Superconducting Electronics

NATO ASI Series

Advanced Science Institutes Series

A series presenting the results of activities sponsored by the NATO Science Committee, which aims at the dissemination of advanced scientific and technological knowledge, with a view to strengthening links between scientific communities.

The Series is published by an international board of publishers in conjunction with the NATO Scientific Affairs Division

A Life Sciences B Physics	Plenum Publishing Corporation London and New York
C Mathematical and Physical Sciences D Behavioural and Social Sciences E Applied Sciences	Kluwer Academic Publishers Dordrecht, Boston and London
F Computer and Systems Sciences G Ecological Sciences H Cell Biology	Springer-Verlag Berlin Heidelberg New York London Paris Tokyo Hong Kong

Series F: Computer and Systems Sciences Vol. 59

Superconducting Electronics

Edited by

Harold Weinstock
Air Force Office of Scientific Research
Washington, DC 20332-6448, USA

Martin Nisenoff
Naval Research Laboratory
Washington, DC 20375-5000, USA

Springer-Verlag Berlin Heidelberg New York
London Paris Tokyo Hong Kong
Published in cooperation with NATO Scientific Affairs Division

Proceedings of the NATO Advanced Study Institute on Superconducting Electronics, held in Il Ciocco, Italy, June 26–July 8, 1988.

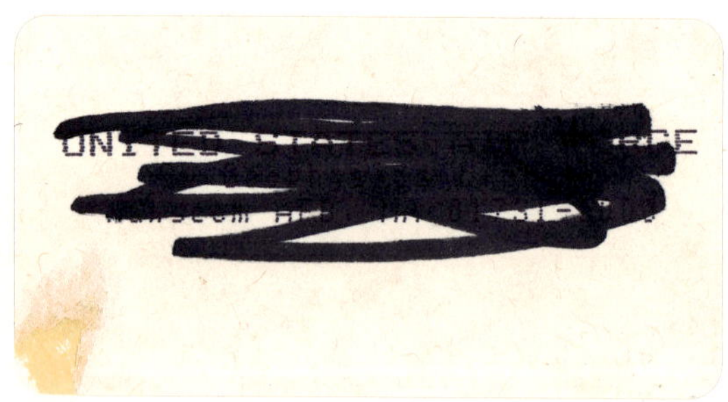

ISBN 3-540-51521-6 Springer-Verlag Berlin Heidelberg New York
ISBN 0-387-51521-6 Springer-Verlag New York Berlin Heidelberg

Library of Congress Cataloging-in-Publication Data. NATO Advanced Study Institute on Superconducting Electronics (1988 : Il Ciocco, Italy). Superconducting electronics / edited by Harold Weinstock, Martin Nisenoff. p. cm.—(NATO ASI series. Series F, Computer and systems science; vol. 59) "Published in cooperation with NATO Scientific Affairs Division." "Proceedings of the NATO Advanced Study Institute on Superconducting Electronics, held in Il Ciocco, Italy, June 26–July 8, 1988"—T.p. verso.
ISBN 0-387-51521-6 (U.S.)
1. Superconductors—Congresses. I. Weinstock, Harold, 1934–. II. Nisenoff, Martin, 1928–. III. North Atlantic Treaty Organization. Scientific Affairs Division. IV. Title. V. Series: NATO ASI series. Series F, Computer and systems sciences; vol. 59. TK7872.S8N23 1988 621.3—dc20 89-21731.

Printed in Germany

Printing: Druckhaus Beltz, Hemsbach; Binding: J. Schäffer GmbH & Co. KG, Grünstadt
2145/3140-543210 – Printed on acid-free-paper

ACKNOWLEDGMENTS

We wish to thank the Scientific Affairs Division of NATO for providing the bulk of the financial support for organizing the Advanced Study Institute (ASI) upon which this volume is based. Additional support for the ASI was provided by the U.S. National Science Foundation and by the organizations which employ us, the Air Force Office of Scientific Research (AFOSR) and the Naval Research Laboratory (NRL).

There are many individuals who contributed to the success of the ASI, but two who contributed immeasurably are Ms. Sandra Chapman of AFOSR, who handled most of the clerical and administrative chores, often working late into the evening, and Mr. Bruno Giannasi of the il Ciocco Hotel, who graciously handled all the special problems that arose and assured that we experienced a most pleasant and trouble-free ambience.

The lecturers, whose contributions are presented on the following pages, provided the intellectual stimulation to an eager and diverse group of participants, which is the foundation for any successful Advanced Study Institute.

Finally, we would like to acknowledge the support of our families in cheerfully putting up with our long absences, even when we were residing in the same building.

Washington, DC
May 1989

Martin Nisenoff
Harold Weinstock

PREFACE

The genesis of the NATO Advanced Study Institute (ASI) upon which this volume is based, occurred during the summer of 1986 when we came to the realization that there had been significant progress during the early 1980's in the field of superconducting electronics and in applications of this technology. Despite this progress, there was a perception among many engineers and scientists that, with the possible exception of a limited number of esoteric fundamental studies and applications (e.g., the Josephson voltage standard or the SQUID magnetometer), there was no significant future for electronic systems incorporating superconducting elements.

One of the major reasons for this perception was the aversion to handling liquid helium or including a closed-cycle helium liquefier. In addition, many critics felt that IBM's cancellation of its superconducting computer project in 1983 was "proof" that superconductors could not possibly compete with semiconductors in high-speed signal processing. From our perspective, the need for liquid helium was outweighed by improved performance, i. e., higher speed, lower noise, greater sensitivity and much lower power dissipation. For many commercial, medical, scientific and military applications, these attributes can lead to either enhanced capability (e.g., compact real-time signal processing) or measurements that cannot be made using any other technology (e.g., SQUID magnetometry to detect neuromagnetic activity).

We also were aware that the IBM action was made as a business decision based on a variety of factors. However, IBM did not abandon its research activities in Josephson junction electronics; the program to manufacture a superconducting supercomputer was postponed until technological advances and market demand produce a more favorable climate for its resumption. Indeed, one observer has likened the IBM project to the equivalent of the Wright brothers trying to build a "747" on their first attempt at powered flight. Since 1983, using this analogy, we have seen the first successful commercial "flight" of a high-speed sampling oscilloscope based upon niobium Josephson-junction technology; additionally, SQUID-based magnetometry has acquired a solid niche as a tool in geomagnetic exploration and biomagnetic diagnostics, as well as showing promise for nondestructive evaluation. Another significant fact is that the 10-year program started about 1980 by the Japanese Ministry of International Trade and Industry (MITI) to develop digital logic and memory chips based on Josephson-junction technology has continued on schedule and with the achievement of some impressive milestones. We are fortunate to have Professor Hayakawa, a former director of this program at the Electrotechnical Laboratory (ETL), as a contributor to this volume.

At the time (January 1987) that we submitted our proposal for the NATO ASI, the scientific community was just beginning to experience the impact of what one Japanese scientist referred to as the "second IBM shock" - the first being the IBM decision discussed earlier. This refers to the discovery of superconductivity in a ceramic material at temperatures above 30 K by Professor K. Alex Müller and Dr. J. Georg Bednorz of the IBM-Zurich Laboratory. The shock over the next year became something more like a major scientific earthquake with the subsequent discovery by others of ceramics with superconducting transition temperatures as high as 125 K. Currently, there are preliminary indications that even higher temperature ceramic superconductors have been fabricated.

Although so far no one has manufactured a Josephson junction using only high-temperature ceramic electrodes, there has been steady progress toward this goal, with recent reports indicating the successful fabrication of (SNS) tunnel junctions and weak-link SQUIDs with low noise capable of operating at 77 K. Thus, there is reason for optimism that, within the next few years, superconducting electronics will become a reality at or even above liquid nitrogen temperature. This leads to new horizons not only because of reduced refrigeration requirements, but because there will be unique opportunities for the development of hybrid superconductor-semiconductor electronic circuitry which was not possible before the recent discoveries.

In this volume we attempt to provide a basic understanding of superconductivity and superconducting electronics, in addition to presenting a comprehensive survey of the major applications of this technology. In the closing session of the ASI several of the lecturers were asked to predict the impact of high-temperature superconductivity on the future of electronics. Each lecturer made many suggestions, but the one common to all of them was that the greatest impact would be derived from some system or device that has yet to be conceived. We hope that our readers will utilize the chapters that follow as a foundation for creating new concepts and inventions that utilize superconducting electronics.

Washington, DC
May 1989

Martin Nisenoff
Harold Weinstock

TABLE OF CONTENTS

SUPERCONDUCTIVITY THEORY

John R. Clem
Ames Laboratory-USDOE and Department of Physics
Iowa State University
Ames, Iowa 50011
USA

INTRODUCTION

Since the primary focus of this volume is superconducting electronics, not superconductivity theory, this chapter will reflect that fact. My goal here is simply to review a few of the key theoretical ideas needed for the development of superconducting electronics and not to give a rigorous theoretical derivation of the existing structure of superconductivity theory. There is an enormous body of literature on superconductivity, and I can provide references to only a tiny fraction of what is available. Many of these references are not the originals, but are review articles that refer to the seminal papers as well as to subsequent work. This discourse will reveal my myopic view of superconductivity, and I apologize to those whose favorite topic is given little or no attention. As for the organization of this chapter, it is simple. I will discuss two main subjects: BCS theory and Ginzburg-Landau theory.

BCS THEORY

Every time I look at the Bardeen-Cooper-Schrieffer paper[1] I am more impressed. It truly is a classic paper, not only presenting a new way of looking at superconductors, but crammed with information. The authors first describe the system before the subtle interactions leading to superconductivity are turned on. The appropriate starting point at zero temperature is a Fermi liquid, in which electrons fill up the lowest energy states, subject to the Pauli principle. Interactions can be thought of as taken into account in an averaged way. Essentially, each electron moves in the average field produced by all the other electrons and ions in the system. There is a well-defined Fermi surface, and the energy of a single-particle state $\mathbf{k}$, measured with respect to the Fermi energy, is $\varepsilon_{\mathbf{k}}$.

NATO ASI Series, Vol. F 59
Superconducting Electronics
Edited by H. Weinstock and M. Nisenoff

Naively one would think that, whatever the interactions among the electrons may be, the many-body wavefunction describing the ground state would have to be some linear combination of the filled single-particle states with $\varepsilon_{\mathbf{k}} < 0$ ($k < k_F$). (It helps to think of a Fermi sphere here, though it is simple to generalize to more complicated Fermi surfaces.) A many-body wavefunction consisting of some states with $\varepsilon_{\mathbf{k}} > 0$ ($k > k_F$) might be expected to have a higher energy than that of the filled Fermi sea. An attractive interaction, however, changes this. As had been shown earlier by Bardeen and Pines,[2] the electron-phonon interaction leads to an attractive effective interaction between electrons in second order, i. e., a term with negative sign in the Hamiltonian. This term can be taken advantage of only if the ground-state wavefunction permits scattering of pairs of electrons from filled states into empty states. In the Fermi sea, however, all states are either completely full ($k < k_F$) or completely empty ($k > k_F$), and thus the attractive term cannot contribute.

The BCS paper shows how to construct the ground-state wavefunction that takes optimum advantage of the attractive pairing interaction, $-V_{\mathbf{kk}'}$. The idea is to consider single-particle states in pairs ($\mathbf{k}\uparrow$, $-\mathbf{k}\downarrow$), which in the ground-state wavefunction are not completely full or completely empty, as in the Fermi sea, but are occupied with a probability $h_{\mathbf{k}}$ (and unoccupied with a probability $1-h_{\mathbf{k}}$). The $h_{\mathbf{k}}$'s are variational parameters to be determined by minimizing the ground-state energy. The result is

$$h_{\mathbf{k}} = (1/2)(1 - \varepsilon_{\mathbf{k}}/E_{\mathbf{k}}), \tag{1}$$

where

$$E_{\mathbf{k}} = (\varepsilon_{\mathbf{k}}{}^2 + \Delta_{\mathbf{k}}{}^2)^{1/2}, \tag{2}$$

and the superconducting energy gap parameter $\Delta_{\mathbf{k}}$ is to be determined from the self-consistency relation

$$\Delta_{\mathbf{k}} = \Sigma_{\mathbf{k}'}\, V_{\mathbf{kk}'}\Delta_{\mathbf{k}'}/2E_{\mathbf{k}'}. \tag{3}$$

The subscript $\mathbf{k}$ on Δ reminds us that in a single crystal the anisotropy of $V_{\mathbf{kk}'}$ also makes $\Delta_{\mathbf{k}}$ anisotropic; i. e., $\Delta_{\mathbf{k}}$ depends on the position of $\mathbf{k}$ on the Fermi surface.

Note that $h_{\mathbf{k}} \simeq 1$ for single-particle energies deep within the Fermi sea ($\varepsilon_{\mathbf{k}} << -\Delta_{\mathbf{k}}$), $h_{\mathbf{k}} = 1/2$ for single-particle energies at the Fermi level ($\varepsilon_{\mathbf{k}} = 0$), and $h_{\mathbf{k}} \simeq 0$ for single-particle energies well above the Fermi energy ($\varepsilon_{\mathbf{k}} >> \Delta_{\mathbf{k}}$). The energy scale over which $h_{\mathbf{k}}$ varies is $\Delta_{\mathbf{k}}$, which is of the order of $k_B T_c$, T_c being the superconducting transition temperature.

In the BCS paper, an isotropic model is assumed, in which $V_{\mathbf{kk}'}$ is replaced by its average value V when $\varepsilon_{\mathbf{k}}$ and $\varepsilon_{\mathbf{k}'}$ are within a characteristic phonon energy $\hbar\omega$ of the Fermi level, and $V_{\mathbf{kk}'} = 0$ otherwise. This permits sums like that of Eq. (3) to be converted to integrals upon the replacement

$$\Sigma_{\mathbf{k}} \rightarrow N(0) \int d\varepsilon_{\mathbf{k}}, \tag{4}$$

where N(0) is the density of states of one spin at the Fermi level. In the weak coupling limit, $N(0)V << 1$, the zero-temperature gap parameter, obtained from Eq. (3), becomes

$$\Delta(0) = 2\hbar\omega \exp[-1/N(0)V], \tag{5}$$

and the energy of the ground state, relative to that of the filled Fermi sea, is

$$W_0 \simeq -N(0)[\Delta(0)]^2/2. \tag{6}$$

Note that, although a kinetic-energy price has to be paid to have partially filled single-particle states within energy $\simeq\Delta$ above E_F in the ground-state wavefunction and partially empty states below E_F, the negative pair-scattering term, which takes optimum advantage of this situation, wins even more energy back, such that the resulting ground-state energy is well below that of the filled Fermi sea.

The ground state is a very special, highly correlated state, as can be seen by investigating the possible excited states. For example, an excited state whose many-body wavefunction is just like that of the ground state, except that the single-particle states $\mathbf{k}''\uparrow$ and $-\mathbf{k}'\downarrow$ are occupied and the states $-\mathbf{k}''\downarrow$ and $\mathbf{k}'\uparrow$ are empty, can be shown to have the excitation energy

$$W_{exc} - W_0 = E_{\mathbf{k}'} + E_{\mathbf{k}''}, \tag{7}$$

where $E_{\mathbf{k}}$ is given by Eq. (2). The main reason for this energy increase is that the pair states $(\mathbf{k}'\uparrow, -\mathbf{k}'\downarrow)$ and $(\mathbf{k}''\uparrow, -\mathbf{k}''\downarrow)$ are now blocked; they no longer can participate in energy-lowering pair-scattering processes. The smallest value this energy difference can take (when $\varepsilon_{\mathbf{k}'} = \varepsilon_{\mathbf{k}''} = 0$) is $\Delta_{\mathbf{k}'} + \Delta_{\mathbf{k}''}$. In conventional anisotropic superconductors there are no places on the Fermi surface where $\Delta_{\mathbf{k}}$ is zero, and there is an energy gap for every possible excitation out of the ground state. For the isotropic case, this energy gap is just $2\Delta(0)$. [The excited state described by Eq. (7) can be thought of as a BCS condensate plus two excited quasiparticles of energy $E_{\mathbf{k}'}$ and $E_{\mathbf{k}''}$.]

As shown in the BCS paper, the case of non-zero temperature can be handled by calculating the free energy using a similar variational approach. The quasiparticle energy $E_{\mathbf{k}}$ takes the same form as in Eq. (2), except that $\Delta_{\mathbf{k}}$, now temperature-dependent, is determined by a modified self-consistency equation,

$$\Delta_{\mathbf{k}} = \Sigma_{\mathbf{k}'} V_{\mathbf{k}\mathbf{k}'}[\Delta_{\mathbf{k}'}/2E_{\mathbf{k}'}] \tanh(E_{\mathbf{k}}/2k_BT). \tag{8}$$

The transition temperature T_c is obtained from this equation as the temperature at which $\Delta_{\mathbf{k}} \to 0$ in Eq. (8). In the weak-coupling limit [$N(0)V \ll 1$], the BCS isotropic-gap model yields[3,4]

$$k_BT_c = 1.134\hbar\omega \exp[-1/N(0)V]. \tag{9}$$

(The numerical coefficient is $2e^C/\pi = 1.134$, where $e^C = 1.78107\ldots$ and $C = 0.5772\ldots$ is Euler's constant.) Combination of Eqs. (5) and (9) yields the famous result,

$$2\Delta(0)/k_BT_c = 3.528. \tag{10}$$

(The numerical coefficient here is $2\pi/e^C$.)

Also calculated in the BCS paper are the electronic contributions to the important thermodynamic functions in the superconducting state: the free energy F_s, entropy S, specific heat C_{es}, and the bulk thermodynamic critical field H_c, as well as their temperature dependences. Note that many of the computed numerical coefficients given in the BCS paper for the isotropic-gap model in the weak-coupling limit are slightly in error. See Refs. 3 and 4 for corrected values.

Anisotropy in $V_{\mathbf{kk}'}$ translates into corresponding anisotropy in $\Delta_{\mathbf{k}}$. The separable model[5] $V_{\mathbf{kk}'} = (1 + a_{\mathbf{k}})V(1 + a_{\mathbf{k}'})$, where $\langle a_{\mathbf{k}} \rangle = 0$ (brackets denote the Fermi-surface average) yields $\Delta_{\mathbf{k}} = \langle \Delta_{\mathbf{k}} \rangle (1 + a_{\mathbf{k}})$. Typically $\langle a_{\mathbf{k}}^2 \rangle$ is about 0.01 or so for conventional superconductors, such that $\Delta_{\mathbf{k}}$ can be expected to vary by about $\pm$ 10% as **k** moves around the Fermi surface. This model, first introduced[5] to study T_c, also has been used to examine the effect of anisotropy upon all the thermodynamic properties.[6] In most cases, the corrections to the BCS results are of order $\langle a_{\mathbf{k}}^2 \rangle$. For example, the correction to the weak-coupling-limit result of Eq. (10) becomes[6]

$$2\langle \Delta_{\mathbf{k}}(0) \rangle / k_B T_c = 3.528\ (1 - 3\langle a_{\mathbf{k}}^2 \rangle /2). \tag{11}$$

When an electron tunnels through an insulating barrier between a superconductor and a normal metal (or another superconductor) at voltage bias V, various kinds of quasiparticle excitations can be generated near the junction. The total excitation energy (by energy conservation) is just the work eV done by the battery. The current-voltage characteristics reveal much about the energy gap parameter $\Delta_{\mathbf{k}}$ characterizing the excitations. For example, anisotropy in the BCS quasiparticle density of states,[6]

$$N_{\mathbf{k}}(E) = N(0)\ \mathrm{Re}\ [E/(E^2 - \Delta_{\mathbf{k}}^2)^{1/2}], \tag{12}$$

is to be expected for different directions of **k** relative to the crystal axes.

Since the energy scale of superconductivity $[\Delta(0) \sim k_B T_c]$ is so small, one might have expected impurity scattering to destroy superconductivity whenever the uncertainty-principle level broadening $\hbar/\tau$ exceeds $\Delta(0)$ or, equivalently, when the normal-state mean-free path $l = v_F \tau$ is smaller than the BCS coherence length $\xi_0 = \hbar v_F / \pi\Delta(0)$. It is found experimentally, however, that the addition of large amounts of nonmagnetic impurities has surprisingly little effect upon T_c, even when $l \ll \xi_0$. This was explained by Anderson's theory of dirty superconductors,[7] in which each pair is formed by a one-electron state and its exact time reverse. This is the generalization of ($\mathbf{k}\uparrow$, $-\mathbf{k}\downarrow$) pairing when scattering makes **k** no longer a good quantum number. When the mean-free path is reduced predominantly by magnetic-impurity scattering, however, superconductivity is destroyed; in terms of the Anderson theory, this is because magnetic scattering destroys time-reversal symmetry.

Another important effect of scattering by nonmagnetic impurities is the reduction of energy-gap anisotropy.[5-9] Anisotropy effects (for example, the anisotropy corrections to BCS of order $\langle a_{\mathbf{k}}^2 \rangle$) are at their strongest for long mean-free paths l such that $l \gg \xi_0$; they are reduced to roughly half their original sizes when $l \sim \xi_0$; and they approach zero as $l/\xi_0 \to 0$. In the latter limit the tunneling density of states again becomes isotropic[8] (independent of the direction of **k**):

$$N(E) = N(0)\ \mathrm{Re}\ [E/(E^2 - \Delta^2)^{1/2}], \tag{13}$$

where $\Delta = \langle \Delta_{\mathbf{k}} \rangle$.

The BCS theory has been extended by various authors in many directions. For example, the BCS treatment of the electron-phonon interaction (parameterized by the interaction constant V and the cutoff $\hbar\omega$) was oversimplified. An elegant treatment of strong electron-phonon coupling and retardation effects has been developed using a Green's function approach (which is state-of-the-art even today), in which the gap parameter $\Delta(\omega) = \Delta_1(\omega) + i\Delta_2(\omega)$ depends upon the energy ω (or frequency, taking $\hbar = 1$) and has both a real and an imaginary part. The self-consistency equation (3 or 8) of the BCS theory is replaced by a pair of nonlinear integral equations, called the Eliashberg[10] equations, which determine $\Delta(\omega)$ and the phonon renormalization parameter $Z(\omega)$. These equations bring in details of the electron-phonon interaction via the quantity $\alpha^2(\omega)F(\omega)$, where $F(\omega)$ is the phonon density of states and $\alpha^2(\omega)$ is the effective electron-phonon coupling function. [The absence of the argument **k** in $\Delta(\omega)$ and $Z(\omega)$ indicates that anisotropy effects are usually ignored; they can be included, but at the expense of additional complications.] As a consequence of the nonlinearity of the Eliashberg equations, both $\Delta(\omega)$ and $Z(\omega)$, viewed as functions of ω, have structure reflecting that of $\alpha^2(\omega)F(\omega)$. This can be revealed experimentally via measurements of the tunneling density of states,

$$N(\omega) = N(0)\ \mathrm{Re}\ \{\omega/[\omega^2 - \Delta^2(\omega)]^{1/2}\}. \tag{14}$$

See Refs. 11-13 for reviews of this subject.

GINZBURG-LANDAU THEORY

The BCS theory and its various extensions justify the interpretation of numerous properties of superconductors by means of a kind of two-fluid picture, which has some features in common with a model introduced by Gorter and Casimir[14] in 1934. There is a superfluid of Cooper pairs, all described by the same macroscopic wavefunction. At zero temperature this wavefunction is just the BCS ground-state wavefunction, and at higher temperature the wavefunction is essentially what is left of the ground-state wavefunction after various quasiparticles have been excited out of it. This BCS condensate wavefunction is quite rigid; how this rigidity leads to screening of applied magnetic fields is best described with the help of the Ginzburg-Landau equations, which we will discuss below. There is a normal fluid of thermally excited quasiparticles, which contribute to the entropy, specific heat, and thermal conductivity. The normal fluid also feels any electric field that penetrates into the superconductor. At low frequencies, however, the superfluid can effectively short out electric fields, or at least cause them to be attenuated sharply as a function of distance from the surface.

The Ginzburg-Landau (GL) theory[15] provides a conceptually simple and elegant way of examining the response of the superfluid to externally applied magnetic fields and currents. Although this theory predated the BCS theory, Gor'kov[16] later showed that the GL theory can be derived from the BCS theory in the vicinity of T_c. Because of its simplicity, the GL theory commonly is used well outside its range of validity, i. e., far from T_c. Fortunately, extensions of the microscopic theory usually show that this is not a bad mistake, the errors seldom being more than 25%.

The Ginzburg-Landau theory describes how magnetic fields penetrate into superconductors and how magnetic structure develops there. The GL approach is to use simple quantum mechanics to derive various contributions to the Helmholtz free energy density F below T_c. There is an order parameter Ψ, which is very similar to the wavefunction of a single particle in ordinary quantum mechanics, but it is normalized such that $|\Psi|^2$ is the local density of Cooper pairs (number per unit volume) in the BCS condensate. For a uniform superconductor in the absence of a magnetic field, the free energy density in the superconducting state F_{s0} relative to that in the normal state F_{n0} is expanded in even powers of Ψ up to the fourth power:

$$F_{s0} - F_{n0} = \alpha|\Psi|^2 + \beta|\Psi|^4/2, \tag{15}$$

where $\alpha = \alpha'(T-T_c)$ and α' and β are positive constants (which we now know are derivable from the BCS theory). A plot of $F_{s0} - F_{n0}$ versus $|\Psi|^2$ exhibits a minimum at the equilibrium value $\Psi = \Psi_e$, where Ψ_e can be taken to be real, and $\Psi_e^2 = -\alpha/\beta$. The value at the minimum is $(F_{s0} - F_{n0})_{min} = - H_c^2/8\pi$, which is just the condensation energy per unit volume of the superconducting state; thus $\alpha^2/2\beta = H_c^2/8\pi$. In this approximation, $H_c \propto (T_c-T)$ and $\Psi_e \propto (T_c-T)^{1/2}$.

When the superconductor is subjected to a magnetic field, the order parameter in general is complex and thus may be written as

$$\Psi = \Psi_e f e^{i\gamma}. \tag{16}$$

Let us consider a particle of mass M and charge Q. When it is in a magnetic field $\mathbf{B} = \nabla \times \mathbf{A}$, its classical Hamiltonian is $(\mathbf{p}-Q\mathbf{A}/c)^2/2M$ (plus any scalar potentials, which we ignore here) and the particle's velocity is $\mathbf{v} = (\mathbf{p}-Q\mathbf{A}/c)/M$. The electrical current density for a collection of such particles with density n, all moving with the same velocity $\mathbf{v}$, is $\mathbf{j} = nQ\mathbf{v}$. For a single particle with wavefunction ψ, the quantum mechanical generalization of this current density is

$$\mathbf{j} = (Q/2M)[\psi^*(\hbar\nabla/i - Q\mathbf{A}/c)\psi + \text{c.c.}], \tag{17}$$

where c.c. denotes the complex conjugate.

We now consider a superconductor in which the local magnetic field is $\mathbf{b} = \nabla \times \mathbf{a}$. To compute the supercurrent density arising from the BCS condensate of Cooper pairs, which have charge $Q = -e^* = -2e$ and mass $M = m^* = 2m$ and are described by the order parameter Ψ, we apply Eq. (17), but with the replacements $Q \to -e^*$, $M \to m^*$, $\psi \to \Psi$, and $\mathbf{A} \to \mathbf{a}$. The result can be written as

$$\mathbf{j} = -(cf^2/4\pi\lambda^2)[\mathbf{a} + (\phi_0/2\pi)\nabla\gamma], \tag{18}$$

where $\phi_0 = hc/e^* = hc/2e = 2.07 \times 10^{-7}$ G-cm^2 is the superconducting flux quantum. Here we have introduced one of the characteristic lengths of the GL theory, the penetration depth $\lambda(T)$, which is defined via

$$\lambda^{-2} = 4\pi\Psi_e^2 e^{*2}/m^*c^2. \tag{19}$$

Note that λ diverges at T_c as $(T_c-T)^{-1/2}$. The quantities **j** and f are gauge-invariant quantities, which remain the same regardless of the choice of the vector potential **a**. This means that, while the phase γ of the order parameter is gauge-dependent, the quantity within brackets on the right-hand side of Eq. (18) is gauge-invariant. The combination $\mathbf{a}_s = \mathbf{a} + (\phi_0/2\pi)\nabla\gamma$ can be thought of as a gauge-invariant vector potential. It has the property that $\nabla \times \mathbf{a}_s = \mathbf{b}$, except for delta-function singularities on the vortex axes, around which the phase changes by a multiple of 2π. The bracketed quantity on the right-hand side of Eq. (18) also is proportional to the superfluid velocity $\mathbf{v}_s$, the local velocity of the Cooper pairs,

$$\mathbf{v}_s = e^*\mathbf{a}_s/m^*c = (\hbar\nabla\gamma + e^*\mathbf{a}/c)/m^*. \tag{20}$$

Equation (18) is known as the second Ginzburg-Landau equation. <u>Fluxoid quantization</u> can be derived immediately by multiplying Eq. (18) by $4\pi\lambda^2/cf^2$ and integrating the result around a closed contour. The result is

$$\int d\mathbf{S}\cdot\mathbf{b} + (4\pi\lambda^2/c)\oint d\mathbf{r}\cdot\mathbf{j}/f^2 = -(\phi_0/2\pi)\oint d\mathbf{r}\cdot\nabla\gamma = N\phi_0, \tag{21}$$

where $N = 0, \pm1, \pm2, \ldots$. In a type-II superconductor, N is the number of vortices minus the number of antivortices (vortices of the opposite sense) included within the integration contour. The left-hand side of Eq. (21) is the fluxoid. Often the line integral of $\mathbf{j}/f^2$ can made very small by appropriate choice of the contour, which also defines the area through which the magnetic flux $\Phi = \int d\mathbf{S}\cdot\mathbf{b}$ passes. In such a case, we have $\Phi \simeq N\phi_0$, which is the source of the commonly heard (but theoretically unsound) expression "flux quantization."

The Helmholtz free energy density in the superconducting state with a magnetic field present, relative to that in the absence of a field, can be written as [see Eq. (13) of Ref. 17]

$$\Delta F = F_s(\mathbf{b}\neq 0) - F_{s0} = \Delta F_c + \Delta F_{kg} + \Delta F_{kj} + \Delta F_f. \tag{22}$$

The first term,

$$\Delta F_c = (H_c{}^2/8\pi)(1 - f^2)^2, \tag{23}$$

represents the cost in condensation energy when the reduced order parameter f is reduced from its value (f = 1) in the absence of the magnetic field, such as in the core of a vortex in a type-II superconductor. The second term,

$$\Delta F_{kg} = (H_c{}^2/4\pi)\xi^2(\nabla f)^2, \qquad (24)$$

is the cost in kinetic energy of having a gradient in the magnitude of the reduced order parameter. The third term,

$$\Delta F_{kj} = f^2 a_s{}^2/8\pi\lambda^2, \qquad (25)$$

is the cost in kinetic energy of having a supercurrent flowing in response to the applied magnetic field. This term is simply the kinetic energy of a Cooper pair, $(1/2)m^* v_s{}^2$, times $|\Psi|^2$, the density of pairs. The fourth term,

$$\Delta F_f = b^2/8\pi, \qquad (26)$$

is just the local energy density of the magnetic field.

The length ξ appearing in Eq. (24) is the Ginzburg-Landau coherence length, which is determined once the two independent quantities H_c and λ are known:

$$\xi = \phi_o/2\pi\sqrt{2}H_c\lambda. \qquad (27)$$

The coherence length ξ also diverges at T_c as $\xi \propto (T_c - T)^{-1/2}$. The Ginzburg-Landau parameter $\kappa = \lambda/\xi$, a material-dependent dimensionless constant, determines the nature of any magnetic structure that penetrates into the superconductor. For $\kappa < 1/\sqrt{2}$, the material is a type-I superconductor, and the magnetic structure normally takes the form of large, multiply-quantized domains. For $\kappa > 1/\sqrt{2}$, the material is a type-II superconductor, and the magnetic structure takes the form of compact, singly-quantized vortices, each carrying flux ϕ_o.

The Ginzburg-Landau theory frequently is reduced to dimensionless form by expressing lengths, magnetic fields, etc. in terms of the corresponding Ginzburg-Landau fundamental units. Table I exhibits the units used for this purpose.

Table I. Ginzburg-Landau fundamental units used to form dimensionless quantities.

Quantity	Ginzburg-Landau fundamental unit
Length	λ
Magnetic field	$\sqrt{2}H_c = \phi_0/2\pi\xi\lambda = \kappa\phi_0/2\pi\lambda^2$
Magnetic flux	$\sqrt{2}H_c\lambda^2 = \phi_0\lambda/2\pi\xi = \kappa\phi_0/2\pi$
Vector potential	$\sqrt{2}H_c\lambda = \phi_0/2\pi\xi = \kappa\phi_0/2\pi\lambda$
Current density	$(c/4\pi)\sqrt{2}H_c/\lambda = c\phi_0/8\pi^2\xi\lambda^2 = \kappa c\phi_0/8\pi^2\lambda^3$
Energy density	$H_c{}^2/4\pi = \phi_0{}^2/32\pi^3\xi^2\lambda^2 = \kappa^2\phi_0{}^2/32\pi^3\lambda^4$
Superfluid velocity	$(e^*/m^*c)\sqrt{2}H_c\lambda = (e^*/m^*c)\phi_0/2\pi\xi = (e^*/m^*c)\kappa\phi_0/2\pi\lambda$
	$= \hbar/m^*\xi = \kappa\hbar/m^*\lambda$

In the literature on this subject, primes often are used to indicate dimensionless quantities formed by normalizing to the Ginzburg-Landau fundamental units. For example, $\mathbf{r}' = \mathbf{r}/\lambda$, $\nabla' = \lambda\nabla$, $\delta(x') = \lambda\delta(x)$, and $\mathbf{b}' = \mathbf{b}/(\sqrt{2}H_c)$. The flux quantum becomes $\phi_0' = \phi_0/(\sqrt{2}H_c\lambda^2) = 2\pi/\kappa$. Equation (22) becomes

$$\Delta F' = (1-f^2)^2/2 + (\nabla' f)^2/\kappa^2 + f^2 a_s'^2 + b'^2. \tag{28}$$

In this equation, f and $\mathbf{a}_s'$ are variational parameters, to be obtained by minimizing the Gibbs free energy in an applied magnetic field. Application of the calculus of variations, integrating by parts, and making use of the Ginzburg-Landau boundary conditions at a superconductor-vacuum interface,

$$\hat{n}\cdot\nabla' f = 0, \tag{29}$$

$$\hat{n}\cdot\mathbf{a}_s' = 0, \tag{30}$$

where $\hat{n}$ is the outward normal to the superconductor, leads to the first and second Ginzburg-Landau equations:

$$\kappa^{-2}\nabla'^2 f = (f^2 - 1)f + a_s'^2 f, \tag{31}$$

$$\mathbf{j}' = -f^2\mathbf{a}_s'. \tag{32}$$

[Equation (32) is simply Eq. (18) written in dimensionless form.] These two equations must be solved subject to the supplemental equations $\mathbf{a}_s' = \mathbf{a}' + \kappa^{-1}\nabla'\gamma = \mathbf{v}_s'$, $\mathbf{b}' = \nabla'\times\mathbf{a}'$, and $\mathbf{j}' = \nabla'\times\mathbf{b}'$, as well as the appropriate boundary conditions describing the experimental arrangement.

For a superconductor initially containing no magnetic flux, the application of a weak applied magnetic field $H_a \ll H_c$ causes f to be suppressed only slightly from its value (f = 1) in the absence of a magnetic field. Substitution of f = 1 into Eq. (32) and taking the curl leads to the London equation,

$$\nabla'^2\mathbf{b}' = \mathbf{b}', \tag{33}$$

which states that **b**, **j**, and $\mathbf{a}_s$ decay exponentially with distance from the surface over the decay length λ.

Increasing the value of the applied field causes $\mathbf{a}_s'$ to increase in magnitude, which, as can be seen from Eq. (31), causes f to be suppressed close to the surface. When the applied field is sufficiently large, an instability occurs at the surface. The order parameter drops to zero there, and a zero of the order parameter then moves in from the surface into the sample's interior. Surrounding this zero is a region of reduced order parameter, and the size of this region depends upon whether the material is a type-I or type-II superconductor. For a type-I superconductor the zero is generally at the center of a large normal domain, while for a type-II superconductor the zero marks the axis of a singly-quantized vortex.

The Ginzburg-Landau equations have been solved numerically for the case of an isolated singly-quantized vortex deep within a type-II superconductor. Cylindrical coordinates (ρ,ϕ,z) are used, and the magnitudes of **b**, **j**, $\mathbf{a}_s$, and **a** depend only upon the value of ρ, the distance from the vortex axis. The phase is $\gamma = -\phi$. When $\xi \ll \lambda$, the solutions are such that f rises linearly with ρ

on the length scale ξ; beyond ξ, $f \simeq 1$. It is therefore said that the vortex has a core of radius ξ. The field magnitude b is nearly constant within the core, but beyond ξ it drops off rapidly. The magnitudes b, j, a_s, and v_s all decay nearly exponentially for ρ larger than λ. Close to the vortex axis, v_s and a_s vary as ρ^{-1}, but it follows from Eq. (32) that j is proportional to ρ there because f^2 is proportional to ρ^2.

A simple variational model for the vortex is given in Ref. 18, where f is assumed to be of the form $f = \rho/R$, where $R = (\rho^2 + \xi_v^2)^{1/2}$, and ξ_v is a variational core radius parameter, expected to be of the order of ξ. Although this Ansatz does not satisfy Eq. (31) exactly, Eq. (32) can be solved exactly. The results for b_z, a_ϕ, and $j_\phi = -(\rho^2/R^2)a_{s\phi}$ are

$$b_z = (\phi_0/2\pi\lambda\xi_v)K_0(R/\lambda)/K_1(\xi_v/\lambda), \quad (34)$$

$$a_\phi = (\phi_0/2\pi\rho)[1 - (R/\xi_v)K_1(R/\lambda)/K_1(\xi_v/\lambda)], \quad (35)$$

$$j_\phi = (c\phi_0/8\pi^2\lambda^2\xi_v)(\rho/R)K_1(R/\lambda)/K_1(\xi_v/\lambda), \quad (36)$$

where $K_n(x)$ is the modified Bessel function of order n. The z component of the flux through a circular area of radius ρ centered on the z axis is $\Phi_z(\rho) = 2\pi\rho a_\phi(\rho)$. These quantities have the same qualitative behavior as found by solving the Ginzburg-Landau equations numerically, but can be calculated in a matter of seconds using a short computer program. The best value of ξ_v is determined by substituting Eqs. (34)-(36) into (22) or (28), integrating over x and y to find the energy $\varepsilon_1 = \phi_0 H_{c1}/4\pi$ per unit length of the vortex, and then minimizing ε_1 with respect to ξ_v. All the integrals can be carried out analytically, and the resulting condition that ξ_v must satisfy in order to minimize ε_1 is

$$\kappa = \sqrt{2}\,[1 - K_0^2(\xi_v')/K_1^2(\xi_v')]^{1/2}/\xi_v', \quad (37)$$

where $\xi_v' = \xi_v/\lambda$. For large κ, we see that $\xi_v' \simeq \sqrt{2}/\kappa \ll 1$, or since $\kappa = \lambda/\xi$, $\xi_v \simeq \sqrt{2}\xi$. For all values of κ corresponding to type-II superconductors, $\xi_v \sim \xi$; the value of ξ_v/ξ ranges from 0.935 at $\kappa = 1/\sqrt{2}$ to 1.414 at $\kappa = \infty$. An excellent approximation to H_{c1} now can be obtained by substituting the solution of Eq. (37) back into the expression for the energy per unit length of vortex. In dimensionless units, the resulting lower critical field is

$$H_{c1}' = H_{c1}/\sqrt{2}H_c = \kappa\xi_v'^2/8 + (8\kappa)^{-1} + K_0(\xi_v')/2\kappa\xi_v' K_1(\xi_v'). \quad (38)$$

Since this "energy" is from a variational calculation, it should not be a surprise that this value of H_{c1} is a few percent higher than the values obtained from numerical solutions of Eqs. (31) and (32). For example, at $\kappa = 1/\sqrt{2}$, Eq. (38) yields $H_{c1}' = 0.732$, about 4% higher than the exact Ginzburg-Landau result of 0.707. This model also gives a simple expression for $\tilde{b}_z(q)$, the two-dimensional Fourier transform of $b_z(\rho)$, which can be measured using small-angle neutron scattering,

$$\tilde{b}_z(q) = \phi_0 K_1(Q\xi_v)/Q\lambda K_1(\xi_v/\lambda), \tag{39}$$

where $Q = (q^2 + \lambda^{-2})^{1/2}$.

A useful approximation for the lower critical field when κ is large is

$$H_{c1} \simeq (\phi_0/4\pi\lambda^2)(\ln \kappa + 0.50), \tag{40}$$

while the upper critical field (at which bulk superconductivity is quenched) is

$$H_{c2} = \phi_0/2\pi\xi^2. \tag{41}$$

In the presence of anisotropy, the Ginzburg-Landau theory can be extended[19-21] with the help of a phenomenological effective mass tensor, with components m_1, m_2, and m_3 associated with the principal axes $\hat{x}_i$ (i = 1, 2, and 3). These are normalized such that $m_1 m_2 m_3 = 1$. Within the Ginzburg-Landau equations, the spatial variation of the order parameter along $\hat{x}_i$ is

$$\xi_i = \xi/\sqrt{m_i}. \tag{42}$$

Similarly, the penetration depth of the component of the screening current that flows along $\hat{x}_i$ is

$$\lambda_i = \lambda\sqrt{m_i}. \tag{43}$$

Note that $(\xi_1\xi_2\xi_3)^{1/3} = \xi$ and $(\lambda_1\lambda_2\lambda_3)^{1/3} = \lambda$, where ξ and λ are related by $H_c = \phi_0/2\pi\sqrt{2}\lambda\xi$, as in the isotropic case. Well outside the core, the field generated by an isolated vortex along the $\hat{x}_3$ axis is

$$b_3 \simeq (\phi_0/2\pi\lambda_1\lambda_2)K_0(r_3/\lambda), \tag{44}$$

where $r_3 = (x_1{}^2/m_2 + x_2{}^2/m_1)^{1/2}$, such that the contours of constant b are ellipses, as are the streamlines of j. The semi-axes of the ellipse defining the vortex core are ξ_1 and ξ_2 along the $\hat{x}_1$ and $\hat{x}_2$ axes, respectively. The lower (H_{c1}) and upper (H_{c2}) critical fields along the $\hat{x}_3$ axis are

$$H_{c13} \approx (\phi_o/4\pi\lambda_1\lambda_2)\{\ln[\lambda/(\xi_1\xi_2)^{1/2}] + 0.50\}, \tag{45}$$

$$H_{c23} = \phi_o/2\pi\xi_1\xi_2. \tag{46}$$

The critical fields along the other principal axes can be obtained by cyclic permutation.

When an array of vortices is present in the specimen, it is always appropriate to think of the net magnetic field in terms of a linear superposition[22-23] of a number of contributions: (a) the Meissner response of the specimen to the applied field, (b) the Meissner response of the specimen to the self field associated with any current applied to the superconductor, and (c) the vortex fields. Each vortex field carries one quantum of flux within an area $\sim\pi\lambda^2$ up along its axis, and then returns this flux in dipole fashion through the non-superconducting space around the sample. All these field contributions decay with the weak-field penetration depth λ (considering isotropic materials for simplicity) if the average flux density B is small by comparison with the upper critical field $B_{c2} = \phi_o/2\pi\xi^2$. However, if the intervortex spacing becomes comparable with ξ (which occurs as B approaches B_{c2}), the average value of f^2 is considerably suppressed. As seen from Eqs. (18) and (32), this has the effect of increasing the effective screening length[22] from λ to $\lambda_B = \lambda/\langle f^2\rangle^{1/2} \approx \lambda/(1-B/B_{c2})^{1/2}$, where the brackets here denote a spatial average. If the specimen thickness d is much thinner than λ_B, the corresponding screening length of the fields along the wide specimen dimensions becomes[24] $\lambda_{eff} = 2\lambda^2/d$. The shape of a vortex threading through a wide film of arbitrary thickness d relative to λ_B is given in Ref. 25.

In most superconducting electronics circuits, any vortex created in some region of the sample during cooldown through the transition temperature will be pinned by various metallurgical imperfections, and normally such a trapped vortex will be benign. However, when the vortex moves under the influence of a fluctuating transport current[23] or thermal activation,[26] noise can be generated in any nearby circuit that is sensitive to the return flux generated by the vortex. Similarly, Johnson noise generated in normal metal close to the superconductor also can be picked up by a superconducting circuit.[27]

The above discussion implicitly has concentrated on the behavior of superconductors at sufficiently low frequencies that the superconductor responds quasistatically. For superconductors, low frequency generally means that $hf = \hbar\omega < 2\Delta(T)$, where the photon energy is too low to break pairs. For $hf = \hbar\omega > 2\Delta(T)$, the behavior is very similar to that in a normal metal. The general theory for this, based on BCS theory, was given by Mattis and Bardeen.[28] A useful description of superconducting electrodynamics at low frequencies was given by London.[29] An electric field $\mathbf{e}$ penetrating into the superconductor induces, by Faraday's law and the second Ginzburg-Landau equation, a supercurrent $\mathbf{j}_s = -(cf^2/4\pi\lambda^2)\mathbf{a}_s$, where $d\mathbf{a}_s/dt = -c\mathbf{e}$. A normal current $\mathbf{j}_n = \sigma_n\mathbf{e}$ is also induced, where σ_n is the effective conductivity of the normal fluid (which is the thermally generated bath of quasiparticles). Note that, because the normal current and supercurrent are 90° out of phase, the frequency-dependent conductivity $\sigma(\omega)$ is complex. [The displacement current usually can be neglected at low frequencies.]

Although the Josephson effect is covered elsewhere in this volume, a lecture on superconductivity theory cannot conclude without some discussion of it. Briefly, the overlap between the wavefunctions of two superconductors a and b, separated by a thin insulating layer, permits a tunneling supercurrent[30] J to flow across the junction between a and b,

$$J = J_o \sin \Delta\gamma, \tag{47}$$

where J_o is the maximum Josephson supercurrent set by the microscopic theory and $\Delta\gamma$ is the gauge-invariant phase difference[30] across the junction,

$$\Delta\gamma = \gamma_a - \gamma_b - (2\pi/\phi_o) \int d\mathbf{r}\cdot\mathbf{a}, \tag{48}$$

the integral being carried out along a straight contour across the junction from a to b. The voltage across the junction is

$$V = V_a - V_b = \int d\mathbf{r}\cdot\mathbf{e} = (\hbar/2e)\, d\Delta\gamma/dt, \tag{49}$$

where the integral is carried out from a to b across the junction.

ACKNOWLEDGMENTS

Ames Laboratory is operated for the Department of Energy by Iowa State University under Contract No. W-7405-Eng-82. This work was supported by the Director for Energy Research, Office of Basic Energy Sciences.

REFERENCES

1. Bardeen, J., Cooper, L. N. and Schrieffer, J. R.: Theory of Superconductivity. Phys. Rev. 108, 1175-1204 (1957)
2. Bardeen, J., Pines, D.: Electron-Phonon Interaction in Metals. Phys. Rev. 99, 1140-1150 (1955)
3. Mühlschlegel, B.: Die thermodynamischen Funktionen des Supraleiters. Z. Phys. 155, 313-327 (1959)
4. Rickayzen, G.: The Theory of Bardeen, Cooper, and Schrieffer. In: Superconductivity (R. D. Parks, ed.), Vol. 1, pp. 51-115. New York: Dekker 1969
5. Markowitz, D., and Kadanoff, L. P.: Effect of Impurities upon Critical Temperature of Anisotropic Superconductors. Phys. Rev. 131, 563-575 (1963)
6. Clem, J. R.: Effects of Energy Gap Anisotropy in Pure Superconductors. Ann. Phys. (N. Y.) 40, 268-295 (1966)
7. Anderson, P. W.: Theory of Dirty Superconductors. J. Phys. Chem. Solids 11, 26-30 (1959)
8. Clem, J. R.: Effects of Nonmagnetic Impurities upon Anisotropy of the Superconducting Energy Gap. Phys. Rev. 148, 392-401 (1966)
9. Clem, J. R.: Impurity Dependence of the Critical Field in Anisotropic Superconductors. Phys. Rev. 153, 449-454 (1967)
10. Eliashberg, G. M.: Interactions between Electrons and Lattice Vibrations in a Superconductor. Sov. Phys. JETP 11, 696-702 (1960; Temperature Green's Function for Electrons in a Superconductor. 12, 1000 (1961)
11. Scalapino, D. J.: The Electron-Phonon Interaction and Strong-Coupling Superconductors. In: Superconductivity (R. D. Parks, ed.), Vol. 1, pp. 449-560. New York: Dekker 1969
12. Schrieffer, J. R.: Theory of Superconductivity. New York: Benjamin 1964
13. McMillan, W. L., and Rowell, J. M.: Tunneling and Strong-Coupling Superconductivity. In: Superconductivity (R. D. Parks, ed.), Vol. 1, pp. 561-614. New York: Dekker 1969
14. Gorter, C. J., and Casimir, H. B. G.: On Supraconductivity I. Physica 1, 306-320 (1934)
15. Werthamer, N. R.: The Ginzburg-Landau Equations and Their Extensions. In: Superconductivity (R. D. Parks, ed.), Vol. 1, pp. 321-370. New York: Dekker 1969
16. Gor'kov, L. P.: Microscopic Derivation of the Ginzburg-Landau Equations in the Theory of Superconductivity. Sov. Phys. JETP 9, 1364-1367 (1959)
17. Fetter, A. L., and Hohenberg, P. C.: Theory of Type II Superconductors. In: Superconductivity (R. D. Parks, ed.), Vol. 2, pp. 817-924. New York: Dekker 1969
18. Clem, J. R.: Simple Model for the Vortex Core in a Type II Superconductor. J. Low Temp. Phys. 18, 427-434 (1975)
19. Klemm, R. A., and Clem, J. R.: Lower Critical Field of an Anisotropic Type-II Superconductor. Phys. Rev. B 21, 1868-1875 (1980)
20. Kogan, V. G.: London Approach to Anisotropic Type-II Superconductors. Phys. Rev. B 24, 1572-1575 (1981)
21. Kogan, V. G., and Clem, J. R.: Uniaxial Type-II Superconductors Near the Upper Critical Field. Phys. Rev. B 24, 2497-2505 (1981)

22. Clem. J. R.: Phenomenological Theory of the Local Magnetic Field in Type-II Superconductors. In: Low Temperature Physics - LT14 (M. Krusius and M. Vuorio, eds.), Vol. 2, pp. 285-288. Amsterdam: North-Holland 1975
23. Clem, J. R.: Flux-Flow Noise in Superconductors. Physics Reports 75, 1-55 (1981)
24. Pearl, J.: Current Distribution in Superconducting Films Carrying Quantized Fluxoids. Appl. Phys. Lett. 5, 65-66 (1964)
25. Clem, J. R.: Vortices in Superconducting Films. In: Inhomogeneous Superconductors - 1979 (D. U. Gubser, T. L. Francavilla, S. A. Wolf, and J. R. Leibowitz, eds.), pp. 245-250. New York: American Institute of Physics 1980
26. Li, P. S., and Clem, J. R.: Johnson Noise in Ideal Type-II Superconducting Films. Phys. Rev. B 23, 2209-2218 (1981)
27. Clem, J. R.: Johnson Noise from Normal Metal Near a Superconducting SQUID Gradiometer Circuit. IEEE Trans. Magn. MAG-23, 1093-1096 (1987)
28. Mattis, D. C., and Bardeen, J.: Theory of the Anomalous Skin Effect in Normal and Superconducting Metals. Phys. Rev. 111, 412-417 (1958)
29. London, F.: Superfluids, Vol 1, pp. 27-33. New York: Dover (1961)
30. Josephson, B. D.: Weakly Coupled Superconductors. In: Superconductivity (R. D. Parks, ed.), Vol. 1, pp. 423-448. New York: Dekker 1969

Quantum Interference in Normal Metals

C. Van Haesendonck and Y. Bruynseraede

Laboratorium voor Vaste Stof-Fysika en Magnetisme
Katholieke Universiteit Leuven, B-3030 Leuven (Belgium)

1 Introduction

In 1924, de Broglie [1] introduced his famous quantum-mechanical wave picture to describe the motion of electrons. Freely moving electrons are described as plane waves $\Psi(\mathbf{r}) \propto \exp(i\,\mathbf{k}.\mathbf{r})$, and their momentum p varies inversely proportional to the wavelength λ :

$$p = \frac{h}{\lambda} = \hbar|\mathbf{k}| \, . \tag{1}$$

The validity of the wave picture can be tested directly by studying the diffraction of an electron beam by a periodic array of atoms. In 1927, Davisson and Germer [2] showed that, after penetration through a crystal, the diffraction patterns for an electron beam and an x-ray beam are identical. Since the electronic wavelength decreases when the electron speed increases, a high energy electron beam will allow imaging of materials on an atomic scale. Indeed, electron microscopy has become one of the most powerful research tools for the identification of microscopic structures.

With the development of the transmission electron microscope, it also became possible to show that the phase of an electron wave has a physical significance. When the electron beam in an electron microscope is split into two parts using an electrostatic biprism, interference fringes will be formed near the point where the two beams converge. Holographic imaging of an object also is possible if this object is placed in one of the interfering beams [3].

When a magnetic field $\mathbf{B}$ is applied, two effects are observed. First of all, the envelope of the interference pattern is shifted because of the Lorentz force $\mathbf{F}_L = -e\mathbf{v} \times \mathbf{B}$, which modifies the path followed by the electrons. At the same time, the position of the interference fringes shifts relative to the envelope of the pattern. This shift is usually referred to as the Aharonov-Bohm effect. As first predicted by Aharonov and Bohm in 1959 [4], the phase shift between the interfering beams depends only upon the total magnetic flux Φ_B enclosed by the interfering beams. Careful experiments confirmed that the Aharonov-Bohm effect occurs even when the flux Φ_B is completely confined to the interior of the electron path so that $F_L = 0$.

It is important to note that the Aharonov-Bohm effect is a direct consequence of the wave character of the electrons and does not have a classical analogue. In order to understand this non-local influence of the magnetic flux, one has to assume that the vector potential $\mathbf{A}$ influences the electron waves even in regions where $\mathbf{B} = \nabla\times\,\mathbf{A} = 0$.

NATO ASI Series, Vol. F 59
Superconducting Electronics
Edited by H. Weinstock and M. Nisenoff

The Aharonov-Bohm effect relies on the fact that the phase of the wave function, which is calculated from the Schrödinger equation, depends upon the specific choice for the vector potential. Contrary to physically observable quantities, the phase of a wave function is not gauge invariant.

As pointed out by Aharonov and Bohm, the wave function of an electron which moves in a region where $\mathbf{A} \neq 0$, will pick up an extra phase φ which is determined by the line integral of the vector potential along the electron path :

$$\varphi = \frac{e}{\hbar} \int \mathbf{A}.d\mathbf{l} \,. \tag{2}$$

For a closed path, Eq. (2) predicts a modulation of the phase by the enclosed magnetic flux :

$$\varphi = \frac{e}{\hbar} \oint \mathbf{A}.d\mathbf{l} = \frac{e}{\hbar} \iint \mathbf{B}.d\mathbf{S} = 2\pi \frac{\Phi_B}{\Phi_o} \,, \tag{3}$$

where the flux quantum $\Phi_o = h/e$. Equation (3) has been successfully used to explain the magnetic-flux quantization within a hollow superconducting cylinder. In a superconductor the electrons form Cooper pairs with charge $2e$. The enclosed magnetic flux will be an integer multiple of the superconducting flux quantum $h/2e = \Phi_o/2$ [5].

Since 1966, the fundamental question has been raised as to whether the Aharonov-Bohm effect also can be observed for the conduction electron waves in a normal metal [6,7]. At high temperatures, the inelastic scattering of the conduction electrons (by other electrons or by phonons) will cause random variations of the electronic wavelength on an atomic length scale and, therefore, destroy the interference between conduction electron waves. At low temperatures, the inelastic scattering time τ_{in} becomes very long and the conduction process is governed by the elastic scattering at lattice defects or impurities. The phase coherence length L_φ, i.e., the length over which the electrons can diffuse elastically without changing their wavelength, can not exceed the inelastic diffusion length $L_{in} = \sqrt{D\tau_{in}}$, where D is the diffusion constant. Although inelastic scattering is the most obvious source of phase incoherence in a metal, other mechanisms such as spin-flip scattering by residual magnetic impurities in the metal may cause a reduction of the phase coherence length L_φ [8].

Typical quantum effects such as the Shubnikov - de Haas - van Alphen magnetoresistance oscillations or the quantum Hall effect, can only be observed in very pure, high mobility samples. One therefore is tempted to believe that elastic disorder scattering destroys the quantum coherence. Here we will show that the quantum coherence between interfering electron waves survives in the presence of elastic defect scattering. The experiments we will discuss in the next sections, prove that elastic scattering enhances even the influence of the interference effects upon the electronic conduction.

From a theoretical point of view, much insight has been gained from the Landauer description [9], which treats the disorder as a tunneling barrier for the conduction electrons. The Landauer model predicts that in an isolated ring of a normal metal (no leads attached), the coherence can be perfect, so that a persistent current, comparable to a superconducting current, can be induced by applying a time-dependent magnetic flux $\Phi_B(t)$. The classical, resistive behaviour appears only when inelastic (phase breaking) scattering is introduced. At the same time, the interference effects are largely damped [10].

2 Interference between elastically scattered electron waves

The basic configuration which is needed to observe the solid-state analogue of the Aharonov- Bohm effect is sketched in Fig. 1 . A conduction electron wave arriving at point L is divided into two partial waves which diffuse elastically along both arms of the square metal loop with size S and interfere at point R. The phase coherence length L_φ is comparable to the distance $2S$ between the points L and R. The black dots in Fig. 1 represent the elastic scattering events where the **k** vector of the electron changes its direction. The probability P_R that the electron arrives at point R, is given by the square of the sum of the two partial wave functions $\Psi_I(R)$ and $\Psi_{II}(R)$ which describe the electron in the upper and lower branch of the metal loop :

$$P_R = |\Psi_I|^2 + |\Psi_{II}|^2 + \Psi_I^*\Psi_{II}e^{i\theta} + \Psi_{II}^*\Psi_I e^{-i\theta} . \qquad (4)$$

The wave character of the conduction electrons causes interference terms whose magnitude depends upon the phase shifts $\delta_I(\mathrm{i})$ and $\delta_{II}(\mathrm{j})$ occurring during each of the elastic scattering events. The phase factor θ is given by :

$$\theta = \sum_i \delta_I(i) - \sum_j \delta_{II}(j) . \qquad (5)$$

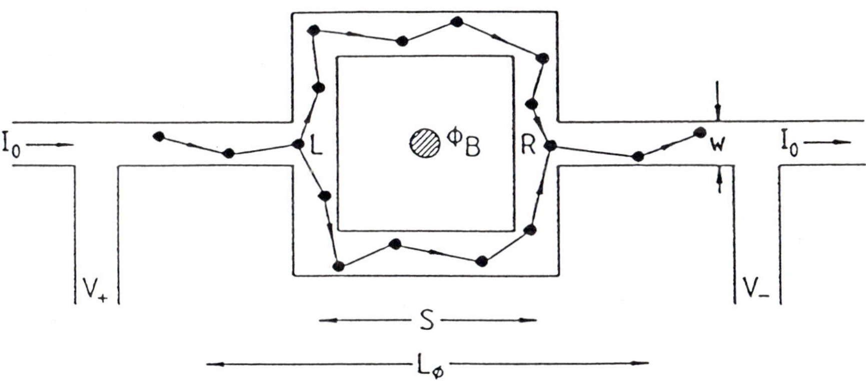

Figure 1 - *Four-terminal measurement of the interference between split conduction electron waves in a disordered metal film loop. The black dots represent the elastic scattering of the electron waves at lattice defects or impurities. The magnetic flux Φ_B is completely confined to the interior of the loop.*

Depending upon the impurity configuration in the upper and lower branch, the interference at point R is either constructive or destructive. When a magnetic flux Φ_B is applied, the Aharonov-Bohm effect induces an additional phase difference φ given by Eq. (3).

Therefore, the probability $P_R(\Phi_B)$ will be an oscillating function of Φ_B. From Eq. (4) we find that:

$$\begin{aligned} P_R(\Phi_B) &= |\Psi_I|^2 + |\Psi_{II}|^2 + \Psi_I^*\Psi_{II}e^{i(\varphi+\theta)} + \Psi_{II}^*\Psi_I e^{-i(\varphi+\theta)} \\ &= 2|\Psi_I|^2\,[1+\cos(\varphi+\theta)] \\ &= 2|\Psi_I|^2\,[1+\cos(2\pi\frac{\Phi_B}{\Phi_\circ}+\theta)]\,, \end{aligned} \tag{6}$$

if we assume that in zero magnetic field the probability to diffuse via the upper and the lower branch is identical ($|\Psi_I(R)|^2 = |\Psi_{II}(R)|^2$).

Until 1981 experimental confirmation of the presence of interference effects in metals seemed impossible. Indeed, Eq. (6) is valid only for a single electron. A real metal loop has a finite width w and a finite thickness d, both being much larger than the Fermi wavelength λ_F for the electrons. This implies that we have to consider the situation where a dense gas of conduction electrons with velocity v_F is moving through the arms of the loop. Each of the electrons will diffuse along a different path with a different phase factor θ. Due to the statistical averaging over the different paths, the interference effects will disappear and the conductivity is proportional to the average elastic mean free path l_{el}, as predicted by classical transport theories. Within the Einstein approach, the conductivity $\sigma_\circ$ is determined by the density of states $N(E_F)$ at the Fermi energy E_F, and the diffusion constant $D = \frac{1}{3}v_F l_{el}$:

$$\sigma_\circ = e^2 N(E_F) D\,. \tag{7}$$

The Einstein model, as well as the more complicated calculations which are based on the Boltzmann transport formalism, completely neglect the interference effects between partial waves of the charge carriers in a metal. The quantum-mechanical nature influences the conductivity only indirectly via an enormous enhancement of the carrier energy to a value $E_F \gg k_B T$.

Since 1981 theoretical calculations and detailed experiments clearly indicated that Eq. (7) must be corrected when one considers the electronic conduction on a length scale which is smaller than the phase coherence length L_φ. Since for pure metals, $L_\varphi \sim 1$ μm at very low temperatures, this regime is usually refered to as the "mesoscopic regime". This regime separates the microscopic, atomic regime where quantum mechanics dominates, from the macroscopic regime where classical mechanics is valid. As first suggested by Gefen, Imry and Azbel [11], the conductance $G = I_\circ/(V_+ - V_-)$ of the mesoscopic loop (shown in Fig. 1), will oscillate about its average value $G_\circ$ (calculated from Eq. (7)) as a function of the magnetic flux Φ_B threading the loop. More detailed calculations based on a Green's function approach allows one to obtain a quantitative estimate of the oscillation amplitudes [12] :

$$\begin{aligned} G = G_\circ \;&+\; A_1 G_{un}\cos(2\pi\frac{\Phi_B}{\Phi_\circ}+\alpha_1)exp(\frac{-2S}{L_\varphi}) \\ &+\; A_2 G_{un}\cos(2\pi\frac{\Phi_B}{\Phi_\circ/2}+\alpha_2)exp(\frac{-4S}{L_\varphi}) + \cdots \end{aligned} \tag{8}$$

In Eq. (8), A_1 and A_2 are sample-dependent correction factors of order unity, and the universal conductance $G_{un} = e^2/h \simeq 4 \times 10^{-5}\Omega^{-1}$. While the second term in Eq. (8) corresponds to the usual Aharonov-Bohm effect, the other terms correspond to electron waves that travel several times around the loop before interfering. Due to the exponential damping of the magnetoconductance oscillations for $2S > L_\varphi$, the amplitude of the higher harmonic $h/2e$ oscillations will be small when compared to the amplitude of the h/e oscillations.

Equation (8) still contains the phase factors α_1 and α_2. Apparently, an arbitrary value for α_1 and α_2 implies a violation of the well-known Onsager principle requiring a magnetoconductance which is symmetric around zero field : $G(B) = G(-B)$. For a two-terminal measurement (common voltage and current probes), one can show that α_1 and α_2 are equal to 0 or π [13]. For a four-terminal measurement on a mesoscopic scale (see Fig. 1), the measured voltage $V_+ - V_-$ also will include an anti-symmetric Hall voltage V_H ($V_H(B) = -V_H(-B)$) which also shows h/e oscillations with an amplitude comparable to the amplitude of the longitudinal voltage oscillations. The presence of the anti-symmetric Hall voltage explains the apparent asymmetry of the magnetoconductance without having to invoke a violation of the Onsager principle [14].

The most surprising consequence of Eq. (8) is the fact that the amplitude of the h/e conductance oscillations is independent of the sample size and the amount of disorder as long as $L_\varphi \simeq 2S$. The conductance oscillations may be considered as statistical fluctuations of the conductance G. Equation (8) implies that the conductance is not a self-averaging property in the mesoscopic regime, and the conductance fluctuations have a "universal" amplitude comparable to $G_{un} = e^2/h$, independent of the amount of impurity averaging [15]. For a macroscopic sample containing a large number N of mesoscopic units, the amplitude of the interference effects will be reduced by a factor $\sqrt{N}$. For the loop shown in Fig. 1, this implies that the constant A_1 in Eq. (8) must be divided by $\sqrt{N}$ if an array of N such loops is considered.

3 Interference processes in the macroscopic regime

Since the development in 1979 of the scaling theory for the metal-insulator transition, it has become clear that a special class of interference processes survives the impurity averaging on a macroscopic scale [16]. These interference processes are caused by two partial waves which travel completely around the loop shown in Fig. 1, and interfere again at point L. Since the two wave functions $\Psi_I(L)$ and $\Psi_{II}(L)$ correspond to time-reversed states, the phase factor $\theta \equiv 0$ in Eq. (6). Consequently, the magnetoconductance oscillations with flux period $\Phi_o = h/2e$ resulting from this coherent back-scattering on a mesoscopic scale, survive in a macroscopic sample.

In a thin metal film, where the loop structure is not present, many different areas S^2 can be encircled by the diffusing electron waves. In the absence of a magnetic field, the enhanced back-scattering probability resulting from the interference will cause the well-known weak electron localization [16]. Due to the rapid increase of L_φ at low temperatures, the resistance $R = G^{-1}$ will diverge logarithmically. In a magnetic field the conductance oscillations are washed out and are replaced by a uniform background. Numerous experiments on thin metal films have shown that an analysis of this background is a unique and extremely powerful tool to study characteristic scattering processes : the inelastic scattering due to the electron-electron or the electron-phonon interaction, the spin-flip scattering caused by the Kondo effect in dilute magnetic alloys [8], the enhancement of the spin-orbit scattering at a disordered film surface [16], etc... .

When a thin-walled metal cylinder with diameter $\phi = 2r$ is placed in an axial magnetic field, all the elecron waves that are elastically back-scattered around the cylinder will enclose the same magnetic flux $\Phi_B = \pi r^2 B$. In 1981, Altshuler, Aronov and Spivak (AAS) [17] used a Green's function perturbation calculation to show that the resistance R measured between the top and the bottom of the cylinder will oscillate with a flux period $h/2e$ and an amplitude ΔR given by :

$$\frac{\Delta R}{R} \simeq G_{un} R_{\square} \cos\left(\frac{2\pi r}{L_\varphi}\right) K_o\left(\frac{2\pi r}{L_\varphi}\right) . \tag{9}$$

$R_{\square}$ is the resistance per square of the metal cylinder wall and $K_o(x)$ is the McDonald function. When the cylinder diameter ϕ becomes larger than the phase coherence length L_φ, the McDonald function results in an exponential damping of the resistance oscillations. Apparently, the amplitude of the $h/2e$ oscillations can be enhanced considerably by increasing the disorder so that the resistance per square $R_{\square}$ becomes comparable to G_{un}^{-1}. Unfortunately, the inelastic diffusion length L_{in} varies inversely proportional to $R_{\square}$ in disordered metal films. The exponential damping caused by the decrease of the phase coherence length L_φ will be much stronger than the enhancement of the oscillation amplitude caused by a larger $R_{\square}$ value. It is important to note that the AAS $h/2e$ oscillations are only a fraction of the higher harmonic $h/2e$ Aharonov-Bohm oscillations given by the third term in Eq. (8).

Since the AAS oscillations result from the interference between time-reversed states, these oscillations will be destroyed by a high magnetic field or by the presence of magnetic impurities [8]. This effect is very similar to the destruction of Cooper pairs in a superconductor. On the other hand, the Aharonov-Bohm oscillations should still be present at high magnetic fields and in loops which are doped with magnetic impurities.

Experimentally, thin-walled metal cylinders with diameter $\phi \sim 1$ μm can be fabricated by evaporating a 20 nm thick metal film onto a quartz fiber which is stretched over a hole (diameter ~ 1 cm) in a glass substrate. During the metal evaporation, the substrate is rotated in a reduced helium atmosphere to ensure a uniform thickness of the cylinder wall. In Fig. 2 we show the low field magnetoresistance (field parallel to the fiber axis) for three Mg cylinders with different diameter ϕ at $T \simeq 1.5$ K. These experimental results, obtained at Leuven in 1983 [18], clearly confirm the existence of the $h/2e$ oscillations which were observed for the first time by Sharvin and Sharvin in 1981 [19].

As shown by the full curves in Fig. 2, the $h/2e$ oscillations can be fitted nicely by the AAS theory when we assume a phase coherence length $L_\varphi \simeq 2.0$ μm. The exponential damping of the oscillation amplitude which occurs when $L_\varphi < \phi$ is clearly present. The disappearance of the $h/2e$ oscillations at higher magnetic fields is mainly due to the magnetic field penetrating into the cylinder walls. This penetration causes a destruction of the time-reversal symmetry between the two partial electron waves which are back-scattered around the cylinder in the clockwise and counter-clockwise direction.

At higher temperatures, the phase coherence length for electron waves traveling around the cylinder rapidly decreases due to the inelastic electron-electron and electron-phonon scattering. This results in an exponential damping of the AAS oscillations above $T = 1.5$ K where $L_{\varphi}(T) \simeq L_{in}(T) < \phi$. This is illustrated in Fig. 3 for a Mg cylinder with diameter $\phi = 1.20$ μm.

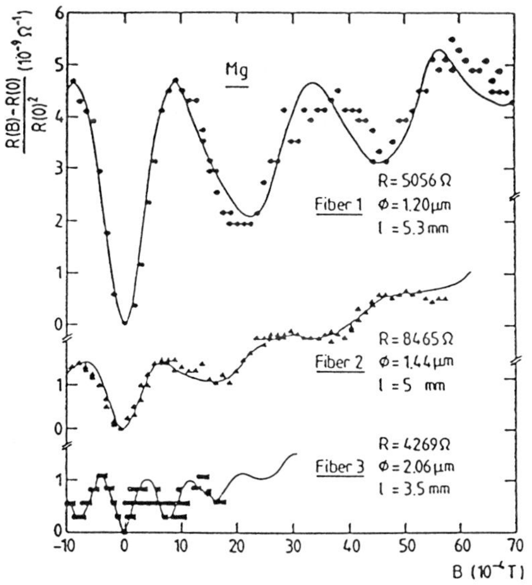

Figure 2 - *Experimental observation of the AAS magnetoresistance oscillations with flux period $\Phi_o = h/2e$ in thin-walled Mg cylinders with different diameter $\phi = 2r$ and length l at $T = 1.5$ K. The field B is parallel to the cylinder axis.(From ref. [18])*

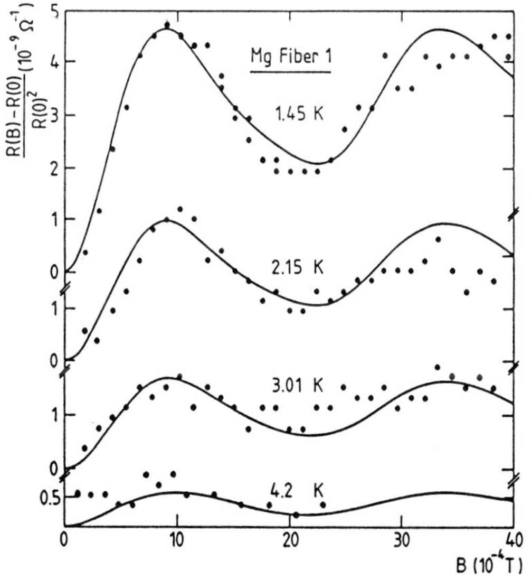

Figure 3 - *Temperature dependence of the AAS magnetoresistance oscillations in a thin-walled Mg cylinder with diameter $\phi = 1.20$ μm.(From ref. [18])*

4 Interference processes in the mesoscopic regime

The experiments performed on metal cylinders clearly have proven that interference processes between electron waves influence the electronic conduction even on a macroscopic scale. The coherent back-scattering, which causes the weak electron localization in a plane metal film, becomes directly observable via the $h/2e$ magnetoresistance oscillations. Since the cylinders are much longer than the phase coherence length L_φ, the h/e oscillations caused by the direct interference between split electron waves can not be observed.

Equation (8) predicts that the h/e oscillations should dominate the $h/2e$ oscillations if one would succeed in preparing a mesoscopic loop having a width and a thickness $w, d \ll L_\varphi$. Recent developments in electron beam lithography made it possible to prepare mesoscopic loops with a diameter $\phi \leq 1$ μm. The first experiments performed by Webb, Washburn, Umbach and Laibowitz [20] failed to show periodic h/e oscillations. Instead, aperiodic (but reproducible) magnetoresistance fluctuations appeared with an amplitude which strongly increased at lower temperatures. As pointed out by Stone [21], the aperiodic oscillations are caused by the magnetic flux penetrating the arms of the loop. Direct interference processes occuring within the arms of the loop will produce sample specific oscillations ("magnetofingerprints") with different flux periods depending on the enclosed area. For a loop with a poor aspect ratio, i.e., w is not much smaller than ϕ, the flux Φ_B through the metal forming the loop is comparable to the flux enclosed by the partial waves which produce the h/e oscillations. The h/e oscillations will therefore be masked by the aperiodic background, since both effects occur with a comparable characteristic field scale.

In 1984, Webb and his coworkers [22] succeeded in fabricating a Au ring with a very good aspect ratio. This loop with a diameter $\phi \simeq 0.8$ μm and $w \simeq d \simeq 0.04$ μm was obtained by "contamination" lithography in a transmission electron microscope. Figure 4 shows a drawing of the mesoscopic loop together with a typical magnetoresistance trace which clearly shows the h/e oscillations. Because of the good aspect ratio, the periodic h/e oscillations can easily be distinguished from the aperiodic fluctuations which appear as a slowly varying background. Figure 4 also shows the Fourier transform of the magnetoresistance trace. The sharp peak corresponding to the h/e effect is much stronger than the second harmonic $h/2e$ effect. This is in agreement with Eq. (8) which predicts an additional exponential damping of the higher harmonic effects when the phase coherence length L_φ is smaller than half the circumference of the loop. Since the magnetoresistance has been measured at high fields, the AAS $h/2e$ oscillations are not present. The peak appearing at low frequencies results from the slowly varying aperiodic background.

Although the ring experiments confirm the importance of the direct interference between partial conduction electron waves in the mesoscopic regime, the presence of a pure Aharonov-Bohm effect is not obvious. Since the magnetic flux also is piercing through the arms of the loop, the **B** field interacts with the electrons via the Lorentz force. The experimental results shown in Fig. 4 therefore can not be considered as an unambiguous proof of the existence of the Aharonov-Bohm effect in normal metals. The clear separation between the h/e signal and the aperiodic background in the Fourier transform does, however, strongly suggest that the electrons experience the presence of the magnetic flux in the interior of the loop via the Aharonov-Bohm effect.

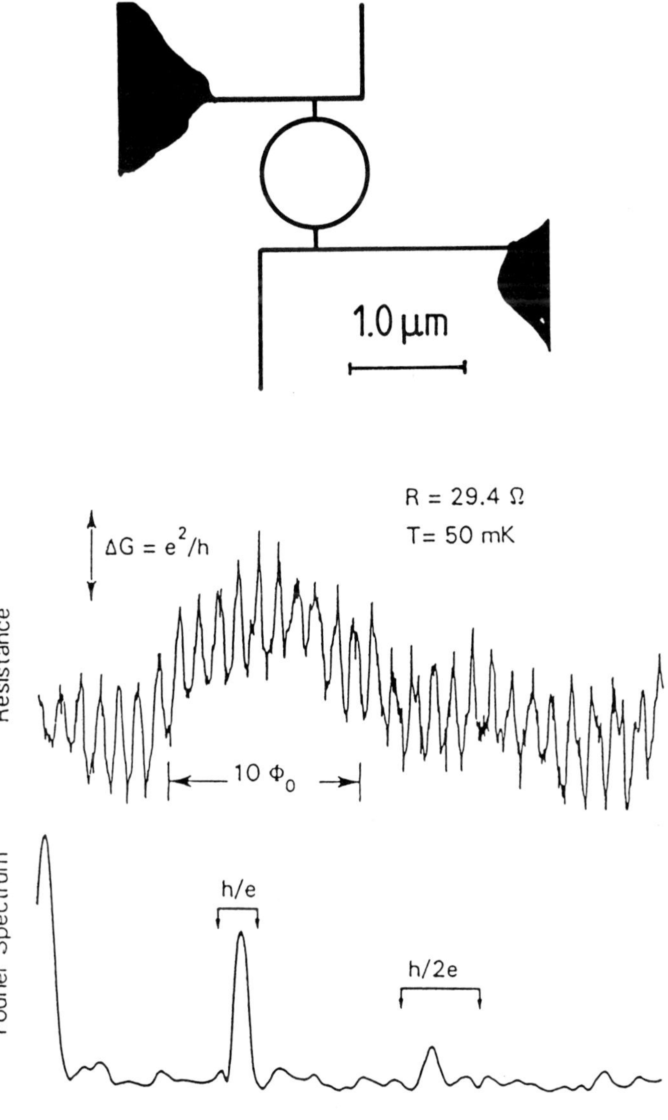

Figure 4 - *Experimental observation of the solid-state Aharonov-Bohm effect in the submicron Au loop shown in the drawing. The upper curve is a typical magnetoresistance trace showing the Aharonov- Bohm oscillations with fundamental flux period* $\Phi_o = h/e$ *and the slower aperiodic oscillations which are caused by the penetration of the magnetic flux into the arms of the loop. The lower curve shows the Fourier spectrum of the magnetoresistance trace.(From ref. [22])*

5 Stochastic averaging of mesoscopic fluctuations

From the experimental results shown in Fig. 2 and Fig. 3, we may conclude that the resistance oscillations which appear for the cylinder geometry are caused by the coherent back-scattering. On the other hand, Fig. 4 indicates that the flux period for the Aharonov-Bohm effect in a mesoscopic loop is determined by the direct interference between split electron waves. These findings confirm that in the mesoscopic regime, impurity averaging is incomplete and at higher magnetic fields, the h/e effect dominates the $h/2e$ effect. In a macroscopic sample, the amplitude of the h/e effect is reduced by a factor $\sqrt{N}$, where N is the number of mesoscopic units within the sample.

In order to estimate this reduction for the cylinder geometry, we can chop the macroscopic cylinder with height H into mesoscopic loops with thickness $d \ll L_\varphi$, as shown in Fig. 5. A cylinder of height $H \sim 1$ cm can be chopped into $N \sim 10^6$ loops with thickness $d \sim 10$ nm. The impurity averaging which occurs on a macroscopic scale will reduce the

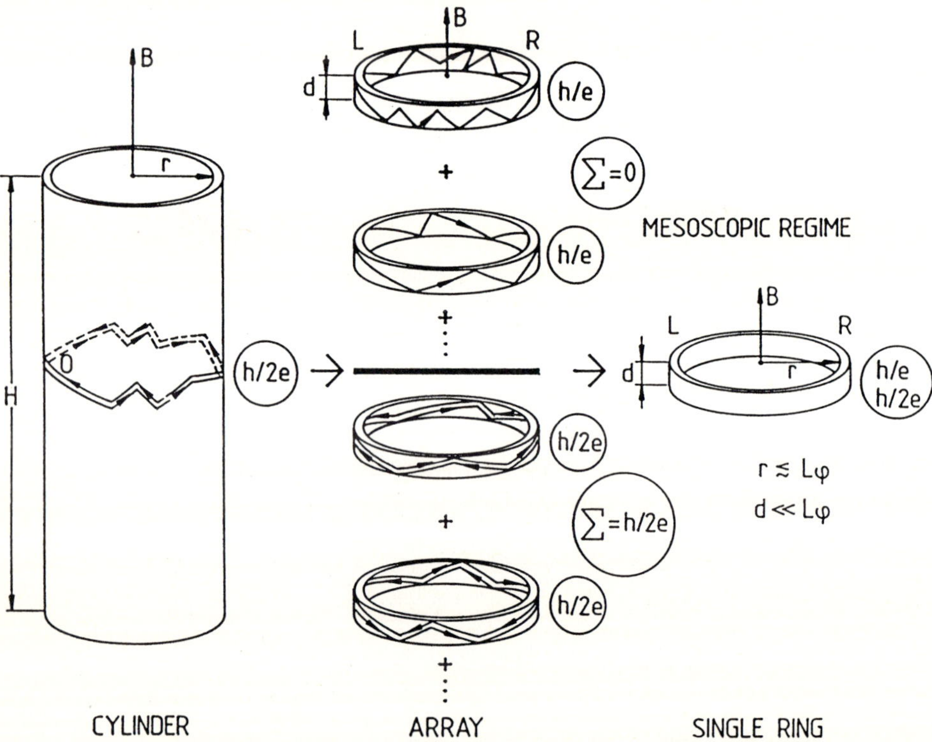

Figure 5 - *Schematic representation of the transition between the "macroscopic" cylinder geometry and the "mesoscopic" loop geometry. The sample specific h/e Aharonov-Bohm oscillations add up incoherently, so that they are no longer observed for a large number N of loops or in a long cylinder. On the other hand, the weaker $h/2e$ AAS oscillations survive for large N since they are independent of the impurity configuration in the sample.*

h/e oscillation amplitude by a factor $\sqrt{N} \sim 10^3$. Since, for the cylinder geometry, the $h/2e$ oscillations can barely be resolved from the intrinsic noise produced by the cylinder resistance, the h/e oscillations will remain invisible.

The equivalence between the cylinder geometry and a large array of mesoscopic loops (see Fig. 5) also has been confirmed experimentally [23]. The magnetoresistance for an array and a cylinder look identical. At low fields the $h/2e$ oscillations dominate. At higher fields the oscillations are damped and replaced by a slowly varying background which is caused by the penetration of the **B** field into the arms of the loop or into the cylinder wall.

When the number N of mesoscopic units in a sample is not very large, the transition toward the macroscopic regime will be incomplete. This is a direct consequence of the slow, statistical character of the impurity averaging. This slow averaging has been beautifully illustrated by experiments with small arrays of Ag loops [24]. These arrays were obtained by a combination of electron-beam lithography in a scanning electron microscope and lift-off techniques.

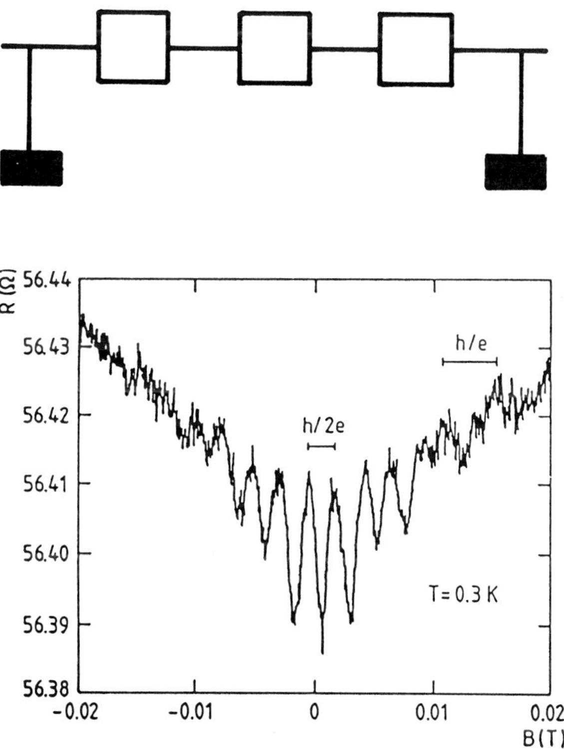

Figure 6 - *Observation of magnetoresistance oscillations for N loops in series. The drawing shows a typical sample with $N = 3$ and a linewidth of* 70 *nm. The low-field magnetoresistance trace has been obtained for a single loop ($N = 1$) and shows the presence of both the Aharonov-Bohm h/e and the AAS $h/2e$ oscillations.(From ref. [24])*

Figure 6 shows the drawing of a typical sample with $N = 3$. The loop structure consists of line segments with a width $w \simeq 70$ nm, a thickness $d \simeq 20$ nm and a length $S \simeq 0.95$ μm. For the Ag material, L_φ is limited to a value of about 2 μm at low temperatures. The loops therefore can be considered as independent mesoscopic units since only interference effects within a single loop will be important. The low-field magnetoresistance which is shown in Fig. 6 has been obtained for a single loop ($N = 1$). The AAS $h/2e$ oscillations which dominate near zero field are gradually replaced by the Aharonov-Bohm h/e oscillations at higher fields.

A typical magnetoresistance trace at higher fields is shown in Fig. 7 for the sample with $N = 30$. Due to the high resolution of the resistance measurements, h/e oscillations can be clearly observed.

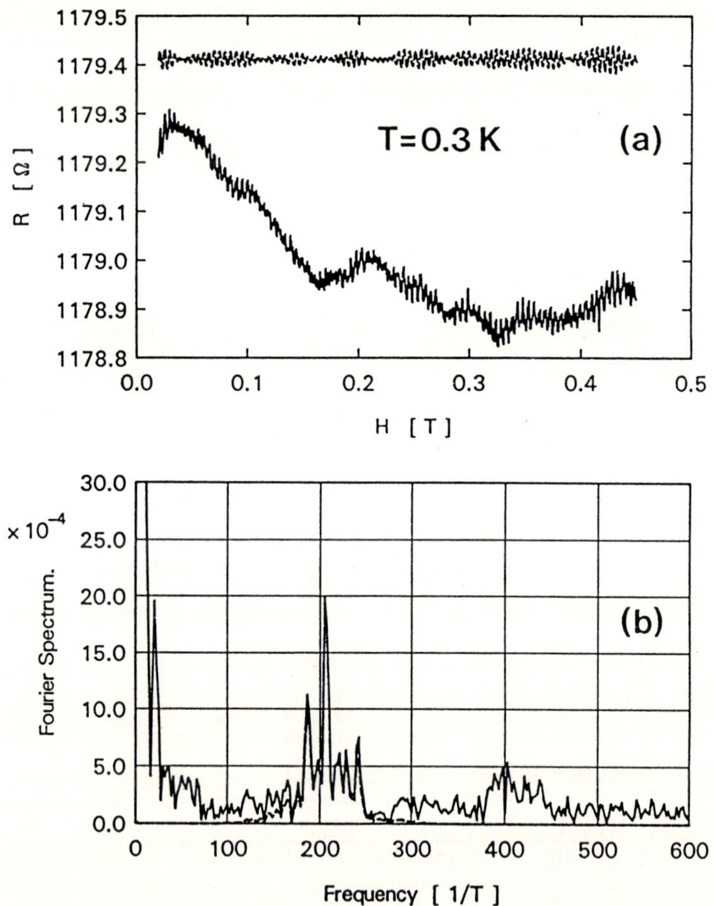

Figure 7 - *(a) Magnetoresistance trace at higher fields for* $N = 30$ *loops in series before (lower curve) and after (upper curve) removal of the aperiodic background by digital filtering. (b) Fourier transform of the data shown in (a). The dashed curve was used to calculate the beating of the* h/e *signal shown by the upper curve in (a).*

Although the total sample length is about 50 μm, the sample-specific h/e oscillations and aperiodic background magnetoresistance are still present. Fig. 7 also demonstrates how the Fourier transform of the magnetoresistance trace can be used for digital filtering of the h/e signal.

The variation of the $h/2e$ (AAS) and the h/e oscillation amplitude has been plotted in Fig. 8 as a function of N. Since the AAS $h/2e$ oscillations do not depend upon the amount of impurity averaging, their amplitude is independent of N. On the other hand, the h/e oscillation amplitude varies inversely proportional to $\sqrt{N}$, confirming the slow, stochastic character of the impurity averaging.

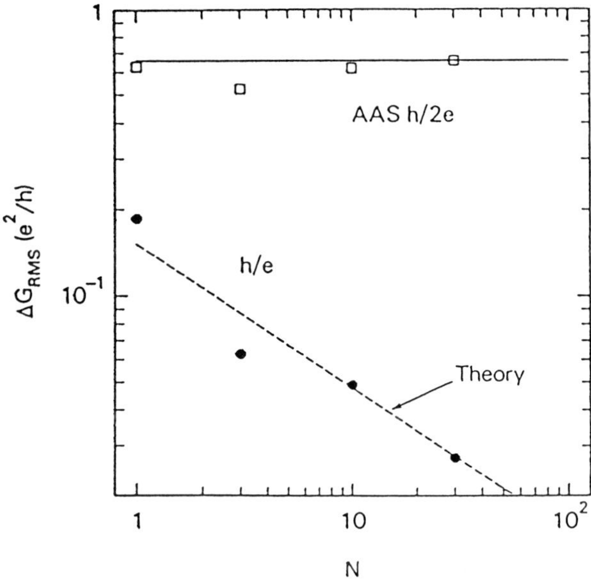

Figure 8 - *The slow ensemble averaging of direct interference processes in a series of N Ag loops. The incoherent addition of the direct interference processes results in a stochastic averaging of the Aharonov-Bohm h/e oscillation amplitude inversely proportional to $\sqrt{N}$. The amplitude of the AAS $h/2e$ oscillations is independent of N.(From ref. [24])*

6 Non-locality of the interference effects

The presence of the solid-state Aharonov-Bohm effect in mesoscopic structures clearly demonstrates that the conduction electrons in a metal have a well-defined wave character. Similar experiments on semiconductor inversion layers [25], which have not been discussed in this review, confirm that the wave character of the charge carriers also influences the conduction process in mesoscopic doped semiconductors. As indicated by Büttiker [26], the presence of the interference effects on a scale L_φ inevitably implies the non-local nature of the flux quantization effects. The electronic properties of mesoscopic samples therefore also will be influenced by regions outside the voltage probes (see Fig. 1).

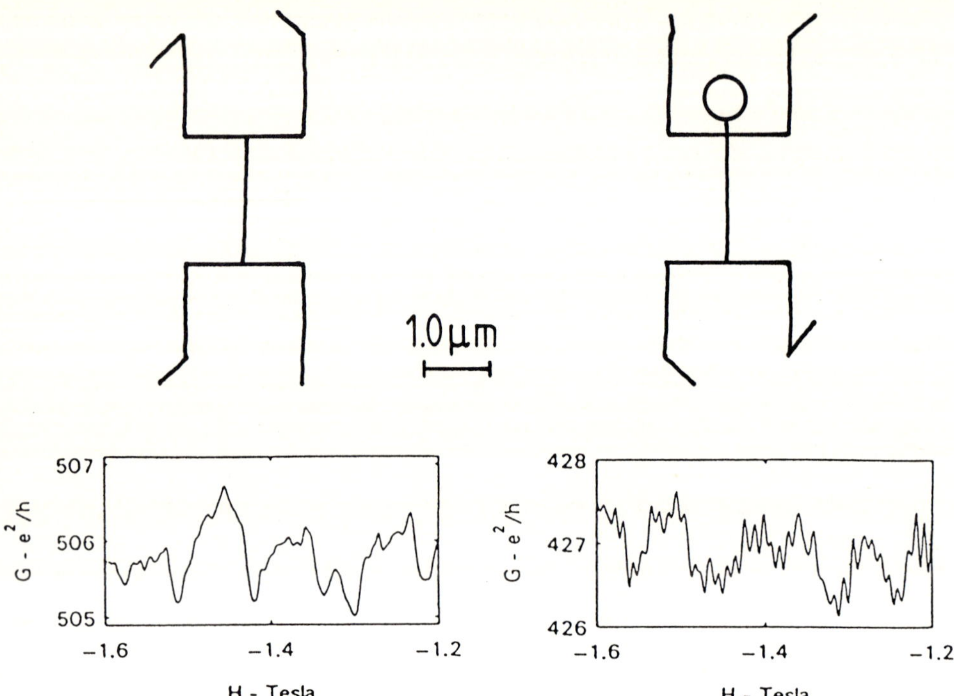

Figure 9 - *Four-terminal measurement of the magnetoconductance of a mesoscopic Au line with and without a loop attached to the outside of the contact probes. The line without the loop shows only aperiodic fluctuations. On the other hand, the line with the loop shows additional oscillations having the Aharonov-Bohm flux period h/e for the loop outside of the contact probes.(From ref. [27])*

Umbach, Santhanam, Webb and one of the authors [27], recently have provided a very graphic and easy to understand demonstration of the non-local interference effects. As shown by the drawing in Fig. 9, two identical mesoscopic Au lines (length 2 μm and linewidth 62 nm) have been prepared by contamination lithography in a transmission electron microscope. A loop with a diameter of 0.7 μm has been connected to one of the lines near the junction of a pair of current and voltage leads. Figure 9 also shows a typical magnetoconductance trace for the two lines. While the single line produces only aperiodic fluctuations which are caused by the penetration of the magnetic field into the line segments, the line with the externally connected loop shows additional h/e oscillations.

The results presented in Fig. 9 imply that a four-terminal measurement fails in the mesoscopic regime. Consequently, the classical "scaling down" approach to modeling small circuits is reaching its limit. The sensitivity of the interference to the specific impurity configuration will result in an important electronic noise at higher temperatures when defects can move in a metal or charge traps are present in a semiconducting device.

However, when the periodic Aharanov-Bohm oscillations can be made large enough, it is also conceivable that devices based on the quantum interference will become important in the future.

Acknowledgements

We are much indebted to M. Büttiker, Y. Gefen, Y. Imry, R. Landauer, R. Laibowitz, P. Santhanam, A. Stone, C. Umbach, S. Washburn and R. Webb for enlightening discussions. This work has been supported by the Belgian Inter-University Institute for Nuclear Sciences (IIKW), the Inter-University Attraction Poles (IUAP) and Concerted Action (GOA) Programmes. C.V.H. is a Research Associate of the Belgian National Fund for Scientific Research (NFWO).

References

[1] L. de Broglie in *Selected Papers on Wave Mechanics* , edited by L. de Broglie and L. Brillouin. Blackie, London (1929).
[2] C. Davisson and L.H. Germer, Phys. Rev. **30**, 705 (1927).
[3] A. Tonomura, N. Osakabe, T. Matsuda, T. Kawasaki, J. Endo, S. Yano and H. Yamada, Phys. Rev. Lett. **56**, 792 (1986).
[4] Y. Aharonov and D. Bohm, Phys. Rev. **115**, 485 (1959).
[5] B.S. Deaver and W.M. Fairbank, Phys. Rev. Lett. **7**, 43 (1961).
[6] R. Landauer, unpublished IBM proposal.
[7] For a recent review see : S. Washburn and R.A. Webb, Adv. Phys. **35**, 375 (1986); A.G. Aronov and Yu.V. Sharvin, Rev. Mod. Phys. **59**, 755 (1987); R.A. Webb and S. Washburn, Physics Today **41**, 46 (1988).
[8] C. Van Haesendonck, J. Vranken, and Y. Bruynseraede, Phys. Rev. Lett. **58**, 1968 (1987).
[9] R. Landauer in *Localization, Interaction and Transport Phenomena* , edited by G. Bergmann, Y. Bruynseraede, and B. Kramer. Springer, Heidelberg (1985).
[10] M. Büttiker, Y. Imry, and M.Ya. Azbel, Phys. Rev. A **30**, 1982 (1984).
[11] Y. Gefen, Y. Imry, and M.Ya. Azbel, Phys. Rev. Lett. **52**, 129 (1984).
[12] P.A. Lee, A.D. Stone, and H. Fukuyama, Phys. Rev. B **35**, 1039 (1987).
[13] A.D. Stone and Y. Imry, Phys. Rev. Lett. **56**, 189 (1986).
[14] M. Büttiker, Phys. Rev. Lett. **57**, 1761 (1986).
[15] P.A. Lee and A.D. Stone, Phys. Rev. Lett. **55**, 1622 (1985).
[16] G. Bergmann, Phys. Rep. **107**, 1 (1984).
[17] B.L. Altshuler, A.G. Aronov, and B.Z. Spivak, JETP Lett. **33**, 94 (1981).
[18] M. Gijs, C. Van Haesendonck, and Y. Bruynseraede, Phys. Rev. Lett. **52**, 2069 (1984).
[19] D.Yu. Sharvin and Yu.V. Sharvin, JETP Lett. **34**, 272 (1981).
[20] C.P. Umbach, S. Washburn, R.B. Laibowitz, and R.A. Webb, Phys. Rev. B **30**, 4048 (1984).
[21] A.D. Stone, Phys. Rev. Lett. **54**, 2692 (1985).
[22] R.A. Webb, S. Washburn, C.P. Umbach, and R.B. Laibowitz, Phys. Rev. Lett. **54**, 2696 (1985).

[23] B. Pannetier, J. Chaussy, R. Rammal, and P. Gandit, Phys. Rev. Lett. **53**, 718 (1984); G.J. Dolan, J.C. Licini, and D.J. Bishop, Phys. Rev. Lett. **56**, 1493 (1986).

[24] C.P. Umbach, C. Van Haesendonck, R.B. Laibowitz, S. Washburn, and R.A. Webb, Phys. Rev. Lett. **56**, 386 (1986).

[25] S. Datta, M.R. Melloch, S. Bandyopadhyay, R. Noren, M. Vaziri, M. Miller, and R. Reifenberger, Phys. Rev. Lett. **55**, 2344 (1985); W.J. Skocpol, P.M. Mankiewich, R.E. Howard, L.D. Jackel, D.M. Tennant, and A.D. Stone, Phys. Rev. Lett. **56**, 2865 (1986).

[26] M. Büttiker, Phys. Rev. B **32**, 1846 (1985).

[27] C.P. Umbach, P. Santhanam, C. Van Haesendonck, and R.A. Webb, Appl. Phys. Lett. **50**, 1289 (1987).

Giaever and Josephson Tunneling

Y. Bruynseraede, C. Vlekken and C. Van Haesendonck

Laboratorium voor Vaste Stof-Fysika en Magnetisme
Katholieke Universiteit Leuven, B-3030 Leuven (Belgium)

1 The tunneling phenomenon: historical review and basic concepts

If the motion of a particle in the neighbourhood of a potential barrier is treated wave-mechanically, it is found that there is a finite probability that the particle can leak through the barrier even though its kinetic energy is less than the potential energy corresponding to the height of the barrier. In other words, a particle impinging on the barrier will not necessarily be reflected, but may pass through the barrier and continue its forward motion.

Already in the late twenties, a number of paradoxial physical phenomena could be successfully explained using the tunneling phenomenon. Gamov [1] showed that the natural decay of a radioactive nucleus is due to tunneling of alpha particles through the Coulomb barrier which surrounds the nucleus. As first proposed by Fowler and Nordheim [2], the field emission of electrons from a cold metal surface under the application of a strong electric field is governed by electron tunneling through an electric field dependent potential barrier. Oppenheimer [3] showed that the ionisation of atomic hydrogen in a strong electric field is also a quantum-mechanical tunneling phenomenon.

The role of tunneling effects in chemical reactions was recognized soon after the introduction of quantum mechanics. In 1927, Hund demonstrated the importance of quantum tunneling for intra-molecular rearrangements in molecules (e.g., ammonia), which is manifested by tunnel splitting of vibrational spectra. In 1929, Bourgin discussed the role of the tunnel effect in chemical kinetics [4]. Since those pioneer days, tunneling has developed in many fields, encompassing biology, crystalline and amorphous solids, electronic devices, tunneling microscopy, etc...[5,6,7]. Recently [8], a search for tunneling phenomena in chemical reactions at very low temperatures was performed. According to the Arrhenius law in classical chemistry, the chemical reaction rate decreases exponentially with temperature and ceases at absolute zero. Experimentally, this law accurately describes the reaction rate at high temperatures, but fails at low temperatures due to the tunneling of entire atoms through barriers caused by the repulsive forces of other atoms.

Before 1940, electron tunneling in solids was mainly studied from a theoretical point of view. At that time a simple model was introduced to describe tunneling between two metal electrodes separated by either a vacuum or an insulating barrier (Sommerfeld and Bethe [9], Frenkel [10]). In 1943, Zener [11] showed that electrons can tunnel from the valence band into the conduction band of a semiconductor by the application of a very large electric field. It was not until the late fifties that experiments clearly demonstrated

NATO ASI Series, Vol. F 59
Superconducting Electronics
Edited by H. Weinstock and M. Nisenoff

that in solid state devices many physical processes are dominated by electron tunneling. One of the key experiments was the measurement by Esaki in 1957 [12] of the tunneling current between two semiconductors. This sparked the development of the tunnel diode.

As pointed out by Harrison [13], tunneling through an insulating barrier between two metal electrodes gives rise to an ohmic behaviour ($I \propto V$) at low voltages - see also Sect. 2. The variation of the density of states N with energy E ($N(E) \propto E^{1/2}$ for a free-electron band) is exactly compensated by the energy dependence of the tunneling probability. Consequently, the band structure of metals near the Fermi level can not be determined via tunneling. Harrison's argument is not valid for deformations of the free-electron band structure ($N_\circ$) resulting from electron-electron interaction effects. The interaction will give rise to a renormalization of the density of states:

$$N(E) = N_\circ(E)\,[1 + d\Sigma/d\varepsilon(E)]^{-1} \ , \tag{1}$$

where Σ represents the self-energy which is the difference between the total energy E and the free-electron energy ε. Tunneling experiments allow one to obtain detailed measurements of the quantity $d\Sigma/d\varepsilon(E)$.

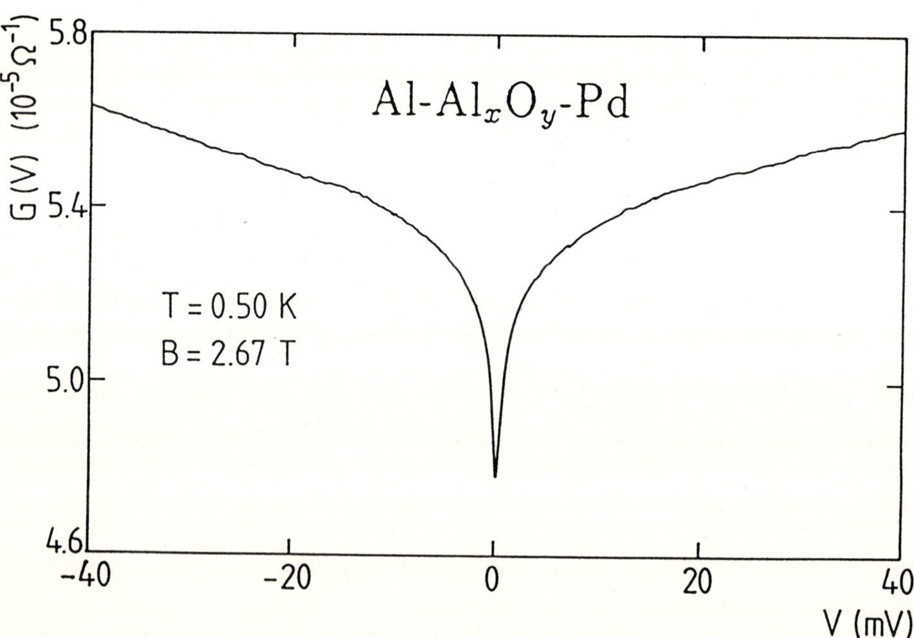

Figure 1 - *Dynamic conductance of an Al-Al$_x$O$_y$-Pd junction at low voltages. The magnetic field $B = 2.67$ T has been applied to suppress the superconductivity of the Al electrode.(From ref. [14])*

In very disordered metals the reduced electron diffusivity will limit the ability of the electrons to screen the Coulomb repulsion. This decrease of the screening effect will induce an anomalous increase of $d\Sigma/d\varepsilon$ near the Fermi level. The resulting dip in the density of states is directly reflected in tunneling measurements, as illustrated in Fig. 1. As will be shown in Sect. 2, the conductance $G(V) = dI/dV$ for tunneling between a clean Al film and a very disordered Pd film is directly proportional to the density of states, $N(E = eV)$, of the disordered Pd film.

Superconductive tunneling - for reviews see [15,16] - was discovered in the early sixties by Giaever [17] and Nicol et al. [18]. Using normal metal-insulator- superconductor (NIS) and superconductor-insulator-superconductor (SIS) sandwich structures, an I-V characteristic was observed which is directly related to the excitation spectrum of the superconducting electrode. In 1962 Josephson [19] predicted that in a SIS junction the macroscopic pair wave functions of both superconductors have a finite overlap, giving rise to tunneling of electron pairs through the oxide barrier. This electron pair tunneling produces a dissipationless current which flows at zero bias voltage. Josephson's prediction was experimentally confirmed for the first time by Anderson and Rowell in 1963 [20] and opened not only a new chapter in solid-state physics, but provided a wide variety of stimulating applications. One such application is a double junction device which acts as a very sensitive magnetic interferometer: the SQUID, an acronym for Superconducting Quantum Interference Device.

In 1982, a new and powerful microscope was developed by Binnig and Rohrer [21] using vacuum tunneling: the Scanning Tunneling Microscope (STM), which enables one to resolve mono-atomic steps on the surface of a crystal (Fig. 2). The fundamental mechanism is very simple and relies on the finite overlap of the electron wave function of a

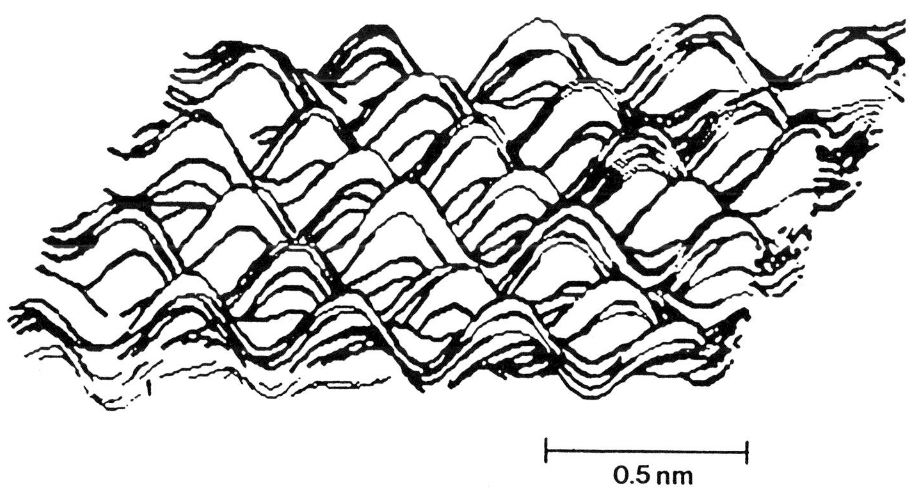

Figure 2 - *Graphite surface as seen by the scanning tunneling microscope.(From ref. [22])*

sharp metal needle and a conducting surface placed within a fraction of a nm of each other. By applying a small voltage between the needle and the surface, one produces a flow of electrons which tunnel through the small vacuum gap. Because the amplitude of the electron wave function decreases exponentially with distance in the vacuum barrier, the tunneling current is extremely sensitive to the separation between the needle tip and the sample surface. By scanning the needle across the sample, surface features of atomic dimensions show up as variations of the tunneling current.

In the tunneling microscope the effective tunneling area is extremely small. This implies that an extra charging energy e^2/C is needed to transfer an individual electron via the tunneling barrier. If we assume that the barrier thickness is 1 nm, and the junction area is about $(25\text{ nm})^2$, we obtain a junction capacity of 5×10^{-18} F. The charging energy will then be about 30 meV. As a result, a "Coulomb blockade" occurs at low temperatures around zero voltage, and a gap appears in the $I(V)$ characteristic when the STM is cooled to liquid-helium temperature [23]. The Coulomb blockade also can be observed at low temperatures in sub-micrometer junctions prepared by electron beam lithography [24] and for ultra-small (10 nm diameter) particles embedded in an insulating matrix. In Fig. 3 we show the experimental result for Ag particles in an oxide film [25]. The I-V characteristic displays not only the Coulomb blockade around zero voltage, but also an entire "staircase" of jumps in the current. Each jump corresponds to the addition of an extra electron on a silver particle.

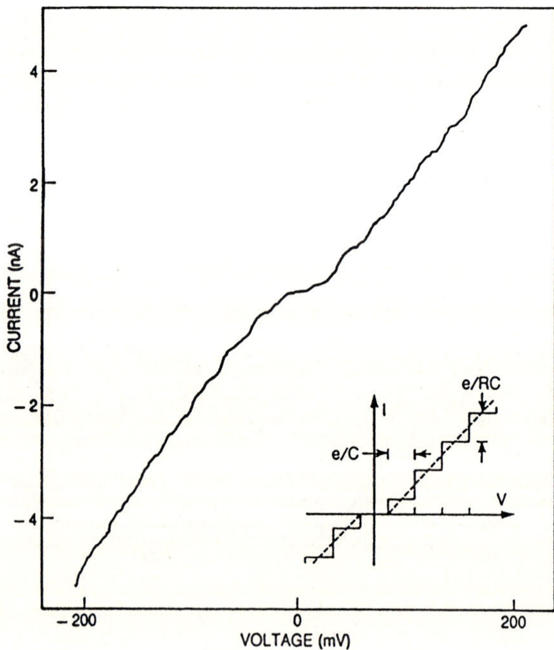

Figure 3 - *Coulomb staircase is a pattern of discrete jumps in the current at fixed intervals of voltage measured for Ag particles embedded in an oxide film.(From ref. [25])*

2 Formulation of the tunneling problem

A tunneling junction connected by electrical leads to a tunable current source and a voltmeter is shown schematically in Fig. 4(a). The two metal electrodes M_L and M_R are separated by a thin ($\sim$ 2 nm) insulating barrier (I). Using the potential barrier model illustrated in Fig. 4(b) and assuming that only a very small bias voltage V is present, we can calculate from the Schrödinger equation the current transmission ratio J_R/J_L through the junction with J_L (J_R) the incoming (transmitted) electron flux :

$$J_R/J_L \propto \frac{|\Psi_R(x)|^2}{|\Psi_L(x)|^2} \propto exp(-2\kappa d) . \qquad (2)$$

$\Psi_L(x)(\Psi_R(x))$ is the wave function in the left (right) electrode; $\kappa \sim \sqrt{\phi}$ is the decay constant for the wave function in the insulating barrier with height ϕ and thickness d. For ϕ =1 eV we obtain a current transmission ratio $J_R/J_L = 10^{-6}$ for a thickness $d = \frac{1}{2\kappa} ln(J_R/J_L) \simeq 1$ nm. Figure 5 shows the experimental results for Al_2O_3 tunneling barriers confirming the exponential decay of the current transmission with increasing barrier thickness d.

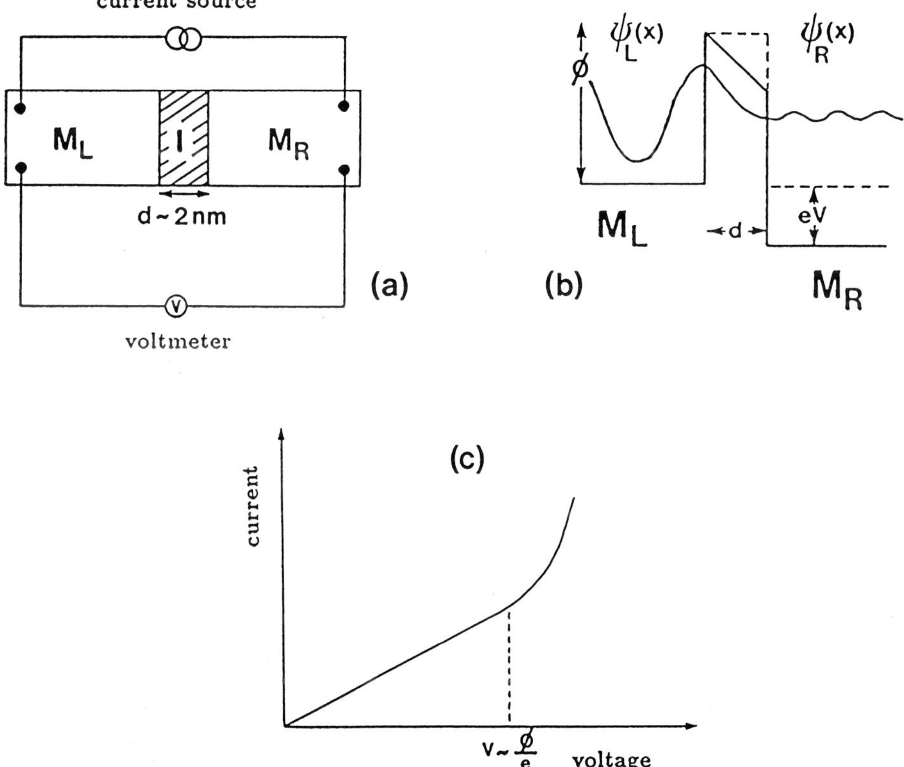

Figure 4 - *(a) Scheme of a tunneling junction; (b) potential barrier model for the NIN structure; (c) current-voltage characteristic for a NIN junction.*

As shown in Fig. 4(c) the tunneling current I versus applied voltage V shows an ohmic behaviour in the region where $eV \ll$ barrier height ϕ. When the applied voltage V is increased such that $eV \sim \phi$ the conductance varies quadratic ($G(V) = G(0) + cV^2 + ...$) and the detailed barrier shape becomes important as illustrated in Fig. 6. For still higher voltages V, the conductance increases exponentially.

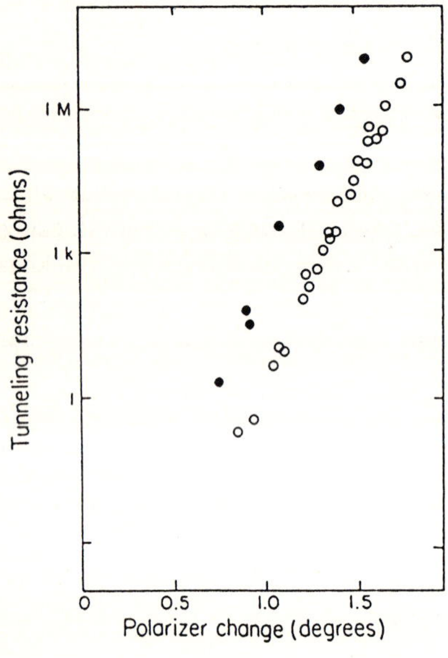

Figure 5 - *The exponential decay of the current transmission with increasing barrier thickness is demonstrated in a plot of the junction resistance versus tunneling barrier thickness (measured with an ellipsometer). A change in polarization of 0.53 ° corresponds to a change of the oxide layer thickness by 1 nm. The base electrode is an Al film and the barrier is Al_2O_3. The solid circles represent measurements obtained for a Pb top electrode, while the open circles are for an Al top electrode.(From ref. [26])*

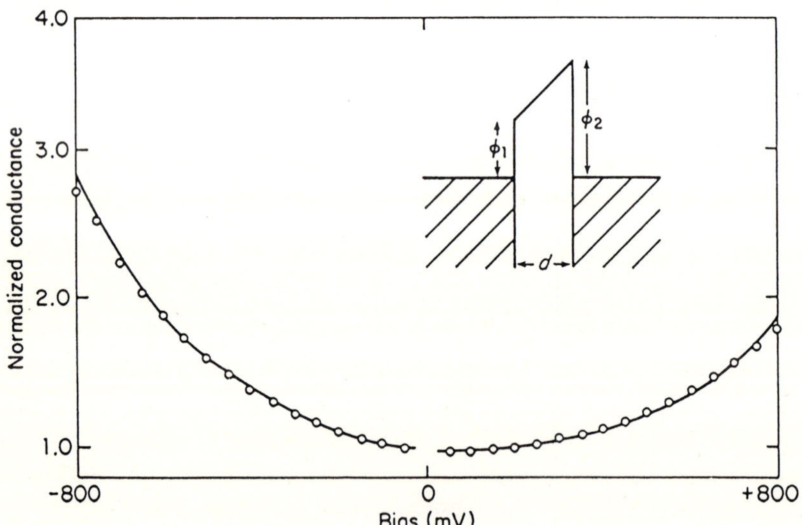

Figure 6 - *Measured conductance of an Al-Al_2O_3-Pb junction (full curve). The open circles have been calculated assuming sharp interfaces and the barrier parameters: ϕ_1= 1.57 V, ϕ_2 =4.07 V, d =1.16 nm.(From ref. [27])*

2.1 Giaever tunneling

A - Visualisation of the tunneling process

To visualize the tunneling process, we adopt a simple representation in terms of energy (E)-momentum (k) diagrams. The normal metal is represented in the E-k plane by the curve shown in Fig. 7(a). The dashed line is the portion of the free- electron parabola below the Fermi energy E_F, representing hole states. The electron-hole pair creation is then visualized with the excitation of two states of energy E_l and E_h.

In the case of a superconductor, all the condensed pairs are at the Fermi level and a minimum threshold energy 2Δ (with 2Δ the energy gap) is required for an excitation of a pair as shown in Fig. 7(b). The total excitation energy of a quasi particle is $E = (\varepsilon^2 + \Delta^2)^{1/2}$, corresponding to a self energy $\Sigma = (\varepsilon^2 + \Delta^2)^{1/2} - \varepsilon$. Using Eq. (1), we can calculate the renormalized density of states for an energy independent gap Δ (in the BCS approximation):

$$\begin{aligned} |E| \geq \Delta : N_S(E) &= N_N(0)\frac{E}{(E^2 - \Delta^2)^{1/2}} \\ |E| < \Delta : N_S(E) &= 0 \, , \end{aligned} \tag{3}$$

with $N_N(0)$ the normal-metal density of states near the Fermi level. In a superconductor, the distinction between electrons and holes is somewhat blurred, and the excitation spectrum in Fig. 7(b) should be understood in terms of superconducting quasiparticles rather than in terms of pure electron or hole excitations.

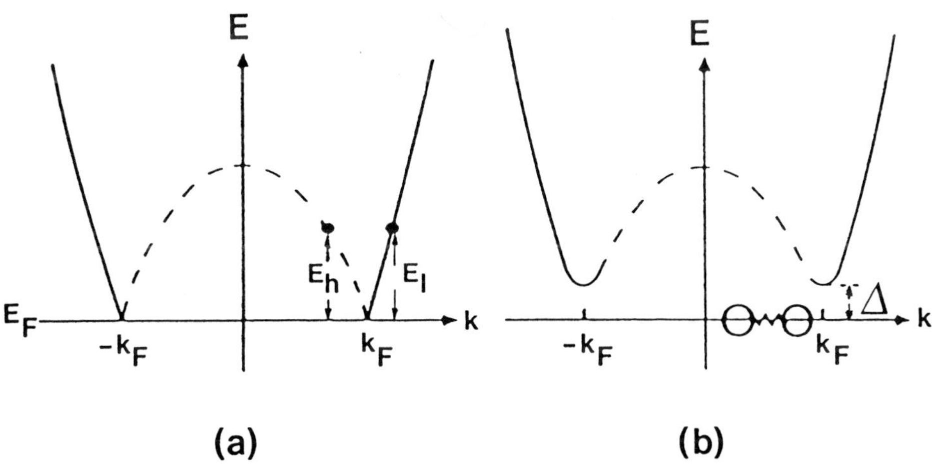

Figure 7 - *The excitation spectrum in the E-k representation for a normal metal (a) and a superconductor (b).*

B - Phenomenological theory of tunneling

Let us now calculate the net current I flowing across the tunneling junction illustrated in Fig. 4(a). The number of electrons which will move from left to right in an energy interval dE is proportional to the number of occupied states on the left, i.e., $N_L(E)f_L(E)dE$, with N_L, the density of states, and f_L, the Fermi distribution function. These electrons can only tunnel to the right if there are unoccupied states available, and the current is therefore also proportional to $N_R(E)(1-f_R(E))$. Finally, the current also is proportional to the tunneling probability $|T_n|^2$, where T_n is the tunneling matrix element describing the overlap between the electronic wave functions of the left and right electrode. After integrating over all allowed energies, the current flowing from left to right is given by:

$$I_{L\to R} = \frac{2\pi}{\hbar}\int_{-\infty}^{+\infty} |T_n|^2 N_L(E) f_L(E) N_R(E)(1-f_R(E))dE \ . \tag{4}$$

Assuming that an electron has an equal probability to tunnel in either direction, the net current is:

$$I = I_{L\to R} - I_{R\to L} = \frac{2\pi}{\hbar}\int_{-\infty}^{+\infty} |T_n|^2 N_L(E) N_R(E)[f_L(E) - f_R(E)]dE \ . \tag{5}$$

Under equilibrium conditions $I = 0$. When we apply a voltage V across the junction, the Fermi energy level on each side will be shifted with respect to one another by an energy eV, and a net current will flow:

$$I = \frac{2\pi}{\hbar}|T_n|^2 \int_{-\infty}^{+\infty} N_L(E) N_R(E+eV)[f_L(E) - f_R(E+eV)]dE, \tag{6}$$

assuming that T_n is independent of energy, which is a good approximation for small bias voltages.

When both metals are in the normal state, we make the assumption that N_L and N_R are constant and equal to the the density of states at the Fermi level. The tunneling current in a NIN junction is:

$$I_{NN} = C \times \int_{-\infty}^{+\infty} [f(E) - f(E+eV)]dE \ . \tag{7}$$

At low voltages, the Fermi function may be expanded into a power series and the Ohmic behaviour of the NIN junction is recovered:

$$I_{NN} = \sigma_N V \ . \tag{8}$$

Next we consider one of the electrodes to be in the superconducting state. The tunneling process in a NIS junction in the E-k plane is represented in Fig. 8. The transfer of an electron from the left (normal metal) to the right (superconductor) can have two possible final states, i.e., an electron-like excitation or a hole-like excitation. According to Eq. (6), the net current in a NIS junction is:

$$I_{NS} = C \times \int_{-\infty}^{+\infty} \frac{E}{|E^2 - \Delta^2|^{1/2}}[f(E) - f(E+eV)]dE \ . \tag{9}$$

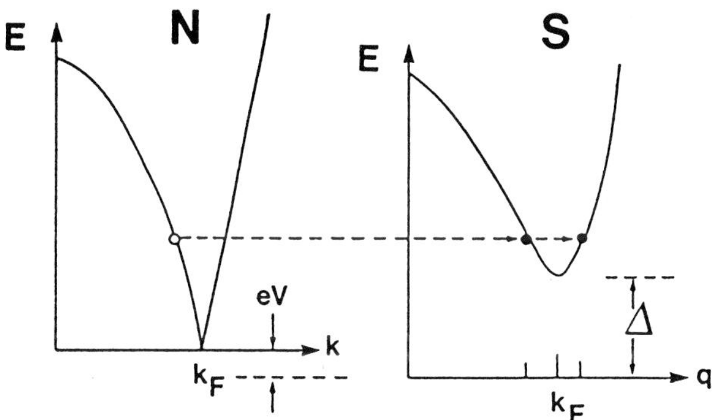

Figure 8 - *Tunneling from a normal metal into a superconductor in the E-k representation.*

For a temperature $T \to 0$, the Fermi function $f(E) = 1$ for $E < E_F$, and $f(E) = 0$ for higher energies. Thus the tunneling current for the NIS junction is given by a step function:

$$I_{NS} = 0 \qquad (eV < \Delta)$$
$$I_{NS} = C \times [(eV)^2 - \Delta^2]^{1/2} \qquad (eV \geq \Delta)\,. \tag{10}$$

For the dynamic conductance G we obtain for $eV \geq \Delta$:

$$G = \frac{dI_{NS}}{dV} = C \times \left(\frac{eV}{|(eV)^2 - \Delta^2|^{1/2}}\right) = C \times N_S(E = eV)\,. \tag{11}$$

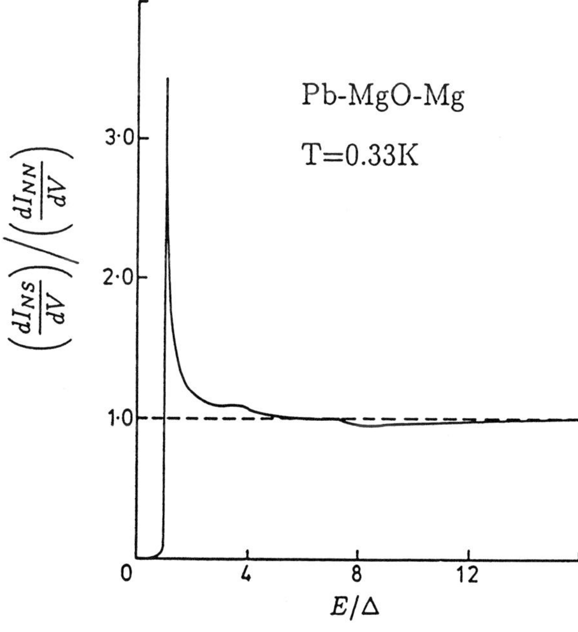

Figure 9 - *The normalized conductance of a Pb-MgO-Mg junction plotted against normalized energy. The small deviations of the density of states from unity in the 4mV to 12 mV range are due to the phonons present in the lead film.(From ref. [28])*

According to Eq. (11), the density of states for the superconductor can be determined experimentally by measuring the normalized conductance at very low temperatures $k_BT \ll \Delta$, as illustrated in Fig. 9.

Tunneling between two superconductors S_L and S_R is illustrated in the excitation representation in Fig. 10. There are three different tunneling processes: in channel A, an existing quasiparticle excitation in the left superconductor tunnels into S_R; channel B is similar, with the exchange of a pair; in channel C, a pair is broken up in S_L, creating quasiparticle excitations in both superconductors. For a SIS junction, Eq. (6) can be written as:

$$I_{SS} = C \times \int_{-\infty}^{+\infty} \frac{|E|}{|E^2 - \Delta_L^2|^{1/2}} \frac{|E + eV|}{|(E + eV)^2 - \Delta_R^2|^{1/2}} [f(E) - f(E + eV)] dE \,. \tag{12}$$

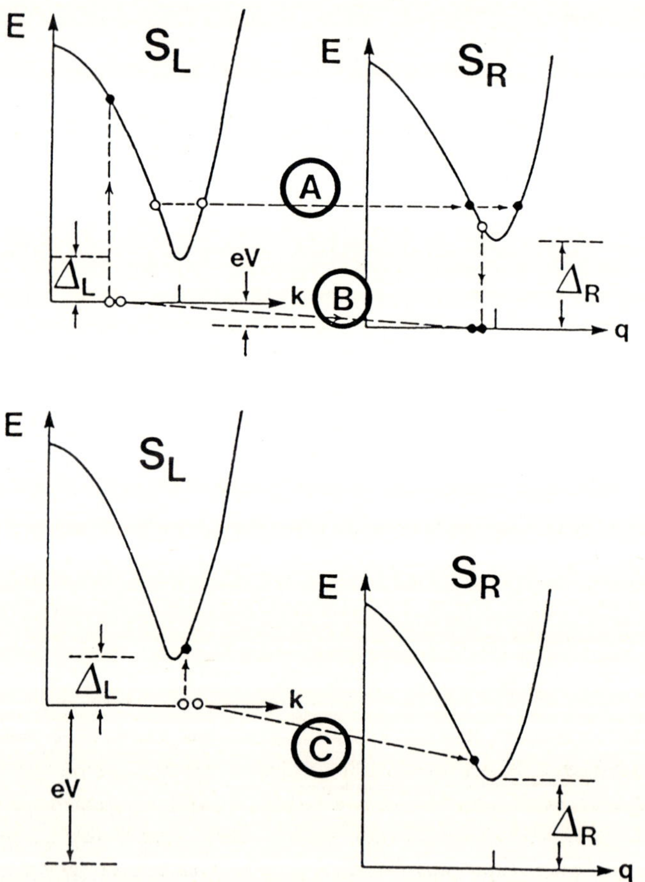

Figure 10 - *Basic processes or channels for quasiparticle tunneling in a S_LIS_R junction in the E-k representation.*

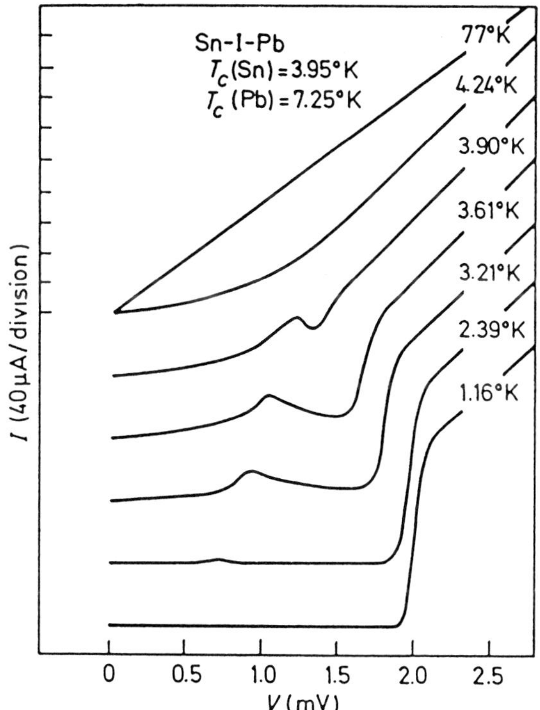

Figure 11 - *I-V characteristics at different temperatures for a Sn-I-Pb junction. (From ref. [30])*

For finite T, this equation was solved numerically by Nicol et al. [18] and Shapiro et al. [29]. They showed that the I-V characteristic contains a logarithmic singularity at $eV = \Delta_R - \Delta_L$, a region of negative resistance occurs for $\Delta_R - \Delta_L < eV < \Delta_L + \Delta_R$, and a discontinuous jump in the current occurs at $eV = \Delta_L + \Delta_R$. These calculations have been confirmed by the experimental results (Fig. 11).

2.2 Josephson tunneling

A - The Josephson equations

Finite overlap of the macroscopic pair wave function between two superconductors can occur when they are weakly coupled, either by (a) a very thin oxide layer, (b) a narrow constriction, (c) a superconducting point contact or (d) a normal metal in a sandwich junction. Josephson [19,31] predicted that a superconducting tunneling current can flow through such a junction without any voltage appearing across the junction. This is illustrated in Fig. 12 in the excitation representation. We discuss a very simple derivation of the Josephson equations, due to Feynman [32], which is based on a macroscopic quantum approach.

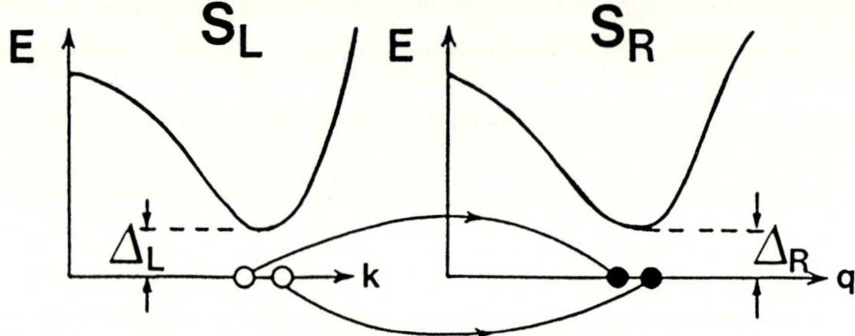

Figure 12 - *E-k representation of Josephson tunneling at V=0 between two superconductors.*

Let us consider a SIS tunneling junction (Fig. 13). We call $\Psi_L(\Psi_R)$ the pair amplitude for the left (right) superconductor. Each superconducting electrode can be described by a single quantum state and the Ψ's can be treated as macroscopic variables, with $|\Psi|^2$ representing the actual Cooper pair density ρ. We indicate with the ket $\mid L >$ ($\mid R >$) the ground state for the left (right) superconductor:

$$
\begin{aligned}
< L \mid \Psi_L^* \Psi_L \mid L >= |\Psi_L|^2 = \rho_L \\
< R \mid \Psi_R^* \Psi_R \mid R >= |\Psi_R|^2 = \rho_R \, .
\end{aligned}
\tag{13}
$$

Due to the weak coupling of the two superconductors, "transitions" (transfer of Cooper pairs) between the two states $\mid L >$ and $\mid R >$ can occur. This coupling is essentially related to the finite overlap of the two pair amplitudes Ψ_L and Ψ_R. A state vector of this system with two ground states can be described as:

$$
\mid \Psi > \; = \Psi_R \mid R > + \Psi_L \mid L > \; . \tag{14}
$$

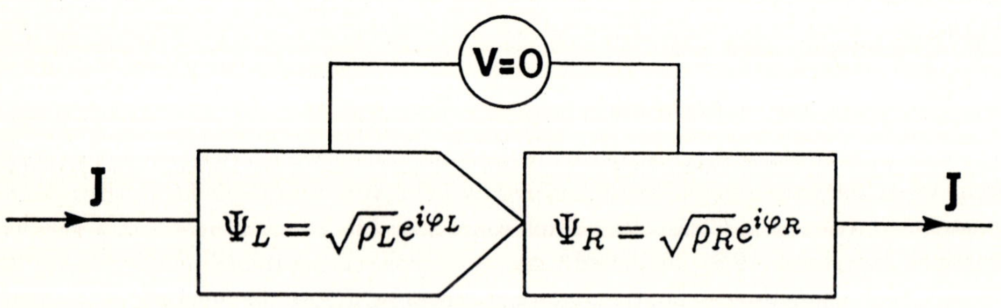

Figure 13 - *Schematic diagram of a single SIS junction.*

So the particle can be either in a "left" or "right" state with amplitude Ψ_L and Ψ_R respectively. The time-dependent Schrödinger equation descibes the time evolution of the system:

$$i\hbar \frac{\partial \mid \Psi >}{\partial t} = H \mid \Psi > , \tag{15}$$

with the Hamiltonian H given by

$$H = H_L + H_R + H_T , \tag{16}$$

where $H_L = E_L \mid L >< L \mid$ and $H_R = E_R \mid R >< R \mid$ are related to the unperturbed states $\mid L >$ and $\mid R >$. E_L and E_R are the ground state energies of the two superconductors and H_T is the tunneling Hamiltonian connecting the two states:

$$H_T = K[\mid L >< R \mid + \mid R >< L \mid] . \tag{17}$$

The coupling amplitude K of the two-state system is a measure of the coupling interaction between the two superconductors and depends on the specific junction structure.

Considering the projections on the two base states, Eq. (15) can be written in terms of the amplitudes Ψ_L and Ψ_R:

$$\begin{aligned} i\hbar(\partial \Psi_L/\partial t) &= E_L \Psi_L + K\Psi_R \\ i\hbar(\partial \Psi_R/\partial t) &= E_R \Psi_R + K\Psi_L . \end{aligned} \tag{18}$$

If there is a potential difference V across the junction, the difference in ground state energies $E_L - E_R = 2eV$. We may, for convenience, define the zero of energy to be halfway between E_L and E_R, and rewrite Eq. (18):

$$\begin{aligned} i\hbar(\partial \Psi_L/\partial t) &= (+eV)\Psi_L + K\Psi_R \\ i\hbar(\partial \Psi_R/\partial t) &= (-eV)\Psi_R + K\Psi_L . \end{aligned} \tag{19}$$

These are the standard equations describing the coupling between two quantum-mechanical states. We now make the substitutions

$$\Psi_L = \sqrt{\rho_L} e^{i\varphi_L} \qquad \Psi_R = \sqrt{\rho_R} e^{i\varphi_R} , \tag{20}$$

where φ_R and φ_L correspond to the macroscopic phase on both sides of the junction. Substituting Eq. (20) into Eq. (19), we obtain:

$$\begin{aligned} \partial \rho_L/\partial t &= +\frac{2}{\hbar} K \sqrt{\rho_L \rho_R} \sin \varphi \\ \partial \rho_R/\partial t &= -\frac{2}{\hbar} K \sqrt{\rho_L \rho_R} \sin \varphi \end{aligned} \tag{21}$$

$$\partial\varphi_L/\partial t = -\frac{K}{\hbar}\sqrt{\rho_R/\rho_L}\cos\varphi - \frac{eV}{\hbar}$$
$$\partial\varphi_R/\partial t = -\frac{K}{\hbar}\sqrt{\rho_L/\rho_R}\cos\varphi + \frac{eV}{\hbar}\,, \tag{22}$$

where the phase difference $\varphi = \varphi_R - \varphi_L$. The pair current density $J \equiv (\partial\rho_L/\partial t) = -(\partial\rho_R/\partial t)$ can be calculated from Eq. (21):

$$J = \frac{2K}{\hbar}\sqrt{\rho_L\rho_R}\sin\varphi\,. \tag{23}$$

Since ρ_L and ρ_R are constants and equal to $\rho_\circ$, the density of Cooper pairs in the superconducting material, we write:

$$J = J_\circ \sin\varphi\,, \tag{24}$$

with $J_\circ = 2K\rho_\circ/\hbar$, a number characteristic of the particular junction. Equation (22) gives the time dependence of the phase difference $\varphi_R - \varphi_L$:

$$\frac{\partial\varphi}{\partial t} = \frac{2eV}{\hbar}\,. \tag{25}$$

Equations (24) and (25) are the famous Josephson relations for a SIS tunneling junction.

Assuming a zero voltage across the junction, the second Josephson relation (25) implies that the phase difference φ is a constant, but not necessarily equal to zero. The first Josephson relation (24) predicts that a finite current density $J_\circ$ can flow through the barrier without a voltage drop appearing across the junction (no dissipation). This is the dc Josephson effect which is illustrated in Fig. 14.

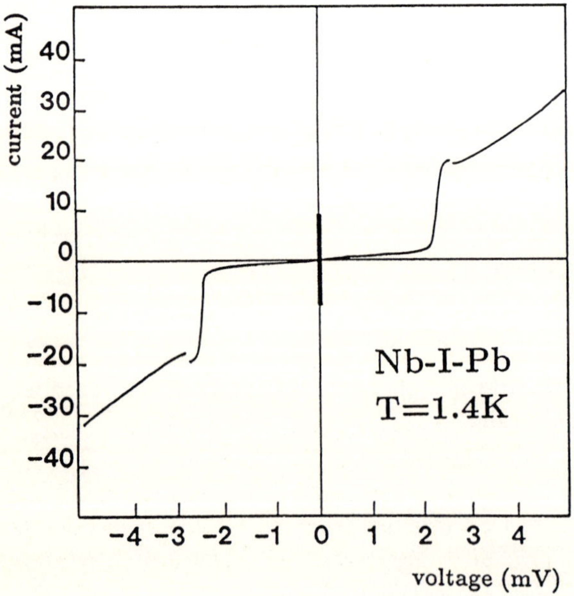

Figure 14 - *I-V characteristic of a Nb-I-Pb Josephson junction clearly showing the Josephson current at zero voltage.(From ref. [33])*

When a dc voltage V is applied, integration of Eq. (25) shows that the argument of the sine function in Eq. (24) becomes $\varphi = \varphi_o + (2e/\hbar)Vt$, and an oscillating supercurrent is generated:

$$J = J_o \sin\left(\varphi_o + \frac{2e}{\hbar}Vt\right) , \tag{26}$$

with the Josephson frequency ν equal to:

$$\nu = \left(\frac{2e}{h}\right) V . \tag{27}$$

These oscillations give rise to the emission of electromagnetic radiation with the same frequency. The existence of the ac Josephson effect was first demonstrated by Shapiro [34] in 1963. Experiments by Finnegan et al. [35] confirmed that a dc voltage of 1 μV across a Josephson junction corresponds to an oscillating current with frequency of 483.6 MHz as predicted by Eq. (27).

Finally, we also note that a junction irradiated with microwaves with frequency ν shows current steps in the I-V characteristic (Fig. 15) at regularly spaced voltages

$$V_n = n\left(\frac{h}{2e}\right)\nu , \tag{28}$$

with n an integer.

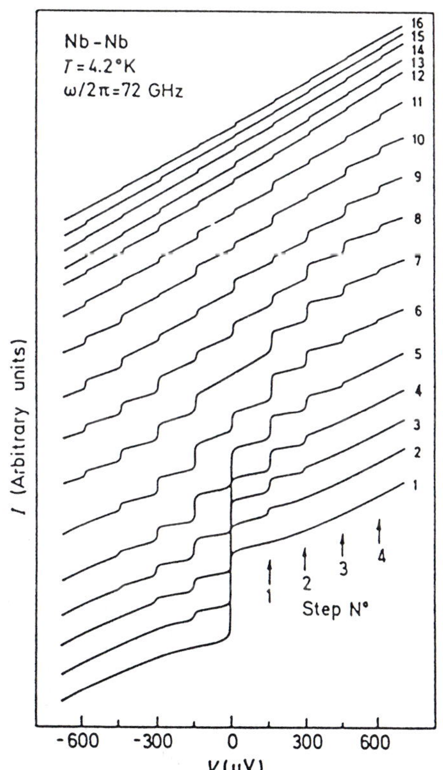

Figure 15 - *I-V characteristics of a Nb-Nb point-contact junction taken by a high impedance source. Curve 1: no microwave power; curves 2-16: microwave power increasing gradually by 26 db. Frequency 72 GHz,* $\hbar\omega/q$*=149* μ*V. The first four steps at multiples of 149* μ*V are clearly discernible.(From ref. [36])*

B - Magnetic field effects

We now calculate the effect of a magnetic field **B** on the Josephson tunneling current. As shown in Fig. 16, we assume that the current flows into the z-direction while the junction plane corresponds to the x-y plane. The magnetic field is parallel to the y-axis. The density of the shielding current (Meissner effect) in a superconductor is given by

$$\mathbf{J}_S = \frac{\rho e}{m}(\hbar\nabla\varphi - 2e\mathbf{A}) \; , \tag{29}$$

from which we calculate the gradient of the phase φ as:

$$\nabla\varphi = \frac{2e}{\hbar}\left(\frac{m}{2e^2\rho}\mathbf{J}_S + \mathbf{A}\right) \; . \tag{30}$$

The vector potential **A** is related to the magnetic field **B** by the usual relation $\nabla \times \mathbf{A} = \mathbf{B}$. Integrating Eq. (30) along the contours C_L and C_R (see Fig. 16) we get:

$$\begin{aligned} \varphi_{R_a}(x) - \varphi_{R_b}(x+\delta x) &= \frac{2e}{\hbar}\int_{C_R}(\mathbf{A} + \frac{m}{2e^2\rho}\mathbf{J}_S).\delta \mathbf{l} \\ \varphi_{L_b}(x+\delta x) - \varphi_{L_a}(x) &= \frac{2e}{\hbar}\int_{C_L}(\mathbf{A} + \frac{m}{2e^2\rho}\mathbf{J}_S).\delta \mathbf{l} \; . \end{aligned} \tag{31}$$

When the thickness of the superconducting films is much larger than the London penetration depths λ_L and λ_R, the contours can be chosen outside the region where the magnetic field penetrates and where the shielding current density $\mathbf{J}_S$ vanishes.

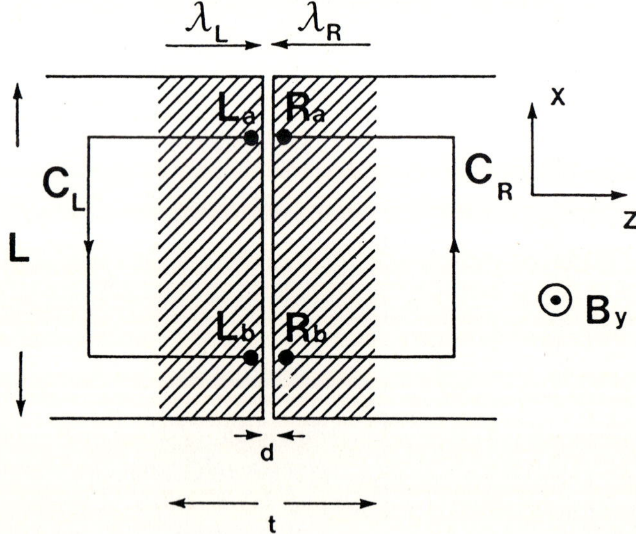

Figure 16 - *Contours of integration C_L and C_R used to derive the magnetic field dependence of the phase difference φ. The dashed zones indicate the regions in which the field penetrates into the superconducting electrodes.*

The portions of the contours in the penetration region can be chosen perpendicular to the $\mathbf{J}_S$ and we have

$$\begin{aligned}\varphi(x+\delta x)-\varphi(x) &= [\varphi_{L_b}(x+\delta x)-\varphi_{R_b}(x+\delta x)]-[\varphi_{L_a}(x)-\varphi_{R_a}(x)] \\ &= \frac{2e}{\hbar}\left[\int_{C_L}\mathbf{A}.\delta\mathbf{l}+\int_{C_R}\mathbf{A}.\delta\mathbf{l}\right] . \end{aligned} \quad (32)$$

Neglecting the barrier thickness $d \ll \lambda_L, \lambda_R$, Eq. (32) can be transformed to:

$$\varphi(x+\delta x)-\varphi(x)=\frac{2e}{\hbar}\oint\mathbf{A}.\delta\mathbf{l}=\frac{2e}{\hbar}B_y(\lambda_R+\lambda_L+d)\delta x \ , \quad (33)$$

where λ_L and λ_R are the London penetration depths for the two superconductors, and d is the barrier thickness. By integration of Eq. (33) we obtain the phase difference across the junction:

$$\varphi=\frac{2e}{\hbar}B_y tx+\varphi_\circ \ , \quad (34)$$

where $t=\lambda_R+\lambda_L+d$ is the length in which the field penetrates into the junction area. With Eq. (24), the pair current density becomes:

$$J=J_\circ\sin(\frac{2e}{\hbar}B_y tx+\varphi_\circ) \ . \quad (35)$$

So the tunneling Josephson current is spatially modulated by B.

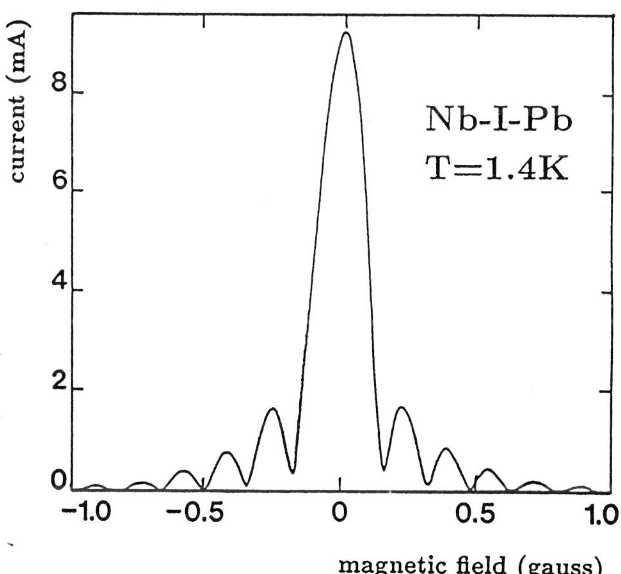

Figure 17 - *Magnetic field dependence of the maximum Josephson current for a Nb-I-Pb Josephson junction. (From ref. [33])*

Taking into account a rectangular junction with width W and length L, we get a supercurrent:

$$I(B) = I_o(0)\left|\frac{\sin \pi(\Phi_B/\Phi_o)}{\pi(\Phi_B/\Phi_o)}\right| , \tag{36}$$

with $I_o(0) = J_o WL$, $\Phi_B = BLt$ (the magnetic flux through the junction), and $\Phi_o = h/2e = 2.07 \times 10^{-15}$ Wb, the "flux quantum". The magnetic field dependence of the Josephson current is identical to diffraction pattern for light waves passing through a small slit. An experimentally observed diffraction pattern is shown in Fig. 17.

3 Quantum interference phenomena

3.1 Superconducting fluxoid

Equation (30) relates the phase φ to $\mathbf{J}_S$ and $\mathbf{A}$. Calculating the integral of this expression along a closed loop and with $\mathbf{B} = \nabla \times \mathbf{A}$, we can define the "fluxoid" as

$$\underbrace{\iint_s \mathbf{B}.d\mathbf{S} + \frac{m}{2e^2\rho}\oint \mathbf{J}_S.d\mathbf{l}}_{\text{“fluxoid”}} = n\Phi_o , \tag{37}$$

which is quantized in units of Φ_o. Inside a superconductor we have $\mathbf{J}_S = 0$ and so

$$\Phi_B = \iint_s \mathbf{B}.d\mathbf{S} = n\Phi_o . \tag{38}$$

This is the quantization of the flux Φ_B enclosed by a thick (that is, thick compared to the penetration length) superconducting loop.

3.2 Superconducting Quantum Interference Device (SQUID)

Let us consider a superconducting loop containing two Josephson junctions. We assume a symmetric configuration as illustrated in Fig. 18. In this configuration, on

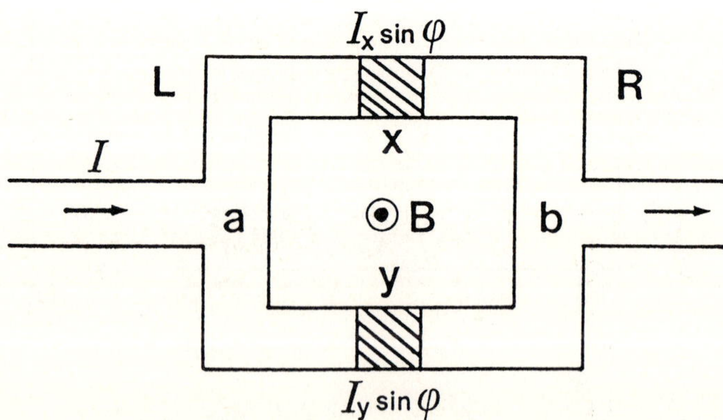

Figure 18 - *Symmetric two junction configuration.*

increasing the current I, a value I_c will be reached for which both junctions x and y switch simultaneously to the normal state. At this point, a voltage V develops across the device. We now investigate the dependence of I_c on the externally applied field B. Throughout the regions a and b (see Fig. 18) we have $\nabla\varphi = (2e/\hbar)\mathbf{A}$. Taking the integral along xby and xay we find:

$$[\varphi_R(x) - \varphi_L(x)] + [\varphi_L(y) - \varphi_R(y)] = \frac{2e}{\hbar}\oint \mathbf{A}.d\mathbf{l} = \frac{2e}{\hbar}\Phi_B = 2\pi(\Phi_B/\Phi_\circ)\,. \tag{39}$$

We now write

$$\begin{aligned} \varphi_R(x) - \varphi_L(x) &= \varphi + \pi(\Phi_B/\Phi_\circ) \\ \varphi_R(y) - \varphi_L(y) &= \varphi - \pi(\Phi_B/\Phi_\circ)\,. \end{aligned} \tag{40}$$

The supercurrent flowing between points a and b will be the sum of the Josephson currents through Josephson junction x and y

$$I = I_\circ \sin\left(\varphi_R(x) - \varphi_L(x)\right) + I_\circ \sin\left(\varphi_R(y) - \varphi_L(y)\right)\ ,$$

or

$$I = 2I_\circ \sin\varphi \cos\pi(\Phi_B/\Phi_\circ)\ , \tag{41}$$

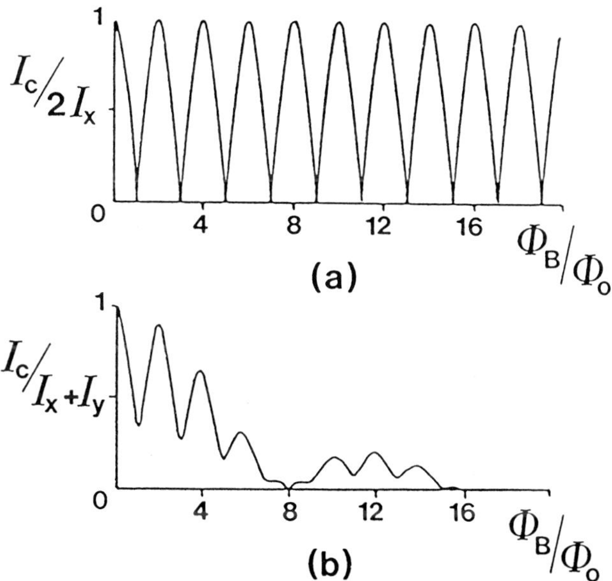

Figure 19 - *Critical current versus external flux for a two junction loop with (a)* $I_x = I_y$, $\Phi_J = 0$; *(b)* $I_x = I_y/2$, $\Phi_J/\Phi_B = 0.2$. *(From ref. [16, p. 372])*

where we assumed that both junctions have the same critical current $I_o = I_x = I_y$. For a given flux Φ_B, the Josephson current is maximized by taking $\sin\varphi = 1$

$$I_c \equiv I_{max} = 2I_o|\cos\pi(\Phi_B/\Phi_o)|\,. \tag{42}$$

This equation is valid only when we assume that the magnetic flux Φ_J threading the individual junctions can be neglected. This result also implies that the SQUID operation is identical to an interference experiment with coherent optical waves (see Fig. 19(a)). The phase difference is tuned by the magnetic flux threading the SQUID loop. The small flux period $\Phi_o = 2.07 \times 10^{-15}$ Wb explains why the SQUID will be an extremely sensitive magnetometer for a typical loop area of 1 mm^2 [15,16]. Figure 19(b) shows the interference pattern when we assume that the critical currents of the individual junctions are not the same and when the magnetic flux Φ_J enclosed by each junction is taken into account. The exponential dependence of the tunneling resistance (and also the critical current) upon the barrier thickness, implies that the fabrication of a reliable SQUID with two identical high- quality and very small Josephson junctions is extremely difficult. From the results shown in Fig. 5 and in Fig. 19(b) we conclude that a difference in barrier thickness on the order of 0.1 nm can largely destroy the interference pattern.

Acknowledgements

This work was supported by the Belgian Inter-University Institute for Nuclear Sciences (IIKW), the Inter-University Attraction Poles (IUAP) and Concerted Action (GOA) Programmes. C.V. is a Research Assistant and C.V.H. a Research Associate of the Belgian National Fund for Scientific Research (NFWO).

References

[1] G. Gamov, Z. Phys. **51**, 204 (1928).
[2] R.H. Fowler and L. Nordheim, Proc. Roy. Soc. **A119**, 173 (1928).
[3] J.R. Oppenheimer, Phys. Rev. **31**, 66 (1928).
[4] P. Hänggi and H. Grabert, Europhysics News **18**, 71 (1987).
[5] For a survey of modern trends in tunneling see: *Tunneling*, edited by J. Jortner and B. Pullman. Reidel, Boston (1986).
[6] J. Clarke and G. Schön, Europhysics News **17**, 94 (1986).
[7] For recent reviews see the articles in Physics Today **39/3**, 22 (1986).
[8] V.I. Goldanskii, Scientific American **254/2**, 38 (1986).
[9] J.A. Sommerfeld and H. Bethe, *Handbuch der Physik* **24/2**, p. 450, edited by S. Flügge. Springer-Verlag, Berlin (1933).
[10] J. Frenkel, Phys. Rev. **36**, 1604 (1930).
[11] C. Zener, Proc. Roy. Soc. **A145**, 523 (1934).
[12] L. Esaki, Phys. Rev. **109**, 603 (1957).
[13] W.A. Harrison, Phys. Rev. **123**, 85 (1961).
[14] M. Gijs and Y. Bruynseraede, Solid State Commun. **57/2**, 141 (1986).
[15] L. Solymar, *Superconductive Tunneling and Applications*. Chapman and Hall Ltd, London (1972).

[16] A. Barone and G. Paterno, *Physics and Applications of the Josephson Effect.* John Wiley & Sons, New York (1982).
[17] I. Giaever, Phys. Rev. Lett. **5**, 147 (1960).
[18] J. Nicol, S. Shapiro, and P.H. Smith, Phys. Rev. Lett. **5**, 461 (1960).
[19] B.D. Josephson, Phys. Lett. **1**, 251 (1962).
[20] P.W. Anderson and J.M. Rowell, Phys. Rev. Lett. **10**, 230 (1963).
[21] G. Binnig, H. Rohrer, Ch. Gerber, and E. Weibel, Phys. Rev. Lett. **49**, 57 (1982).
[22] L. Stockman, C. Van Haesendonck, and Y. Bruynseraede, unpublished.
[23] P.J.M. van Bentum, H. van Kempen, L.E.C. van de Leemput, and P.A.A. Teunissen, Phys. Rev. Lett. **60**, 369 (1988).
[24] T.A. Fulton and G.J. Dolan, Phys. Rev. Lett. **59**, 109 (1987).
[25] J.B. Barner and S.T. Ruggerio, Phys. Rev. Lett. **59**, 807 (1987).
[26] K. Knorr and J.D. Leslie, Solid State Commun. **12**, 615 (1973).
[27] D.G. Walmsley, R.B. Floyd, and W.E. Timms, Solid State Commun. **22**, 497 (1977).
[28] I. Giaever, H.R. Hart, and K. Megerle, Phys. Rev. **126**, 941 (1962).
[29] S. Shapiro, P.H. Smith, J.L. Miles, and P.F. Strong, IBM J. Res. Develop. **6**, 34 (1962).
[30] B.N. Taylor, J. Appl. Phys. **39**, 2490 (1968).
[31] B.D. Josephson, Rev. Mod. Phys. **36**, 216 (1964).
[32] R.P. Feynman, R.B. Leighton, and M. Sands, *The Feynman Lectures on Physics* **3**, chap.21. Addison Wesley, New York (1966).
[33] L. Van den Dries, Ph.D. thesis, K.U.Leuven, 1981 (unpublished).
[34] S. Shapiro, Phys. Rev. Lett. **11**, 80 (1963).
[35] T.F. Finnegan, A. Denenstein, and D.N. Langenberg, Phys. Rev. **B4**, 1487 (1971); Rev. Mod. Phys. **41**, 375 (1969).
[36] C.C. Grimes and S. Shapiro, Phys. Rev. **169**, 397 (1968).

Fabrication of Tunnel Junction Structures

G.B. Donaldson
Department of Physics and Applied Physics
University of Strathclyde, Glasgow G4 0NG
Scotland

1. INTRODUCTION

The fundamental element characterising a *superconducting* electronic device is almost always a weak electrical link between two pieces of metal, one or both of them superconducting. Device operation is based on the non-linear DC and AC electrical properties of such links, and can involve either superconducting pairs and phase coherence (usually through the Josephson effect [1,2]) or quasiparticles (usually through Giaever tunnelling [3,4]).

The tunnel junction is only one of several types of weak link. Others include the point contact and the microbridge, and have significant roles in various SQUIDs (Superconducting Quantum Interference Devices) and RF detectors. An extensive literature is available on the physics and applications of each [5,6,7,8]. However, for reasons ranging from fundamental noise processes to the problems of reproducible microfabrication, tunnel junctions are, at present, the preferred structures for most devices.

We therefore concentrate here on superconductive tunnel junctions and on methods of making them. We leave to other chapters the details of their incorporation into complete devices. This review looks back to the 1960's when Duke [9] found that "tunneling is an art and not a science" and claimed the available literature to be "not worth retrieving since it provides neither useful understanding on which to base future research nor reliable data which can be reproduced in another laboratory". However, complete accounts of earlier methods have been given by Solymar [5] and by Barone and Paterno [6], and so our main emphasis is on more recent times when in 1985 Klapwijk could write [10] that "new deposition facilities and improved surface analytical tools allow fairly accurate control and knowledge of barrier composition. This provides hope for a systematic analysis which was unthinkable in the past and might prevent us from slipping back from science into art".

We also briefly discuss current high T_c tunnel junction problems.

NATO ASI Series, Vol. F 59
Superconducting Electronics
Edited by H. Weinstock and M. Nisenoff

2. Early work

For most purposes, whether used as tools in physics or as devices, tunnel junctions must exhibit high quality Giaever or Josephson properties. This means that electron transfer should be describable by the complete tunnelling Hamiltonian [11]:

$$H_T = \Sigma[T_{\underline{k}q} c^+_{\underline{k}\sigma} d_{q\sigma} + T^*_{\underline{k}q} d^+_{\underline{q}\sigma} c_{\underline{k}\sigma}] \quad , \tag{1}$$

where $c^+_{\underline{k}\sigma}(c_{\underline{k}\sigma})$ are creation and destruction operators for electrons of wave-vector $\underline{k}$ and spin σ in one electrode, while $d^+_{\underline{q}\sigma}$ and $d_{\underline{q}\sigma}$ refer to the other electrode (wave-vector $\underline{q}$). $T_{\underline{kq}}$ is a matrix element for the probability of transition from ($\underline{k}\sigma$) to ($\underline{q}\sigma$), and is given in a WKB approximation by

$$|T_{\underline{kq}}|^2 \; \alpha \; k_z q_z \; \exp(-[2mU]^{\frac{1}{2}} d/h) \; \delta(k_y q_y) \; \delta(k_x q_x) \quad , \tag{2}$$

where U and d are the energy height and spatial width of the barrier, and the delta functions account for the conservation of wave-number components (momentum) parallel to the barrier.

Thus the barrier must be thin enough to allow reasonable current flow at biases of a few millivolts, but not so thin as to allow distortion of the independent quasiparticle densities of states in the two electrodes, which contribute to the summation in Eq. 1. The barrier height must be large enough to maintain normal tunnelling linearity over a bias range of several millivolts. Also, the junction capacitance must often be very small: DC SQUIDs, for example, require values as small as 1-10pF, pointing to very small area junctions. Finally, there should be no competing transport processes, such as leakage paths or traps in the barrier. This has been expressed in the Rowell criterion [12] which requires the sub-gap leakage conductance (σ_L) to be less than 0.1% of the above-gap conductance (σ_{NN}) before a junction is acceptable for tunnelling spectroscopy More recently this has been expressed for SIS junctions as a requirement to maximise $V_m - (\sigma_{NN}/\sigma_L) \times 2\Delta$, where 2Δ is the superconducting energy gap. Most workers are less stringent than Rowell: $V_m = 15 \times 2\Delta$ is usually regarded as good.

The electrodes themselves present special problems. Tunnelling samples the superconducting properties of the electrodes only to a depth of about one coherence length (ξ). This has implications for surface

Table 1: Superconducting parameters at 4.2K of materials used in tunnel junctions

	Critical Temperature (K)	Coherence Length (nm)	Energy Gap Parameter Δ (meV)
In	3.4		0.52
Sn	3.7	230	0.56
PbIn(Au)	6.8		1.15
Pb	7.2	83	1.3
Nb	9.2	38	1.5
NbN	14	4-7	2.8
V_3Si	15	3-4	2.6
YBaCuO	92	0.2-3	12-27
		(see Table 2)	

stoichiometry, especially for superconducting alloys and ceramics where ξ is very short. (See Table 1). Also, Eq. 2 depends on tunnelling being one-dimensional, and restricted essentially to electrons travelling in a single direction [13]. Reasonably <u>plane</u> surfaces are therefore required.

Early experiments met these requirements easily. They concentrated on Type I materials such as Pb and Al, for which ξ is large. Usually the first electrode was evaporated through a mechanical mask on to a glass substrate. A judicious period of thermal or dc glow discharge oxidation of the surface followed. The native oxide so produced was then covered by evaporating the counterelectrode as a cross stripe. Finally, contacts were made to the films using indium.

Such junctions were made in S-I-N, S-I-S and S_1-I-S_2 forms, with many metals (Solymar [Ref. 5 in Chap 23] gives a useful directory). They were used for early Giaever and Josephson developments and brought nee insights into superconductivity, ranging from measurement of energy gaps and their temperature variation to phonon spectroscopy and lifetime effects. Quantum interference and work on the value of h/e were among other applications.

However these methods were quite inadequate for the development of electronic device technology. Junction yields were small. Resistances could change rapidly during storage at room temperature. Few junctions survived more than a few thermal cyclings to 4.2K. Worse, since tight control of I_c (to better than 5%) is needed for a practical Josephson logic technology, these methods were very unpredictable. Thermal oxidation depended strongly on local conditions such as temperature, humidity and small concentrations of other molecular species. (A nearby dish of formic acid speeds the oxidation of indium, for example.) These "conditions" could also cover contaminants in the deposition system, and even details of how it was pumped down. Recipes were rarely transferable between individuals in the same research group, let alone between different laboratories.

Fig 1 illustrates some of the problems raised by these early evaporation and native oxidation methods:

A. *Substrate to film thermal mismatch*
Stress induced by differential thermal contraction between electrodes and substrate can lead to film tearing or recrystallisation.

B. *Film surface*
Though films deposited in poor vacua were amorphous and had smooth surfaces, diffusion or electromigration (of included impurities) to the surface led to unstable junctions.

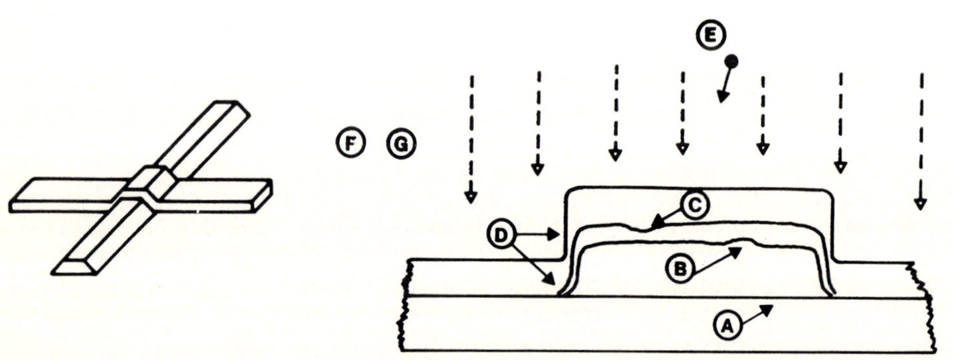

Figure 1 : Weaknesses of traditional junction methods– A Substrate mismatch; B Film surface; C Non-uniform oxide; D Edge effects; E Mechanical masks; F No passivation; G Non-refractory electrodes

Better vacuum systems were employed; purer films were deposited. Unfortunately, the results were polycrystalline structures with rough surfaces. Gap spreading in the I-V characteristics (due to anisotropy) was common. Dendrite ("hillock") growth during thermal cycling and room temperature storage led to electrical shorts, especially in Pb-based films, due to oxide barrier rupture.

C. Non-uniform oxide layers

Both thermal and DC glow discharge (activated oxygen) give limiting barrier thicknesses which depend on the point-to-point properties of the underlying film and its crystallites. Flaws are not self healing. Barriers then have variable thickness, with unpredictable J_c's. Junction conductance (exponentially weighted to regions of small *d*) ceases to be simply related to junction capacitance (depending on average *d*).

D. Edge effects

Film thinning near edges can lead to altered T_c and Δ values. For example T_c's as high as 2.2K can be seen for the edges of aluminium films with bulk T_c = 1.2K. Poor I-V characteristics result, accentuated perhaps by thin oxide layers and by thinning of the counterelectrode on a steeply rising edge. The last problem can even lead to film discontinuity, making the junction useless.

E. Mechanical Mask effects

For good definition without penumbra, mechanical masks must be positioned close to the film. There are alignment and clearance problems, and risks of sputtering of mask material into the junction during film evaporation.

F. Lack of passivation

Atmospheric degradation of junctions can be serious. Pb films can totally disappear into $Pb(OH)_3$ in the presence of dampness. Protection, usually with dried layers of thinned glue or varnish, was very limited.

G. Lack of rugged and refractory junctions

The first devices used soft metals such as Pb and Sn. Nb and other transition metals and alloys (e.g. V_3Si) make harder and more rugged films, and have larger T_c's (9K for Nb). But they are sensitive to oxygen impurities, and so before about 1970, it was not easy to produce films clean enough to show superconductivity. Stable native oxides would not form. Coherence lengths (Table 1) presented surface stoichiometry problems.

The rest of this chapter describes how these problems have been overcome. It is a success story, culminating in structures so reproducible and reliable that complete Josephson microprocessors are now practical [14,15]. But it is a *qualified* story: the solutions apply to particular structures, most notably the refractory system Nb-I-Nb, regarded at this time as the most promising for low T_c electronics. The solutions are *not* universal: the difficulties in developing high T_c tunnel junctions remind us of that.

2. DEPOSITION OF ELECTRODES AND OTHER LAYERS

Tunnel junction electrodes are always vacuum deposited - by Joule evaporation, electron beam evaporation, or by DC or RF sputtering usually in an inert gas such as argon. Evaporated atoms arrive at the substrate with energies appropriate to their melting temperature ($kT_m \approx 0.1$eV for $T_m \approx 600°$C); in sputtering, however, atoms leave the target with energies of $eV_{sputter}$, where $V_{sputter} \approx 200$eV , and in spite of scattering on the way to the substrate, will still arrive with energies of tens of volts.

2.1 Soft metals and alloys

The non-transition metal elements used in tunnelling devices (typically Pb, Sn, In, and Al as the superconductors and Au, Ag, Cu and Mg as the normal metals) can all be deposited from Joule heated boats or filaments.

Alloying of soft metals (as in PbBi) can reduce the gap anisotropy which causes smearing of I-V characteristics. In the dirty limit (short electron mean free paths), Cooper pairs contain wavefunction components from all round the Fermi surface [16], resulting in a single gap for all crystal directions.

Alloys were also useful in solving the mechanical "hillock" problem of stress, leading to dendrite growth and junction destruction on thermal cycling. This problem had (for pure metals) only been partly helped by choosing the substrate material (e.g. Corning 7059 glass or Si wafer) to reduce expansion mismatch. Alloying Pb films with 5% of In greatly reduced hillock growth [17], while the further addition of about 0.1 atomic % of Au essentially eliminated it. The alloys are easily made by premelting the components and subsequently evaporating the mixture from a single boat, or by successively depositing layers of the component metals: these interdiffuse over 100nm in a few minutes at room temperature.

Indium also has an important role in oxidation (see Sec. 4.3).

2.2 Refractory metals and alloys

Some transition metal superconductors (notably Nb and Ta) and certain alloys (NbN, V_3Si) are very attractive as electrodes: they have high adhesion to substrates, and good thermal and chemical stability. They also have large energy gaps, and at T = 4.2K, a value of T/T_c small enough to minimise residual changes (say in I_c) when temperature fluctuates.

Their high melting temperature rules out Joule evaporation of these materials. Electron beam evaporation can be a difficult alternative, because of strong gettering. Partial pressures of O_2 or H_2O in excess of 3×10^{-6}Pa (2×10^{-8} torr) can produce steep reductions in T_c. Such partial pressures will occur, by outgassing of evaporator components due to radiative heating, unless deposition rates are high and deposition times are short. The necessary high power guns (say 10KW) and cryopumps are usually not available. Where they are, however, and especially with rate-controlled multigun systems [18], excellent films of both metals and alloys can be made.

Laboratories without such advanced facilities rely on RF or DC sputtering, or on variations such as ion-beam sputtering, for deposition of refractory films. Prolonged pre-pumping is still needed to remove oxygen and water, and the sputter gas must have less than 1ppm of significant contaminants. But heating and outgassing is nearly eliminated. At Strathclyde, for example, we use a DC magnetron sputter gun for niobium. We pre-pump to 1.3×10^{-6}Pa (10^{-8} torr) in an oil-pumped, LN_2 trapped chamber, then back-fill with argon to about 10-100 millitorr. Deposition on to unheated pre-cleaned Si wafers at about 100nm-min^{-1} is achieved with DC currents of about 4A. The films have excellent adhesion, good residual resistance ratio ($R_{300K}:R_{4.2K} \approx 8$), and $T_c > 9$K.

NbN is attractive for its high T_c (up to 16K), and can be produced by sputtering Nb in nitrogen gas on to heated substrates [19]. Other alloys can be made by co-sputtering on metals (c.f. high T_c ceramics (Sec. 6)), but are of little device interest.

NbN is *un*-attractive for some magnetic applications because of its large penetration depth, involving large kinetic inductances and increased sensitivity to stray fields. Double layered Nb-NbN films overcome these problems by using the Nb to screen the NbN [20].

2.3 Insulators

Electrically insulating layers are required to keep wiring levels apart, or to separate junction electrodes away from tunnel barrier regions. SiO or SiO_x can be deposited by sputtering or evaporation and yield pinhole-free layers down to about 200nm thickness. Alternatively, selective anodisation of Al, Nb and other materials is available (see Sec. 5 and Fig 8).

3. JUNCTION GEOMETRY; LITHOGRAPHY

3.1 Patterning

Mechanical masks present difficulties with alignment and material sputtering. In any case they are impractical for structures which include many junctions or complex interwiring. Even for simple devices such as DC SQUIDS, they are usually unsuitable because of the need for small junction sizes.

For optimum noise behaviour, a DC SQUID must satisfy a number of conditions (see Ref. 21), involving the SQUID loop inductance (L_s), the junction critical current (I_c), the junction capacitance (C) and shunt resistance (R_s). (Note that I_c can be changed by orders of magnitude without affecting C by more than a few percent simply by changing the oxide barrier thickness by a few Å in about 20-30Å .) The optimal flux noise spectral density is

$$\Phi^2_{NW} = 16k_BT(\pi L_s{}^3C)^{1/2} \quad , \qquad (3)$$

so that C should have the smallest value possible. With L_s = 1.0nH (good for coupling to the SQUID), C = 1.5pF is needed to achieve $\Phi_{NW} \approx 4 \times 10^{-6}\Phi_0\text{-Hz}^{-1/2}$. Since oxide barriers of typical thicknesses have capacitances of order 0.2 - 0.03 pF-μm^{-2}, it follows that junction <u>areas</u> down to 1-10$\mu m^2$are needed.

These sizes are too small for simple mechanical masks (though some bimetal structures have given good results [22]). Occasionally, one may have to use them and accept poorer resolution, for example, to avoid impasses which can be created if photoresist is rendered insoluble by high energy argon beams used for cleaning niobium surfaces (see Sec. 4.3).

Resolutions below 0.5μm can be achieved with e-beam or deep-UV lithography. However, 2.5μm conventional photolithography with a standard mask-aligner is usually adequate. Projection or contact chrome-on-glass masks are used, with standard photoresists (e.g. Shipley AZ 1450). Material-specific wet etchants are rarely used for the unusual metals employed in superconductive technology. Instead, wet processing uses lift-off methods, where acetone dissolves and removes photoresist and unwanted overlying layers (Fig 2). To prevent tearing, a chloro-benzene surface toughening procedure is used to ensure photoresist undercutting.

Dry patterning is also used, often to produce differential etching between different materials, as in edge junction fabrication (Sec. 3.3). Typical processes include RF plasma etching in CF_4 mixtures (usually with oxygen or argon) , and reactive ion etching (e.g. CCl_2F_2 with O_2 or CF_4).

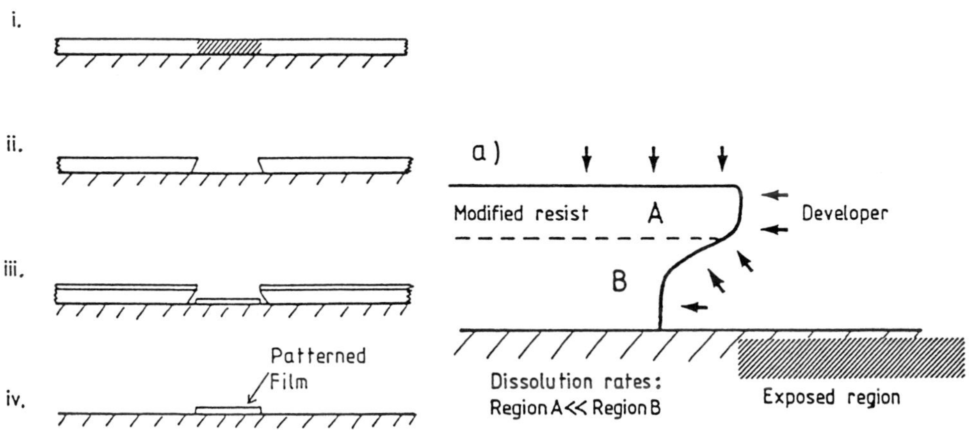

Figure 2: Lift-off: (i) Exposure of photoresist; (ii) Development and removal of exposed resist; (iii) Deposition of metal film (iv) Lift off by immersion in acetone.

Figure 2a:Photoresist is pre-immersed in chloro-benzene before development to ensure undercutting at stage (ii).

3.2 Window structures

Difficulties with thin edges (Fig 1-D) are avoided by insulating a base electrode with SiO and lithographically opening a "window" of appropriate area in this overlayer. After cleaning (Sec. 4.2) and possible processes associated with self alignment (Sec. 5.3), the barrier is formed on the main film rather than its edges. Junction diameters down to 3μm (area 7μm^2) are easy to produce in this way.

3.3 Edge structures

Still smaller junction areas (for lower noise SQUIDs) can be obtained by using the film *thickness* ($\approx$0.5μm typically), rather than the lithographic limit (2.5μm, say), as one of the junction dimensions (see Fig 3). A niobium film is covered by an insulating overlayer. When exposed to a CF_4-O_2 RF plasma, the overlayer (chosen for this purpose) is etched more rapidly than the niobium. At dynamic equilibrium the niobium has an edge-face sloping at an angle which can be adjusted to about 45° by varying the oxygen content of the plasma. The face is then used for barrier formation and counter-electrode coverage. In this way junction areas down to 1μm^2 are obtainable with photolithography [23] and 0.1μm^2 with e-beam lithography [24].

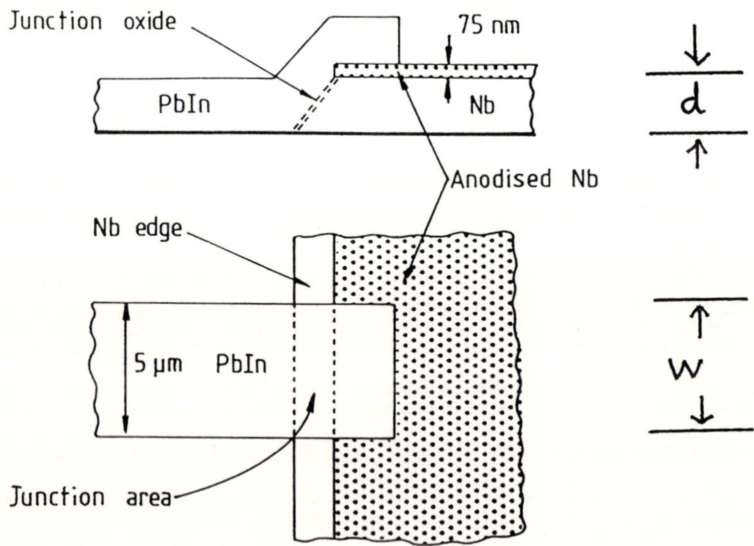

Figure 3 : Edge junction: the junction area is approximately $\sqrt{2} \times d \times w$, where d ($\approx$ 0.4μm) is the base electrode thickness and w ($\approx$ 2.5μm) the counterelectrode width.

4 TUNNEL BARRIERS AND COUNTER ELECTRODES

Making tunnel barriers has been central to the development of superconducting electronics. Controlled and uniform barrier thicknesses are needed to ensure that the Josephson current densities satisfy the conditions necessary for different devices. RF properties are involved too, because many devices (including the *DC* SQUID) depend on the presence of Josephson *AC* currents. Barrier dielectric constants and loss factors are therefore important also.

Dominant, however, is the property with the ugly name "ruggedness". The barrier must survive repeated thermal cycling, so that there must be no differential thermal stresses between barrier and electrode. The junction must not change during room temperature storage: thus, after barrier formation there must be no further reaction (e.g. oxidation) at either barrier-electrode interface. This involves electrochemical questions which are reviewed by Braginski and others [25].

All our methods involve oxide layers a few nanometres thickness *(*d) with large barrier heights (U in Eq. 2). Eq 2 suggests that thicker layers with lower barrier heights such as semiconductors also might be usable. They could be deposited uniformly by evaporation, and the "dangerous" interfacial regions would represent only a small proportion of the total thickness. In fact, with the exception of α-Si (see Sec. 5), semiconductor barriers, such as C, CdS and ZnS, have generated interesting physics, but few devices [26]. They show pinholes, competing current paths, and sensitivity to light. Even if these could be removed, traps and RF losses would probably rule out these materials for low noise, high integration structures.

4.1 Thermal oxidation; DC glow discharge oxidation

Some control over simple thermal oxide growth was developed in the 1960's for Pb, Sn, and other soft elements, making junction resistances predictable to within factors of 3-10. A typical exposure might be 100-760 torr (P) of air or oxygen for a period from 10 sec to days. Often the ultimate junction resistance was a linear function of time implying (Eq 1) a $d \propto \ln(time)$ growth during oxidation. This fits the Cabrera-Mott oxidation mechanism [27], which describes the early oxide growth on many metals, either by the migration of oxygen through the barrier to the metal surface, or by the passage of metal ions to the oxide-oxygen interface.

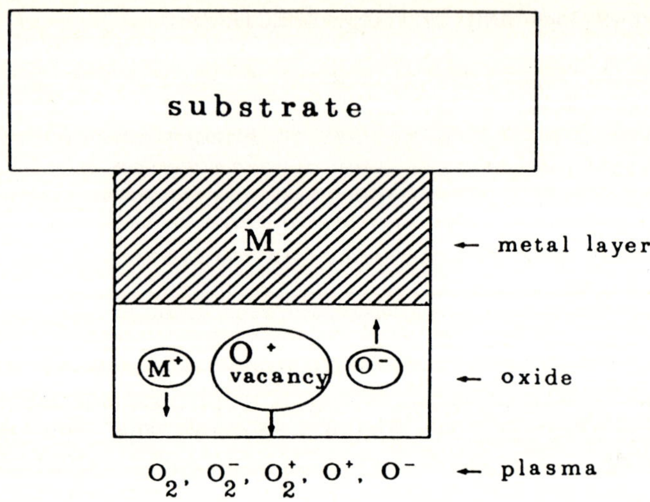

Figure 4 : Plasma oxidation processes (from Ref 6)

Cabrera-Mott oxidation depends on natural activation potentials which drive the various ions through the oxide. The associated fields, however, decrease as the oxide thickens, leading to the self-limiting logarithmic growth. Adequate tunnelling thicknesses may not grow within reasonable times.

DC plasma oxidation overcomes these problems (Fig 4). A discharge is created in a pressure of 1.5-15Pa (10-100μm) of oxygen using negative potentials of a few hundred volts applied to an electrode in the oxidation chamber. The base electrode floats, but in practice its potential is fairly close to ground. Plasma oxidation is faster than thermal oxidation, and thicker layers can be grown, for two reasons. First, the discharge produces O^+ ions, which have a smaller physical size than O_2 molecules, so that the diffusion process within oxide layers is enhanced. Secondly, the cross-barrier fields are now determined in part by the field within the discharge itself, which does not decrease as the oxide layer thickens.

However, the method can still produce junctions with pinholes or regions of thin barriers because the process does not deal with flaws or chemical impurities on the base electrode before oxide growth.

4.2 RF sputter etching and oxidation

Barriers on lead-based electrodes

A more satisfactory process was developed for Pb, PbIn and PbAuIn by Greiner and others [28] by using RF (13.6MHz) discharges rather than DC (see Fig 5). Power from an RF oscillator is fed *via* a matching network to a shrouded water cooled RF anode which carries the base electrode on its substrate. The discharge is struck between the anode and its grounded shroud in a continuous flow of either argon or a premixed argon-oxygen mixture. The chamber pressure is held at 1.5-15Pa (10-100μm). RF amplitudes of 100-400 volts are usually employed, and the specimen itself self-biases to a positive voltage equal to this amplitude; under these conditions the power incident on the film is then a few milliwatts-cm^{-2}.

Barrier fabrication is in two phases. A preliminary cleaning process using a discharge in pure argon sputters

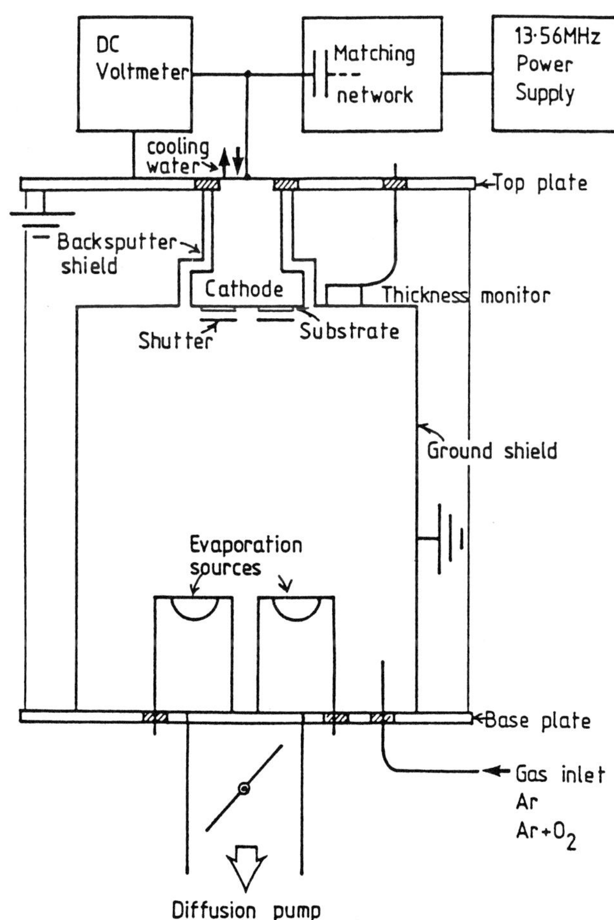

Figure 5 :
RF oxidation apparatus
(from D Hutson,
PhD Thesis,
Univ. of Strathclyde)

away surface "impurities" which can range from a few atoms desorbed from the chamber walls after evaporation of the electrode, up to complete oxide layers if the film was grown in another chamber. Provided care is taken to prevent deposition of material sputtered from other surfaces in the chamber, the cleaning process will produce atomically clean base electrodes.

The second phase is oxide growth. The plasma gas is now oxygen. Two processes compete - sputtering and activated oxidation. Oxygen ions and electrostatic fields produce accelerated oxide growth as in the dc plasma method. Sputtering by O^+ ions, however, constantly removes atoms at the oxide surface. The governing equation for the oxide thickness is:

$$dx/dt = -R + K \exp(-x/x_o) \quad , \qquad (4)$$

where R corresponds to an etching rate, K is a constant and x_o describes diffusion of ions through the oxide to the oxide-metal interface. R is found to vary as $(V)^{1.3}$, where V = RF voltage amplitude; K varied as $(P)^{1.5}$, where P = oxygen pressure.

For $t \rightarrow \infty$, the limiting thickness,

$$x_L = x_o \ln(K/R) \quad , \qquad (5)$$

is determined only by varying parameters such as P and V, and not by initial oxide thickness.

Using this approach and tools such as optical ellipsometry to monitor barrier composition and x(t), oxide growth on Pb, PbIn, and PbAuIn was made highly reliable at IBM [29] and elsewhere [30]. Figure 6 shows how varying the sputtering parameters can be used to either increase or decrease the thickness of a barrier which has already formed. The best barriers process developed was on $Pb_{0.84}In_{0.12}Au_{0.04}$.

On oxidised alloys, the relative proportions of the oxides usually differ from those of the underlying metals. Thus, on Pb-5%In the equilibrium concentrations are 30% PbO: 70% In_2O_3; the concentration reaches 100% In_2O_3 for only 35 atomic % of In in the base electrode [31].

Counterelectrodes of Pb or PbIn deposited on these barriers formed the basis for the once-promising IBM logic family.

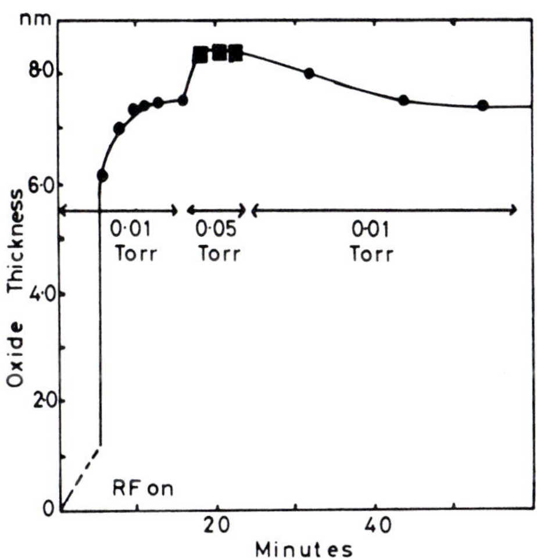

Figure 6 : RF sputter oxidation - control of limiting oxide thickness by varying oxygen pressure (Ref 30)

High critical current densities (> 10^4 A-cm^{-2}) with uniformities of 10%, and up to 1 million junction-thermal cycles between room temperature and 4.2K between failures were achieved with sets of several thousand junctions.

Barriers on niobium electrodes

RF oxidation was not easily extended to niobium, because of gettering. During argon ion cleaning, outgassing products from the chamber can react with the film itself. Often the surface is no cleaner at the end of A^+ sputter etching than at the beginning.

With good sputter-gun design and careful system management, acceptable control of barrier growth was obtained [32], though simple relations such as Eq. 5 have not been established. In fact, most methods may involve a monatomic diffusion barrier of NbCO deposited due to pump-oil contamination during plasma cleaning [33]. This would provide protection similar to that produced by the "wetting" methods described in the next section.

If a soft metal counterelectrode is deposited after NbO_x formation, good results are obtained: indeed, an entire SQUID technology was built on Nb-PbIn junctions [34]. The same was not true for Nb counterelectrodes. Attempts to make all-refractory junctions (Nb-NbO_x-Nb) by sputtering Nb after RF oxidation always resulted in leaky junctions (low V_m) and often in complete shorts. Only if the Nb was e-beam evaporated could reasonable tunnel junctions be made [35], but even here reproducibility was poor.

These problems with NbO_x might be due to sputtered Nb^{+++} ions arriving at the oxide surface with several eV of energy and rupturing the insulator layer; evaporated Nb atoms, on the other hand, with $k_BT_{2500K} \approx 0.25$eV might carry too little momentum to penetrate far into the oxide. However, a full explanation may rely on the electrochemistry of the structure, as can be seen from work [36,37] in which ultrathin layers of Cu, Ag, or Au were deposited on top of freshly produced NbO_x. Subsequent sputtering of Nb produced non-shorted junctions with well behaved critical currents. Since less than a monolayer of noble metal was enough to provide some "protection", a simple mechanical barrier cannot be the answer. Bain and Donaldson [38] proposed that the thin metallic layer establishes electrostatic potentials which eliminate strong barrier fields and reduce the forces driving Nb^{+++} ions through the barrier.

These results have not been followed up because V_m was too low, and devices formed from the junctions were too noisy, perhaps indicating electron traps in whatever barrier was formed. However, the use of thin "protective" layers has persisted, as we shall now see.

4.3 Niobium "wetting" methods

The key to all refractory technology was found by Rowell *et al.* [39]. If a fresh Nb film is overcoated with 1-4nm of sputtered aluminium, the Al fully "wets" the niobium surface, and suppresses the niobium oxidation properties. Upon thermal oxidation at room temperature, a very uniform layer of Al_2O_3 grows. The aluminium thickness is usually such that after oxidation a small residual layer of Al remains at the Nb surface. This scavenges oxygen from the niobium, so that XPS data shows no evidence of NbO_x and a sharp interface is created between the metal and the Al_2O_3; thus the "coherence length" condition is satisfied even though ξ_{Nb} is only 10-30nm.

When a Nb counterelectrode is sputtered, the Al_2O_3 (unlike NbO_x) retains its integrity. Excellent tunnel junctions

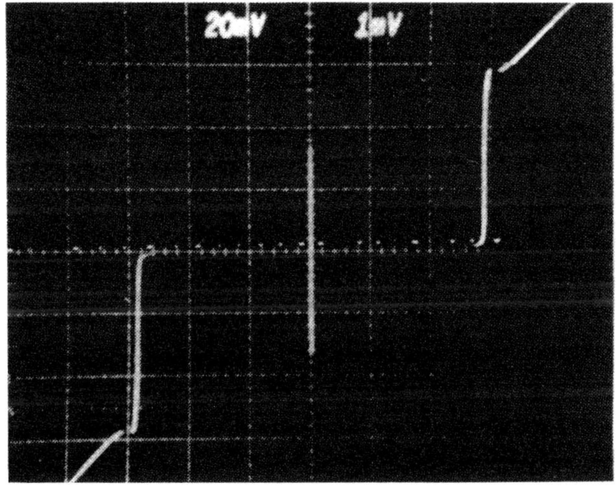

Figure 7: I-V characteristic of Nb-Al_2O_3-Nb junction (Ref 40)

result, with some of the highest V_m's ever obtained, and with very sharp gap structure - see Fig 7 [40] . Predeposition of another ultrathin Al layer on top of the Al_2O_3 can produce even further improvement.

The "wetting" approach works also for NbN, V and A15 compounds such as V_3Si; likewise, the Al layer can be replaced by Si, Ge, Mg, and Y. Direct deposition of Al_2O_3 and MgO has also been demonstrated [41]. The technique could possibly be extended to epitaxial growth of artificial oxide barriers, with low concentrations of defects and traps which are believed to contribute to 1/f noise. For full details of film and barrier formation, and of studies on barrier composition and uniformity, we refer the reader to a review by Braginski *et al.* [25] We merely remark here on the irony that a method, which signally failed to work in the early days of tunnelling (oxidation of an Al overlayer on a Pb electrode does <u>not</u> lead to good barriers), should 20 years later be the basis of the success of all-refractory Josephson fabrication technology.

The "artificial barrier" principle has been exploited in a variety of *whole-wafer* processes, intended for the production of large numbers of junctions with small spreads of critical current densities. They use a *trilayer* structure, from which individual junctions are later defined (Section 5). The trilayer itself is a vast Josephson junction - an extended sandwich composed of an Nb, NbN or Nb/NbN lower layer (perhaps deposited over sub-layers intended as buffers, ground planes, shunts or insulated spacers), oxidised by one of the "wetting" techniques, and completed by a second refractory electrode.

5 JUNCTION DEFINITION IN TRILAYERS; COMPLETE CIRCUITS

To form complete devices and circuits, individual trilayer junctions must be isolated from the whole wafer structures (Sec. 4.3) and interconnected. Since the trickiest part- growing the oxide layer- was already complete when the sandwich was produced, all the methods adopted maintain the tunnel barrier unexposed and protected by its overlayer throughout.

5.1 Selective Niobium Anodisation Process (SNAP)

In this process, which was pioneered by Kroger *et al.* [42], junctions (see Figure 8) are defined from a trilayer by using anodisation to convert the upper layer of Nb to insulating oxide, except at the desired junction sites, which were protected by photoresist. Anodisation in SNAP involves an ammonium pentaborate and ethylene glycol electrolyte and current densities of order 0.5mA-cm^{-2} for 5-30min. The photoresist is removed before a wiring layer of Nb is deposited to provide interconnects. The anodised niobium can be further protected by SiO if necessary,. Contacts to the base electrode are made, not by trying to expose it directly, but by forming either large area tunnel junctions or superconducting shorts.

The first SNAP junctions [42] were Nb/α-Si:H/Nb , but the process now extends to Ge [43] and native oxide [44] barriers, and to NbN electrodes[45,43,44].

5.2 Selective Niobium Etching Process (SNEP)

SNEP [46] uses dry reactive ion etching (RIE) of Nb in a 90% CF_4-10% O_2 RF plasma (see Fig 9). Anodisation is not involved so that junctions can be defined even when the trilayer has been deposited on an insulator (covering a ground plane, for example). A photoresist mask protects the

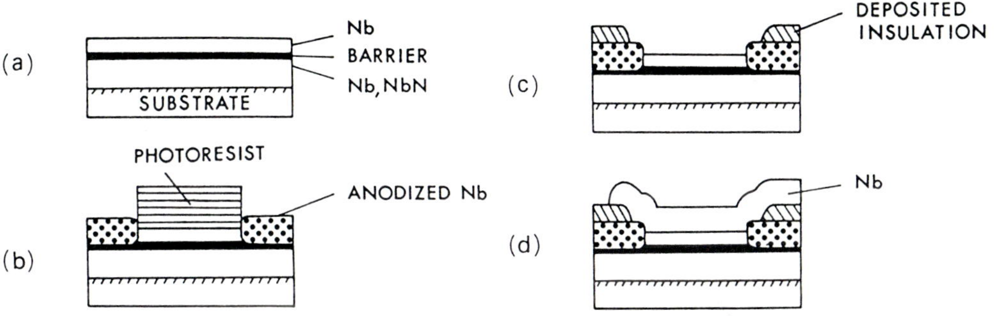

Figure 8 : Selective Niobium Anodisation Process (from Ref 48 - Shoji, *SQUID '85*, p633)

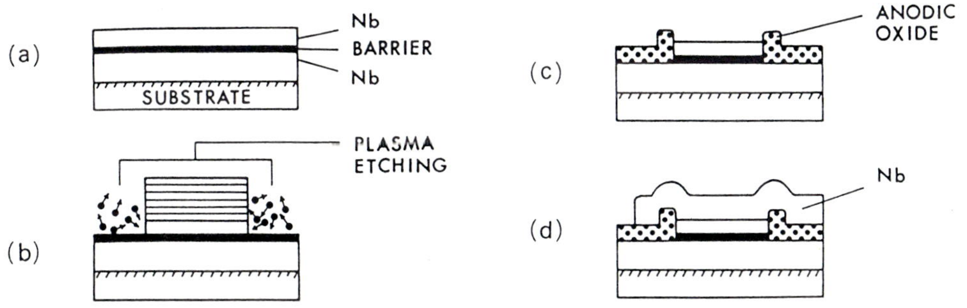

Figure 9 : Selective Niobium Etching Process (from Ref 48 - Shoji, *SQUID '85*, p638)

junction region, while the plasma etches away the upper layer of Nb in a Nb-Al_2O_3-Nb trilayer. The Al_2O_3 barrier is a good etch stop and protects the underlayer of Nb. SiO insulator is then deposited, and after photoresist removal, a wiring layer deposition interconnects the junctions.

RIE is also used to make contact holes through SiO insulator layers. However, with RIE, care is needed to ensure that excessive sputtering of other components such as resistors and even the substrate itself does not occur during niobium removal.

5.3 Self aligned processes-SNIP and others

Self alignment typically involves depositing an insulating layer after junction definition, but before removal of the resist protecting the upper niobium layer. It ensures that junction edges are protected by insulatng layers, and eases the masking problem for the wiring layer. There is excellent control over junction area, and reduced edge effects on I_c.

In SNIP (Self-aligned Niobium-nitride Isolation Process [47]) (see Fig 10), SiO covers both the junction and the sides of the counterelectrode before the photoresist is removed and the wiring procedure carried out.

These processes are now well developed. A set of 49,152 Nb-Al_2O_3-Nb junctions, first produced in 1985 with only 8 shorted junctions, has shown no further failures or even changes in critical current since then [47]. Nb/NbN-MgO-NbN/Nb junctions, with their larger gaps and higher T_c, have also been successfully produced [48]. Good linearity (<3%) has been obtained between I_c and junction area (Fig 11). Residual scatter and gap smearing has been identified with stress relief in niobium films as individual junctions are "excavated"; this becomes relatively larger as the junction area decreases. Stress can be either compressive or tensile, depending on deposition conditions: by choosing appropriate underlayers and a particular argon sputtering pressure, zero stress can be achieved [49]. Trilayer junction areas down to $(0.5\mu m)^2$ should become possible soon and will improve cell density (there are typically 4 junctions per cell) in logic applications.

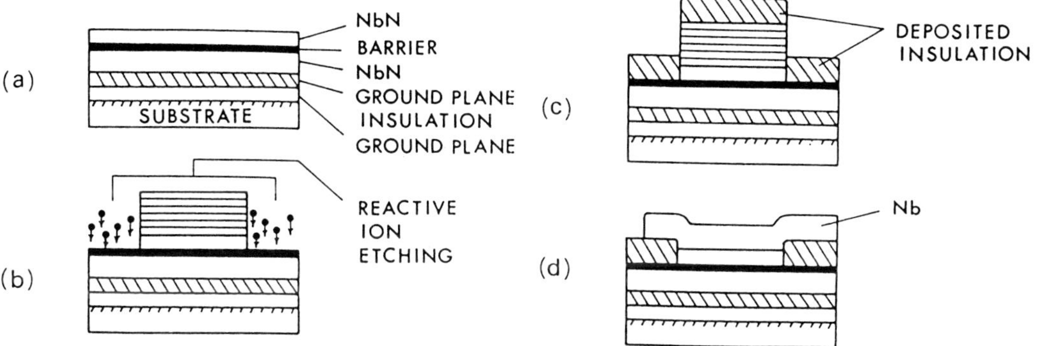

Fig 10: Self-aligned Niobium-nitride Isolation Process
(from Ref 48 - Shoji, *SQUID '85*, p635)

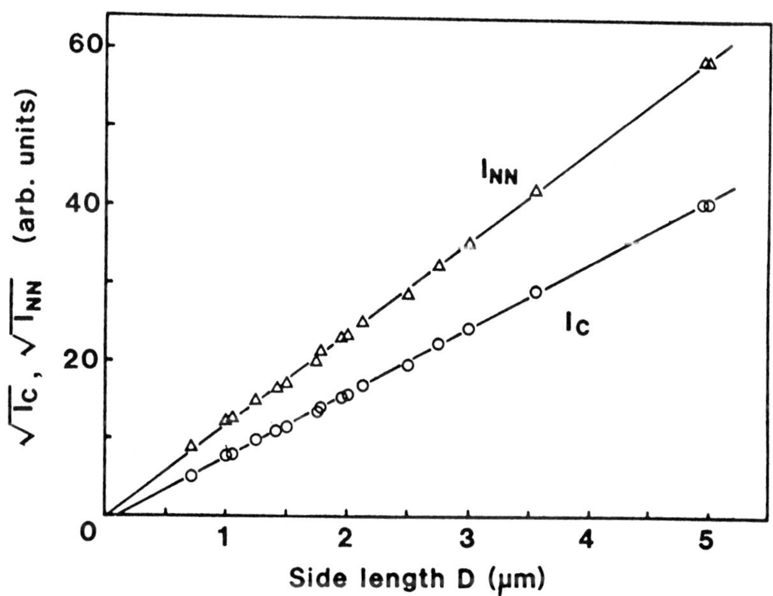

Figure 11: Correlation between self-aligned refractory junction dimensions and electrical properties

5.4 Passivation, planarisation and multilevel devices

These refractory junctions are very rugged; stories are told of boiling in water without damage! Nevertheless, it is usual to protect (passivate) devices with an overlayer of insulator. SiO can be used, but if sputtered may have to be very thick to avoid breaks at sharp vertical edges in underlying structures. "Spun-on-glass" avoids this by filling "hollows" preferentially and producing a smooth surface [50]. Still better results are obtained with DC-biased RF sputtering of SiO in which dynamic equilibrium between SiO deposition and removal is reached. Plane surfaces are obtained, which can be flat enough to carry further structures. In this way the Fujitsu group has produced a vertically stacked junction structure (Fig 12) in which the upper junctions are of as good quality as the lower ones [51].

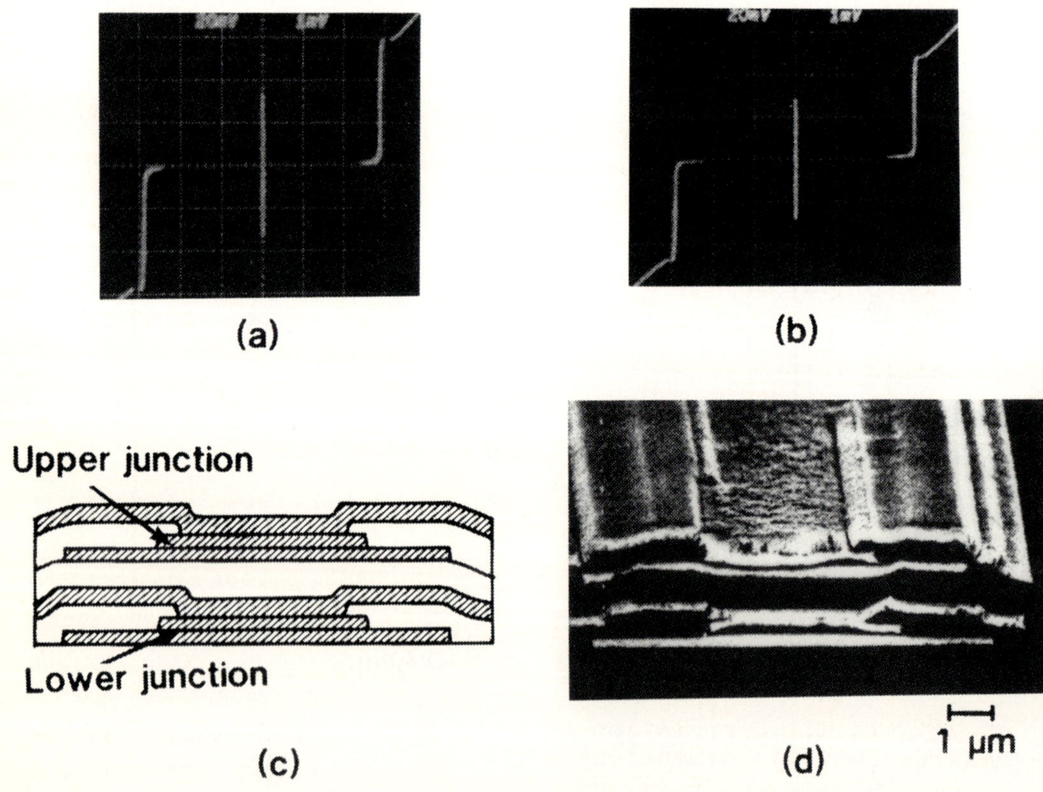

Figure 12: Vertically stacked junction structure (Ref 51)
(a) upper junction (b) lower junction

5.5 Applications of trilayer devices

The MITI supercomputer project in Japan has been the spur for much of the work on trilayer processes (see Ref. 52 for a full account of Japanese work in all aspects of Josephson electronics to 1987). Many interesting structures have been made, of which the Fujitsu Josephson microprocessor is perhaps the most impressive [53]. Designed to replicate all the functions of a recent very fast semiconductor device (AMD 2901), the superconducting version runs 10 times faster (clock rate 770MHz, basic junction switching time 2.5ps) and dissipates 500 times less power (5mW total for its 5011 junctions).

Although developed for LSI logic structures all-refractory trilayer methods can also be used for simpler structures with small junction counts, such as SQUIDs, A-D converters, S-I-S RF detectors and Josephson voltage sources. Also the use of a whole wafer process need not exclude some prepatterning. Thus at Strathclyde we predeposit and pattern DC SQUID shunts, followed by patterned photoresist on which the whole wafer $Nb-Al_2O_3-Nb$ trilayer is then formed. After lift off, we have a patterned trilayer, corresponding to the SQUID washers and junctions. Self aligned etch processing and interconnect deposition follows. The final devices [54] have white noise below $2 \times 10^{-5}\Phi_0\text{-Hz}^{-1/2}$.

It also seems certain that trilayer technology will form the basis for a growing variety of devices of intermediate complexity (up to a few hundred junctions). An example is an integrated SQUID magnetometer [55] in which the feedback circuitry consists of Josephson logic switches deposited on the SQUID chip itself.

6 HIGH-Tc SUPERCONDUCTORS

Most proposals on device applications of high-T_c superconductors (HTS) have been based on simple extrapolations of principles already demonstrated with low-T_c materials. They therefore need high quality tunnel junctions, usually of the S-I-S type, where both superconductors are high-T_c. Since these are not yet available, those SQUIDS [56] and thin film RF detectors [57] that *have* been made all rely on adventitious intergranular weak links in bulk or thin-film material.

In this section, therefore, we briefly review the problems and the progress in making HTS tunnel junctions.

6.1 Films

Deposition of HTS films is not straightforward. Early difficulties were presented by the high processing temperatures needed for annealing and oxygenation (up to 950°C in early work), and by the strong reactivity of the perovskite materials with water and with simple substrate materials (such as silicon). These have been partly overcome, and now many deposition processes, involving both DC and RF sputtering, Joule and e-beam evaporation, and even MBE and MOCVD, have been described for the production of "good" films on substrates such as $SrTiO_3$, at temperatures down to as little as 550°C [58]. Methods involving enhanced oxygenation using ions (produced for example by microwave discharges) have been described [59] and may further reduce processing temperatures.

Though BiSrCaCuO and TlBaCaCuO have been studied, most emphasis so far has been on YBaCuO. High and sharp transition temperatures have been achieved, with good adhesion and control of grain size. Equiaxy and even epitaxy have been demonstrated and provide enhanced critical properties. Critical current densities exceeding 1.5×10^6A-cm^{-2} in fields up to 1T have been demonstrated [60].

Patterning methods involving lift-off, as well as wet and dry etch, now yield feature sizes below 1μm [61]. This has allowed the development of simple dual-weak-link SQUIDs, which though neither reproducible nor reliable, have white noise quite close to the theoretical limits [62]. Elementary thin film passive microwave devices have also been demonstrated with such films [63].

6.2 Tunnel Junctions

Attempts have been made to produce tunnel junctions on HTS films, with at least partial success when the counterelectrode was a normal metal or low-T_c superconductor;. One method [64] was to deposit Y_2O_3 to form an artificial barrier on top of a sputtered film of YBCO. The counterelectrode was lead. The results were typical of many similar attempts: a clear evidence of gap structure at temperatures up to 77K but with gap smearing and leakage currents which were too great for meaningful

Table 2: Anisotropy in YBaCuO

	a-b plane	along c-axis
Coherence Length (nm)	1.6-3.0	0.2-0.4
Penetration Depth (nm)	270	1800

measurements, and certainly far from satisfying Rowell criterion (Section 2) for junctions which are acceptable for measuring gap functions or phonon spectra.

Strong anisotropy of properties (Table 2) in typical ceramic superconductors should contribute to gap smearing unless the base electrode is epitaxially grown. However, the small size and the anisotropy of the coherence lengths is probably much more significant. Even along the *a*-axis, the clean limit value is only 3nm, which will set the coherence length criterion (Section 2) for surface stoichiometry and cleanliness. This is clearly difficult to meet, and there is further discouragement in XPS studies [65] which show that oxygen disaggregation may produce a natural non-superconducting layer at a free YBCO surface. Against this, Braginski *et al* [66] have produced YBCO/Au-MgO/Nb junctions on *a*-oriented films which, although they gave disappointing tunnelling results because of the patchiness of the MgO overlayer, did at least show superconducting shorts. This suggests that the YBCO surface <u>was</u> superconducting and thus, in principle, amenable to eventual junction formation.

Even when good tunnel barriers *are* formed, successful deposition of <u>HTS</u> counterelectrodes may be impossible because of the high processing temperatures needed. They seem bound to lead to diffusive destruction of the sharp interlayer boundaries needed. Epitaxy of the counterelectrode would also be a problem. All these problems would be repeated with each successive HTS level in a complex device such as an tunnel junction SQUID with a superimposed flux transformer or control lines.

All is <u>not</u> gloom, however. Shiota *et al.* [67] have produced a YBCO/barrier/YBCO junction with good S-I-N (though not yet S-I-S) characteristics. The leakage conductivity was much less than 5% of the above-gap conductance. No special deposition precautions were taken with either electrode, and processing temperatures up to 700°C were involved in each case. The barrier was formed by an RF discharge in CF_4-O_2 plasma. Since CF_4 does not form a gaseous etch product on YBCO, it may be that the effect of the discharge is to re-oxygenate the surface layer, making it superconducting, and simultaneously to form an insulating barrier above it.

The barrier surface of the counterelectrode has still to be made fully superconducting. Until it is, all-HTS Josephson devices will not be possible. But the success of Shiota *et al* is clear encouragement that junctions will eventually be made, and the way opened to new devices.

7 CONCLUSIONS; FUTURE WORK

Junction fabrication has progressed from the *ad hoc* metal mask methods of the 1960's to the highly reliable microelectronic trilayer technologies of today. But success has been achieved within the limited context of specialisation to a few refractory metals - the trilayer processes are not of universal application. Faced with a new superconducting material to tunnel into, the basic scientist really has to start again, as the problems presented by HTS tunnelling clearly show.

Future developments are likely to include a reduction in junction sizes to well below 1 micron. This will reduce cell sizes, and hence propagation delays, in Josephson logic circuits. Recent developments in semiconductor technology also open the way to new research on barriers. The use of MBE or MOCVD to produce trap-free barriers may further reduce 1/f noise in Josephson devices.

Structures with shaped and resonant barriers, based on multiple oxide layers of different U and d (Eq. 2), will also be studied. An introductory review by Klapwijk [10] indicated interesting possibilities, though few experimental results were available when he was writing. Since then the beautiful quantum-well properties of low-dimensional semiconductor structures have been demonstrated. Their superconducting analogues await discovery, while the possible applications, especially for semiconductor-HTS hybrid structure, are pure speculation at this moment.

The future is promising, but, as always, depends on improved understanding of tunnel-barrier physics.

REFERENCES

1. B D Josephson, Phys. Letts., 1, 251 (1962).
2. P W Anderson and J M Rowell, Phys. Rev. Letts., 10, 230 (1963).
3. I Giaever, Phys. Rev. Letts., 5, 147 (1960)
4. M H Cohen, L M Falicov and J C Phillips, Phys. Rev. Letts., 8, 316 (1962).

For information on device physics and applications, see, for example

5. L Solymar, *Superconductive Tunnelling and Applications* Chapman and Hall, London, 1972.
6. A Barone and G Paterno, *Physics and Applications of the Josephson Effect*, John Wiley, London and New York, 1982.
7. Proceedings of successive Applied Superconductivity Conferences, published biennially in IEEE Trans Magnetics eg in MAG 25 to be published in 1989, MAG 23 (1987),.....
8. Proceedings of successive International Conferences on Superconducting Quantum Interference Devices in Berlin: *SQUID '76, SQUID '80, SQUID '85*, (H D Hahlbohm and H Lubbig, eds), de Gruyter, Berlin
9. C B Duke, *Tunnelling in Solids*, Academic Press, New York and London, 1962.
10. T Klapwijk SQUID '85, pp1-29, Superconducting Quantum Interference Devices and their Applications, (H D Hahlbohm and H Lubbig, eds), de Gruyter, Berlin, 1985.
11. See Chapter by Y Bruynserade (this volume) or Chapter 2 of Ref 6.
12. W L MacMillan and J M Rowell, "Tunneling and strong-coupling superconductivity", in *Superconductivity* (R D Parks,Ed) Vol 1, Marcel Dekker, New York, Chap. 11 pp 561-613: see p585.
13. W A Harrison, Phys. Rev. 123, 85 (1961).
14. S Kotani, N Fujimaki, T Imamura and S Hasuo, ISSCC88 Digest 150 (1988).
15. S Hasuo, to be published in IEEE Trans Magnetics MAG 25, (1989).
16. P W Anderson, J. Phys. Chem. Solids, 11, 26 (1959).
17. S K Lahiri, J. Vac. Sci. Tech., 13, 148 (1976).
18. R H Hammond, J. Vac. Sci. Tech., 15, 382 (1978).
19. S A Wolf, I L Singer, E J Cukauskas, T L Francavilla and E F Skelton, J. Vac. Sci. Tech., 17, 411 (1980); D D Bacon, A T English, S Nakahara, F G Peters, H Schreiber, W R Sinclair and R B van Dover, J. Appl. Phys., 55, 6509 (1983).
20. S Kosaka, K Kojima, F Shinoki, A Shoji and H Hayakawa, Jpn. J. Appl. Phys., 21, Supp. 21-1, 319 (1982).

21. C D Tesche and J Clarke, J. Low Temp. Phys., 29, 301 (1984).
22. A Cucolo (University of Salerno), private communication.
23. R F Broom, A Oosenbrug and W Walter, Appl. Phys. Lett., 37, 237 (1980); A Tugwell, C M Pegrum, G B Donaldson and M Wicks, IEEE Trans Magnetics MAG-23, 1036 (1987).
24. R T Wakai and D J van Harlingen, Appl. Phys. Lett., 52, 1182 (1988).
25. A I Braginski, J R Gavaler, M A Janocko and J Talvacchio SQUID '85- p591-629, Superconducting Quantum Interference Devices and their Applications, (H D Hahlbohm and H Lubbig, eds), de Gruyter, Berlin, 1985; see also Reference 6, Chapter 8.
26. A G Barone, G Paterno, M Russo and R Vaglio, p88, Proc. 14th Int. Conf on Low Temp. Phys. (M Krusius and M Vurio, Eds.), North-Holland, Amsterdam, 1975; see also Reference 6, Chap 8.
27. K Hauffe, pp 125-8, *"Oxidation of Metals"*, Plenum Press, New York (1965).
28. J H Greiner, J Appl. Phys., 42, 5151 (1971); J H Greiner, J Appl. Phys., 45, 32 (1974); P C Karlukar and J E Nordman, J Appl. Phys., 50, 7051 (1979).
29. J H Greiner, S Basavaiah and I Ames, J. Vac. Sci. Tech., 11, 81 (1974).
30. G B Donaldson and H Faghihi-Nejad, IEEE Trans Electron Devices ED-27, 1988 (1980).
31. J M Eldridge, D W Dong and K L Komarek, J. Electron Mat., 4, 1191 (1974); N J Chou, S Lahiri, R Hammer and K L Komarek, J. Chem. Phys., 63, (1975); A Emmanuel, G B Donaldson, W T Band and D Dew-Hughes, IEEE Trans Magnetics MAG-11, 763 (1975).
32. R F Broom, S I Raider, A Oosenbrug, R E Drake and W Walter, IEEE Trans Electron Devices ED-27, 1998 (1980); D Hutson PhD Thesis, University of Strathclyde (1988).
33. T S Kuan, S I Raider and R E Drake, J Appl. Phys., 53, 7464 (1982).
34. J Clarke, W M Goubau, and M B Ketchen, J. Low Temp. Physics, 25, 99 (1976); M B Ketchen, IEEE Trans Magnetics MAG-21, 543 (1985).
35. S I Raider and R E Drake, IEEE Trans Magnetics MAG-17, 299 (1981).
36. G Hawkins and J Clarke, J Appl. Phys., 47, 1616 (1976).
37. R J P Bain and G B Donaldson, IEEE Trans Magnetics MAG-23, 1650 (1987).
38. R J P Bain and G B Donaldson, J Phys.C, 18, 2539 (1985).

39. J M Rowell, M Gurvitch and J Geerk, Phys. Rev. B, 24, 2278 (1981).
40. S Morohashi and S Hasuo, J. Appl. Phys. 81, 4835 (1987).
41. A Shoji, M Aoyagi, S Kosaka, F Shinoki and H Hayakawa, Appl. Phys. Lett., 46, 1098 (1985).
42. H Kroger, L N Smith and D W Jillie, Appl. Phys. Lett., 39, 280 (1981).
43. H Kroger, D W Jillie, L N Smith, L E Phaneuf,C N Potter, D M Shaw, Appl. Phys. Lett., 44, 562 (1984).
44. T Iwata, K Takei, M Igarashi, Jpn. J. Appl. Phys., 23, L327 (1984).
45. D W Jillie, H Kroger, L N Smith, E J Cukauskas and M Nisenoff, Appl. Phys. Lett., 40, 747 (1983).
46. M Gurvitch, M A Washington and H A Huggins, Appl. Phys. Lett., 42, 472 (1983).
47. H Hoko, A Yoshida, H Tamura, T Imamura and H Hasuo, Int. Conf. on Solid State Devices and Materials, Tokyo 1986, 447.
48. A Shoji, F Shinoki, S Kosaka, M Aoyagi and H Hayakawa, Appl. Phys. Lett., 41, 1097 (1982); A Shoji, SQUID '85-p631-657, Superconducting Quantum Interference Devices and their Applications, (H D Hahlbohm and H Lubbig, eds), de Gruyter, Berlin, 1985.
49. T Imamura and H Hasuo, to be published in IEEE Trans Magnetics MAG 25, (1989).
50. S Kosaka, A Shoji, M Aoyagi, F Shinoki, S Tahara, H Ogihashi, H Nakagawa, S Takada and H Hayakawa, IEEE Trans Magnetics MAG-21, 102 (1985).
51. H Hoko, T Imamura and S Hasuo, Proc. Int. Electron Devices Meeting, Washington, December 1987; see also Ref. 15.
52. *Superconductivity Electronics* (K Hara, Ed.), Ohmsha (Tokyo), and Prentice Hall (Englewood Cliffs, NJ), 1987.
53. S Kotani, N Fujimaki, S Morohashi, S Ohara and S Hasuo, IEEE J. Solid State Circuits, SC-22, 98 (1988); see also Ref.15.
54. S Yano, W Lea, C M Pegrum and G B Donaldson, to be published.
55. N Fujimaki, H Tamura, T Imamura and S Hasuo, Digest of San Francisco ISSC 1988, p40; see also Ref.15.
56. R H Koch, C P Umbach, G J Clark, P Chaudhari and R B Laibowitz, Appl. Phys. Lett., 51, 200 (1987); U Kawabe, Physica C, 153-155, 1586 (1988).
57. S Kita, H Tanabe and T Kobayashi, to be published in IEEE Trans Magnetics MAG 25, (1989).
58. Y Bando, T Terashima, K Iijima, K Yamomoto, K Hirata and H Mazaki, p11, High T_c Superconducting Devices Meeting, Miyagi-Zao, 1988.
59. T Nojima, S Awaji, Y Maeno and T Fujita, Paper II-24, High T_c Superconducting Devices Meeting, Miyagi-Zao, 1988.

60. H Itizakji, S Tanaka, K Harada, K Higaki, N Fujimori and S Yasu, p149 (Paper II-24), High T_c Superconducting Devices Meeting, Miyagi-Zao, 1988.
61. M.R.Scheuermann, C C Chi, C C Tsuei, D S Yee, J J Cuomo, R B Laibowitz, B Braren, R Srinivasan and M M Plechaty, Appl. Phys. Lett. 51, 1951 (1988); G C Hilton, E B Harris and D J van Harlingen, Appl. Phys. Lett. 53, 1107 (1988).
62. C M Pegrum and G B Donaldson, Proc. European Workshop on High.-Tc Superconductors and Potential Applications, p 125 (Commission of the European Communities, Brussels, 1987).
63. D R Dykaar, J M Chwalek, R Sobaliewski, J F Whitaker, T Y Hsiang, G A Mourou,D K Lathrop, S E Russek and R Burhman, Appl. Phys. Lett. May 1988 and to be published in IEEE Trans Magnetics MAG 25, (1989).
64. M G Blamire, G W Morris, R E Somekh and J E Evetts, J. Phys. D, 20, 1330 (1987).
65. Many XPS studies have been reported: a typical example is T Kachel, P Sen, B Dauth and M Campagna, Z. Phys. B., 70, 137 (1988).
66. A I Braginski, M G Forrester, J Talvacchio and J R Wagner, Proc. Future Electron Devices Meeting, Miyagi-Zao, Japan, June 1988, p171.
67. T Shiota, K Takechi, Y Takai and H Hayakawa, to be published in Proc. ISS88, Nagoya, September 1988.

SQUID CONCEPTS AND SYSTEMS

John Clarke

Department of Physics, University of California
and
Materials and Chemical Sciences Division, Lawrence Berkeley Laboratory
Berkeley, California 94720

I. INTRODUCTION

Superconducting QUantum Interference Devices (SQUIDs) are the most sensitive detectors of magnetic flux currently available. They are amazingly versatile, being able to measure any physical quantity that can be converted to a flux, for example, magnetic field, magnetic field gradient, current, voltage, displacement, and magnetic susceptibility. As a result, the applications of SQUIDs are wide ranging, from the detection of tiny magnetic fields produced by the human brain and the measurement of fluctuating geomagnetic fields in remote areas to the detection of gravity waves and the observation of spin noise in an ensemble of magnetic nuclei.

NATO ASI Series, Vol. F 59
Superconducting Electronics
Edited by H. Weinstock and M. Nisenoff

SQUIDs combine two physical phenomena, flux quantization, the fact that the flux Φ in a closed superconducting loop is quantized[1] in units of the flux quantum $\Phi_0 \equiv h/2e \cong 2.07 \times 10^{-15}$ Wb, and Josephson tunneling[2]. There are two kinds of SQUIDs. The first[3], the dc SQUID, consists of two Josephson junctions connected in parallel in a superconducting loop, and is so named because it can be operated with a steady current bias. The second[4,5], the rf SQUID, involves a single Josephson junction interrupting the current flow around a superconducting loop, and is operated with a radiofrequency flux bias. In both cases, the output from the SQUID is periodic with period Φ_0 in the magnetic flux applied to the loop. One generally is able to detect an output signal corresponding to a flux change of much less than one flux quantum.

In this chapter I try to give an overview of the current state of the SQUID art. I cannot hope to describe all of the SQUIDs that have been made or, even less, all of the applications in which they have been successfully used. I begin, in Sec. II, with a brief review of the resistively-shunted Josephson junction, with particular emphasis on the effects of noise. Section III contains a description of the dc SQUID: how these devices are made and operated, and the limitations imposed by noise. Section IV contains a similar description of the properties of rf SQUIDs, but because there has been little development of these devices in the 1980's, I shall keep this section relatively brief. In Sec.V, I describe a selection of instruments based on SQUIDs and mention some of their applications. Section VI contains a discussion of the impact of high temperature superconductivity on SQUIDs, and of future prospects in this area, while Sec.VII contains a few concluding remarks.

II. THE RESISTIVELY SHUNTED JUNCTION

A Josephson junction[2] consists of two superconductors separated by a thin insulating barrier. Cooper pairs of electrons (or holes) are able to tunnel through the barrier, maintaining phase coherence in the process. The applied current, I, controls the difference $\delta = \phi_1 - \phi_2$ between the phases of the two superconductors according to the current-phase relation

$$I = I_0 \sin \delta, \tag{2.1}$$

where I_0 is the critical current, that is, the maximum supercurrent the junction can sustain. When the current is increased from zero, initially there is no voltage across the junction, but for $I > I_0$ a voltage V appears, and δ evolves with time according to the voltage frequency relation

$$\dot{\delta} = 2eV/\hbar = 2\pi V/\Phi_0. \tag{2.2}$$

A high quality Josephson tunnel junction has a hysteretic current-voltage (I - V) characteristic. As the current is increased from zero, the voltage switches abruptly to a nonzero value when I exceeds I_0, but returns to zero only when I is reduced to a value much less than I_0. This hysteresis must be eliminated for SQUIDs operated in the conventional manner, and one does so by shunting the junction with an external shunt resistance. The "resistively shunted junction" (RSJ) model [6,7] is shown in Fig.1(a). The junction has a critical current I_0 and is in parallel with its self-capacitance C and with its shunt resistance R, which has a current noise source $I_N(t)$ associated with it. The equation of motion is

$$C\dot{V} + I_0 \sin\delta + V/R = I + I_N(t). \tag{2.3}$$

Neglecting the noise term for the moment and setting $V = \hbar\dot{\delta}/2e$, we obtain

$$\frac{\hbar C}{2e}\ddot{\delta} + \frac{\hbar}{2eR}\dot{\delta} = I - I_0 \sin\delta = -\frac{2e}{\hbar}\frac{\partial U}{\partial \delta}, \tag{2.4}$$

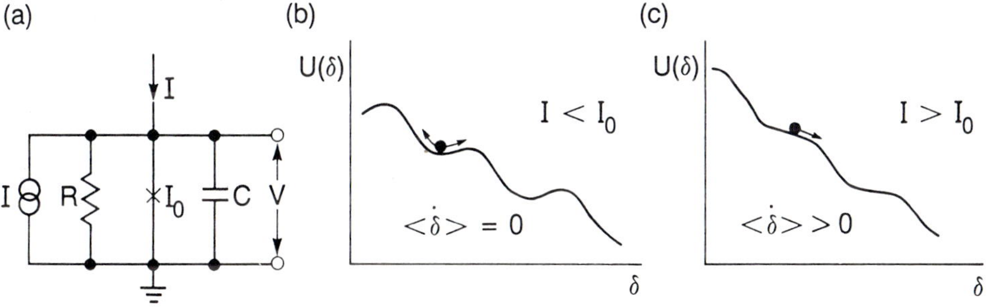

Fig. 1(a) The resistively shunted Josephson junction; (b) and (c) show the tilted washboard model for $I < I_0$ and $I > I_0$.

where

$$U = -\frac{\Phi_0}{2\pi}(I\delta + I_0 \cos \delta). \qquad (2.5)$$

One obtains considerable insight into the dynamics of the junction by realizing that Eq. (2.4) also describes the motion of a ball moving on the "tilted washboard" potential U. The term involving C represents the mass of the particle, the l/R term represents the damping of the motion, and the average "tilt" of the washboard is proportional to -I. For values of $I < I_0$, the particle is confined to one of the potential wells [Fig. 1(b)], where it oscillates back and forth at the plasma frequency[1] $\omega_p = (2\pi I_0/\Phi_0 C)^{1/2}\,[1-(I/I_0)^2]^{1/4}$. In this state $<\dot{\delta}>$ and hence the average voltage across the junction are zero (< > represents a time average). When the current is increased to I_0, the tilt increases, and when I exceeds I_0, the particle rolls down the washboard; in this state $<\dot{\delta}>$ is nonzero, and a voltage appears across the junction [Fig.1(c)]. As the current is increased further, $<\dot{\delta}>$ increases, as does V. For the nonhysteretic case, as soon as I is reduced below I_0 the particle becomes trapped in one of the wells, and V returns to zero. In this, the overdamped case, we require[6,7]

$$\beta_C \equiv (2\pi I_0 R/\Phi_0)RC = \omega_J RC \lesssim 1; \qquad (2.6)$$

$\omega_J / 2\pi$ is the Josephson frequency corresponding to the voltage I_0 R.

We introduce the effects of noise by restoring the noise term to Eq. (2.4) to obtain the Langevin equation

$$\frac{\hbar C}{2e}\ddot{\delta} + \frac{\hbar}{2eR}\dot{\delta} + I_0 \sin \delta = I + I_N(t). \qquad (2.7)$$

In the thermal noise limit, the spectral density of $I_N(t)$ is given by the Nyquist formula

$$S_I(f) = 4k_B T/R, \qquad (2.8)$$

where f is the frequency. It is evident that $I_N(t)$ causes the tilt in the washboard to fluctuate with time. This fluctuation has two effects on the junction. First, when I is less than I_0, from time to time fluctuations cause the total current $I + I_N(t)$ to exceed I_0, enabling the particle to roll out of one potential minimum into the next. For the underdamped junction, this process produces a series of voltage pulses randomly spaced in time. Thus, the time average of the voltage is nonzero even though $I < I_0$, and the I - V characteristic is "noise-rounded" at low voltages.[8] Because this thermal activation process reduces the observed value of the critical current, there is a minimum value of I_0 for which the two sides of the junction remain coupled together. This condition may be written as

$$I_0\Phi_0/2\pi \gtrsim 5k_BT, \tag{2.9}$$

where $I_0\Phi_0/2\pi$ is the coupling energy of the junction[2] and the factor of 5 is the result of a computer simulation[9]. For $T = 4.2K$, we find $I_0 \gtrsim 0.9\mu A$.

The second consequence of thermal fluctuations is voltage noise. In the limit $\beta_C << 1$ and for $I > I_0$, the spectral density of this noise at a measurement frequency f_m that we assume to be much less than the Josephson frequency f_J is given by[10,11]

$$S_V(f_m) = \left[1+\frac{1}{2}\left(\frac{I_0}{I}\right)^2\right]\frac{4k_BTR_D^2}{R} \quad . \quad \left\{\begin{matrix}\beta_C << 1\\ I > I_0\\ f_m << f_J\end{matrix}\right\} \tag{2.10}$$

The first term on the right-hand side of Eq. (2.10) represents the Nyquist noise current generated at the measurement frequency f_m flowing through the dynamic resistance $R_d \equiv dV/dI$ to produce a voltage noise - see Fig. 2. The

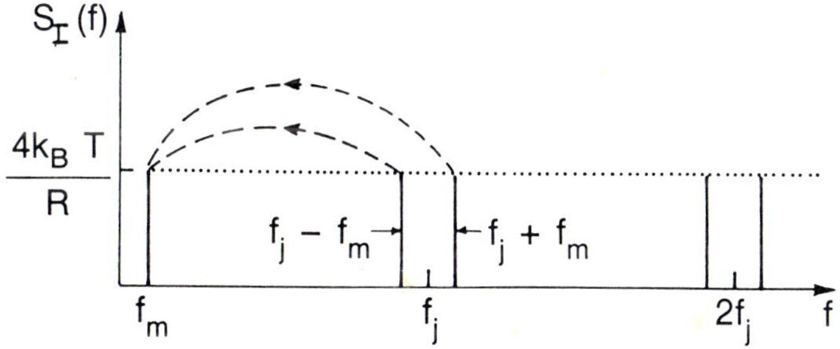

Fig. 2 Schematic representation for the noise terms in Eq.(2.10). The Nyquist noise generated in the resistor at frequency f_m contributes directly at f_m; that generated at $f_J \pm f_m$ is mixed down to f_m .

second term, $(1/2)(I_0 / I)^2 (4k_BT/R)\, R_d^2$, represents Nyquist noise generated at frequencies $f_J \pm f_m$ mixed down to the measurement frequency by the Josephson oscillations and the inherent nonlinearity of the junction. The factor $(1/2)(I_0 / I)^2$ is the mixing coefficient, and it vanishes for sufficiently large bias currents. The mixing coefficients for the Nyquist noise generated near harmonics of the Josephson frequencies, $2f_J, 3f_J$, ---, are negligible in the limit $f_m / f_J << 1$.

At sufficiently high bias current, the Josephson frequency f_J exceeds k_BT/h, and quantum corrections[12] to Eq. (2.10) become important, provided the term $(1/2)(I_0 / I)^2$ is not too small. The requirement for observing significant quantum corrections is $eI_0R / k_BT >> 1$. The spectral density of the voltage noise becomes

$$S_V(f_m) = \left[\frac{4k_BT}{R} + \frac{2\,e\,V}{R}\left(\frac{I_0}{I}\right)^2 \coth\left(\frac{eV}{k_BT}\right)\right] R_d^2, \quad \left\{\begin{matrix} \beta_C << 1 \\ I > I_0 \\ f_m << f_J \end{matrix}\right\} \qquad (2.11)$$

where we have assumed that $hf_m / k_BT << 1$, so that the first term on the right-hand side of Eq. (2.11) remains in the thermal limit. In the limit $T \to 0$, the second term, $(2eV/R)\,(I_0 / I)^2 R_d^2$, represents noise mixed down from zero point fluctuations near the Josephson frequency.

This concludes our review of the RSJ, and we now turn our attention to the dc SQUID.

III. THE DC SQUID

A. A First Look

The essence of the dc SQUID[3] is shown in Fig. 3(a). Two junctions are connected in parallel on a superconducting loop of inductance L. Each junction is resistively shunted to eliminate hysteresis on the I -V characteristics, which are shown in Fig. 3(b) for $\Phi = n\Phi_0$ and $(n + 1/2)\Phi_0$, where Φ is the external flux applied to the loop and n is an integer. If we bias the SQUID with a constant current ($> 2\,I_0$), the voltage across the SQUID oscillates with period Φ_0 as we steadily increase Φ, as indicated in Fig. 3(c). The SQUID is generally operated on the steep part of the $V - \Phi$ curve where the transfer

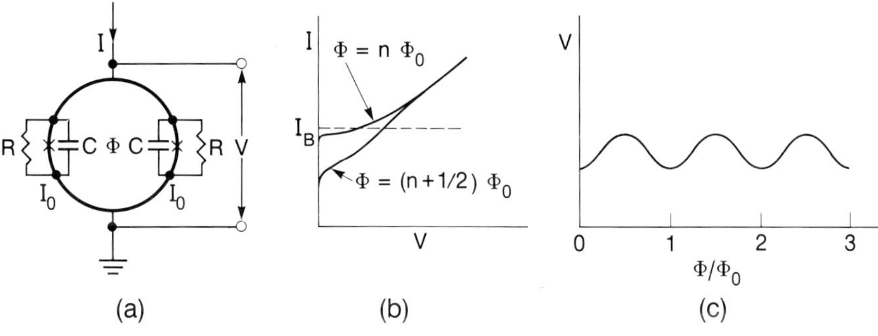

Fig. 3(a) The dc SQUID; (b) I-V characteristics; (c) V vs. Φ/Φ_0 at constant bias current I_B.

coefficient, $V_\Phi \equiv |(\partial V/\partial \Phi)_I|$, is a maximum. Thus, the SQUID produces an output voltage in response to a small input flux $\delta\Phi$ ($<< \Phi_0$), and is effectively a flux-to-voltage transducer.

Before we give a detailed description of the signal and noise properties of the SQUID, it may be helpful to give a simplified description that, although not rigorous, gives some insight into the operation of the device. We assume the two junctions are identical and arranged symmetrically on the loop. We further assume, for simplicity, that the bias current is swept from zero to a value above the critical current of the two junctions at a frequency much higher than $d\Phi/\Phi_0 dt$. In the absence of any applied flux (or with $\Phi = n\,\Phi_0$), there is no current circulating around the loop and the bias current divides equally between the two junctions. The measured critical current is $2I_0$ (if we ignore noise rounding). If we apply a magnetic flux, Φ, the flux in the loop will be quantized and will generate a current $J = -\Phi / L$, where we have neglected the effects of the two junctions [Figs. 4(a) and (b)]. The circulating current adds to the bias current flowing through junction 1 in Fig. 4(a) and subtracts from that flowing through 2. In this naive picture, the critical current of junction 1 is reached when $I/2 + J = I_0$, at which point the current flowing through junction 2 is $I_0 - 2J$. Thus, the SQUID switches to the voltage state when $I = 2I_0 - 2J$. As Φ is increased to $\Phi_0 / 2$, J increases to $\Phi_0 / 2L$ [Fig. 4(b)], and the critical current falls to $2I_0 - \Phi_0 / L$ [Fig. 4(c)]. As the flux is increased beyond $\Phi_0 / 2$, however, the SQUID makes a transition from the flux state n = 0 to n = 1, and J changes sign [Fig. 4(b)]. As we increase Φ to Φ_0, J is reduced to zero and the critical current

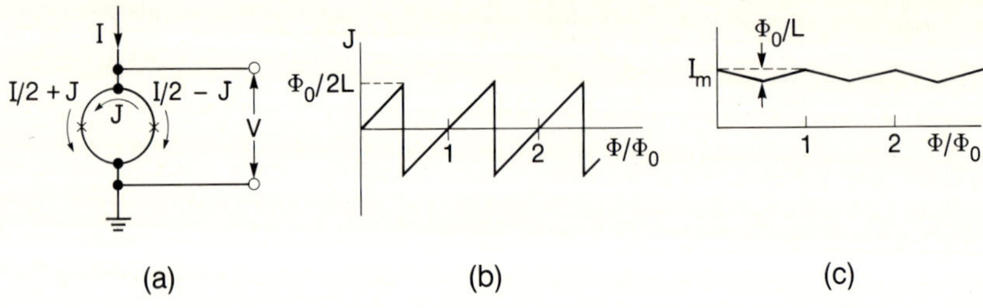

Fig. 4 Simplistic view of the dc SQUID: (a) a magnetic flux Φ generates a circulating current J that is periodic in Φ as shown in (b); as a result (c), the maximum supercurrent I_m is also periodic in Φ.

is restored to its maximum value $I_m = 2I_0$ [Fig. 4(c)]. In this way the critical current oscillates as a function of Φ.

Continuing with our simplified model, we see that the voltage change across the SQUID (at the peak of the current sweep) as we change Φ from 0 to $\Phi_0 / 2$ is $\Delta V = (\Phi_0 / L)R/2$, where R/2 is the parallel resistance of the two shunts. Hence, $V_\Phi = \Delta V / (\Phi_0 / 2) = R / L$.

We also can estimate the equivalent flux noise of the SQUID. If the noise voltage across the SQUID is $V_N(t)$ with a spectral density $S_V(f)$, the corresponding flux noise referred to the SQUID loop is just

$$S_\Phi(f) = S_V(f)/ V_\Phi^2. \tag{3.1}$$

A convenient way of characterizing the flux noise is in terms of the noise energy per unit bandwidth,

$$\varepsilon(f) = S_\Phi(f) / 2L. \tag{3.2}$$

If we assume that the noise in the SQUID is just the Nyquist noise in the shunt resistor with spectral density $4k_BT$ (R/2), we find $\varepsilon(f) = k_BTL/R$. Although these results are not quantitatively correct, they do give the correct scaling with the various parameters. For example, we see that to lower $\varepsilon(f)$ we should reduce T and L while using the largest possible value of R subject to the I - V characteristic remaining nonhysteretic.

Exact results for the signal and noise can be obtained only from computer simulation. The results show that the plots of the circulating supercurrent and the critical current vs. Φ become smoothed. Furthermore, the noise voltage is higher than Nyquist noise because of mixed-down noise; unfortunately the magnitude of this noise cannot be obtained analytically.

One final remark is appropriate at this point. To observe quantum interference effects, we require the modulation depth of the critical current, Φ_0 / L, to be much greater than the root mean square noise current in the loop, $\langle I_N^2 \rangle^{1/2} = (k_B T / L)^{1/2}$. This condition can be written $L \lesssim \Phi_0^2 / 5k_B T$, where the factor of 1/5 is the result of a computer analysis[9]. For T = 4.2K, we find $L \lesssim 15$ nH.

B. Thermal Noise in the SQUID : Theory

A model for noise calculations is shown in Fig. 5. This figure shows two independent Nyquist noise currents, $I_{N1}(t)$ and $I_{N2}(t)$, associated with the two shunt resistors. The phase differences across the junctions, $\delta_1(t)$ and $\delta_2(t)$, obey the following equations:[13-15]

$$V = \frac{\hbar}{4e}\left(\dot{\delta}_1 + \dot{\delta}_2\right), \tag{3.3}$$

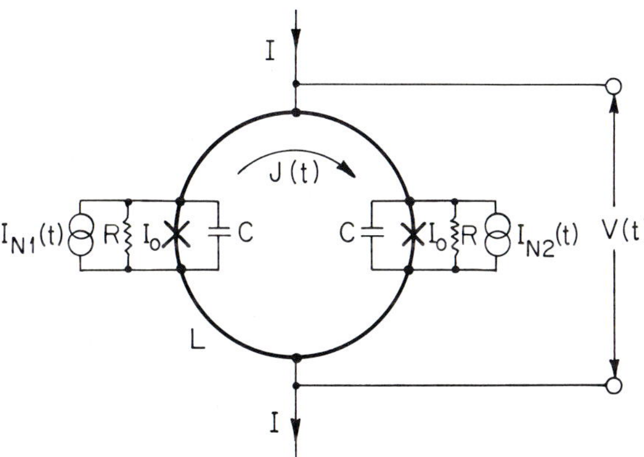

Fig.5 Model of dc SQUID showing noise sources associated with the shunt resistors.

$$J = \frac{\Phi_0}{2\pi L}\left(\delta_1 - \delta_2 - \frac{2\pi\Phi}{\Phi_0}\right), \tag{3.4}$$

$$\frac{\hbar C}{2e}\ddot{\delta}_1 + \frac{\hbar}{2eR}\dot{\delta}_1 = \frac{I}{2} - J - I_0 \sin\delta_1 + I_{N1}, \tag{3.5}$$

and

$$\frac{\hbar C}{2e}\ddot{\delta}_2 + \frac{\hbar}{2eR}\dot{\delta}_2 = \frac{I}{2} + J - I_0 \sin\delta_2 + I_{N2}\,. \tag{3.6}$$

Equation (3.3) relates the voltage to the average rate of change of phase; Eq. (3.4) relates the current in the loop, J, to $\delta_1 - \delta_2$ and to Φ; and Eqs. (3.5) and (3.6) are Langevin equations coupled via J. These equations have been solved numerically for a limited range of values of the noise parameter $\Gamma = 2\pi k_B T/I_0\Phi_0$, reduced inductance $\beta = 2\,LI_0 / \Phi_0$ and hysteresis parameter β_C. For typical SQUIDs in the ^{4}He temperature range, $\Gamma = 0.05$. One computes the time-averaged voltage V vs. Φ, and hence finds V_Φ, which, for a given value of Φ, peaks smoothly as a function of bias current. The transfer function exhibits a shallow maximum around $(2n + 1)\,\Phi_0 / 4$. One computes the noise voltage for a given value of Φ as a function of I, and finds that the spectral density is white at frequencies much less than the Josephson frequency. For each value of Φ, the noise voltage peaks smoothly at the value of I where V_Φ is a maximum. From these simulations, one finds that the noise energy has a minimum when $\beta \approx 1$. For $\beta = 1$, $\Gamma = 0.05$, $\Phi = (2n + 1)\,\Phi_0 / 4$ and for the value of I at which V_Φ is a maximum, the results can be summarized as follows:

$$V_\Phi \approx R/L, \tag{3.7}$$

$$S_V(f) \approx 16 k_B T R, \tag{3.8}$$

and

$$\varepsilon(f) \approx 9 k_B T L/R. \tag{3.9}$$

We see that our rough estimate of V_Φ in Sec. IIIA was rather accurate, but that the assumption that the noise spectral density was given by the Nyquist result underestimated the computed value by a factor of about 8.

It is often convenient to eliminate R from Eq. (3.9) using the expression $R = (\beta_C \Phi_0 / 2\pi I_0 C)^{1/2}$. We find

$$\varepsilon(f) \approx 16\, k_B T(LC/\beta_C)^{1/2}. \quad (\beta_C \lesssim 1) \tag{3.10}$$

Equation (3.10) gives a clear prescription for improving the resolution: one should reduce T, L and C. A large number of SQUIDs with a wide range of parameters have been tested and found to have white noise energies generally in good agreement with the predicted values. It is common practice to quote the noise energy of SQUIDs in units of $\hbar$ ($\approx 10^{-34}$J sec $= 10^{-34}$JHz^{-1}).

In closing this discussion, we emphasize that although $\varepsilon(f)$ is a useful parameter for characterizing the resolution of SQUIDs with different inductances, it is not a complete specification because it does not account fully for the effects of current noise in the SQUID loop. We defer a discussion of this point to Sec. V.D.

C. Practical dc SQUIDs

Modern dc SQUIDs are invariably made from thin films with the aid of either photolithography or electron beam lithography. A major concern in the design is the need to couple an input coil inductively to the SQUID with rather high efficiency. This problem was elegantly solved by Ketchen and Jaycox,[16,17] who introduced the idea of depositing a spiral input coil on a SQUID in a square washer configuration. The coil is separated from the SQUID with an insulating layer. The version[18] of this design made at UC Berkeley is shown in Fig. 6. These devices are made in batches of 36 on 50mm diameter oxidized silicon wafers in the following way. First, a 30nm thick Au (25 wt % Cu) film is deposited and patterned to form the resistive shunts. Next, we sputter a 100nm thick Nb film and etch it to form the SQUID loop and a strip that eventually contacts the inner end of the spiral coil. The third film is a 200 nm SiO layer with 2 μm diameter windows for the junctions, a larger window to give access to the CuAu shunt, and a window at each end of the Nb strip to provide connections to the spiral coil. The next step is to deposit and lift off the 300 nm thick Nb spiral coil, which has 4, 20 or 50 turns. At this point, we usually dice the wafer into

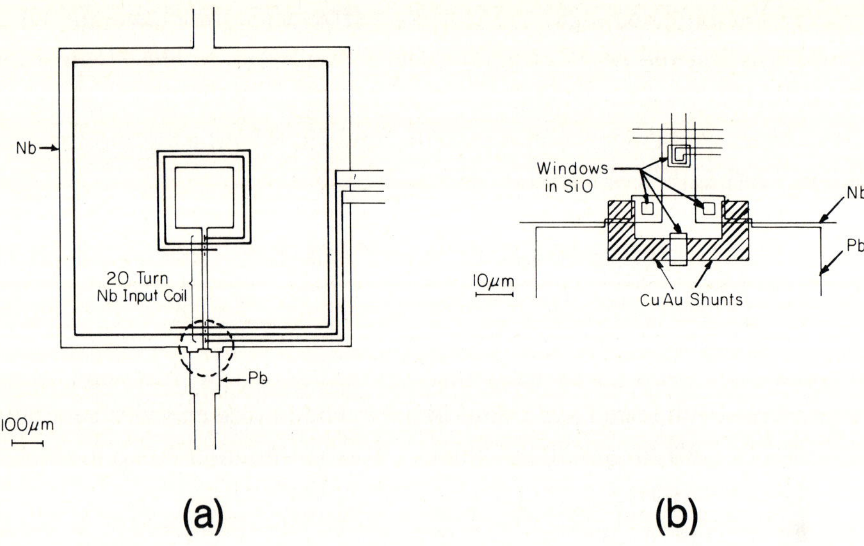

Fig.6 (a) Configuration of planar dc SQUID with overlaid spiral input coil; (b) expanded view of junctions and shunts.

chips, each with a single SQUID which is completed individually. The device is ion milled to clean the exposed areas of Nb and CuAu. We have two procedures for forming the oxide barrier. In one, we oxidize the Nb in a rf discharge in Ar containing 5 vol % O_2, and deposit the 300 nm Pb (5wt % In) counterelectrode which completes the junctions and makes contact with the shunts. In the other process, we deposit approximately 6 nm of Al and form Al_2O_3 by exposing[19] it to O_2. A photograph of the completed SQUID and a scanning electron micrograph of the junctions is shown in Fig. 7. The shunt resistance R is typically 8 Ω, and the estimated capacitance C about 0.5 pF.

Jaycox and Ketchen[17] showed that a square washer (with no slit) with inner and outer edges d and w has an inductance L (loop) = $1.25\mu_0 d$ in the limit w >> d. They gave the following expressions for the inductances of the SQUID, L, and of the spiral coil, L_i, and for the mutual inductance, M_i, and coupling coefficient, α^2, between the spiral coil and the SQUID:

$$L = L\,(\text{loop}) + L_j, \tag{3.11}$$

$$L_i = n^2\,(L-L_j) + L_s, \tag{3.12}$$

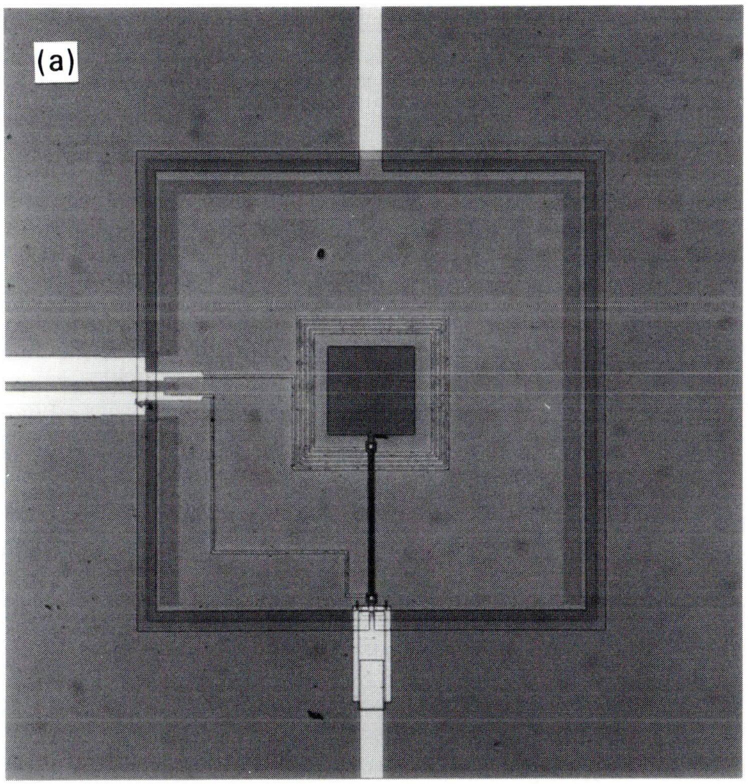

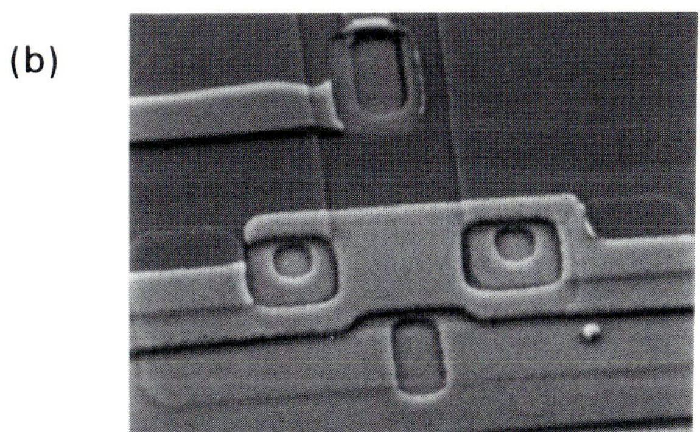

Fig.7(a) Photograph of planar dc SQUID made at UC Berkeley, with 4-turn input coil; the square washer is about 1mm across. (b) Electron micrograph of junctions prior to deposition of counterelectrode; each junction is about 3μm across.

$$M_i = n(L-L_j) \tag{3.13}$$

and

$$\alpha^2 = (1-L_J/L)/ [1+L_s/n^2(L-L_j)]. \tag{3.14}$$

Here, L_j is the parasitic inductance associated with the junctions, n is the number of turns of the input coil and L_s is the stripline inductance of this coil. For the SQUID just described with a 50-turn input coil, one measures $L_i \approx 800$ nH, $M_i \approx 16$ nH and $\alpha^2 \approx 0.75$. These results are in good agreement with the predictions of the above expressions if one takes the predicted value L(loop) ≈ 0.31 nH and assumes $L_j \approx 0.09$ nH to give L ≈ 0.4 nH. The stripline inductance (~10 nH) is insignificant for a 50-turn coil.

References 20-25 are a selection of papers describing SQUIDs fabricated on the basis of the Ketchen-Jaycox design. Some of the devices involve edge junctions in which the counterelectrode is a strip making a tunneling contact to the base electrode only at the edge. This technique enables one to make junctions with a small area and thus a small self-capacitance without resorting to electron-beam lithography. However, stray capacitances are often critically important. As has been emphasized by a number of authors, parasitic capacitance between the square washer and the input coil can produce resonances that, in turn, induce structure on the I-V characteristics and give rise to excess noise. One way to reduce these effects is to lower the shunt resistance in order to increase the damping. A different approach is to couple the SQUID to the signal via an intermediary superconducting transformer[24], so that the number of turns on the SQUID washer and the parasitic capacitance are reduced. Knuutila et al.[25] successfully damped the resonances in the input coil by terminating the stripline with a matched resistor. An alternate coupling scheme has been adopted by Carelli and Foglietti[26], who fabricated thin-film SQUIDs with many loops in parallel. The loops are coupled to a thin-film input coil surrounding them.

D. Flux-locked Loop

In most, although not all, practical applications one uses the SQUID in a feedback circuit as a null detector of magnetic flux[27]. One applies a modulating flux to the SQUID with a peak-to-peak amplitude $\Phi_0/2$ and a frequency f_m usually between 100 and 500 kHz, as indicated in Fig. 8. If the quasistatic flux in the SQUID is exactly $n\Phi_0$ the resulting voltage is a rectified version of the input signal, that is, it contains only the frequency $2f_m$ [Fig. 8(a)]. If this voltage is sent through a lock-in detector referenced to the fundamental frequency f_m, the output will be zero. On the other hand, if the quasistatic flux is $(n + 1/4)\Phi_0$, the voltage across the SQUID is at frequency f_m [Fig. 8(b)], and the output from the lock-in will be a maximum. Thus, as one increases the flux from $n\Phi_0$ to $(n + 1/4)\Phi_0$, the output from the lock-in will increase steadily; if one reduces the flux from $n\Phi_0$ to $(n - 1/4)\Phi_0$, the output will increase in the negative direction [Fig. 8(c)].

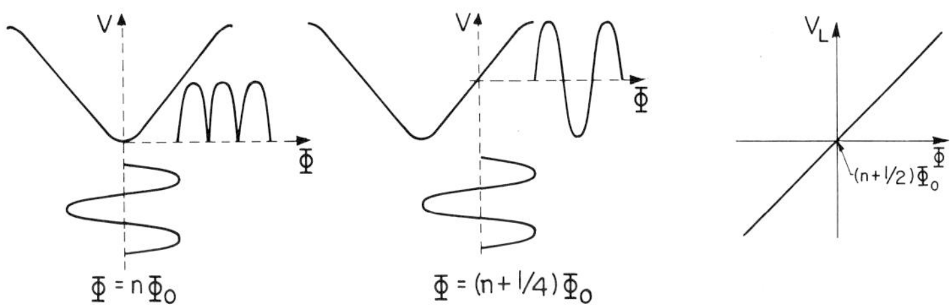

Fig. 8 Flux modulation scheme showing voltage across the SQUID for (a) $\Phi = n\Phi_0$ and (b) $\Phi = (n+1/4)\Phi_0$. The output V_L from the lock-in detector vs. Φ is shown in (c).

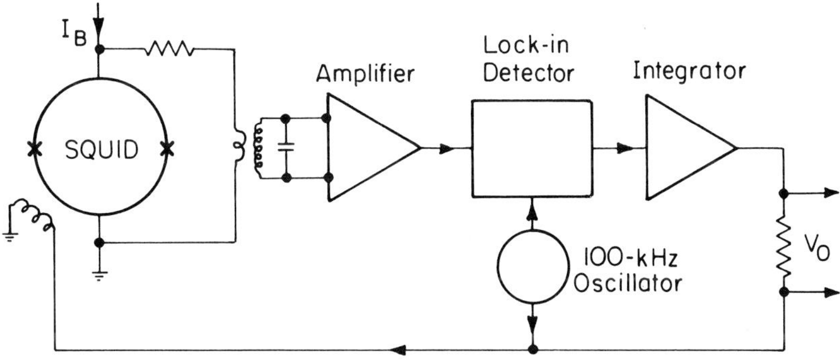

Fig. 9 Modulation and feedback circuit for the dc SQUID.

The alternating voltage across the SQUID is coupled to a low-noise preamplifier, usually at room temperature, via either a cooled transformer[28] or a cooled LC series-resonant circuit[27]. The first presents an impedance N^2R_d to the preamplifier, and the second, an impedance Q^2R_d, where R_d is the dynamic resistance of the SQUID at the bias point, N is the turns ratio of the transformer, and Q is the quality factor of the tank circuit. The value of N or Q is chosen to optimize the noise temperature of the preamplifier; with careful design, the noise from the amplifier can be appreciably less than that from the SQUID at 4.2 K.

Figure 9 shows a typical flux-locked loop in which the SQUID is coupled to the preamplifier via a cooled transformer. An oscillator applies a modulating flux to the SQUID. After amplification, the signal from the SQUID is lock-in detected and sent through an integrating circuit. The smoothed output is connected to the modulation and feedback coil via a large series resistor R_f. Thus, if one applies a flux $\delta\Phi$ to the SQUID, the feedback circuit will generate an opposing flux - $\delta\Phi$, and a voltage proportional to $\delta\Phi$ appears across R_f. This technique enables one to measure changes in flux ranging from much less than a single flux quantum to many flux quanta. The use of a modulating flux eliminates 1/f noise and drift in the bias current and preamplifier. Using a modulation frequency of 500 kHz, a double transformer between the SQUID and the preamplifier, and a two-pole integrator, Wellstood et al.[18] achieved a dynamic range of $\pm 2 \times 10^7$ $Hz^{1/2}$ for signal frequencies up to 6 kHz, a frequency response from 0 to 70 kHz (±3 dB), and a maximum slew rate of $3 \times 10^6\ \Phi_0\ sec^{-1}$.

E. Thermal Noise in the dc SQUID : Experiment

One determines the spectral density of the equivalent flux noise in the SQUID by connecting a spectrum analyzer to the output of the flux-locked loop. A representative power spectrum[29] is shown in Fig. 10: above a 1/f noise region, the noise is white at frequencies up to the roll-off of the feedback circuit. In this particular example, with L = 200pH and R = 8Ω, the measured flux noise was $S_\Phi^{1/2} = (1.9 \pm 0.1) \times 10^{-6}\Phi_0 Hz^{-1/2}$, in reasonable agreement with the predictions of Eqs. (3.7) and (3.8). The corresponding flux-noise energy was 4×10^{-32} $JHz^{-1} \approx 400\ \hbar$. Many groups have achieved noise energies that are comparable or, with lower values of L or C, somewhat better.

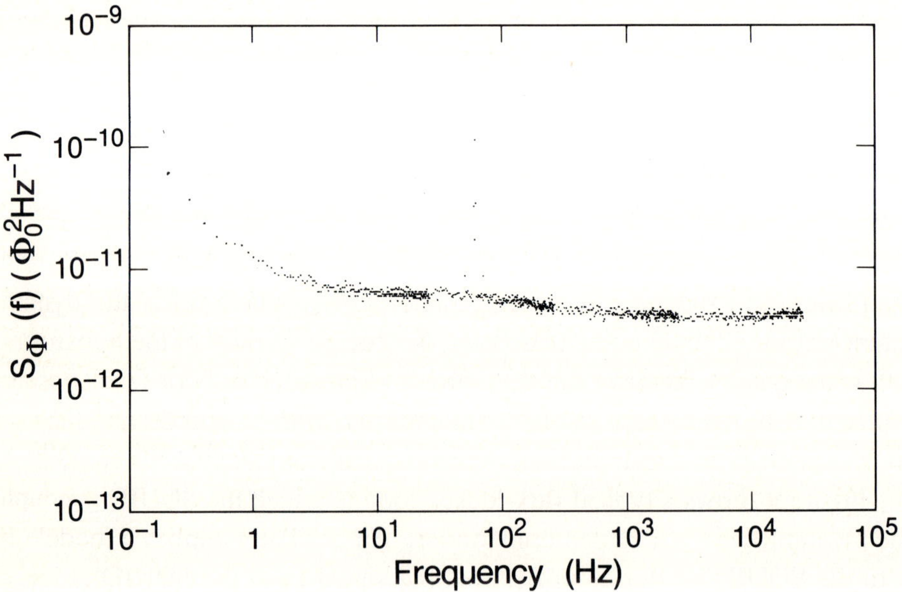

Fig.10 Spectral density of equivalent flux noise for dc SQUID with a Pb body : L = 0.2nH, R = 8Ω, and T = 4.2K (courtesy F.C. Wellstood).

Rather recently, Wellstood et al.[30] have operated SQUIDs in a dilution refrigerator at temperatures T below 1K , using a second dc SQUID as a preamplifier. They found that the noise energy scaled accurately with T at temperatures down to about 150 mK, below which the noise energy became nearly constant. This saturation was traced to heating in the resistive shunts, which prevented them from cooling much below 150mK. This heating is actually a hot-electron effect:[31,32] the bottleneck in the cooling process is the rate at which the electrons can transfer energy to the phonons which, in turn, transfer energy to the substrate. The temperature of the shunts was lowered by connecting each of them to a CuAu "cooling fin" of large volume. The hot electrons diffuse into the fins where they rapidly transfer energy to other electrons. Since the "reaction volume" is now greatly increased, the numbers of electrons and phonons interacting are also increased, and the electron gas is cooled more effectively. In this way, the effective electron temperature was reduced to about 50mK when the SQUID was at a bath temperature of 20mK, with a concomitant reduction in ε to about $5\hbar$. Very recently Ketchen et al.[33] have achieved a noise energy of about $2\hbar$ at 0.3 K in a SQUID with L = 100 pH and C = 0.14 pF.

F. l/f Noise in dc SQUIDs

The white noise in dc SQUIDs is well understood. However, some applications of SQUIDs, for example neuromagnetism, require good resolution at frequencies down to 0.1 Hz or less, and the level of the l/f or "flicker" noise becomes very important.

There are at least two separate sources of l/f noise in the dc SQUID[34]. The first arises from l/f fluctuations in the critical current of the Josephson junctions, and the mechanism for this process is reasonably well understood[35]. In the process of tunneling through the barrier, an electron becomes trapped on a defect in the barrier and is subsequently released. While the trap is occupied, there is a local change in the height of the tunnel barrier and hence in the critical current density of that region. As a result, the presence of a single trap causes the critical current of the junction to switch randomly back and forth between two values, producing a random telegraph signal. If the mean time between pulses is τ, the spectral density of this process is a Lorentzian,

$$S(f) \propto \frac{\tau}{1+(2\pi f\tau)^2}, \qquad (3.15)$$

namely white at low frequencies and falling off as $1/f^2$ at frequencies above $1/2\pi\tau$. In many cases, the trapping process is thermally activated, and τ is of the form

$$\tau = \tau_0 \exp(E/k_0 T), \quad (3.16)$$

where τ_0 is a constant and E is the barrier height.

In general, there may be several traps in the junction, each with its own characteristic time τ_i. One can superimpose the trapping processes, assuming them to be statistically independent, to obtain a spectral density[36]

$$S(f) \propto \int dE\, D(E) \left[\frac{\tau_0 \exp(E/k_B T)}{1+(2\pi f \tau_0)^2 \exp(2E/k_B T)} \right], \quad (3.17)$$

where D(E) is the distribution of activation energies. The term in square brackets is a strongly peaked function of E, centered at $\tilde{E} \equiv k_B T \ln(1/2\pi f \tau_0)$, with a width $\sim k_B T$. Thus, at a given temperature, only traps with energies within a range $k_B T$ of $\tilde{E}$ contribute significantly to the noise. If one now assumes D(E) is broad with respect to $k_B T$, one can take $D(\tilde{E})$ outside the integral, and carry out the integral to obtain

$$S(f,T) \propto \frac{k_B T}{f} D(\tilde{E}). \quad (3.18)$$

In fact, one obtains a 1/f-like spectrum from just a few traps.

The magnitude of the 1/f noise in the critical current depends strongly on the quality of the junction as measured by the current leakage at voltages below $(\Delta_1 + \Delta_2)/e$, where Δ_1 and Δ_2 are the energy gaps of the two superconductors. Traps in the barrier enable electrons to tunnel in this voltage range, a process producing both leakage current and 1/f noise. Thus, for a given technology, junctions with low subgap leakage currents will have low 1/f noise. Figure 11 shows an example of a Nb-Al_2O_3-Nb junction with a single trap[37]. The junction was resistively shunted and voltage biased at typically 1.5 μV; the noise currents were measured with a SQUID. At 4.2 K [Fig.11(a)], the noise is approximately Lorentzian; the switching process producing the noise is shown in the inset. Figure 11(b) shows that at 1.5 K the noise is substantially reduced as the trap freezes out. By measuring the temperature dependence of the random telegraph signal, Savo et al.[37] found that τ obeyed Eq. (3.16) with τ_0 = 10 s and

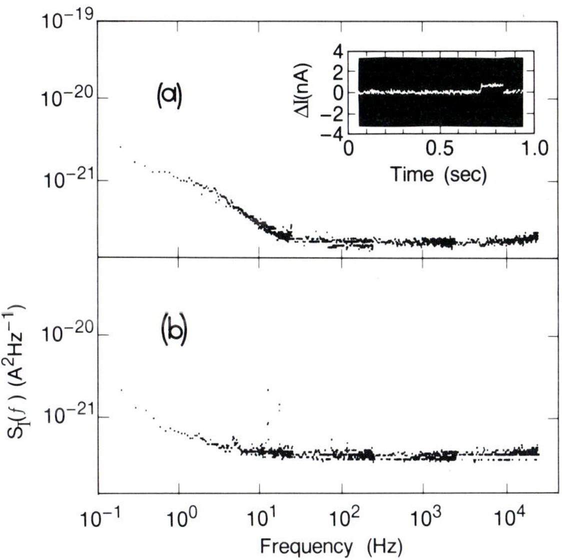

Fig.11 Spectral density of fluctuations in the critical current of a single Nb-Al_2O_3-Nb tunnel junction at (a) 4.2K and (b) 1.5K. Inset in (a) shows fluctuations vs. time (from ref.37).

E = 1.8 meV. Furthermore, τ was exponentially distributed, as expected, with an average value of 107 ms at 4.2 K.

The second source of 1/f noise in SQUIDs appears to arise from the motion of flux lines trapped in the body of the SQUID[34], and is less well understood than the critical current noise. This mechanism manifests itself as a flux noise; for all practical purposes the noise source behaves as if an external flux noise were applied to the SQUID. Thus, the spectral density of the 1/f flux noise scales as V_Φ^2, and, in particular, vanishes at $\Phi = (n \pm 1/2)\Phi_0$ where $V_\Phi = 0$. By contrast, critical current noise is still present when $V_\Phi = 0$, although its magnitude does depend on the applied flux.

The level of 1/f flux noise appears to depend strongly on the microstructure of the thin films. For example, SQUIDs fabricated at Berkeley with Nb loops sputtered under a particular set of conditions show 1/f flux noise levels of typically[34] $10^{-10}\ \Phi_0^2\,Hz^{-1}$ at 1 Hz. On the other hand, SQUIDs with Pb loops in exactly the same geometry exhibit a 1/f noise level of about $2x10^{-12}\ \Phi_0^2 Hz^{-1}$ at 1 Hz, arising from critical current fluctuations. Tesche <u>et al</u>.[38]

reported a l/f noise level in Nb-based SQUIDs of about $3x10^{-13}\,\Phi_0^2\,Hz^{-1}$, while Foglietti et al.[39] found a critical current l/f noise corresponding to $2x10^{-12}\,\Phi_0^2\,Hz^{-1}$, also in Nb based devices. Thus, we conclude that the quality of the Nb films plays a significant role in the level of l/f flux noise. It is of considerable fundamental and practical interest to understand the mechanism in detail.

There is an important practical difference between the two sources of l/f noise: critical current noise can be reduced by a suitable modulation scheme, whereas flux noise cannot. To understand how to reduce critical current l/f noise, we first note that at constant current bias the spectral density of the l/f voltage noise across the SQUID can be written in the approximate form

$$S_V(f) \approx \frac{1}{2}\left[(\partial V/\partial I_0)^2 + L^2 V_\Phi^2\right] S_{I_0}(f). \qquad (3.19)$$

In Eq. (3.19), we have assumed that each junction has the same level of critical current noise, with a spectral density $S_{I_0}(f)$. The first term on the right is the "in-phase mode", in which each of the two junctions produces a fluctuation of the same polarity. This noise is eliminated (ideally) by the conventional flux modulation scheme described in Sec. III D, provided the modulation frequency is much higher than the l/f noise frequency. The second term on the right of Eq. (3.19) is the "out-of phase" mode in which the two fluctuations are of opposite polarity and, roughly speaking, result in a current around the SQUID loop. This term appears, therefore, as a flux noise, vanishing for $V_\Phi = 0$, but is not reduced by the usual flux modulation scheme. Fortunately, there are schemes by which this second term, as well as the first, can be reduced.

One such scheme was described by Koch et al.[34] and is also available on the dc SQUID manufactured by BTi[40]. An alternate scheme, second harmonic detection (SHAD), has recently been developed by Foglietti et al.[39], and we shall briefly describe it. In Fig. 12(a) we see that both I and Φ are switched among the three states A: $(+I, \Phi+\Phi_0/4)$, B: $(0, \Phi+\Phi_0/2)$ and C: $(-I, \Phi+3\Phi_0/4)$. If the static flux is $n\Phi_0$ [dashed curve and pulses in Fig.12(a)], the positive and negative voltage pulses across the SQUID are of equal magnitude, and if one detects this signal at twice the modulation frequency, the output is zero. If we apply an additional flux $\delta\Phi$, however, the V-Φ curve is shifted along the Φ-axis [solid curve and pulses in Fig.12(a)], and the pulse heights become unequal. Lock-in detection at twice the modulation frequency produces an output proportional to $\delta\Phi$. The output from the lock-in detector is integrated and fedback to flux-lock

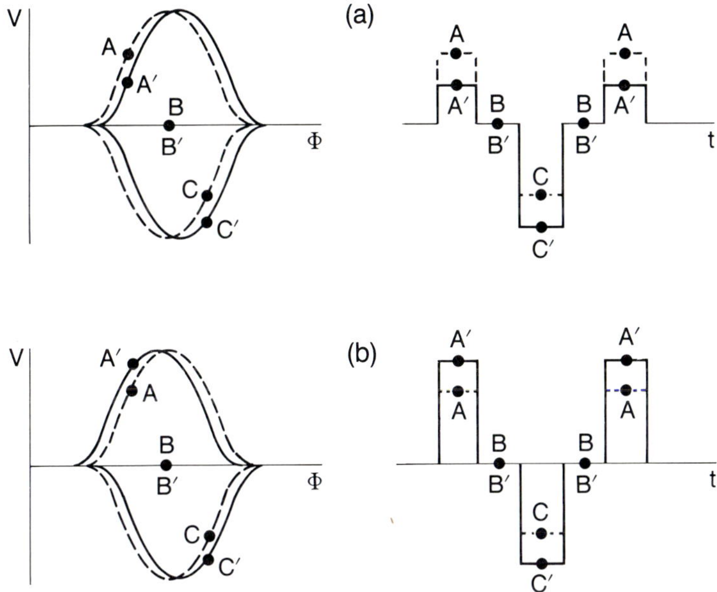

Fig.12 SHAD modulation scheme for a dc SQUID to reduce l/f noise due to critical current fluctuations: (a) shows the response to an external flux δΦ, and (b) the response to a critical current fluctuation (re-drawn from ref. 39).

the SQUID in the usual way. We note that in-phase fluctuations in critical current increase or decrease the magnitude of the voltage across the SQUID, but do not shift the V-Φ curve along the Φ axis. The magnitudes of the voltage pulses grow or shrink in a symmetric way, and still contain no second harmonic component. Thus, the in-phase fluctuations in critical current produce no response. Similarly, the out-of-phase fluctuations produce no response, as we see in Fig. 12(b). These fluctuations produce an effective flux in the SQUID, with a polarity that depends on the direction of I. Thus, the V-Φ curve shifts to the left for positive bias currents and to the right for negative bias currents. The resulting voltage pulses shrink in a symmetric way, producing no component at the second harmonic of the modulation frequency. Finally, we emphasize that any l/f flux noise produces an output from the lock-in detector in precisely the same way as the applied flux in Fig. 12(a).

Figure 13 shows an example of the reduction in l/f noise achieved by Foglietti et al.[34] Figure 13(a) shows the noise as measured with a rf SQUID, and Fig. 13(b), with the conventional flux modulation scheme. The reduction in l/f noise indicates that in-phase critical current fluctuations have been removed.

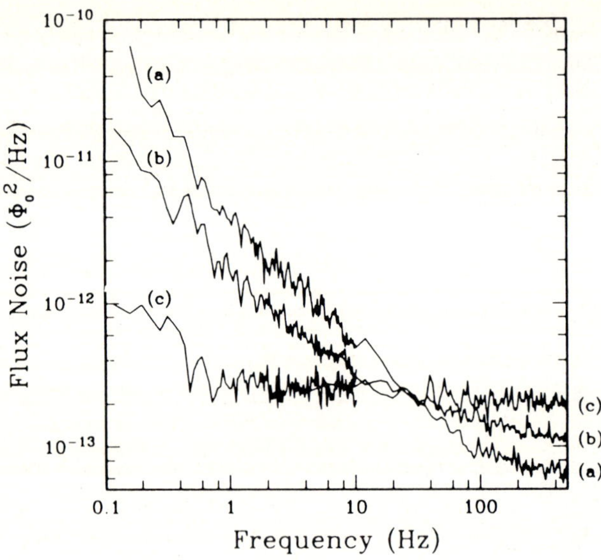

Fig.13 Flux noise spectral density for dc SQUID measured (a) with an rf SQUID small-signal readout scheme, (b) with the flux modulation scheme of Figs.8 and 9 and (c) with SHAD (Fig.12) (from ref. 39).

Figure 13(c) shows the considerable reduction in the l/f noise resulting from SHAD: the measured l/f noise was $10^{-12}\,\Phi_0^2\mathrm{Hz}^{-1}$ at 0.1 Hz. We note that SHAD increases the spectral density of the white noise by a factor of 2: V_Φ^2 is reduced by a factor of 4, while the noise is measured only during the 50% of the modulation cycle in which the SQUID is biased.

For measurements at low frequencies, it is clearly advantageous to use a l/f noise reduction scheme for SQUIDs in which the l/f noise is dominated by critical current fluctuations. As stressed repeatedly, however, nothing can be done to reduce l/f flux noise. As we shall see in Sec. VI, l/f flux noise is a particularly serious problem for high-T_C SQUIDs.

G. Alternate Read-Out Schemes

Although the flux modulation method described in Sec. III.D has been used successfully for many years, alternate schemes recently have been developed. These efforts have been motivated, at least in part, by the need to simplify the electronics required for the multichannel systems used in neuromagnetism - see Sec. V.A. Fujimaki and co-workers[41] and Drung and co-workers[42] have devised schemes in which the output from the SQUID is sensed digitally and fed back as an analog signal to the SQUID to flux-lock the loop. Fujimaki et al.[41] used Josephson digital circuitry to integrate their feedback system on the same chip as the SQUID so that the flux-locked signal was available directly from the cryostat. The system of Drung and co-workers, however, is currently the more sensitive, with a flux resolution of about $10^{-6}\,\Phi_0\mathrm{Hz}^{-1/2}$ in a 50 pH SQUID. These workers also were able to reduce the l/f noise in the system using a modified version of the modulation scheme of Foglietti et al.[39] Although they need further development, cryogenic digital feedback schemes offer several advantages: they are compact, produce a digitized output for transmission to room temperature, offer wide flux-locked bandwidths, and need not add any noise to the intrinsic noise of the SQUID.

In yet another system, Mück and Heiden[43] have operated a dc SQUID with hysteretic junctions in a relaxation oscillator. The oscillation frequency depends on the flux in the SQUID, reaching a maximum at $(n+1/2)\Phi_0$ and a minimum at $n\Phi_0$. A typical frequency modulation is 100 kHz at an operating frequency of 10 MHz. This technique produces large voltages across the SQUID so that no matching network to the room temperature electronics is required. The room temperature electronics is simple and compact, and the resolution at 4.2 K is about $10^{-5}\Phi_0\mathrm{Hz}^{-1/2}$ with an inductance estimated to be about 80 pH.

IV. THE RF SQUID

A. Principles of Operation

Although the rf SQUID is still the more widely used device because of its long-standing commercial availability, it has seen very little development in recent years. For this reason I will give a rather brief account of its principles and noise limitations, following rather closely descriptions in earlier reviews.[44,45]

The rf SQUID[4,5] shown in Fig.14 consists of a superconducting loop of inductance L interrupted by a single Josephson junction with critical current I_0 and a nonhysteretic current-voltage characteristic. Flux quantization[1] imposes the constraint

$$\delta + 2\pi\Phi_T/\Phi_0 = 2\pi n \qquad (4.1)$$

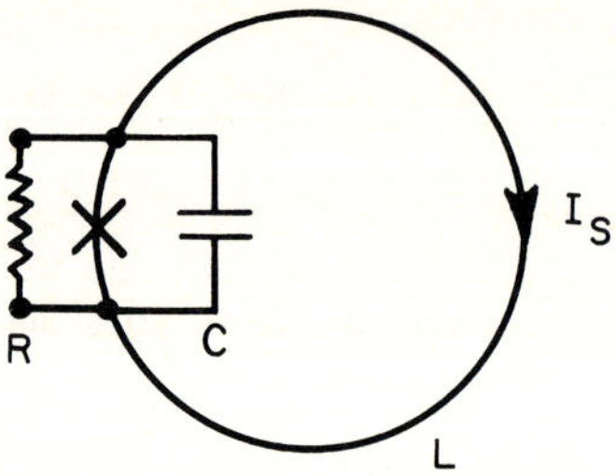

Fig.14 The rf SQUID.

on the total flux Φ_T threading the loop. The phase difference δ across the junction determines the supercurrent

$$I_s = -I_0 \sin(2\pi\Phi_T/\Phi_0) \tag{4.2}$$

flowing in the ring. A quasistatic external flux Φ thus gives rise to a total flux

$$\Phi_T = \Phi - LI_0 \sin(2\pi\Phi/\Phi_0). \tag{4.3}$$

The variation of Φ_T with Φ is sketched in Fig.15(a) for the typical value $LI_0 = 1.25\ \Phi_0$. The regions with positive slope are stable, whereas those with negative slope are not. A "linearized" version of Fig.15(a) showing the path traced out by Φ and Φ_T is shown in Fig.15(b). Suppose we slowly increase Φ from zero. The total flux Φ_T increases less rapidly than Φ because the response flux $-LI_S$ opposes Φ. When I_S reaches I_0, at an applied flux Φ_C and a total flux Φ_{TC}, the junction switches momentarily into a nonzero voltage state and the SQUID jumps from the $k = 0$ to the $k = 1$ quantum state. If we subsequently reduce Φ from a value just above Φ_C, the SQUID remains in the $k = 1$ state until $\Phi = \Phi_0 - \Phi_C$, at which point I_S again exceeds the critical current and the SQUID returns to the $k = 0$ state. In the same way, if we lower Φ to below $-\Phi_C$ and then increase it, a second hysteresis loop will be traced. We note that this hysteresis occurs provided $LI_0 > \Phi_0/2\pi$; most practical SQUIDs are operated in this regime. For $LI_0 \approx \Phi_0$, the energy ΔE dissipated when one takes the flux around a single hysteresis loop is its area divided by L:

$$\Delta E \approx I_0\Phi_0. \tag{4.4}$$

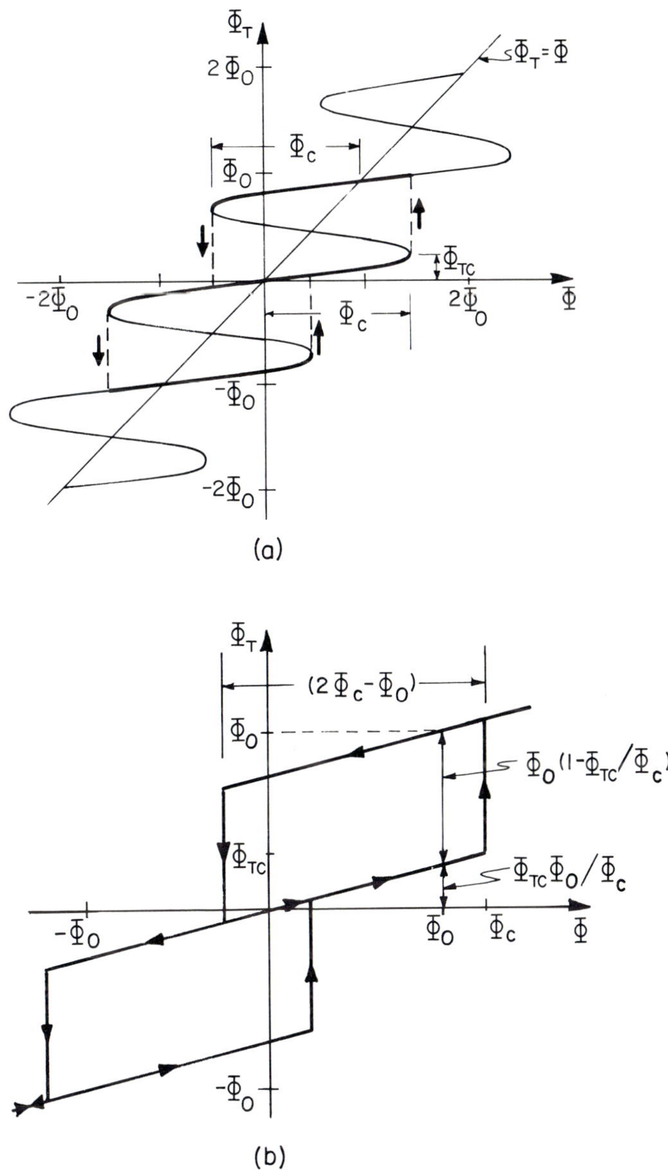

Fig.15 The rf SQUID: (a) total flux Φ_T vs. Φ for $LI_0 = 1.25\ \Phi_0$; (b) values of Φ_T as Φ is quasistatically increased and then decreased.

We now consider the radio frequency (rf) operation of the device. The SQUID is inductively coupled to the coil of an LC-resonant circuit with a quality factor $Q = R_T/\omega_{rf}L_T$ via a mutual inductance $M = K(LL_T)^{1/2}$ - see Fig. 16. Here,

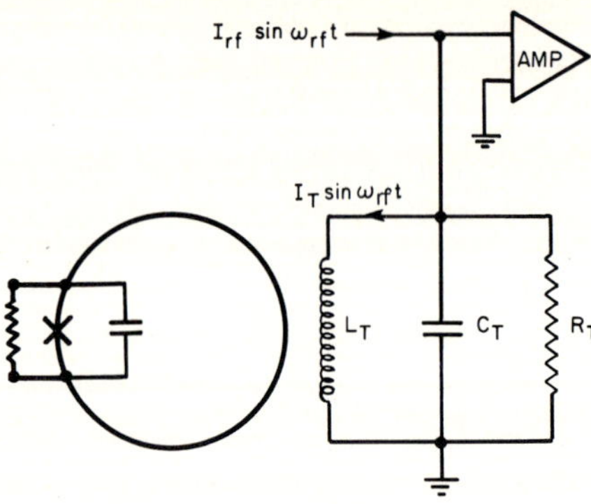

Fig.16 The rf SQUID inductively coupled to a resonant tank circuit.

L_T, C_T, and R_T are the inductance, capacitance and (effective) parallel resistance of the tank circuit, and $\omega_{rf}/2\pi$ is its resonant frequency, typically 20 or 30 MHz. The tank circuit is excited at its resonant frequency by a current I_{rf} sinω_{rf}t, which generates a current of amplitude $I_T = QI_{rf}$ in the inductor. The voltage V_T across the tank circuit is amplified with a preamplifier having a high input impedance. First, consider the case $\Phi = 0$. As we increase I_{rf} from zero, the peak rf flux applied to the loop is $MI_T = QMI_{rf,}$ and V_T increases linearly with I_{rf}. The peak flux becomes equal to Φ_C when $I_T = \Phi_C/M$ or $I_{rf} = \Phi_C/ MQ$, at A in Fig.17. The corresponding peak rf voltage across the tank circuit is

$$V_T^{(n)} = \omega_{rf} L_T \Phi_C/M, \tag{4.5}$$

where the superscript (n) indicates $\Phi = n\Phi_0$, in this case with n=0. At this point the SQUID makes a transition to either the k = +1 state or the k = -1 state. As the SQUID traverses the hysteresis loop, energy ΔE is extracted from the tank circuit. Because of this loss, the peak flux on the next half cycle is less than Φ_C, and no transition occurs. The tank circuit takes many cycles to recover sufficient energy to induce a further transition, which may be into either the k = +1 or -1 states. If we now increase I_{rf}, transitions are induced at the same values of I_T and V_T but, because energy is supplied at a higher rate, the stored energy builds up more rapidly after each energy loss ΔE, and transitions occur more frequently. In the absence of thermal fluctuations (Sec. IV.B), the "step" AB in Fig.17 is at constant voltage. At B, a transition is induced on each positive and negative rf

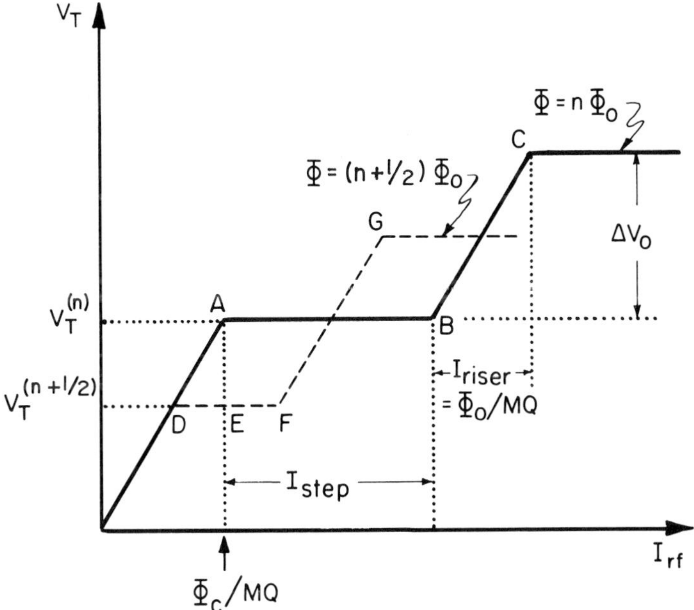

Fig.17 V_T vs. I_{rf} in the absence of thermal noise for $\Phi = n\Phi_0$, $(n+1/2)\Phi_0$.

peak, and a further increase in I_{rf} produces the "riser" BC. At C, transitions from the $k = \pm1$ to the $k = \pm2$ states occur, and a second step begins. As we continue to increase I_{rf}, we observe a series of steps and risers.

If we now apply an external flux $\Phi = \Phi_0/2$, the hysteresis loops in Fig.15(b) are shifted by $\Phi_0/2$. Thus, a transition occurs on the positive peak of the rf cycle at a flux $(\Phi_c - \Phi_0/2)$, whereas on the negative peak the required flux is $-(\Phi_c + \Phi_0/2)$. As a result, as we increase I_{rf} from zero, we observe the first step at D in Fig.17 at

$$V_T^{(n+1/2)} = \omega_{rf}L_T(\Phi_C-\Phi_0/2)/M. \tag{4.6}$$

As we increase I_{rf} from D to F, the SQUID traverses only one hysteresis loop, corresponding to the $k = 0$ to $k = +1$ transition at $(\Phi_c - \Phi_0/2)$. A further increase in I_{rf} produces the riser FG, and at G, transitions begin at a peak rf flux $-(\Phi_C + \Phi_0/2)$. In this way, we observe a series of steps and risers for $\Phi = \Phi_0/2$, interlocking those for $\Phi = 0$ (Fig.17). As we increase Φ from zero, the voltage at which the first step appears will drop to a minimum (D) at $\Phi_0/2$ and rise to its maximum value (A) at $\Phi = \Phi_0$. The change in V_T as we increase Φ from 0 to $\Phi_0/2$, found by subtracting Eq. (4.6) from Eq.(4.5), is $\omega_{rf}L_T\Phi_0/2M$. Thus, for a small change in flux near $\Phi = \Phi_0/4$ we find the transfer function

$$V_\Phi = \omega_{rf}L_T/M. \tag{4.7}$$

At first sight, Eq.(4.7) suggests that we can make V_Φ arbitrarily large by reducing K sufficiently. However, we obviously cannot make K so small that the SQUID has no influence on the tank circuit, and we need to establish a lower bound on K. To operate the SQUID, we must be able to choose a value of I_{rf} that intercepts the first step for all values of Φ : this requirement is satisfied if the point F in Fig. 17 lies to the right of E, that is, if DF exceeds DE. We can calculate DF by noting that the power dissipation in the SQUID is zero at D and $\Delta E(\omega_{rf}/2\pi) \approx I_0\Phi_0\omega_{rf}/\, 2\pi$ at F. Thus, $(I_{rf}^{(F)}-I_{rf}^{(D)})V_T^{(n+1/2)}/2 = I_0\Phi_0\omega_{rf}/\, 2\pi$ (I_{rf} and V_T are peak, rather than rms values). Furthermore, we easily can see that $I_{rf}^{(E)} - I_{rf}^{(D)} = \Phi_0/\, 2MQ$. Assuming $LI_0 \approx \Phi_0$ and using Eq.(4.5), we find that the requirement $I_{rf}^{(E)} > I_{rf}^{(D)}$ can be written in the form

$$K^2Q \gtrsim \pi/4. \tag{4.8}$$

If we set $K \approx Q^{-1/2}$, Eq.(4.7) becomes

$$V_\Phi \approx \omega_{rf}(QL_T/L)^{1/2}. \tag{4.9}$$

To operate the SQUID, one adjusts I_{rf} so that the SQUID remains biased on the first step - see Fig.17 - for all values of Φ. The rf voltage across the tank circuit is amplified and demodulated to produce a signal that is periodic in Φ. A modulating flux, typically at 100kHz and with a peak-to-peak amplitude of $\Phi_0/2$, also is applied to the SQUID, just as in the case of the dc SQUID. The voltage produced by this modulation is lock-in detected, integrated, and fed back as a current into the modulation coil to flux-lock the SQUID.

B. Theory of Noise in the rf SQUID

A detailed theory has been developed for noise in the rf SQUID;[46-54] in contrast to the case for the dc SQUID, noise contributions from the tank circuit and preamplifier also are important. We begin by discussing the intrinsic noise

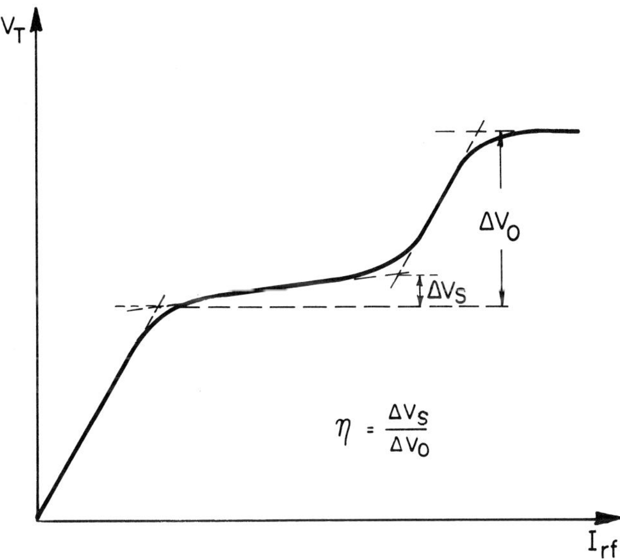

Fig.18 V_T vs. I_{rf} showing the effects of thermal noise.

in the SQUID. In the previous section we assumed that transitions from the $k = 0$ to the $k = 1$ state occurred precisely at $\Phi = \Phi_C$. In fact, thermal activation causes the transition to occur stochastically, at lower values of flux. Kurkijärvi[46] calculated the distribution of values of Φ at which the transitions occur; experimental results[55] are in good agreement with his predictions. When the SQUID is driven with an rf flux, the fluctuations in the value of flux at which transitions occur have two consequences. First, noise is introduced on the peak voltage V_T, giving an equivalent intrinsic flux noise spectral density[47,51]

$$S_{\Phi}^{(i)} \approx \frac{(LI_0)^2}{\omega_{rf}} \left(\frac{2\pi k_B T}{I_0 \Phi_0} \right)^{4/3} . \tag{4.10}$$

Second, the noise causes the steps to tilt (Fig.18), as we easily can see by considering the case for $\Phi = 0$. In the presence of thermal fluctuations the transition from the $k = 0$ to the $k = 1$ state (for example) has a certain probability of occurring at any given value of the total flux $\Phi + \Phi_{rf}$. Just to the right of A in Fig.17 this transition occurs at the peak of the rf flux once in many rf cycles. Thus, the probability of the transition occurring in any one cycle is small. On the other hand, at B a transition must occur at each positive and negative peak of the rf flux, with unity probability. To increase the transition probability, the peak value of the rf flux and hence V_T must increase as I_{rf} is increased from A to B. Jackel and Buhrman[48], introduced the slope parameter η defined in Fig.18, and showed that it was related to $S_\Phi^{(i)}$ by the relation

$$\eta^2 \approx S_\Phi^{(i)}\omega_{rf}/\pi\Phi_0^2, \tag{4.11}$$

provided η was not too large. This relation is verified experimentally.

The noise temperature T_a of typical rf amplifiers operated at room temperature is substantially higher than that of amplifiers operated at a few hundred kilohertz, and is therefore not negligible for rf SQUIDs operated at liquid ^{4}He temperatures. Furthermore, part of the coaxial line connecting the tank circuit to the preamplifier is at room temperature. Since the capacitances of the line and the amplifier are a substantial fraction of the capacitance of the tank circuit, part of the resistance damping the tank circuit is well above the bath temperature. As a result, there is an additional contribution to the noise which we combine with the preamplifier noise to produce an effective noise temperature T_a^{eff}. The noise energy contributed by these extrinsic sources can be shown to be[48,52] $2\pi\eta k_B T_a^{eff}/\omega_{rf}$. Combining this contribution with the intrinsic noise, one finds

$$\varepsilon \approx \frac{1}{\omega_{rf}}\left(\frac{\pi\eta^2\Phi_0^2}{2L} + 2\pi\eta k_B T_a^{eff}\right). \tag{4.12}$$

Equation (4.12) shows that ε scales as $1/\omega_{rf}$, but one should bear in mind that T_a tends to increase with ω_{rf}. Nonetheless, improvements in performance have been achieved by operating the SQUID at frequencies[56,57] much higher than the usual 20 or 30 MHz. One also can reduce the T_a^{eff} by cooling the preamplifier,[56,58] thereby reducing T_a and reducing the temperature of the tank circuit to that of the bath. However, the best noise energies achieved for the rf SQUID are substantially higher[59] than those routinely obtained with thin-film dc SQUIDs, and for this reason workers requiring the highest possible resolution almost invariably use the latter device.

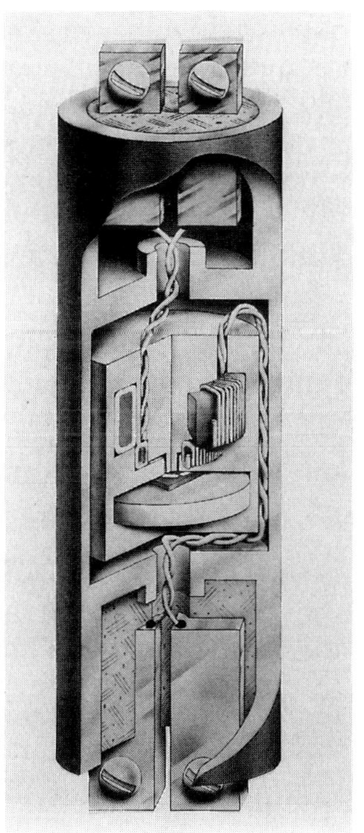

Fig.19 Cut-away drawing of toroidal SQUID (courtesy BTi, inc.).

C. Practical rf SQUIDs

Although less sensitive than the dc SQUID, the rf SQUID is entirely adequate for a wide range of applications. It is therefore more widely used than the dc SQUID, for the simple reason that reliable, easy-to-operate devices have been commercially available since the early 1970's, notably from BTi (formerly SHE). We therefore shall confine ourselves to a brief description of the device available from this company.

Figure 19 shows a cut-away drawing of the BTi rf SQUID[40], which has a toroidal configuration machined from Nb. One way to understand this geometry is to imagine rotating the SQUID in Fig.14 through 360° about a line running through the junction from top to bottom of the page. This procedure produces a toroidal cavity connected at its center by the junction. If one places a toroidal coil in this cavity, a current in the coil produces a flux that is tightly coupled to the SQUID. In Fig.19, there are actually two such cavities, one containing the tank circuit-modulation-feedback coil and the other the input coil. This separation eliminates cross-talk between the two coils. Leads to the two coils are brought out via screw-terminals. The junction is made from thin films of Nb. This device is self-shielding against external magnetic field fluctuations, and has proven to be reliable and convenient to use. In particular, the Nb input terminals enable one to connect different input circuits in a straightforward way. A typical device has a white noise energy of $5x10^{-29}$ JHz^{-1}, with a l/f noise energy of perhaps 10^{-28} JHz^{-1} at 0.1 Hz.

V. SQUID-BASED INSTRUMENTS

Both dc and rf SQUIDs are used as sensors in a far-ranging assortment of instruments. I here briefly discuss some of them: my selection is far from exhaustive, but does include the more commonly used instruments.

Each instrument involves a circuit attached to the input coil of the SQUID. We should recognize from the outset that, in general, the presence of the input circuit influences both the signal and noise properties of the SQUID while the SQUID, in turn, reflects a complex impedance into the input. Because the SQUID is a nonlinear device a full description of the interactions is complicated, and we shall not go into the details here. However, one aspect of this interaction, first pointed out by Zimmerman,[60] is easy to understand. Suppose we connect a superconducting pick-up loop of inductance L_P to the input coil of inductance L_i to form a magnetometer, as shown in Fig.20(a). It is easy to show that the SQUID inductance L is reduced to the value

$$L' = L[1-\alpha^2 L_i/(L_i+L_P)], \tag{5.1}$$

where α^2 is the coupling coefficient between L and L_i. We have neglected any stray inductance in the leads connecting L_i and L_P, and any stray capacitance. The reduction in L tends to increase the transfer coefficient of both the dc SQUID [Eq.(3.7)] and the rf SQUID [Eq.(4.9)]. The reduction in L and the change in noise properties can affect the results of optimization schemes appreciably under certain circumstances,[29] but we shall not discuss these issues here.

A. Magnetometers and Gradiometers

One of the simplest instruments is the magnetometer [Fig.20(a)]. A pick-up loop is connected across the input coil to make a superconducting flux transformer. The SQUID and input coil are generally enclosed in a superconducting shield. If one applies a magnetic flux, $\delta\Phi^{(p)}$, to the pick-up loop, flux quantization requires that

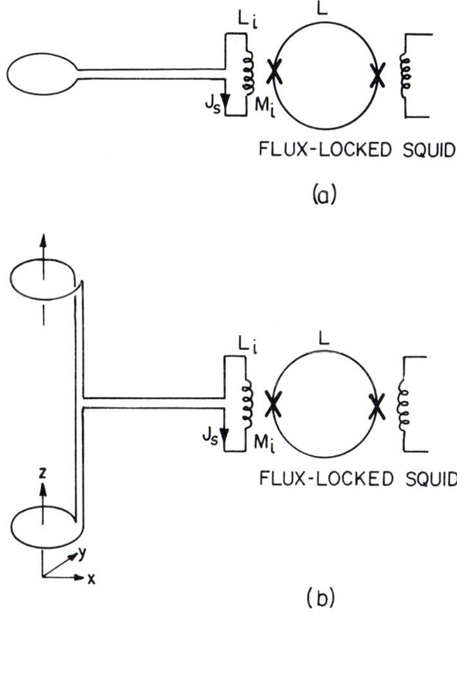

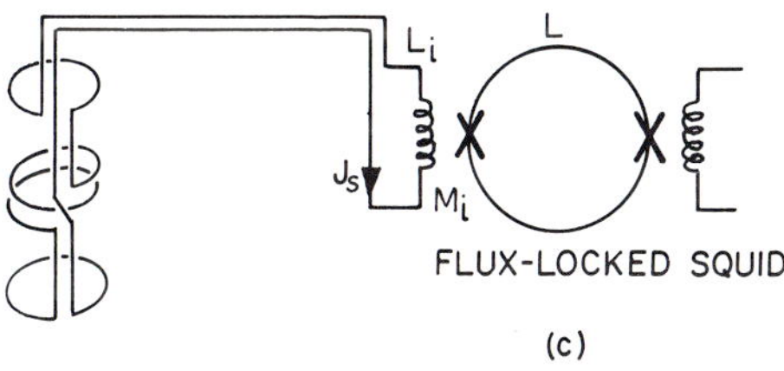

Fig.20 Superconducting flux transformers:(a) magnetometer, (b) first-derivative gradiometer, (c) second-derivative gradiometer.

$$\delta\Phi^{(p)} + (L_i + L_p)J_S = 0, \tag{5.2}$$

where J_S is the supercurrent induced in the transformer. We have neglected the effects of the SQUID on the input circuit. The flux coupled into the SQUID, which we assume to be in a flux-locked loop, is $\delta\Phi = M_i|J_S| = M_i\delta\Phi^{(p)}/(L_i + L_P)$. We find the minimum detectable value of $\delta\Phi^{(p)}$ by equating $\delta\Phi$ with the equivalent flux noise of the SQUID. Defining $S_\Phi^{(p)}$ as the spectral density of the equivalent flux noise referred to the pick-up loop, we find

$$S_\Phi^{(p)} = \frac{(L_p+L_i)^2}{M_i^2} S_\Phi. \quad (5.3)$$

Introducing the equivalent noise energy referred to the pick-up loop, we obtain

$$\frac{S_\Phi^{(p)}}{2L_p} = \frac{(L_p+L_i)^2}{L_i\, L_p}\, \frac{S_\Phi}{2\alpha^2 L}\,. \quad (5.4)$$

We observe that Eq.(5.4) has the minimum value

$$S_\Phi^{(p)}/2L_p = 4\varepsilon(f)/\alpha^2 \quad (5.5)$$

when $L_i = L_p$. Thus, a fraction $\alpha^2/4$ of the energy in the pick-up loop is transferred to the SQUID. In this derivation we have neglected noise currents in the input circuit arising from noise in the SQUID, the fact the the input circuit reduces the SQUID inductance, and any possible coupling between the feedback coil of the SQUID and the input circuit.

Having obtained the flux resolution for $L_i = L_p$, we can immediately write down the corresponding magnetic field resolution $B_N^{(p)} = (S_\Phi^{(p)})^{1/2}/\pi r_p^2$, where r_p is the radius of the pick-up loop:

$$B_N^{(p)} = 2\sqrt{2}L_p^{1/2}\varepsilon^{1/2}/\pi r_p^2\alpha\,. \quad (5.6)$$

For a loop made from wire of radius r_0, one finds[61] $L_p = \mu_0 r_p[\ln(8r_p/r_0) - 2]$, where $\mu_0 = 4\pi \times 10^{-7}$ henries/meter; for a reasonable range of values of r_p/r_0 we can set $L_P \approx 5\mu_0 r_p$. Thus, we obtain $B_N^{(p)} \approx 2(\mu_0\varepsilon)^{1/2}/\alpha r_P^{3/2}$. This indicates that one can, in principle, improve the magnetic field resolution indefinitely by increasing r_p, keeping $L_i = L_p$. Of course, in practice, the size of the cryostat will impose an upper limit on r_p. If we take $\varepsilon = 10^{-28}$ JHz^{-1} (a somewhat conservative value for an rf SQUID), $\alpha = 1$, and $r_p = 25$ mm, we find $B_N^{(p)} \approx 5 \times 10^{-15}$ tesla Hz$^{-1/2}$ = 5×10^{-11} gauss Hz$^{-1/2}$. This is a much higher sensitivity than that achieved by any nonsuperconducting magnetometer.

Magnetometers have usually involved flux transformers made of Nb wire. For example, one can make the rf SQUID in Fig.19 into a magnetometer merely by connecting a loop of Nb wire to its input terminals. In the case of the thin-film dc SQUID, one can make an integrated magnetometer by fabricating a Nb loop across the spiral input coil. In this way, Wellstood et al.[18] achieved a magnetic field white noise of 5×10^{-15} tesla $Hz^{-1/2}$ using a pick-up loop with a diameter of a few millimeters.

Magnetometers with typical sensitivities of 0.01 pT $Hz^{-1/2}$ have been used in geophysics in a variety of applications,[62] for example, magnetotellurics, active electromagnetic sounding, piezomagnetism, tectonomagnetism, and the location of hydrofractures. Although SQUID-based magnetometers are substantially more sensitive than any other type, the need to replenish the liquid helium in the field has restricted the extent of their applications. For this reason, the advent of high-temperature superconductors may have considerable impact on this field - see Sec.VI.

An important variation of the flux transformer is the gradiometer. Figure 20(b) shows an axial gradiometer that measures $\partial B_z/\partial z$. The two pick-up loops are wound in opposition and balanced so that a uniform field B_z links zero net flux to the transformer. A gradient $\partial B_z/\partial z$, on the other hand, does induce a net flux and thus generates an output from the flux-locked SQUID. Figure 20(c) shows a second-order gradiometer that measures $\partial^2 B_z/\partial z^2$; Fig.21(a) is a photograph of a practical version.

Thin-film gradiometers based on dc SQUIDs were made as long ago[28] as 1978, and a variety of devices[25,63-67] have been reported since then. To my knowledge, all of the thin film gradiometers made to date have been planar, and therefore measure an off-diagonal gradient, for example, $\partial B_z/\partial x$ or $\partial^2 B_z/\partial x \partial y$. A representative device is shown in Fig.21(b)-(d).

The most important application of the gradiometer thus far is in neuromagnetism[68], notably to detect weak magnetic signals emanating from the human brain. The gradiometer discriminates strongly against distant noise sources, which have a small gradient, in favor of locally generated signals. One can thus use a second-order gradiometer in an unshielded environment, although the present trend is toward using first-order gradiometers in a shielded room of aluminum and mu-metal that greatly attenuates the ambient magnetic noise. In this application, axial gradiometers of the type shown in Fig.20(a) actually sense magnetic field, rather than the gradient, because the distance from the signal source to the pick-up loop is less than the baseline of the gradiometer. The magnetic field sensitivity referred to one pick-up loop is typically 10 fT $Hz^{-1/2}$. Although great progress in this field has been made in recent years, it is generally

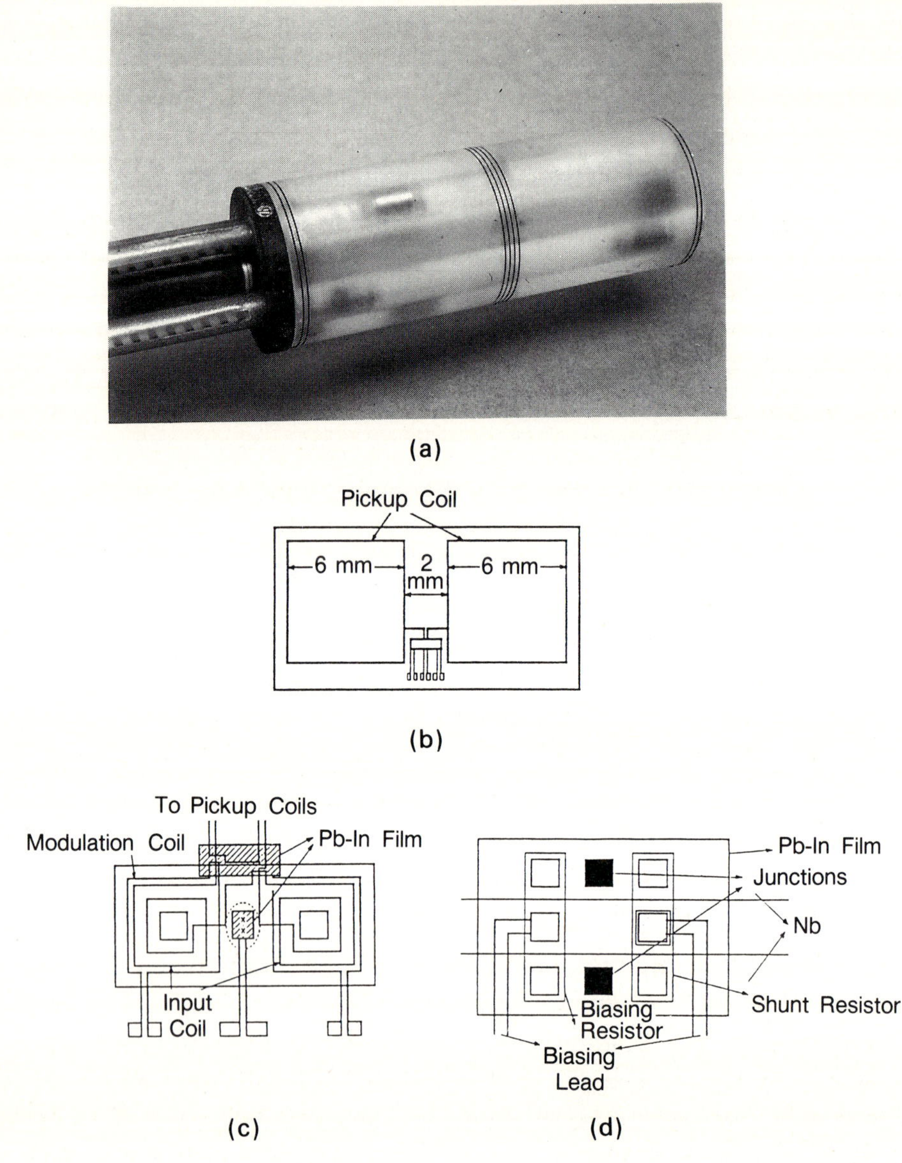

Fig.21 (a) Photograph of wire-wound second-derivative gradiometer for biomedical applications (courtesy BTi, inc.); thin-film first-derivative gradiometer: (b) pick-up loops, (c) two-hole SQUID with spiral input coils, and (d) expanded view of the dotted circle in (c) showing junctions and resistive shunts (from ref. 66).

agreed that one needs an array of 50 to 100 gradiometers to make a clinically viable system. This requirement has greatly spurred the development of integrated, thin-film devices. For example, Knuutila[69] has reported that a 24-channel first-derivative array is under construction.

There are two basic kinds of measurements made on the human brain. In the first, one detects spontaneous activity: a classic example is the generation of magnetic pulses by subjects suffering from focal epilepsy[70]. The second kind involves evoked response: for example, Romani et al.[71] detected the magnetic signal from the auditory cortex generated by tones of different frequencies. Romani has given a extensive review of this work elsewhere in these proceedings.

There are several other applications of gradiometers. One kind of magnetic monopole detector[72] consists of a gradiometer: the passage of a monopole would link a flux h/e in the pick-up loop and produce a step-function response from the SQUID. Gradiometers have recently been of interest in studies of corrosion and in the location of fractures in pipelines and other structures.

B. Susceptometers

In principle, one easily can use the first-derivative gradiometer of Fig.20(b) to measure magnetic susceptibility χ. One establishes a static field along the z-axis and lowers the sample into one of the pick-up loops. Provided χ is nonzero, the sample introduces an additional flux into the pick-up loop and generates an output from the flux-locked SQUID. A very sophisticated susceptometer is available commercially[73]. Room temperature access enables one to cycle samples rapidly, and one can measure χ as a function of temperature between 1.8 K and 400 K in fields up to 5.5 tesla. The system is capable of resolving a change in magnetic moment as small as 10^{-8} emu.

Novel miniature susceptometers have been developed by Ketchen and co-workers[33,74,75]. One version is shown schematically in Fig.22. The SQUID loop incorporates two pick-up loops wound in the opposite sense and connected in series. The two square pick-up loops, 17.5 μm on a side and with an inductance of about 30 pH, are deposited over a hole in the ground plane that minimizes the inductance of the rest of the device. The SQUID is flux biased at the maximum of V_Φ by means of a control current I_C in one of the pick-up loops. One can apply a magnetic field to the two loops by means of the current I_F; by passing a fraction of this current into the center tap I_C, one can achieve a high degree of electronic balance between the two loops. The sample to be studied is placed over one of the loops, and the output from the SQUID when the field is applied is directly proportional to the magnetization. At 4.2 K, the susceptometer is capable of detecting the magnetization due to as few as 3000 electron spins.

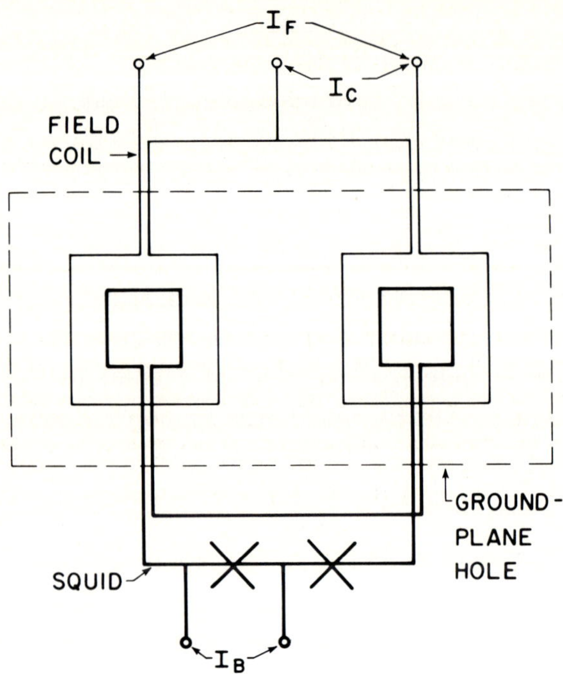

Fig.22 Thin-film miniature susceptometer (from ref. 74).

Awschalom and co-workers[33,75], have used a miniature susceptometer to perform magnetic spectroscopy of semiconductors with picosecond time-resolution. Linearly polarized pulses 4 ps in length are generated with a dye laser and split into a pump train and a weaker probe train. The time delay between the two trains can be varied, and each train is converted to circular polarization by a quarter-wave plate. The beams are chopped at 197 Hz and passed down an optical fiber to the sample in the cryostat. The pump pulses induce a magneto-optical susceptibility χ_{op} which is subsequently measured by means of the much weaker probe pulses of intensity δI that induce a magnetization $\chi_{op}\delta I$. The magnetization is detected by the SQUID at the chopping frequency, and its output is lock-in detected. By varying the time delay between the pump and probe pulses, one can investigate the dynamics of the induced magnetization. One also can vary the dye laser frequency through the red region of the visible spectrum to study the energy dependence of the magnetization. This technique recently has been extended to temperatures down to 0.3K[75].

C. Voltmeters

Probably the first practical application of a SQUID was to measure tiny, quasistatic voltages.[76] One simply connects the signal source -- for example a low resistance through which a current can be passed -- in series with a known resistance and the input coil of the SQUID. The output from the flux-locked loop is connected across the known resistance to obtain a null-balancing measurement of the voltage. The resolution is generally limited by Nyquist noise in the input circuit, which at 4.2 K varies from about 10^{-15} V $Hz^{-1/2}$ for a resistance of $10^{-8}\,\Omega$ to about 10^{-10} V $Hz^{-1/2}$ for a resistance of 100 Ω.

Applications of these voltmeters range from the measurement of thermoelectric voltages and of quasiparticle charge imbalance in nonequilibrium superconductors to noise thermometry and the high precision comparison of the Josephson voltage-frequency relation in different superconductors.

D. The dc SQUID as a Radiofrequency Amplifier

Over recent years, the dc SQUID has been developed as a low noise amplifier for frequencies up to 100 MHz or more[77]. To understand the theory for the performance of this amplifier, we need to extend the theory of Sec.III by taking into account the noise associated with the current J(t) in the SQUID loop. For a bare SQUID with $\beta = 1$, $\Gamma = 0.05$ and $\Phi = (2n+1)\Phi_0/4$, one finds the spectral density of the current to be[78]

$$S_J(f) \approx 11\, k_BT/R. \tag{5.7}$$

Furthermore, the current noise is partially correlated with the voltage noise across the SQUID, the cross-spectral density being[78]

$$S_{VJ}(f) \approx 12\, k_BT. \tag{5.8}$$

The correlation arises, roughly speaking, because the current noise generates a flux noise which, in turn, contributes to the total voltage noise across the junction, provided $V_\Phi \neq 0$.

If one imagines coupling a coil to the SQUID, the coil will "see" an impedance Z in the SQUID loop that can be written in the form[79]

$$\frac{1}{Z} = \frac{1}{j\omega L} + \frac{1}{R}\,. \tag{5.9}$$

The dynamic inductance $\mathcal{L}$ and dynamic resistance $\mathcal{R}$ are not simply related to L and R, but vary with bias current and flux; for example, $1/\mathcal{L}$ is zero for certain values of Φ.

One can make a tuned amplifier, for example, by connecting an input circuit to the SQUID, as shown in Fig.23. In general, the presence of this circuit modifies all of the SQUID parameters and the magnitude of the noise spectral densities.[80] Furthermore, the SQUID reflects an impedance $\omega^2 M_i^2/\mathcal{Z}$ into the input circuit. Fortunately, however, one can neglect the mutual influence of the SQUID and input circuit, provided the coupling coefficient α^2 is sufficiently small, as it is under certain circumstances. For the purpose of illustration, we shall derive the noise temperature of the amplifier in Fig.23.

In the weak coupling limit, the noise current $J_N(t)$ induces a voltage $-M_i\dot{J}_N$ into the input circuit, and hence a current $-M_i\dot{J}_N / Z_i$, where

$$Z_i \approx R_i + j\omega L_i + 1/j\omega C_i \ . \tag{5.10}$$

Here, Z_i is the impedance of the input circuit and L_i and C_i are the series inductance and capacitance. The noise current in the input circuit, in turn, induces a flux $-M_i^2\dot{J}_N/Z_i$ in the SQUID loop and finally a voltage $-M_i^2\dot{J}_N V_\Phi/Z_i$ across the SQUID. Thus, the noise voltage across the SQUID in the presence of the input circuit is[77]

$$V_N'(t) = V_N(t) - M_i^2\dot{J}_N V_\Phi/Z_i \ , \tag{5.11}$$

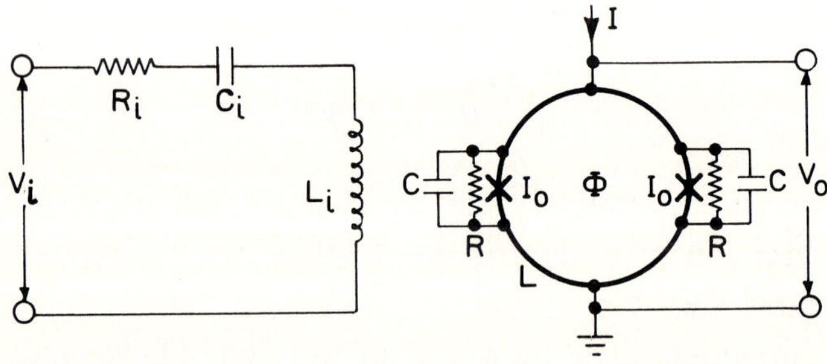

Fig.23 Tuned radiofrequency amplifier based on dc SQUID (from ref.77).

where $V_N(t)$ is the noise voltage of the bare SQUID, which we assume to be unchanged by the input circuit in the limit $\alpha^2 \to 0$. The spectral density of $V'_N(t)$ is easily found to be

$$S'_V(f) = S_V(f) + \frac{\omega^2 M_i^4 V_\Phi^2 S_J(f)}{|Z_i|^2} - \frac{2\omega M_i^2 V_\Phi(\omega L_i - 1/\omega C_i) S_{VJ}(f)}{|Z_i|^2} \cdot \quad (5.12)$$

We now suppose that we apply a sinusoidal input signal at frequency $\omega/2\pi$, with a mean-square amplitude $<V_i^2>$. The mean-square signal at the output of the SQUID is

$$<V_0^2> = M_i^2 V_\Phi^2 <V_i^2>/|Z_i|^2. \quad (5.13)$$

The signal-to-noise ratio is

$$S/N = <V_0^2>/S'_V(f)B \quad (5.14)$$

in a bandwidth B. It is convenient to introduce a noise temperature T_N for the amplifier by setting $S/N = 1$ with $<V_i^2> = 4k_B T_N R_i B$. This procedure implies that the output noise power generated by the SQUID is equal to the output noise power generated by the resistor R_i when it is at a temperature T_N. We then can optimize T_N with respect to R_i and C_i for a given value of L_i, and find

$$R_i^{(opt)} = \frac{\alpha^2 \omega L_i L V_\Phi}{S_V} (S_V S_J - S_{VJ}^2)^{1/2}, \quad (5.15)$$

$$\frac{1}{\omega C_i^{(opt)}} = \omega L_i \left(1 + \frac{\alpha^2 S_{VJ} L V_\Phi}{S_V}\right). \quad (5.16)$$

and

$$T_N^{(opt)} = \frac{\pi f}{k_B V_\Phi} (S_V S_J - S_{VJ}^2)^{1/2}. \quad (5.17)$$

We note from Eq.(5.16) that the optimum noise temperature occurs off-resonance. It often is more convenient in practice to use the amplifier at the resonant frequency of the tank circuit, given by $\omega^2 L_i C_i = 1$ (neglecting reflected components from the SQUID). In that case, one finds optimum values[77]

$$R_i^{(res)} = \alpha^2 \omega L_i L V_\Phi (S_J / S_V)^{1/2} \tag{5.18}$$

and

$$T_N^{(res)} = \frac{\pi f}{k_B V_\Phi} (S_V S_J)^{1/2}. \tag{5.19}$$

Using the results of Eqs.(3.7), (3.8), (5.7) and (5.8), we can write Eq.(5.18) in the form

$$\alpha^2 \omega L_i / R_i^{(res)} = \alpha^2 Q \approx 1. \tag{5.20}$$

This result shows that high-Q input circuits imply that α^2 is small, thereby justifying the assumption made at the beginning of this section. One also finds

$$T_N(f) \approx 18 f T / V_\Phi \approx 2 f \varepsilon(f) / k_B. \tag{5.21}$$

Thus, although $\varepsilon(f)$ does not fully characterize an amplifier, as noted earlier, within the framework of the model, it does enable one to predict T_N.

One can easily calculate the gain on resonance. For $\alpha^2 \ll 1$, an input signal V_i produces an output voltage $V_0 \approx (V_i / R_i^{(res)}) M_i V_\Phi$. The power gain is thus $G = (V_0^2/R_D)/(V_i^2/R_i)$, where R_d is the dynamic output resistance $(\partial V/\partial I)_\Phi$ of the SQUID. If we take $R_d \approx R$, we find

$$G \approx V_\Phi / \omega. \tag{5.22}$$

Hilbert and Clarke[77] made several radiofrequency amplifiers with both tuned and untuned inputs, flux biasing the SQUID near $\Phi = (2n + 1)\Phi_0/4$. There was no flux-locked loop. The measured parameters were in good agreement with predictions. For example, for an amplifier with $R \approx 8\ \Omega$, $L \approx 0.4$ nH, $L_i \approx 5.6$

nH, $M_i \approx 1$ nH and $V_\Phi \approx 3 \times 10^{10}$ sec^{-1} at 4.2 K, they found G = 18.6 ± 0.5 dB and T_N = 1.7 ± 0.5 K at 93 MHz. The predicted values were 17 dB and 1.1 K, respectively.

We emphasize that in this theory and these measurements one is concerned only with the noise temperature of the amplifier itself. Nyquist noise from the resistor adds a contribution which, in the example just given, exceeds the amplifier noise. Thus, the optimization procedure just outlined does not necessarily give the lowest system noise, and one would use a different procedure when the value of T_N in Eq.(5.17) or Eq.(5.19) is well below T.

In concluding this section, we comment briefly on the quantum limit for the dc SQUID amplifier. At T = 0, Nyquist noise in the shunt resistors should be replaced with zero point fluctuations [Eq.(2.11]. Koch et al.[81] performed a simulation in this limit and concluded that, within the limits of error, the noise temperature of a tuned amplifier in the quantum limit should be given by

$$T_N \approx hf/k_B \ln 2. \tag{5.23}$$

This is the result for any quantum-limited amplifier. The corresponding value for ε was approximately ħ, but it should be emphasized that quantum mechanics does not impose any precise lower limit on ε.[82] A number of SQUIDs have obtained noise energies of 3ħ or less, but there is no evidence as yet that a SQUID has attained quantum-limited performance as an amplifier.

E. Magnetic Resonance

SQUIDs have been used for two decades to detect magnetic resonance.[83] Most of the experiments involved the detection of magnetic resonance at low frequencies or the change in the static susceptibility of a sample induced by a resonance at high frequency. However, the development of the radiofrequency amplifier described in the previous section enables one to detect pulsed magnetic resonance directly at frequencies up to ~300 MHz.

Clarke, Hahn and co-workers have used the radiofrequency amplifier to perform nuclear quadrupole resonance[83] (NQR) and nuclear magnetic resonance[84] (NMR) experiments. They observed NQR in ^{35}Cl, which, in zero magnetic field, has two doubly-degenerate nuclear levels with a splitting of 30.6856 MHz. The experimental configuration is shown in Fig.24. The sample is placed in a superconducting pick-up coil, in series with which is an identical, counterwound coil. These coils are in series with an adjustable tuning capacitor

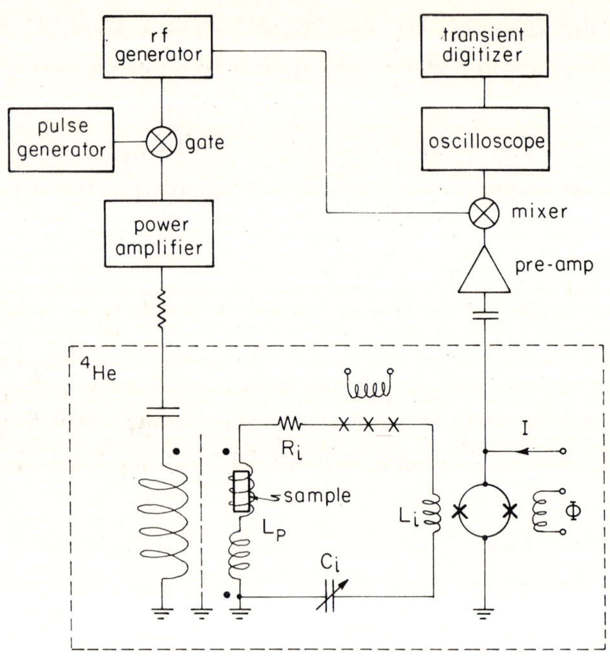

Fig.24 Circuit for NQR with dc SQUID amplifier (from ref. 83).

C_i, the 4-turn input coil of a planar dc SQUID and 20 unshunted Josephson junctions. The resistor R_i represents contact resistance and losses in the capacitor. Radiofrequency pulses applied to the transmitter coil cause the nuclear spins to precess; after each pulse is turned off, the amplifier detects the precessing magnetization. The amplified signal is mixed down with a reference provided by the rf generator, and the mixed-down signal is passed through a low-pass filter, observed on an oscilloscope, and recorded digitally for further analysis.

The major difficulty with this technique, and indeed with other pulsed methods, is the saturation of the amplifier by the very large rf pulse. In the present experiments, the effects of this pulse are reduced in two ways. First, the gradiometer-like configuration gives a common-mode rejection that can be as high as 3×10^4. Second, the series of junctions in the input circuit acts as a Q-spoiler[83]. As the current begins to build in the tuned circuit, the junctions switch to the resistive state with a total resistance of about 1 kΩ, thereby reducing the Q to ~ 1. When the pulse is turned off, the transients die out very quickly and the junctions revert to their zero voltage state, rapidly restoring Q to its full value, usually several thousand. In this way, one can combine the benefits of a high-Q tuned circuit and a sensitive amplifier while retaining a relatively short dead-time after each pulse. In their initial experiments, Hilbert et al.[83] achieved a resolution for a single pulse of ~2 x 10^{16} spins (~2 x 10^{16} nuclear Bohr magnetons) in a bandwidth of 10 kHz.

Subsequently, the Q-spoiler and SQUID amplifier were used to detect atomic polarization induced by precessing nuclear electric quadrupoles.[85] In this experiment, the $NaClO_3$ sample was placed in a capacitor that formed part of the tuned input circuit, and NQR induced in the usual way by radiofrequency pulses. The precessing electric quadruple moments induce a net electric dipole moment in the neighboring atoms, provided the crystal is non-centrosymmetric. These dipole moments, in turn, produce an oscillating electric polarization in the crystal and hence a voltage on the capacitor that is amplified in the usual way. This technique yields information on the location and polarization of atoms near nuclear quadruple moments.

The Q-spoiler and amplifier also have been used to detect nuclear magnetic resonance[84]. In these experiments one applies a magnetic field with an amplitude of several tesla to the crystal, and places the superconducting circuitry some distance away in a relatively low field. In yet another experiment, Sleator et al.[86] observed "spin noise" in ^{35}Cl. An rf signal at the NQR frequency equalized the populations of the two nuclear spin levels, and then was turned off to leave a zero-spin state. A SQUID amplifier (without a Q-spoiler) was able to detect the photons emitted spontaneously as the upper state decayed, even though the lifetime per nucleus for this process was ~ 10^6 centuries. The detected power was about 5 x 10^{-21}W in a bandwidth of about 1.3 kHz.

F. Gravity Wave Antennas

A quite different application of SQUIDs is the detection of minute displacements, such as those of the bar in a gravity wave antenna.[87,88] About a dozen groups worldwide are using these antennas to search for the pulse of gravitational radiation that is expected to be emitted when a star collapses. The radiation induces longitudinal oscillations in the large, freely suspended bar, but because the amplitude is very tiny, one requires the sensitivity of a dc SQUID to detect it. As an example, we briefly describe the antenna at Stanford University, which consists of an aluminum bar 3 meters long (and weighing 4800 kg) suspended in a vacuum chamber at 4.2 K. The fundamental longitudinal mode is at $\omega_a/2\pi \approx 842$ Hz, and the Q is 5 x 10^6. The transducer is shown schematically in

Fig.25. A circular niobium diaphragm is clamped at its perimeter to one end of the bar, with a flat spiral coil made of niobium wire mounted on each side. The two coils are connected in parallel with each other and with the input coil of a SQUID; this entire circuit is superconducting. A persistent supercurrent circulates in the closed loop formed by the two spiral coils. The associated magnetic fields exert a restoring force on the diaphragm so that by adjusting the current, one can set the resonant frequency of the diaphragm equal to that of the bar. A longitudinal oscillation of the bar induces an oscillation in the position of the diaphragm relative to the two coils, thereby modulating their inductances. As a result of flux quantization, a fraction of the stored supercurrent is diverted into the input coil of the SQUID, which detects it in the usual way.

The present Stanford antenna has a root-mean-square strain sensitivity $<(\delta\ell)^2>^{1/2}/\ell$ of 10^{-18}, where ℓ is the length of the bar, and $\delta\ell$ is its longitudinal displacement. This very impressive sensitivity, which is limited by thermal noise in the bar, is nonetheless adequate to detect events only in our own galaxy. Because such events are rare, there is very strong motivation to make major improvements in the sensitivity.

If the bar could be cooled sufficiently, the strain resolution would be limited only by the bar's zero-point motion and would have a value of about 3×10^{-21}. At first sight one might expect that the bar would have to be cooled to

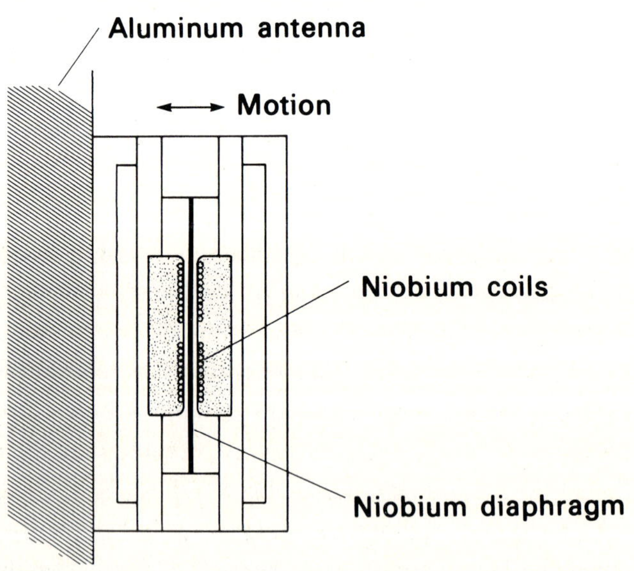

Fig.25 Transducer for gravity wave antenna (courtesy P.F. Michelson).

an absurdly low temperature to achieve this quantum limit, because a frequency of 842 Hz corresponds to a temperature $\hbar\omega_a/k_B$ of about 40 nK. However, it turns out that one can make the effective noise temperature T_{eff} of the antenna much lower than the temperature T of the bar. If a gravitational signal in the form of a pulse of length τ_S interacts with an antenna that has a decay time Q/ω_a, then the effective noise temperature is given approximately by the product of the bar temperature and the pulse length divided by the decay time: $T_{eff} \approx \tau_S\omega_a T/Q$. Thus, one can make the effective noise temperature much less than the temperature of the bar by increasing the bar's resonant quality factor sufficiently. To achieve the quantum limit, in which the bar energy $\hbar\omega_a$ is greater than the effective thermal energy $k_B T_{eff}$, one would have to lower the temperature T below $Q\hbar/k_B\tau_S$, which is about 40 mK for a quality factor Q of 5×10^6 and a pulse length τ_S of 1 msec. One can cool the antenna to this temperature with the aid of a large dilution refrigerator.

Needless to say, to detect the motion of a quantum-limited antenna, one needs a quantum-limited transducer, a requirement that has been the major driving force in the development of ultra-low-noise dc SQUIDs. As we have seen, however, existing dc SQUIDs at low temperatures are now within striking distance of the quantum limit, and there is every reason to believe that one will be able to operate an antenna quite close to the quantum limit within a few years.

G. Gravity Gradiometers

The gravity gradiometer, which also makes use of a transducer to detect minute displacements, has been pioneered by Paik[89] and Mapoles[90]. The gradiometer consists of two niobium proof masses, each constrained by springs to move along a common axis (Fig.26). A single-layer spiral coil of niobium wire is attached to the surface of one of the masses so that the surface of the wire is very close to the opposing surface of the other mass. Thus, the inductance of the coil depends on the separation of the two proof masses, which, in turn, depends on the gravity gradient. The coil is connected to a second superconducting coil which is coupled to a SQUID via a superconducting transformer. A persistent supercurrent, I, maintains a constant flux in the detector circuit. Thus, a change in the inductance of the pick-up coil produces a change in I, and hence, a flux in the SQUID that is related to the gravity gradient. More sophisticated versions of this design enable one to balance the restoring forces of the two springs electronically,[90] thereby eliminating the response to an

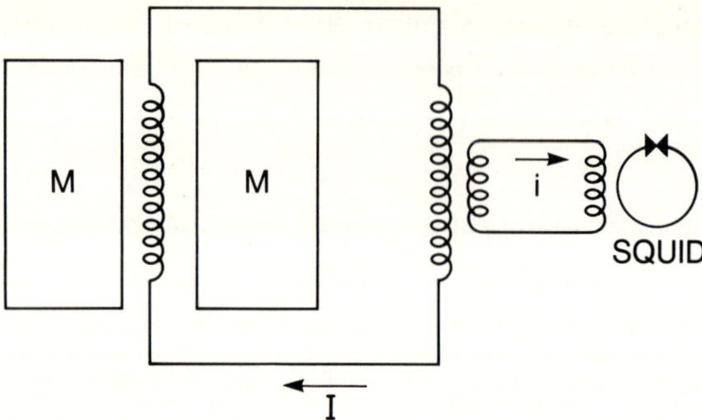

Fig.26 Gravity gradiometer showing two proof masses (M) on either side of a planar spiral coil (from ref. 90).

acceleration (as opposed to an acceleration gradient). Sensitivities of a few Eötvös $Hz^{-1/2}$ have been achieved at frequencies above 2 Hz.

Instruments of this kind could be used to map the earth's gravity gradient, and may be used to test the inverse gravitational square law and in inertial navigation.

VI. THE IMPACT OF HIGH TEMPERATURE SUPERCONDUCTIVITY

The advent of the high-transition-temperature (T_C) superconductors[91] has stimulated great interest in the prospects for superconducting devices operating at liquid nitrogen (LN) temperature (77 K). Indeed, a number of groups have already successfully operated such SQUIDs. In this section I shall give a brief overview of this work.

A. Predictions for White Noise

In designing a SQUID for operation at LN temperature, one must bear in mind the constraints imposed by thermal noise on the critical current and inductance, $I_0 \gtrsim 10\pi k_B T/\Phi_0$ and $L \lesssim \Phi_0^2/5k_B T$. For T =77 K, we find $I_0 \gtrsim 16\ \mu A$ and $L \lesssim 0.8$ nH. If we take as arbitrary but reasonable values, L = 0.2 nH and $I_0 = 20\ \mu A$, we obtain $2LI_0/\Phi_0 = 4$ for the dc SQUID and $LI_0/\Phi_0 = 2$ for the rf SQUID. These values are not too far removed from optimum, and to a first approximation, we can use the equations for the noise energy given in Secs. III and IV.

For the case of the dc SQUID, the noise energy is predicted by either Eq.(3.9) or Eq.(3.10). However, since nobody has yet made a Josephson tunnel junction with high-T_C materials, it is somewhat unrealistic to use Eq.(3.10), which involves the junction capacitance, and instead we use Eq.(3.9). The value of R is an open question, and we rather arbitrarily adopt 5 Ω, which is not too different from values achieved experimentally for high-T_C grain boundary junctions.[92] With L = 0.2 nH, T = 77 K and R = 5 Ω, Eq.(3.9) predicts $\varepsilon \approx 4 \times 10^{-31}$ J Hz^{-1}. This value is only about one order of magnitude higher that that found at 4.2 K for typical Nb-based, thin-film dc SQUIDs, and is somewhat better than that found in commercially available toroidal SQUIDs. These various values are summarized in Fig.27. If one could actually achieve the predicted resolution in a SQUID at 77 K at frequencies down to 1 Hz or less, it would be adequate for most of the applications discussed in Sec. V.

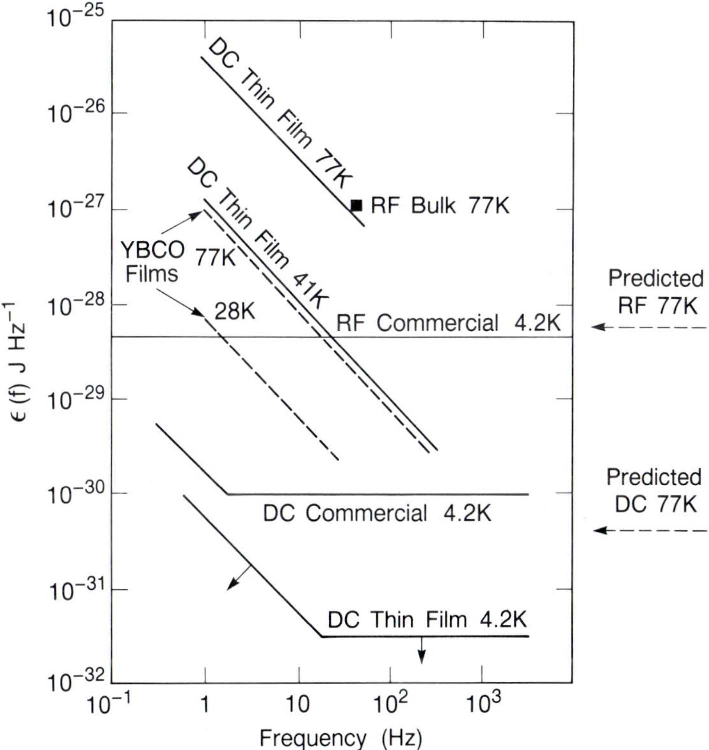

Fig.27 Noise energy ε(f) vs. frequency for several SQUIDs and for a YBCO film.

For the rf SQUID, Eq.(4.10) predicts an intrinsic noise energy of about 6×10^{-29} J Hz^{-1} for $I_0 = 20\ \mu A$, $LI_0 = 2\Phi_0$, $\omega_{rf}/2\pi = 20$ MHz and T = 77 K. This value is comparable with the overall value obtained experimentally with 4.2 K devices where the effective noise temperature T_a^{eff} of the preamplifier and tank circuit is much higher than the bath temperature for the case in which the preamplifier is at room temperature - see Sec.IV. B. However, when one operates a SQUID at 77 K, there is no reason for T_a^{eff} to increase and the system noise energy should be comparable with that at 4.2K.

With regard to l/f noise, in general, one might expect both critical current noise and flux noise to contribute. However, it seems impractical to make any a priori predictions of the magnitude of these contributions.

B. Practical Devices

Although a number of dc and rf SQUIDs have been made from YBCO, I shall describe just one of each type. It appears that the first dc SQUID was made by Koch et al.[92] In their devices, they patterned the films by covering the regions of YBCO destined to remain superconducting with a gold film, and ion implanted the unprotected regions so that they became insulators at low temperatures. The configuration is shown in Fig.28. The estimated inductance is 80 pH. The two microbridges exhibited Josephson-like behavior, which actually arose from junctions formed by grain boundaries between randomly oriented grains of YBCO. As the quality of the films has improved, conventional patterning techniques such as lift off and ion etching have become possible. The I-V characteristics of these devices are modulated by an applied flux, although the

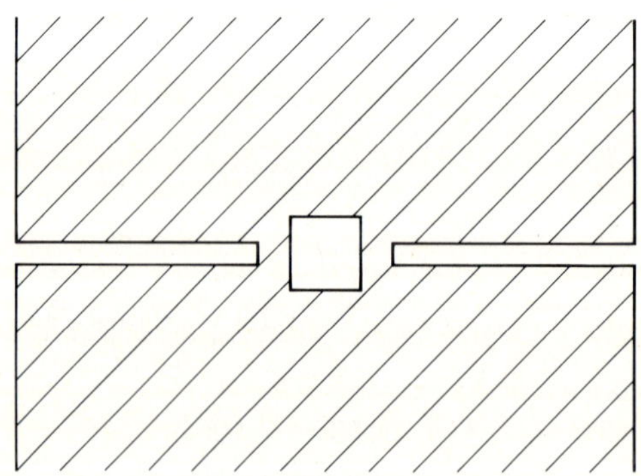

Fig.28 Planar thin-film dc SQUID fabricated from YBCO (re-drawn from ref. 92).

V-Φ curves are often hysteretic and nonperiodic, probably because of flux trapped in the YBCO films. The noise energy scaled approximately as l/f over the frequency range investigated, usually 1 to 10^3 Hz. The lowest noise energies achieved to date at 1 Hz are 4 x 10^{-27} J Hz^{-1} at 41 K and, in a different device, 2 x 10^{-26}J/Hz at 77 K. These values are plotted in Fig.27.

The best characterized rf SQUID reported so far is that of Zimmerman et al.[93] They drilled a hole along the axis of a cylindrical pellet of YBCO, and cut a slot part way along a radius (Fig.29). The pellet was glued into an aluminum holder, also with a slot, and the assembly was immersed in LN. A taper pin forced into the slot in the mount caused the YBCO to break in the region of the cut; when the pin was withdrawn slightly, the YBCO surfaces on the two sides of the crack were brought together, forming a "break junction". The rf SQUID so formed was coupled to a resonant circuit and operated in the usual way. The best flux resolution was 4.5 x 10^{-4} Φ_0 $Hz^{-1/2}$ at 50 Hz, corresponding to a noise energy of 1.6 x 10^{-27} J Hz^{-1} for L = 0.25 nH - see Fig.27.

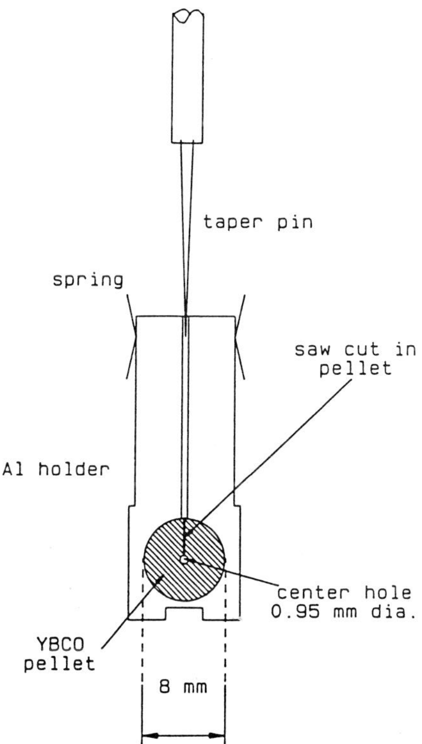

Fig. 29 Break-junction rf SQUID (from ref. 93).

C. Flux Noise in YBCO Films

It is evident that the l/f noise level in YBCO SQUIDs is very high compared with that in Nb or Pb devices at 4.2K. Ferrari et al.[94] have investigated the source of this noise by measuring the flux noise in YBCO films. Each film, deposited on a $SrTiO_3$ chip, was patterned into a loop and mounted parallel and very close to a Nb-based SQUID (with no input coil) so that any flux noise in the YBCO loop could be detected by the SQUID. The assembly was enclosed in a vacuum can immersed in liquid helium. The SQUID was maintained at 4.2 K, while the temperature of the YBCO film could be increased by means of a resistive heater. Below T_C, the spectral density of the flux noise scaled as l/f over the frequency range 1 to 10^3 Hz, and increased markedly with temperature. Three films were studied, with microstructure improving progressively with respect to the fraction of grains oriented with the c-axis perpendicular to the substrate. The critical current density correspondingly increased, to a value of 2×10^6 Acm^{-2} at 4.2 K in the best film. The spectral density of the noise measured at 1Hz is shown vs. temperature in Fig.30. We see that in each case the noise increases rapidly as the temperature approaches T_C, and that, at a given

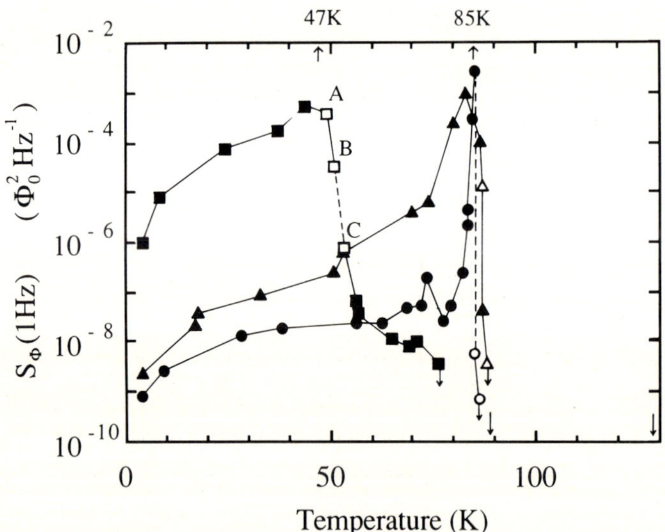

Fig.30 Spectral density of flux noise at 1Hz vs. temperature for three YBCO films: polycrystalline (squares), mixed a- and c-axis (triangles) and >90% c-axis (circles). Solid symbols indicate the noise is l/f at 1Hz, open that it is white or nearly white (from ref. 94).

temperature, the noise decreases dramatically as the quality of the films is improved. The noise energy estimated at 28 K and 77 K with an assumed inductance of 400 pH is shown in Fig.27.

These results demonstrate that YBCO films are intrinsically noisy. The noise presumably arises from the motion of flux quanta trapped in the films, possibly at grain boundaries. This mechanism is almost certainly the origin of the l/f noise observed in YBCO SQUIDs, and, in general terms, is similar to the origin of l/f flux noisc in Nb SQUIDs. It is cncouraging that thc noisc is rcduccd as the microstructure of the films is improved, and it should be emphasized that there is no reason to believe the lowest noise measured so far represents a lower bound. The implications are that SQUIDs and flux transformers coupled to them should be made of very high quality films.

D. Future Prospects for High-T_c SQUIDs

One rather obvious application of a high-T_C SQUID is as a geophysical magnetometer - see Sec.V.A. At the moment, however, it is not entirely straightforward to predict the performance, since the devices are still evolving, and it is evident from the noise measurements on YBCO films that thin-film flux transformers may introduce considerable levels of low frequency noise. To make an estimate, we assume that we can optimally couple the 77 K dc SQUID with the noise energy shown in Fig.27 to a <u>noiseless</u> flux transformer with a thin-film pick-up loop with a diameter of 50 mm. The estimated loop inductance is about 150 nH. Using Eq.(5.6), we find a magnetic field resolution of roughly 0.1 pT $Hz^{-1/2}$ at 1 Hz, improving to 0.01 pT $Hz^{-1/2}$ at 100 Hz. Although this performance is quite good, one should realize that commercially available coils operated at room temperature offer a resolution of about 0.03 pT $Hz^{-1/2}$ over this frequency range. Furthermore, our assumption of a noise-free flux transformer is rather optimistic. Nonetheless, given the short time over which the high-T_C materials have been available, one should be rather encouraged: a relatively modest reduction in the l/f noise that might be gained from improving the quality of YBCO films or even from using alternate materials might yield a useful geophysical device.

One might note here that the real advantage in using liquid nitrogen as opposed to liquid helium for field applications, is not really the reduction in cost, a savings which is negligible compared with the cost of mounting a field operation, but rather is the very much slower boil-off rate of liquid nitrogen. The latent heat of vaporization of liquid N_2 is about 60 times that of liquid ^4He, so that one should be able to design cryostats of modest size with hold-times of up to a year. It also is noteworthy that liquid Ne, which boils at 27 K, has a latent heat roughly 40 times that of liquid He, and its use also would greatly extend the running time over that of liquid He, for roughly the same cost per day. We see from Fig.30 that the l/f noise in YBCO films can be considerably lower at 27 K than at 77 K, so that the lower temperature operation could be a considerable advantage.

The likely impact of high-T_C SQUIDs on the more demanding neuromagnetic applications is much smaller, at least for the near future. Here, one needs very high sensitivity at frequencies down to 0.1 Hz or less, but is not particularly concerned with the cost of liquid ^{4}He or the need to replenish it every day or two. Furthermore, low-noise, closed-cycle refrigerators are just becoming available that obviate the need to supply liquid cryogens in environments where electrical power is readily available. Thus, it is difficult to imagine that high-T_C SQUIDs will have a significant impact in this area unless there is a major reduction in the l/f noise.

In concluding this section, we note that two key problems must be solved before high-T_C SQUIDs are likely to become technologically important. The first is the development of a reproducible and reliable Josephson junction. Although great progress has been made with grain boundary junctions, it is not clear that one can base a technology on this technique. Shiota et al.[95] have reported YBCO-insulator-YBCO junctions formed by fluorination of the base electrode, but the I-V characteristics revealed that one of the surfaces was normal. It is hoped that it will be possible to produce all-YBCO junctions exhibiting Josephson tunneling in the near future. An alternative might be a superconductor-normal metal-superconductor junction.[96] The second problem is concerned with the reduction of hysteresis and noise in thin films of high-T_C material. The motion of magnetic flux in the films is responsible for both effects, and one has to learn to produce films with lower densities of flux lines or higher pinning energies. Given the world-wide effort on the new superconductors, there is every reason to be optimistic about the long-term future of SQUIDs based on these materials.

VII. CONCLUDING REMARKS

In this chapter I have tried to give an overview of the current status of dc and rf SQUIDs. I make no pretence that this account is comprehensive. There are many SQUID designs and applications that I have not mentioned, but I hope that I have given some flavor of the amazing versatility of these devices. I find it remarkable that SQUIDs are the basis of both the most sensitive magnetometer available at 10^{-4} Hz and the lowest noise radio frequency amplifier at 10^{8} Hz.

It is somewhat ironic that the best developed devices, thin-film dc SQUIDs, are still not commercially available. This seemingly lamentable state of affairs simply reflects supply and demand. If thin-film SQUIDs were cheaper, people would buy them, and if people would buy them thin-film SQUIDs would be cheaper. The basic problem is that the demand for these devices has not yet been sufficient to convince a company to produce them on a large enough scale to bring the price down to an affordable level. However, some 14 years after low noise dc SQUIDs were first demonstrated, I believe that the situation is finally about to change. After all, one needs only a single major application to make reasonably-priced SQUIDs available for any number of applications, and there now seem to be two such major applications on the horizon. The first is in neuromagnetism: if this application is to become a clinical reality, one will need systems with as many as 100 channels, and the need for 100 channels will inevitably lead to the production of thin-film SQUIDs on a large scale. The second is the advent of high-T_C thin-film SQUIDs. If these devices attain sufficient sensitivity and reliability for geophysical applications, not to mention laboratory-based applications such as voltmeters, they will be in sufficient demand to justify production on a commercial basis.

Either way, I think the next few years are going to be particularly interesting.

ACKNOWLEDGMENTS

I am indebted to D. Crum for supplying Fig. 21(a), P.F. Michelson Fig. 25, D. Paulsen Fig. 19, F. C. Wellstood Fig. 10, and J. E. Zimmerman Fig. 29. For that part of the work carried out at Berkeley, I wish to thank the following for their hard work and dedication: Gordon Donaldson, Wolf Goubau, Mike Heaney, Claude Hilbert, Cristof Heiden, Mark Ketchen, Roger Koch, John Martinis, Bonaventura Savo, Claudia Tesche, Cristian Urbina,

Dale Van Harlingen and Fred Wellstood. I thank in particular Roger Koch and Fred Wellstood for helpful conversations during the preparation of this manuscript. I am also grateful to Marty Nisenoff and Harold Weinstock for their careful editing. This work was supported by the Director, Office of Energy Research, Office of Basic Energy Sciences, Materials Sciences Division of the U.S. Department of Energy under contract number W-7405-ENG-48.

REFERENCES

1. London. F.: Superfluids. Wiley, New York 1950
2. Josephson. B.D.: Possible new effects in superconductive tunneling. Phys. Lett. **1**. 251-253 (1962); Supercurrents through barriers. Adv. Phys. **14**. 419-451 (1965)
3. Jaklevic. R.C., Lambe. J., Silver. A.H., and Mercereau. J.E.: Quantum interference effects in Josephson tunneling. Phys. Rev Lett. **12**. 159-160 (1964)
4. Zimmerman. J.E., Thiene. P., Harding. J.T.: Design and operation of stable rf-biased superconducting point-contact quantum devices, and a note on the properties of perfectly clean metal contacts. J. Appl. Phys. **41**. 1572-1580 (1970)
5. Mercereau, J.E.: Superconducting magnetometers. Rev. Phys. Appl. **5**. 13-20 (1970); Nisenoff. M.: Superconducting magnetometers with sensitivities approaching 10^{-10} gauss. Rev. Phys. Appl. **5**. 21-24 (1970)
6. Stewart. W.C.: Current-voltage characteristics of Josephson junctions. Appl. Phys. Lett. **12**. 277-280 (1968)
7. McCumber. D.E.: Effect of ac impedance on dc voltage-current characteristics of Josephson junctions. J. Appl. Phys. **39**. 3113-3118 (1968).
8. Ambegaokar. V. and Halperin. B. I.: Voltage due to thermal noise in the dc Josephson effect. Phys. Rev. Lett. **22**. 1364-1366 (1969)
9. Clarke. J. and Koch. R. H.: The impact of high-temperature superconductivity on SQUIDs. Science **242**. 217-223 (1988)
10. Likharev. K. K. and Semenov. V. K.: Fluctuation spectrum in superconducting point junctions. Pis'ma Zh. Eksp. Teor. Fiz. **15.** 625-629 (1972). [JETP Lett. **15**. 442-445 (1972)]
11. Vystavkin. A. N., Gubankov. V.N., Kuzmin. L.S., Likharev. K.K., Migulin. V.V. and Semenov. V.K.: S-c-s junctions as nonlinear elements of microwave receiving devices. Phys. Rev. Appl. **9**. 79-109 (1974)
12. Koch. R.H., Van Harlingen. D.J. and Clarke. J.: Quantum noise theory for the resistively shunted Josephson junction. Phys. Rev Lett. **45**. 2132-2135 (1980)
13. Tesche. C.D. and Clarke. J.: dc SQUID: Noise and Optimization. J. Low. Temp. Phys. **27**. 301-331 (1977)

14. Bruines. J.J.P., de Waal. V.J. and Mooij. J.E.: Comment on "dc SQUID noise and optimization" by Tesche and Clarke. J. Low. Temp. Phys. **46**. 383-386 (1982)
15. De Waal. V.J., Schrijner. P. and Llurba. R. Simulation and optimization of a dc SQUID with finite capacitance. J. Low. Temp. Phys. **54**. 215-232 (1984)
16. Ketchen. M.B. and Jaycox. J.M.: Ultra-low noise tunnel junction dc SQUID with a tightly coupled planar input coil. Appl. Phys. Lett. **40**. 736-738 (1982)
17. Jaycox J.M. and Ketchen M.B.: Planar coupling scheme for ultra low noise dc SQUIDs. IEEE Trans. Magn., **MAG- 17**. 400-403 (1981)
18. Wellstood. F.C., Heiden. C. and Clarke. J.: Integrated dc SQUID magnetometer with high slew rate. Rev. Sci. Inst. **55**. 952-957 (1984)
19. Gurvitch. M., Washington. M.A. and Huggins. H.A.: High quality refactory Josephson tunnel junction utilizing thin aluminum layers. Appl. Phys. Lett. **42**. 472-474 (1983)
20. De Waal. V.J., Klapwijk. T.M. and Van den Hamer. P.: High performance dc SQUIDs with submicrometer niobium Josephson junctions. J. Low. Temp. Phys. **53**. 287-312 (1983)
21. Tesche. C.D., Brown. K.H., Callegari. A.C., Chen. M.M., Greiner. J.H., Jones. H.C., Ketchen. M.B., Kim. K.K., Kleinsasser. A.W., Notarys. H.A., Proto. G., Wang. R.H. and Yogi. T.: Practical dc SQUIDs with extremely low l/f noise. IEEE Trans. Magn. **MAG-21.** 1032-1035 (1985)
22. Pegrum. C.M., Hutson. D., Donaldson. G.B. and Tugwell. A.: DC SQUIDs with planar input coils. ibid. 1036-1039 (1985)
23. Noguchi. T., Ohkawa. N. and Hamanaka. K.: Tunnel junction dc SQUID with a planar input coil. SQUID 85 Superconducting Quantum Interference Devices and their Applications. Ed. Hahlbohm. H. D. and Lubbig. H. (Walter de Gruyter, Berlin, 1985) 761-766
24. Muhlfelder. B., Beall. J.A., Cromar. M.W. and Ono. R.H.: Very low noise tightly coupled dc SQUID amplifiers. Appl. Phys. Lett. **49**. 1118-1120 (1986)
25. Knuutila. J., Kajola. N., Seppä. H., Mutikainen. R. and Salmi. J.: Design, optimization and construction of a dc SQUID with complete flux transformer circuits. J. Low. Temp. Phys. **71**. 369-392 (1988)
26. Carelli. P. and Foglietti. V.: Behavior of a multiloop dc superconducting quantum interference device. J. Appl. Phys. **53**. 7592-7598 (1982)
27. Clarke. J., Goubau. W.M. and Ketchen. M.B.: J. Low. Temp. Phys. **25**. 99-144 (1976)
28. Ketchen. M.B., Goubau. W.M., Clarke. J. and Donaldson. G.B.: Superconducting thin-film gradiometer. J. Appl. Phys. **44**. 4111-4116 (1978)
29. Wellstood. F.C., and Clarke. J.: unpublished.

30. Wellstood. F.C., Urbina. C. and Clarke. J.: Low-frequency noise in dc superconducting quantum interference devices below 1K. Appl. Phys. Lett. **50**. 772-774 (1987)
31. Roukes. M. L., Freeman. M. R., Germain. R. S., Richardson. R. C. and Ketchen. M. B.: Hot electrons and energy transport in metals at millikelvin temperatures. Phys. Rev. Lett. **55.** 422-425 (1985)
32. Wellstood. F.C., Urbina. C. and Clarke. J.: Hot electron effect in the dc SQUID. IEEE Trans. Magn. **MAG-25.** 1001-1004 (1989); Appl. Phys. Lett. (to be published)
33. Ketchen. M.B., Awschalom. D.D., Gallagher. W.J., Kleinsasser. A.W., Sandstrom. R.L., Rozen. J.R. and Bumble. B.: Design, fabrication and performance of integrated miniature SQUID susceptometers. IEEE Trans. Magn. **MAG-25.** 1212-1215 (1989)
34. Koch. R.H., Clarke. J., Goubau. W.M., Martinis. J.M., Pegrum. C.M. and Van Harlingen. D.J.: Flicker (l/f) noise in tunnel junction dc SQUIDs. J. Low. Temp. Phys. **51**. 207-224 (1983)
35. Rogers. C.T. and Buhrman. R.A.: Composition of l/f noise in metal-insulator-metal tunnel junctions. Phys. Rev. Lett. **53**. 1272-1275 (1984)
36. Dutta. P. and Horn. P.M.: Low-frequency fluctuations in solids: l/f noise. Rev. Mod. Phys. **53**. 497-516 (1981)
37. Savo. B., Wellstood. F.C. and Clarke. J.: Low-frequency excess noise in Nb-Al_2O_3-Nb Josephson tunnel junction. Appl. Phys. Lett. **50**. 1757-1759 (1987)
38. Tesche. C.D., Brown. R. H., Callegari. A. C., Chen. M. M., Greiner. J. H., Jones. H. C., Ketchen. M. B., Kim. K. K., Kleinsasser. A. W., Notarys. H. A., Proto. G., Wang. R. H. and Yogi. T.: Well-coupled dc SQUID with extremely low l/f noise. Proc. 17th International Conference on low temperature physics LT-17. (North Holland, Amsterdam 1984) 263-264
39. Foglietti. V, Gallagher. W. J., Ketchen. M. B., Kleinsasser. A. W., Koch. R. H., Raider. S. I. and Sandstrom. R. L.: Low-frequency noise in low l/f noise dc SQUIDs. Appl. Phys. Lett. **49**. 1393-1395 (1986).
40. Biomagnetic Technologies Inc. 4174 Sorrento Valley Blvd., San Diego, CA 92121.
41. Fujimaki. N., Tamura. H., Imamura. T. and Hasuo. S.: A single-chip SQUID magnetometer. Digest of Tech. papers of 1988 International Solid-state conference. (ISSCC) San Francisco. pp. 40-41. A longer version with the same title is to be published.
42. Drung. D.: Digital Feedback loops for dc SQUIDs. Cryogenics **26**. 623-627 (1986). Drung. D., Crocoll. E., Herwig. R., Neuhaus. M. and Jutzi. W.: Measured performance parameters of gradiometers with digital output. IEEE Trans. Magn. **MAG-25.** 1034-1037 (1989)
43. Mück. M. and Heiden. C.: Simple dc SQUID system based on a frequency modulated relaxation oscillator. IEEE Trans. Magn. **MAG-25.** 1151-1153 (1989)

44. Clarke. J.: Superconducting QUantum Interference Devices for Low Frequency Measurements. Superconductor Applications : SQUIDs and Machines, Ed. Schwartz. B. B.and Foner. S. (Plenum New York 1977). pp 67-124.
45. Giffard. R. P., Webb. R.A. and Wheatley. J.C.: Principles and methods of low-frequency electric and magnetic measurements using rf-biased point-contact superconducting device. J. Low. Temp. Phys. **6**. 533-610 (1972)
46. Kurkijärvi. J.: Intrinsic fluctuations in a superconducting ring closed with a Josephson junction. Phys. Rcv. B **6**. 832-835 (1972)
47. Kurkijärvi. J. and Webb. W.W.: Thermal noise in a superconducting flux detector. Proc. Applied Superconductivity Conf. (Annapolis, MD.) 581-587 (1972)
48. Jackel. L.D. and Buhrman. R.A.: Noise in the rf SQUID. J. Low. Temp. Phys. **19**. 201-246 (1975)
49. Ehnholm. G.J.: Complete linear equivalent circuit for the SQUID. SQUID Superconducting Quantum Interference Devices and their Applications. Ed. Hahlbohm. H.D. and Lubbig. H. (Walter de Gruyter, Berlin, 1977) 485-499; Theory of the signal transfer and noise properties of the rf SQUID. J. Low. Temp. Phys. **29**. 1-27(1977)
50. Hollenhorst. H.N. and Giffard. R.P.: Input noise in the hysteretic rf SQUID: theory and experiment. J. Appl. Phys. **51**. 1719-1725 (1980)
51. Kurkijärvi. J.: Noise in the superconducting flux detector. J. Appl. Phys. **44**. 3729-3733 (1973)
52. Giffard. R.P., Gallop. J.C. and Petley. B.N.: Applications of the Josephson effects. Prog. Quant. Electron **4**. 301-402 (1976)
53. Ehnholm. G.J., Islander. S.T., Ostman. P. and Rantala. B.: Measurements of SQUID equivalent circuit parameters. J. de Physique **39**. colloque C6. 1206-1207 (1978)
54. Giffard. R.P. and Hollenhorst. J.N.: Measurement of forward and reverse signal transfer coefficients for an rf-biased SQUID. Appl. Phys. Lett. **32**. 767-769 (1978)
55. Jackel. L. D., Webb. W. W., Lukens. J. E. and Pei. S. S.: Measurement of the probability distribution of thermally excited fluxoid quantum transitions in a superconducting ring closed by a Josephson junction. Phys. Rev. **B9**. 115-118 (1974)
56. Long. A., Clark. T. D., Prance. R. J. and Richards. M. G.: High performance UHF SQUID magnetometer. Rev. Sci. Instrum. **50**. 1376-1381 (1979)
57. Hollenhorst. J. N. and Giffard. R. P.: High sensitivity microwave SQUID. IEEE Trans. Magn. **MAG-15**. 474-477 (1979)
58. Ahola. H., Ehnholm. G. H., Rantala. B. and Ostman. P.: Cryogenic GaAs-FET amplifiers for SQUIDs. J. de Physique **39**. colloque C6. 1184-1185 (1978); J. Low Temp. Phys. **35**. 313–328 (1979)
59. For a review, see Clarke. J.: Advances in SQUID Magnetometers. IEEE Trans. Election Devices. **ED-27**.1896-1908 (1980)

60. Zimmerman. J. E.: Sensitivity enhancement of Superconducting Quantum Interference Devices through the use of fractional-turn loops. J. Appl. Phys. **42**. 4483-4487 (1971)
61. Shoenberg. D.: Superconductivity (Cambridge University Press 1962) 30.
62. For a review, see Clarke. J.: Geophysical Applications of SQUIDs. IEEE Trans. Magn. **MAG-19**. 288-294 (1983)
63. De Waal. V. J. and Klapwijk. T. M.: Compact Integrated dc SQUID gradiometer. Appl. Phys. Lett. **41**. 669-671 (1982)
64. Van Nieuwenhuyzen G. J. and de Waal. V. J.: Second order gradiometer and dc SQUID integrated on a planar substrate. Appl. Phy. Lett. **46**. 439-441 (1985)
65. Carelli. P. and Foglietti. V.: A second derivative gradiometer integrated with a dc superconducting interferometer. J. Appl. Phys. **54.** 6065-6067 (1983)
66. Koyangi. M., Kasai. N., Chinore. K., Nakanishi. M. and Kosaka. S.: An integrated dc SQUID gradiometer for biomagnetic application. IEEE Trans. Magn. **MAG-25.** 1166-1169 (1989)
67. Knuutila. J., Kajola. M., Mutikainen. R., Salmi. J.: Integrated planar dc SQUID magnetometers for multichannel neuromagnetic measurements. Proc. ISEC '87 p. 261
68. For reviews, see Romani. G. L., Williamson. S. J. and Kaufman. L.: Biomagnetic instrumentation. Rev. Sci. Instrum. **53**. 1815-1845 (1982); Buchanan. D. S., Paulson. D. and Williamson. S. J.: Instrumentation for clinical applications of neuromagnetism. Adv. Cryo. Eng. (to be published)
69. Knuutila. J.: European Physical Society Workshop "SQUID: State of Art, Perspectives and Applications". Rome, Italy June 22-24, 1988 (unpublished)
70. Barth. D. S., Sutherling. W., Engel. J. Jr. and Beatty J.: Neuromagnetic evidence of spatially distributed sources underlying epileptiform spikes in the human brain. Science **223.** 293-296 (1984)
71. Romani. G. L., Williamson. S. J. and Kaufman. L.: Tonotopic organization of the human auditory cortex. Science **216.** 1339-1340 (1982)
72. Cabrera. B.: First results from a superconductive detector for moving magnetic monopoles. Phys. Rev. Lett. **48**. 1378-1381 (1982)
73. Quantum Design, 11568 Sorrento Valley Road, San Diego, CA 92121.
74. Ketchen. M. B., Kopley. T. and Ling. H.: Minature SQUID susceptometer. Appl Phys. Lett. **44.** 1008-1010 (1984)
75. Awschalom. D. D. and Warnock. J.: Picosecond magnetic spectroscopy with integrated dc SQUIDs. IEEE Trans. Magn. **MAG-25.** 1186-1192 (1989)
76. Clarke. J.: A superconducting galvanometer employing Josephson tunneling. Phil. Mag. **13.** 115-127 (1966)

77. Hilbert. C. and Clarke. J.: DC SQUIDs as radiofrequency amplifiers. J. Low Temp. Phys. **61**. 263-280 (1985)
78. Tesche. C. D. and Clarke. J.: DC SQUID: current noise. J. Low Temp. Phys. **37.** 397-403 (1979)
79. Hilbert. C. and Clarke. J.: Measurements of the dynamic input impedance of a dc SQUID. J. Low Temp. Phys. **61.** 237-262 (1985)
80. Martinis. J. M. and Clarke. J.: Signal and noise theory for the dc SQUID. J. Low Temp. Phys. **61.** 227-236 (1985), and references therein.
81. Koch. R.H., Van Harlingen. D. J. and Clarke. J.: Quantum noise theory for the dc SQUID. Appl. Phys. Lett. **38**. 380-382 (1981)
82. Danilov. V. V., Likharev. K. K. and Zorin. A. B.: Quantum noise in SQUIDs. IEEE Trans. Magn. **MAG-19**. 572-575 (1983)
83. Hilbert. C., Clarke. J., Sleator. T. and Hahn. E. L.: Nuclear quadruple resonance detected at 30MHz with a dc superconducting quantum interference device. Appl. Phys. Lett. **47.** 637-639 (1985). (See references therein for earlier work on NMR with SQUIDS).
84. Fan. N. Q., Heaney. M.B., Clarke. J., Newitt. D., Wald. L. L., Hahn. E. L., Bielecke. A. and Pines. A.: Nuclear magnetic resonance with dc SQUID preamplifiers. IEEE Trans. Magn. **MAG-25.** 1193-1199 (1989)
85. Sleator. T., Hahn. E. L., Heaney, M.B., Hilbert. C. and Clarke. J.: Nuclear electric quadrupole induction of atomic polarization. Phys. Rev. Lett. **57.** 2756-2759 (1986)
86. Sleator. T., Hahn. E. L., Hilbert, C. and Clarke. J.: Nuclear-spin noise and spontaneous emission. Phys. Rev. B. **36**. 1969-1980 (1987)
87. For an elementary review on gravity waves, see Shapiro. S. L., Stark. R.F. and Teukolsky. S. J.: The search for gravitational waves. Am. Sci. **73.** 248-257 (1985)
88. For a review on gravity-wave antennae, see Michelson. P. F., Price. J. C. and Taber. R. C.: Resonant-mass detectors of gravitational radiation. Science **237**.150-157 (1987)
89. Paik. H. J.: Superconducting tensor gravity gradiometer with SQUID readout. SQUID Applications to Geophysics. Ed. Weinstock H. and Overton. W. C., Jr. (Soc. of Exploration Geophysicists, Tulsa, Oklahoma, 1981) 3-12
90. Mapoles. E.: A superconducting gravity gradiometer. ibid. 153-157
91. Bednorz. J. G. and Muller. K. A.: Possible high T_c superconductivity in the Ba-La-Cu-O system. Z. Phys. **B64.** 189-193 (1986)
92. Koch. R. H., Umbach. C. P., Clark. G. J., Chaudhari. P. and Laibowitz. R. B.: Quantum interference devices made from superconducting oxide thin films. Appl. Phys. Lett. **51.** 200-202 (1987)
93. Zimmerman. J. E., Beall. J. A., Cromar. M. W. and Ono. R. H.: Operation of a Y-Ba-Cu-O rf SQUID at 81K. Appl. Phys. Lett. **51.** 617-618 (1987)

94. Ferrari. M. J., Johnson. M., Wellstood. F. C., Clarke. J., Rosenthal. P. A., Hammond. R. H. and Beasley. M. R.: Magnetic flux noise in thin film rings of $YBa_2Cu_30_{7-\delta}$. Appl. Phys. Lett. **53**. 695-697 (1988)
95. Shiota. T., Takechi. K., Takai. Y and Hayakawa. H.: An observation of quasiparticle tunneling characteristics in all Y-Ba-Cu-0 thin film tunnel junctions (unpublished)
96. Mankiewich. P. M., Schwartz. D. B., Howard. R. E., Jackel. L. D., Straughn. B. L., Burkhardt. E. G. and Dayem. A. H.: Fabrication and characterization of an $YBa_2Cu_30_7$/Au/$YBa_2Cu_30_7$ S-N-S microbridge. Fifth International Workshop on Future Electron Devices - High Temperature Superconducting Devices. June 2-4, 1988, Miyaki-Zao, Japan. 157-160

Added Note: Recently, Koch et al. [Appl. Phys. Lett. **54.** 951-953 (1989)] operated dc SQUIDs fabricated from Tℓ Ba Ca Cu O at 77 K, and found that a few showed an appreciably lower noise level than YBCO devices. At low frequencies, the spectral density of the noise was l/f. At 10Hz, the best values of the noise energy were 5×10^{-29} JHz^{-1} and 2×10^{-29} JHz^{-1} for devices with inductances of about 80pH and 5pH, respectively. The low-frequency noise was neither the flux noise nor the critical noise described in Sec. III F, but appeared to arise from the re-arrangements of flux vortices pinned in the microbridges. The low level of flux noise was possibly due to the large grain size of the Tℓ Ba Ca Cu O films (10-40μm) compared with those of the YBCO films (~ 1μm). Most of the Tℓ-based SQUIDs were approximately in the configuration of Fig. 28 and exhibited V-Φ curves that were rather aperiodic and hysteretic. Fortunately, when the devices were operated in a flux-locked loop (Sec. III D), the flux in the SQUID was kept constant and the hysteresis had little or no effect on the performance. Subsequently, a device fabricated with a narrow loop showed periodic V-Φ curves with negligible hysteresis.

Although these Tℓ-based SQUIDs relied on grain boundary junctions, their noise level would be low enough for many applications if a suitable low-noise flux transformer could be made. The outlook for high-T_c SQUIDs seems to be extremely promising.

THE USE OF SQUIDS IN THE STUDY OF BIOMAGNETIC FIELDS

Gian Luca Romani*
Istituto di Fisica Medica, Universita' "G. D'Annunzio"
Via dei Vestini, 56100 Chieti
Italy

1. INTRODUCTION

The investigation of biomagnetic fields, i.e. fields associated with bioelectrical activity in the human body, has marked impressive progress in the last few years and is proving to be a unique tool to achieve functional imaging of fundamental mechanisms in the heart and in the brain. In particular, the neuromagnetic approach to the study of cerebral functions provided definitive evidence on specific organizations of neural networks located in primary areas, i.e., those devoted to the first analysis of input signals from peripheral sensory systems. Last, but not least, some important pathologies of the heart and of the brain are being investigated by many groups in the world, and the results so far achieved have raised the enthusiasm of exponents from the clinical side. As a unique example we mention the identification of epileptic foci, in cases of partial (focal) epilepsy. The importance of this possibility is well focused if we remember that this disease affects an impressively large percentage of inhabitants in highly industrialized countries.

Several motivations stand behind such impressive and rapid progress in biomagnetic research: the strongest of them, however, is the rapid development of SQUID-based instrumentation that has permitted the achievement of unrivaled performance in magnetic field sensitivity, as well as in reduction of unwanted signals, currently defined as magnetic noise. It is probably worth remembering that approximately a dozen years ago the state of the art of biomagnetic instrumentation was reviewed during another NATO Advanced Study Institute on superconductivity. The performances of

* and: Istituto di Elettronica dello Stato Solido - CNR
Via Cineto Romano 42, 00156 Roma, Italy

NATO ASI Series, Vol. F 59
Superconducting Electronics
Edited by H. Weinstock and M. Nisenoff

biomagnetic detectors have since improved by much more than one order of magnitude, and the miniaturization of sensing units, which may even include the detection coil(s) in a single chip, has permitted several research groups to start projects to develop large multichannel instrumentation for real-time functional brain imaging.

In this article we will describe the most studied biomagnetic fields and the primary sources of magnetic "noise". Then a simple outline of the problem of modeling biomagnetic sources will serve as an introduction to the instrumentation chapter, in which we will dwell on different designs for detecting circuitry, disregarding any mention of SQUIDs proper, as they are described in detail elsewhere in this volume. A glimpse at the state of the art of multichannel systems will precede a brief overview of some of the most recent and important findings which have significantly contributed to the understanding of cerebral functions.

2. FIELDS AND SOURCES

In 1963 Baule and McFee[1] succeeded in measuring magnetic fields associated with bioelectric activity in the human heart, namely a *magnetocardiogram*. The technological state of the art of the mid 1960's did not permit using a superconducting magnetometer. It was only a few years later that a SQUID sensor, located inside a magnetically shielded room, produced a magnetocardiogram[2], thus creating a new application of superconducting devices based on the Josephson effect. The first measurements of cerebral magnetic signals were reported a little later, and involved both spontaneous brain activity[3] and evoked fields[4].

The following period was devoted to a broad investigation of fields related to various bioelectric activities in the human body - see next section- while the number of research groups getting involved with the new technique rapidly increased. At the beginning of the 1980's the time was ripe for a new, fundamental step forward, i.e., the use of a relatively simple model of a bioelectric source to confront the inverse problem and achieve *source localization*. This feature, as described in a section below, undoubtedly represents the major advantage of the biomagnetic approach to the investigation of the brain and the heart. A localization procedure in use since the early 1980's has produced important results, such as the experimental demonstration of the tonotopic organization of the human auditory cortex[5], and the identification of epileptic foci in partial epilepsies[6-9]. The progress achieved in the source identification procedure during the last few years has marked impressive results, which will be partially described in the last section of this article. Before that, we must establish a comprehensive view of the phenomena under study and a brief description of the modeling problem.

2.1 Biomagnetic fields and noise sources

Figure 1 provides a comprehensive representation of the most extensively studied biomagnetic signals. The field amplitudes are expressed in femtotesla, 1 fT = 10^{-15} T, and should be compared with the intensity of the earth magnetic field, which is approximately 5×10^{-5} T. Other unwanted ambient magnetic fields, which are commonly defined as magnetic noise, are generated by micropulsations of the earth's field and by pumps, fans, elevators and other instruments located in proximity to the experimental area. The effect of these fields can be characterized in terms of a frequency spectrum. Considering that the experimental bandwidth required for all biomagnetic measurements - similar to what happens with any bioelectric phenomenon - is limited to the range from about 10^{-1} to 10^{3} Hz, the micropulsations of the earth's field affect the lower part of this spectral range, featuring an approximately 1/f behavior with circadian variations peaked in intensity during daytime. The other sources of *urban* noise are mainly characterized by specific components usually below 10^{2} Hz, and to a typical contribution at the line frequency and its harmonics which may extend well above the upper limit of the recording bandwidth and seriously affect the measurement.

Not all the fields reported in Fig.1 have the same bioelectric origin. Indeed, strong signals may be produced by magnetic particles contaminating the lung of specific workers, such as asbestos miners or arc welders.

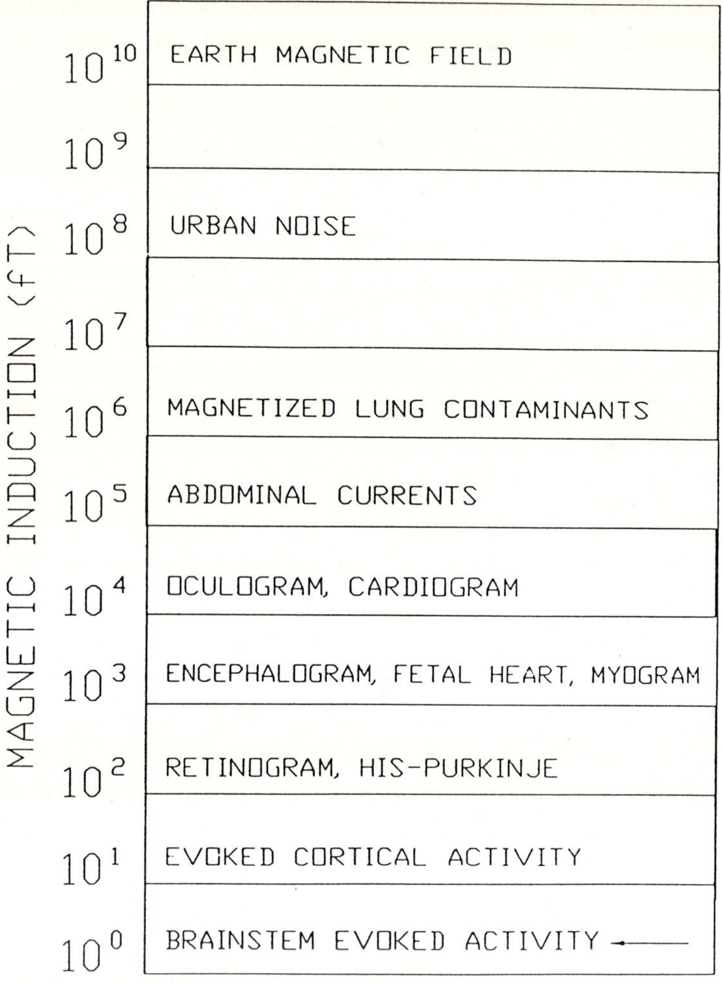

Fig.1 Typical values of the amplitude of the most commonly studied biomagnetic fields, as compared with those of the earth's field and of the environmental field noise in working areas. The intensity of the magnetic induction is expressed in femtoteslas (1 fT = 10^{-15} T). The sensitivity of the best SQUID systems is at the level indicated by the arrow.

Similarly, accumulation of iron compounds (like, for instance, hemoferritine) in the liver, spleen and myocardial tissue of patients affected by specific endemic diseases (thalassemia major, hemocromatosis, etc.) may result in a net paramagnetic signal measurable over the involved portion of the body. Also in these areas the application of the biomagnetic method has provided, or is in the process of providing, important results, and the interested reader is directed to appropriate references[10-13]. Spatial constraints force us to limit this discourse to fields associated with bioelectric activity and generated in the human brain. We infer from the figure that we are dealing with the weakest signals of the entire range. These signals originate from deep cerebral sources, like the brain stem, and are only a few femtoteslas in amplitude. Nevertheless, the SQUID has the necessary sensitivity to investigate all the mentioned phenomena, and the problem of reducing the ambient noise to levels comparable with, or possibly smaller than, the amplitude of signals to be measured is now solvable with the use of a gradiometric design for the detection coils- see Section 3 - and, possibly, with the additional help of a radiofrequency and magnetically shielded room. Indeed, it is an accepted procedure[14] to screen the experimental area by alternating layers of materials with high permeability and high conductivity, in order to get an attenuation of ambient fields about 100 below 1 Hz, and an attenuation increasing with frequency above that threshold, with an effective depletion of line frequency noise larger than 90 dB[15]. Equally important, the available space in these rooms for experimental measurements can be as large as about 30 m^3, more than adequate for any clinical study.

2.2 Modeling of neural activity

The challenge experimentalists must cope with to achieve source localization is commonly defined as the *inverse problem*, that is, the identification of a specific source configuration from a measured distribution of magnetic fields and electric potentials at the body surface. The inverse problem does not have a unique solution, as an infinite number of equivalent source distributions may account for the measured patterns. Thus, the most convenient procedure involves calculating the theoretical field and potential patterns as generated by a suitable model source, i.e., a source physiologically meaningful but yet mathematically tractable. Successively, the theoretical and experimental distributions are compared in a fit, and eventually the location of the source(s) may be identified.

Two events occur in the nervous system, and specifically at its higher level, the cerebral cortex: i) action potentials, and ii) post-synaptic currents. Both are related to ionic flows inside, across and outside the neural membrane - see Fig.2a. There is no space here to dwell on the physiology of these phenomena and on the generation of magnetic fields and electric potentials by these currents. The interested reader is directed to a recent article on the theory of neuroelectric and neuromagnetic fields[16].

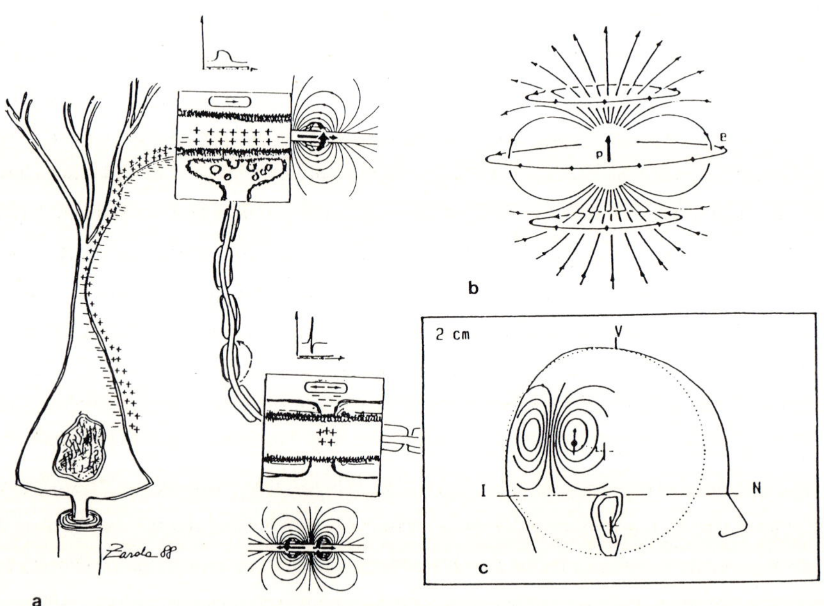

Fig.2 a) Schematic modeling of neural activity in terms of current dipoles (postsynaptic activity), and of oppositely directed dipoles (action potentials). The latter might also be considered as a magnetic quadrupole. b) A current dipole immersed in an infinite medium with homogeneous conductivity. The thin lines around the dipole are the volume currents, whereas the transverse circumferences are the magnetic field lines. c) Scalp distribution of the normal component of the magnetic field generated by a current dipole (immersed in a homogeneously conducting sphere).

We recall here only that the overall pattern of the current flow may reasonably be accounted for by that of a short element of current I, of length L and neglegible cross-section, namely a *current dipole*, with a "moment" Q = IL, when it is immersed in an infinite medium with homogeneous conductivity. It has been shown[16] that one current dipole may represent the pattern of post-synaptic currents, whereas two oppositely directed dipoles, traveling along the neural axon, may account for an action potential. It should be remarked that the majority of magnetic fields which are measured outside the head are associated with the relatively slow (10-100 ms) post-synaptic currents flowing in the cerebral cortex, at the level of the apical dendrites of pyramidal neurons and, consequently, may be accounted for by simple current dipoles or combinations of current dipoles - see Fig.2a.

The Biot-Savart law permits calculation of the magnetic field associated with a current dipole when it is immersed in an infinite medium with homogeneous conductivity. The magnetic field has an axial symmetry with respect to the direction of the current dipole, as shown in Fig.2b. The current lines flowing in the medium outside the dipole are called *volume currents*, and represent the extracellular ionic flows around the neuron. One significant feature of this configuration is that the magnetic field at a point P at a certain distance from the dipole, depends only on the primary (i.e., intracellular) current, whereas the potential at the same point depends both on the primary and volume currents. Thus, the axial symmetry of the field provides a unique tool to obtain information directly on the actual sources of cerebral activity, independently of the smearing and spreading effect due to the outside medium. This fortunate situation is partially maintained also in more realistic cases, where a "boundary" is present to constrain the conducting medium. In particular, we consider a half space with homogeneous conductivity, a sphere with homogeneous conductivity, and a set of concentric spheres with homogeneous conductivity within each shell. In all these cases, the independence of volume currents is saved for the component of the field perpendicular to the bounding surface. For this reason all biomagnetic measurements of fields associated with bioelectric phenomena are performed almost invariably by sensing the normal component of the field.

We should remark that there is a magnetically "silent" situation, where the dipole is oriented normal to the surface. This statement can easily be verified by applying Ampere's law, and the reader is again directed to the cited reference[16]. An important consequence is that the magnetic measurement is somewhat blind to part of the cerebral activity, in that it cannot detect fields produced by current flows oriented perpendicularly to the scalp. Only flows with a tangentially-oriented component produce measurable fields outside the head. We will soon see that this "disadvantage" may become a real advantage in actual measurements. The pattern of the normal component of the field from a current dipole over the bounding surface, as shown in Fig.2c, features two regions of maximum field with opposite polarity, symmetrically shaped over the dipole. This distribution, referring to the case of a tangential dipole, shows only a decrease in relative amplitude when the dipole is tilted toward a normal orientation. A relatively simple relationship links the pattern shape to the three-dimensional location of the source in the medium below the surface. It is, therefore, clear that a simple fit can be performed between the measured distribution and the theoretical one, any time that the experiment provides a "dipolar" shape, such as the one depicted in the figure.

It is worth dwelling on a last point. We have considered, in a first approximation, the medium around the model source as a sphere. The real case is not exactly so, since the geometry of real human heads varies significantly from this ideal shape. Nevertheless, it has been demonstrated repeatedly[16-18] that even in the real case, the contribution from volume currents to the normal field is of minor importance, being limited to a few percent. Scalp potentials always depend on volume currents and are inevitably dependent on different conductivities of various layers interposed between the source and the electrodes. Indeed, we should consider that the skull conductivity is about 80 times lower than that of the scalp and of the cerebral fluid. As a consequence the potential distribution over the scalp is typically more widespread and smeared, and a localization procedure is much more complex, yielding poor results. The mentioned blindness to normally oriented current flows, typical of the magnetic measurement, may even be an advantage, in the sense that a "simplified" situation is presented to the experimenter, making it easier to distinguish specific cerebral events. We should finally consider that in the five millimeters of grey matter constituting the cerebral cortex, at least one population of neurons, the pyramidal cells, are preferentially aligned perpendicularly to the cortex surface. This means that we should expect to be most sensitive to activities occurring in the fissures rather than in convolutions. Fortunately, most of the primary cortical areas are located inside brain fissures, and indeed, the most significant results achieved by means of the magnetic approach regard these areas.

3. INSTRUMENTATION

As mentioned in the previous section, the amplitude of biomagnetic fields spans several orders of magnitude, so it is not necessary for a SQUID to be used in all applications. Non-superconducting induction coils have been used in several occasions to detect magnetic heart signals, even in a vectorial form[19]. The actual limitation for this kind of sensor is due to Nyquist noise associated with the resistance of the windings. This problem can be only removed partially by cooling the coils to liquid nitrogen temperature[11]. Again, only limited applications have been found for fluxgate magnetometers[11], the performance of which is limited by Barkausen noise in the ferrite core. In practice, fluxgates have been used to set up a lung scanner for the evaluation of lung contamination in workers exposed to occupational pollution. An extensive clinical investigation has been performed with this kind of instrument and has produced excellent results[20].

Also mentioned earlier is the fact that the SQUID has sufficient sensitivity to investigate all the other biomagnetic fields, especially those associated with brain activity. Rf biased SQUIDs have been used first, due to the higher reliability these devices showed for many years (relative to dc SQUIDs). Nevertheless, over the last few years a new generation of microfabricated ultra-low noise dcSQUIDs has demonstrated greater reliability and sensitivity. Thus, these devices are slowly replacing rfSQUIDs as the detector of choice. SQUIDs have been extensively described in another chapter of this volume, therefore we shall turn our attention directly to different geometries for the detection coil to be coupled to the SQUID in order to achieve the best reduction of ambient noise and improve *spatial discrimination*. Indeed, SQUIDs alone are not likely to be used directly for magnetic field sensing, as their field sensitivity is not particularly high. It therefore is most fruitful to couple the SQUID to an external detection coil, the shape of which can be adjusted to suit the experimental requirements and to increase field sensitivity. It is worth stressing that the operation of coupling a detection coil to the input coil of the SQUID, i.e., the construction of a *flux transformer or transporter*, should be handled carefully in order to achieve the best signal energy transfer from the primary to the secondary element of the flux transformer, and at the same time to provide acceptable noise reduction.

3.1 Detecting circuitry

It has been shown elsewhere in this volume that a flux transformer is "matched" when the inductance of the detection coil L_d is made equal to that of the SQUID input coil L_i. Under this condition the ratio between the flux measured by the SQUID and the external flux, namely the flux transformer ratio, is proportional to the number of turns of the detection coil and to the mutual indutance M_i between the input coil and the SQUID, which is generally a fixed parameter. Then, in the simplest case of a detection coil with just a few turns N_d of wire spaced fairly close together, enclosing an area A_d, the minimum detectable external field B_n is

$$B_n = \Phi_n(L_d + L_i)/A_dN_dM_i \tag{1}$$

where Φ_n is the SQUID flux noise, and $L_d = L_i$ for optimal matching. As the area of the detection coil cannot be made larger than a few square centimeters - typical detection coil diameters span from 1 to 2 cm - the unique parameter to be handled is N_d. Furthermore, it is desirable to separate the individual turns of the detection coil so to reduce their mutual inductance. By using a spacing between adjacent coils of 1 mm or more, it is possible to make L_d about proportional to N_d rather than to N_d^2, and consequently, to achieve the inductance match with a larger value of N_d.

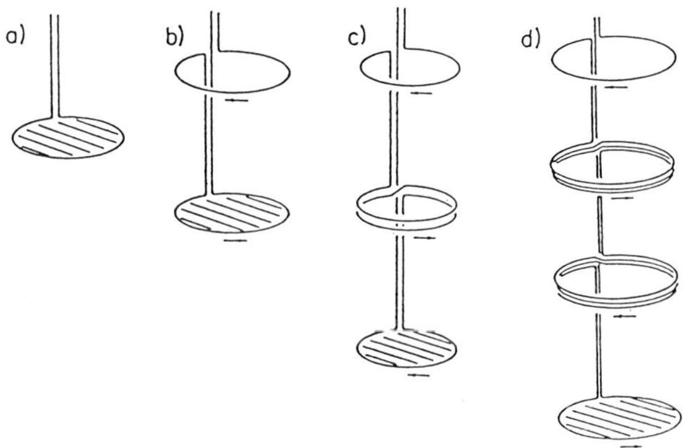

Fig.3 Most commonly used geometries for the detection coil to be coupled to a SQUID for biomagnetic measurements: a) magnetometer, b) first-order gradiometer, c) second-order gradiometers, and d) third-order gradiometer.

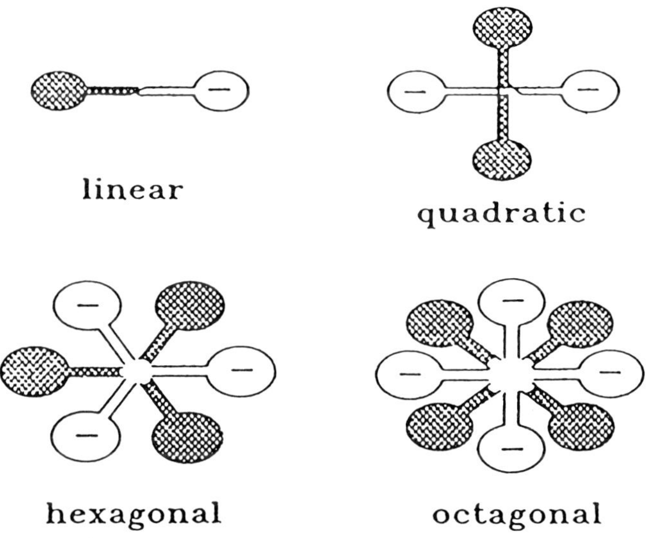

Fig.4 Possible configurations for planar gradiometers, to be directly integrated in the SQUID loop (not shown in the figure).

A detection coil consisting of a single coil, referred to as a *magnetometer coil*, responds to the applied field regardless of the distance of the source. However, the detection coil geometry can be altered, as illustrated in Fig.3, in order to achieve greater insensitivity to far sources. A *first-order gradiometer* is insensitive to fields uniform in space, whereas a *second-order gradiometer* is insensitive to both spatially uniform fields and gradients. The basic advantage in using these geometries is that a "noise" field produced by a far source, relative to the distance between the subcoils, or *baseline* of the gradiometer, can be substantially cancelled**, whereas a high sensitivity is maintained for sources close enough to one of the extreme subcoils of the gradiometer. Hence, the need for helium cryostat is needed with a "tail" providing a small distance- about 1 cm - from the inside to the outside of the dewar***. The penalty we have to pay in return for ambient noise reduction is a reduction in the overall sensitivity, as the signal energy is shared among the subcoils of the gradiometer. This effect[11] limits the performance of systems for unshielded environments to a sensitivity which is about a factor of 3 worse than that obtainable in heavily shielded rooms, where magnetometers can be used straightforwardly. It should be noted that for many years systems based on second-order gradiometers have been routinely used in many laboratories around the world. Nevertheless, the relatively low cost of a new generation of shielded rooms, and the greater complexity in balancing multi-gradiometric systems are convincing experimentalists to combine the rejection effect of gradiometers with the shielding effect of rooms.

The gradiometric geometries so far used for biomagnetic measurements have

---

** The balance of the gradiometer versus constant fields and gradients is in practice limited by the mechanical accuracy during construction. Therefore, additional balancing procedures - by means of superconducting trim tabs, and fine baseline adjustements[11] - are usually required to achieve satisfactory insensitivity to ambient fields.

***The technological state of the art for cryogenic dewars is more than satisfactory. Continuous operation for about a week is guaranteed not only for single-channel sensors, but also for larger systems providing a "concave" tail for a few adjacent channels to ensure perpendicularity with respect to the head. Looking forward to imaging systems with 100 channels or so, further improvements are still to be made, particularly with regard to helium boil-off rate and magnetic noise associated with the thermal shielding in the tail.

the common feature of a vertical symmetry axis****. A different approach consists in using a *planar* configuration and integrating the gradiometer directly in the same chip that contains a planar SQUID. The motivation for this choice relates to the obvious difficulty that integration of a large number of magnetic sensors in a complex multichannel system (see below) presents, particularly with respect to accurate balancing for each channel. Figure 4 shows some examples of planar geometries for the detection coil which can be obtained with standard photolitographic techniques. Each configuration should be regarded as directly connected to a microfabricated SQUID, or inductively coupled to the SQUID input coil via a flux transformer also integrated in the same chip[21]. This feature is important in that it permits elimination of superconducting connections between each gradiometer and the respective SQUID. The increasing number of loops provides higher and higher spatial discrimination, and at the same time, the overall sensitivity is reduced as expected. Early attempts to build experimental devices trace back to the end of the 1970's and to the beginning of the 1980's[22-24], whereas none of these geometries has been used yet for practical measurements. Several computer simulations performed by many authors[25-28] have pointed out various advantages of these devices with respect to vertical ones, among which are a higher discriminating

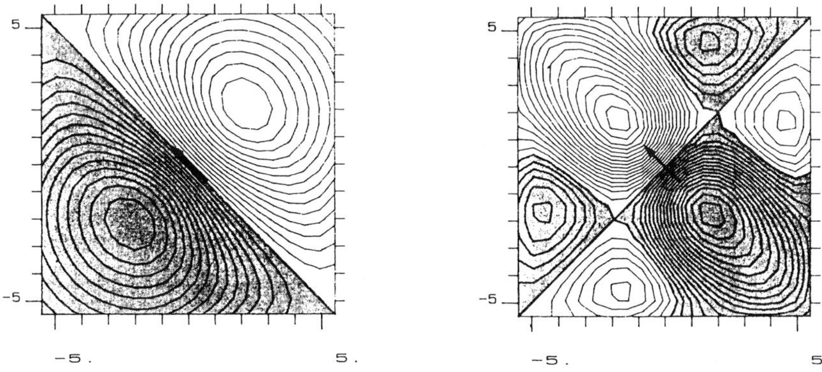

Fig.5 Simulated iso-contour maps illustrating the distribution of the field generated by a current dipole immersed in a homogeneously conducting half space at a depth of 4 cm, as measured (left) by an array of vertical gradiometers, 1.5 cm diameter detection coil, 5 cm baseline, and (right) by an array of planar (quadratic) gradiometers, .5 cm loop diameter, 1.5 cm baseline. The spacing in both directions of the sensors is 3 cm. Units are expressed in centimeters.

****measurements of the alpha activity with an off-diagonal, or 2-D gradiometer, were performed by Cohen in 1978[20] factor versus ambient noise, a larger sensitivity to higher spatial frequencies, and a calculatable quite satisfactory "intrinsic" balancing. It should be remarked, however, that a vertical gradiometer with a baseline longer than about 4 to 5 cm, essentially measures brain fields, rather than gradients. By contrast, a planar gradiometer, the baseline of which cannot be made larger than about 1-2 cm for practical reasons, always measures field differences. As a consequence, the detected patterns are complex- see Fig.5 - and do not show a symmetry for different orientations of the source. On balance, planar coil geometries are favored, particularly for integration into large multi-channel assemblies.

3.2 Multichannel systems

The goal toward which many groups around the world are moving is the achievement of neuromagnetic images through large systems with a number of channels of the order of 100. Later in this section we will see the basic requirements for these systems: the present state of the art is limited to instruments with a limited number of adjacent sensors[28-31]. A common feature for the systems which allow seven adjacent measuring sites is that they place six gradiometers equally spaced on a circle, and a seventh one in the center equidistant from the other six. It should be remarked that this configuration ensures maximum "packing" of sensors in dewars with a cylindrical tail and that, consequently, a sort of "magic numbers" sequence is established also for larger systems according to the progression $[1 + 3n(n + 1)]$, with n = an integer. On this basis the successive steps would be 19, 37, 61,... It should be pointed out that the spacing between adjacent channels is strictly related to the depth of the source under study[32], in that it determines the rate of spatial sampling of the field. Small spacing is useless, as the minimum distance of a cerebral source from the detection coil is typically larger than 3 cm. Large spacing is dangerous and may require replication of the measurement at shifted sites. A spacing of about 3 cm seems optimal given dewar constraints and the depth of the neuromagnetic sources.

The experimental procedure so far adopted to measure neuromagnetic fields and to obtain their spatial distribution over the scalp consists in accurately positioning the concave tail of the dewar over a specific scalp area and in recording activity related to underlying sources. The dewar positioning must be carefully checked with respect to anatomical reference points to ensure reliability. Since the present state of the art requires successive positioning over adjacent scalp areas, errors in this procedure may dramatically affect the validity of localization[32]. It is clear that this problem will gradually disappear when larger systems become available, and when magnetic activity related to a specific source is measurable in a single "shot".

Finally, the shape of the subject's head must be satisfactorily known before applying the localization algorithm. This can be achieved by direct craniometric measurements, by x-ray pictures or, better, by MRI images. This last method, currently under study, would permit identification of the internal structures of the subject's brain, matching the neuromagnetic localization with actual anatomy. The best "local" sphere, fitting the curvature of the head over the region of the measuring sites, should then be used to evaluate the theoretical distribution from the model source by fitting a least-squares algorithm to the experimental pattern. Statistical tests also may be used to determine the level of significance of the fit and 95% confidence intervals[32]. A further step, currently under development, consists in calculating the forward problem directly in the actual head of the subject, as reconstructed by a finite element method[17]. If successful, this procedure will definitely improve results, but at the cost of longer computing time.

4. OVERVIEW OF NEUROMAGNETIC RESULTS

We have seen at the beginning of this article that the most promising results of the biomagnetic method have been obtained in the study of cerebral functioning and pathology. This does not exclude that important results have been achieved in other areas. For example, we note that magnetic investigation of some heart pathologies, such as ventricular arrythmias and abnormal accessory excitation pathways, is proving to be a significant tool for preoperative studies. A recently identified application[33] probably will permit a follow-up of transplanted hearts and possibly predict rejection phenomena. More complete review articles are available[10,12,34,14,13,35].

4.1 Epilepsy

The analysis of epileptic activity by means of magnetoencephalographic (MEG) recordings represents one of the most promising applications of the neuromagnetic method in the clinical field. A systematic study of this disease was initiated in 1980 at the Istituto di Elettronica dello Stato Solido in Roma[36,7,9,37,38] and immediately pointed out the possibility of the new approach. An "historic" recording in Fig.6 shows a magnetic trace detected simultaneously with seven electrical traces. The measurement was of a patient with a calcification in the temporal lobe (as shown by the CT image), and the interesting finding was that a "paroxismal" activity - a sequence of sharp waves, namely *spikes* combined with a slow frequency modulation - was clearly identifiable in the MEG, with no electrical counterpart. The abnormal magnetic activity was quite evident also with respect to the "standard" activity detected over different regions of the scalp, and was most probably associated with an abortive seizure provoked by compression of neural tissue by the intrusive mass.

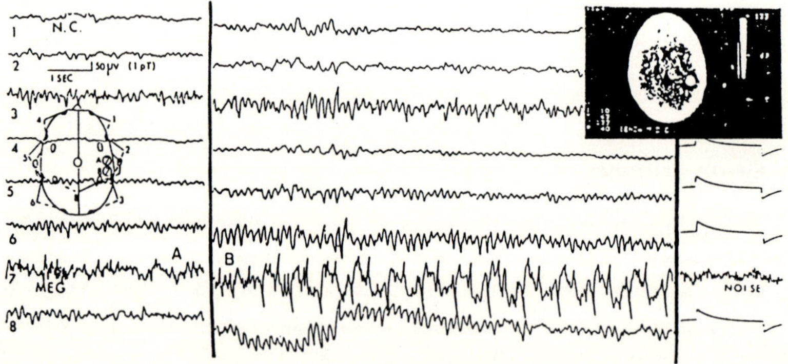

Fig.6 One of the first simultaneous EEGs and MEG (seventh trace) recording from a patient affected by focal epilepsy produced by the calcification revealed by the CT picture. The central magnetic tracing clearly shows pathological signals, probably related to an abortive seizure, not evident in the EEGs.

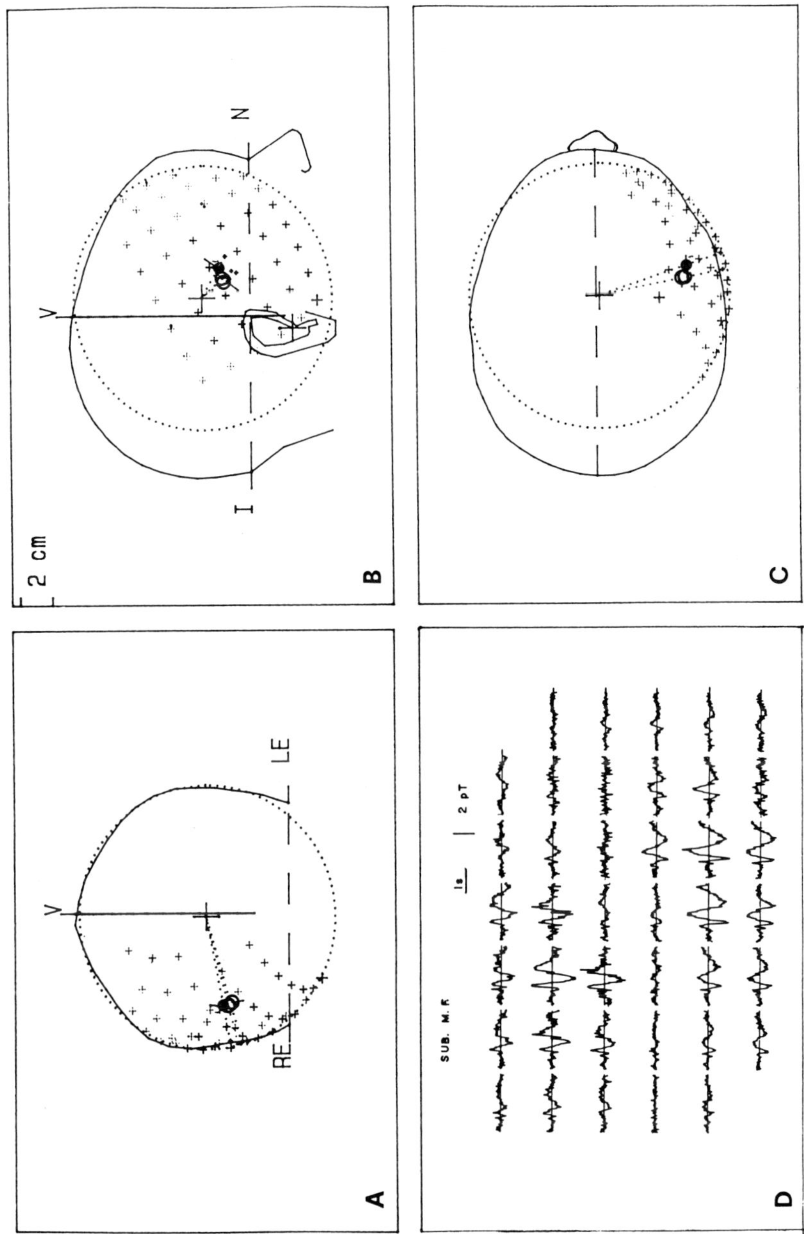

Fig.7 Map of pathological magnetic signals recorded over the right hemisphere (temporal lobe) of a patient affected by focal epilepsy. Each trace was recorded at the corresponding site identified by the cross in the three profiles of the patient's head. The neuromagnetic localization was carried out for different conponents of the pathological signals. The equivalent generators are identified by circles.

In general, the study of "interictal" spikes, i.e., activity occurring between seizures, has shown that magnetic signals often display a dipolar distribution over the scalp[6,7,8]. The localization of the relatively concentrated neural tissue which is firing abnormally is, therefore, often simple. The same procedure based on measured electric potentials would be much more difficult, in fact, even impossible in many cases, due to the interference of extracellular currents and the smearing effect of different layers interposed between the source and the probe. In practice, the magnetic method permitted in many cases a three-dimensional localization of the epileptic focus that was confirmed by intra-operative findings[7,38,39]. One of these examples is illustrated in Fig.7, where the map of spike signals clearly shows polarity reversal. The neuromagnetic localization carried out for the different components of the abnormal signal is reported in the three profiles of the patient's head. This result was confirmed during surgery where the extension of the "active" area was found to involve approximately 7 cm^2 of cortex. Relatively extensive spreading of pathological tissue is typical of this kind of disease and may cause problems for possible non-invasive tissue destruction by means of radiographic techniques. An additional problem may be the insensitivity of the magnetic measurement to radially oriented current flows. A combined mapping of magnetic and electric signals is probably desirable for large-scale clinical use[39].

4.2 Evoked fields

The procedure to study functions related to primary areas of the brain consists in stimulating a peripheral sensory system (visual, somatosensory, auditory) and in measuring the evoked response over the appropriate region of the scalp. The evoked field is generally quite small and, consequently, is by no means distinguishable from the background cerebral activity, which can be regarded as "noise". If the brain is considered as a "stationary"

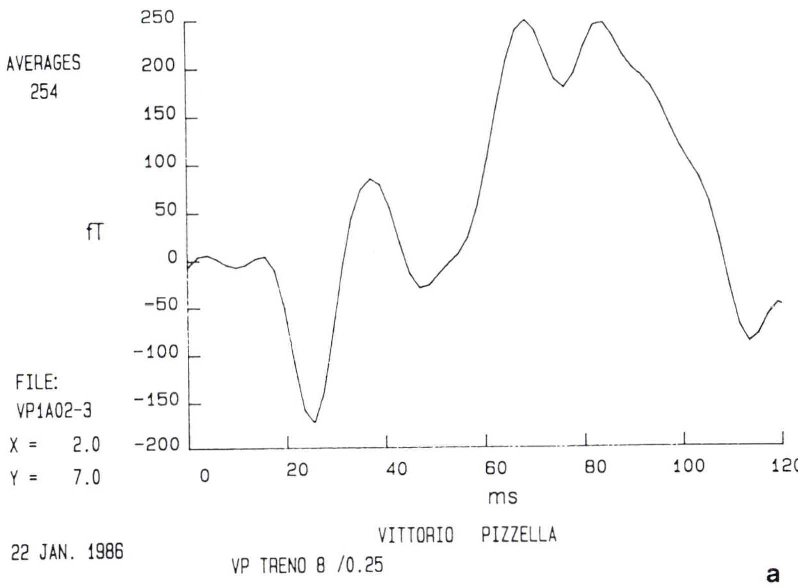

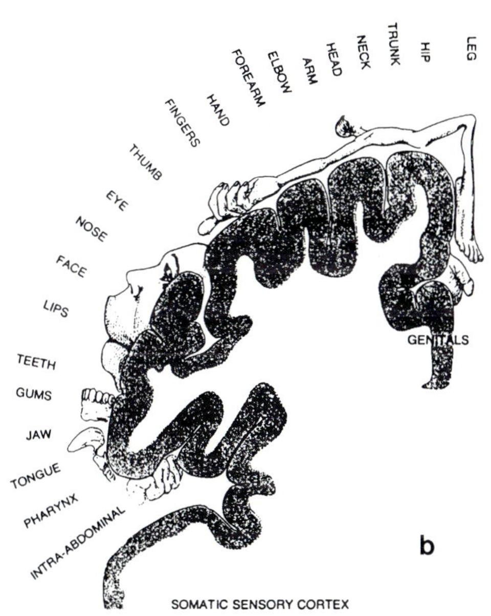

Fig.8 a) Evoked magnetic field under median nerve stimulation at the wrist, as recorded over the side of the scalp contralateral to the stimulated limb. b) Somatosensory homunculus, representing the one-to-one mapping of the peripheral somatosensory system onto the cortex.

system*****, each individual response is "time locked" to the stimulus and is identically repeated after each stimulation. The signal-to-noise ratio can then be increased by averaging many epochs, with a gain proportional to the square root of the number of trials. The result is an *evoked field* like the one reported in Fig.8a. In this case, the stimulation was delivered at the wrist of a normal subject, and consisted of a short electric pulse, 0.1 ms in duration, applied to the median nerve so as to produce an appreciable twisting of the thumb. The response recorded over a specific site of the scalp contralateral to the stimulated limb is structured in a sequence of "components" with different polarity, each one identified by a respective delay, or "latency", from the stimulus onset. By replicating the measurement, other responses can be obtained at many positions of the scalp, typically 40 to 50. From all these time traces it is then possible to obtain field maps corresponding to the latencies of different components. Finally, if the structure of the field distribution is dipolar, a source localization procedure can be applied to identify equivalent generators responsible for the recorded activity. This procedure, applied to the study of the somatosensory cortex, provides results that are in quite good agreement with the so-called somatosensory "homunculus", i.e., the projection of the soma onto the cortex - see Fig.8b - and that has been established many years ago in a highly invasive way during surgery[40]. Fig.9 shows the isofield contour maps obtained for the first component of the responses evoked by stimulation of the median nerve at the wrist and of the tibial nerve at the ankle, respectively. The two maps are both dipolar in shape and the results of the localization algorithm are illustrated in the three plots on the right, representing the three profiles of the subject's head. The equivalent generators, inserted with their respective 95% confidence interval, fit positions in the brain which agree well with the location of the hand and foot area in the homunculus.

A similar procedure can be used to investigate other cortical areas, such as the auditory one. For this case the stimulus usually consists of tone

---

*****the hypothesis of brain stationarity is very questionable, and indeed, has been criticized by many authors. Nevertheless, it has been commonly accepted in all clinical measurements of evoked potentials. Accurate selection of individual responses to each stimulus usually shows different cerebral responses, ranging from a maximum amplitude to even no response at all. We should consider the assumption only as a "working" hypothesis, to get a true "averaged" response.

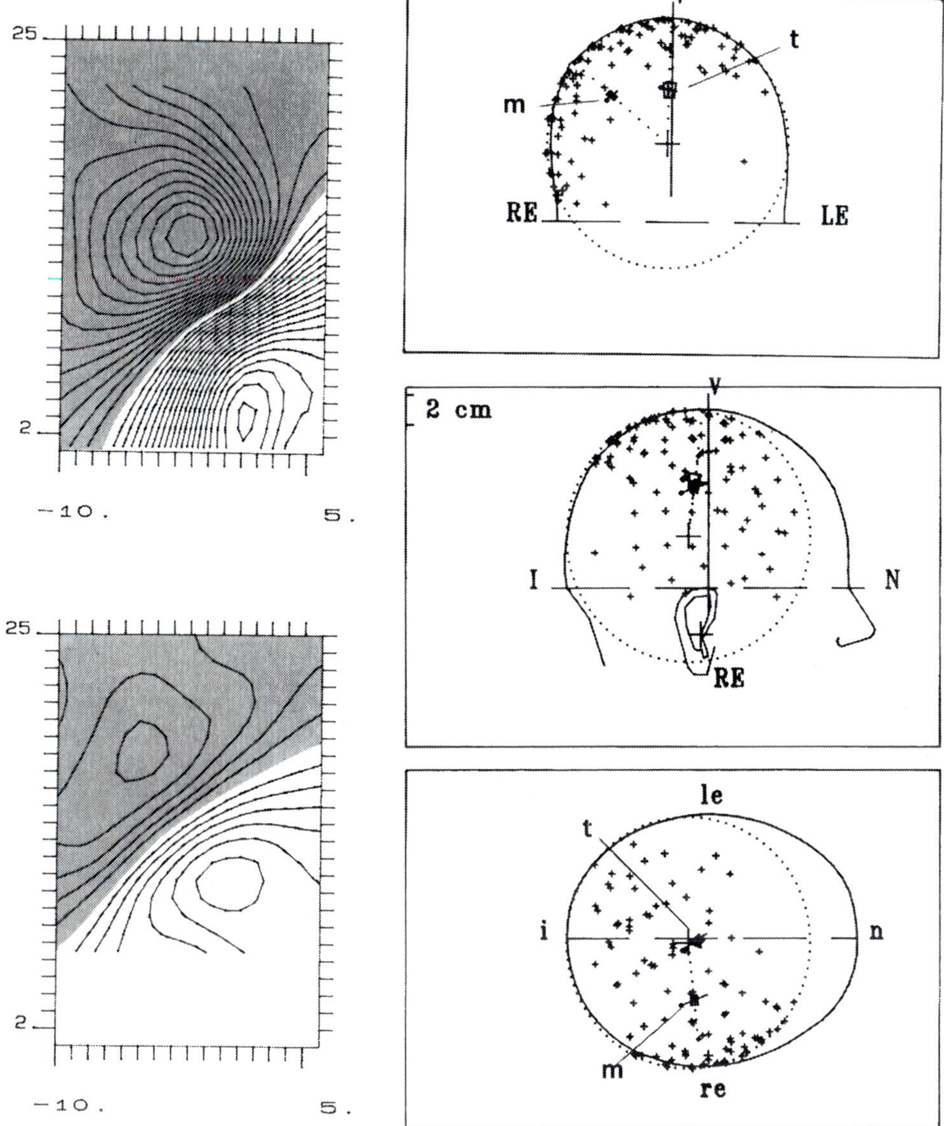

Fig.9 Left: The iso-field contour maps illustrating the scalp distribution of the magnetic field evoked by median nerve stimulation at the wrist (top, m) and by tibial nerve stimulation at the ankle (bottom, t). The iso-line step is 7 fT, and the shaded areas identify negative field polarity. Right: Three-dimensional localization in the actual profiles of the subject's head of the equivalent generators (m, t) accounting for the dipolar distributions reported on the left. The crosses identify the measuring sites and the 95% confidence intervals are also shown.

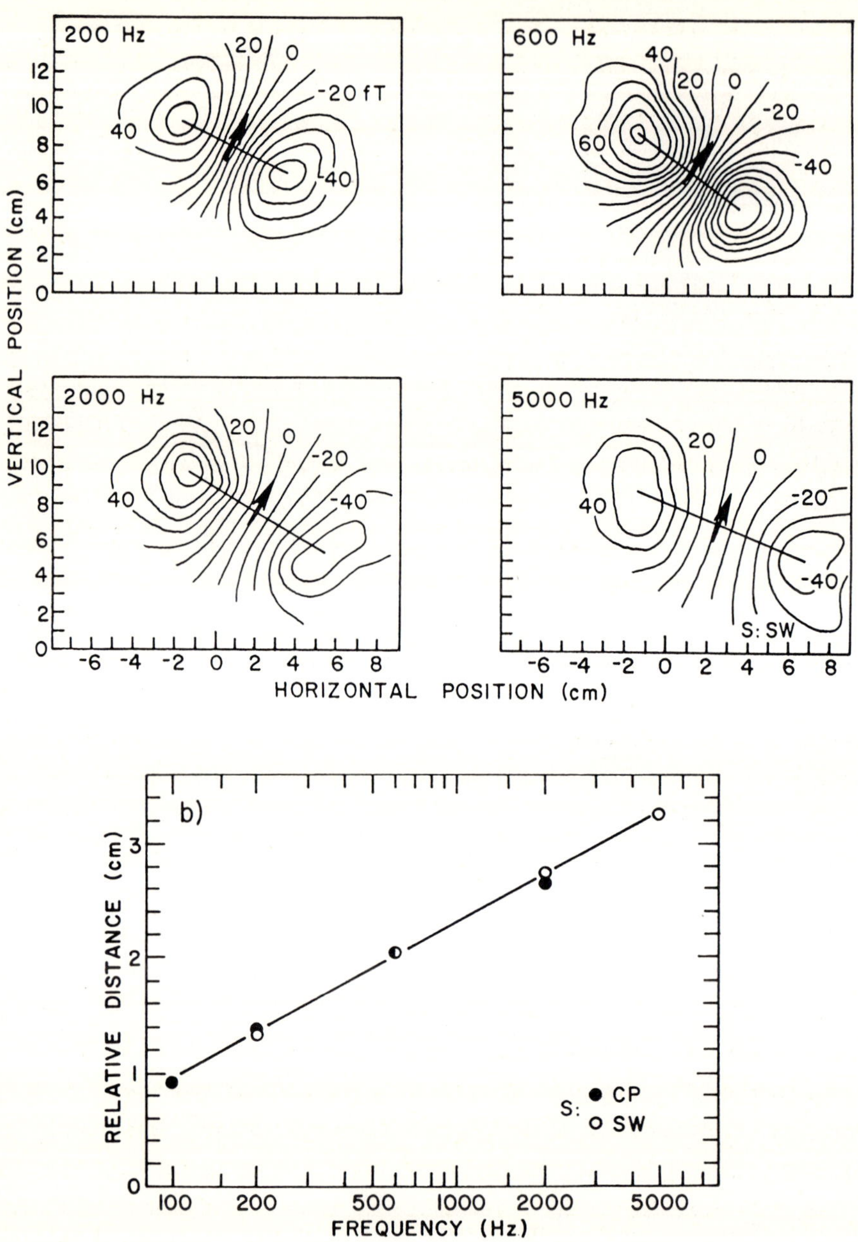

Fig.10 Top: Sequence of iso-field contour maps illustrating the distribution of the magnetic field over the temporal region of the scalp in response to stimulation of the auditory system by means of pure tones with increasing frequency. Bottom: The graph represents the relative straight-line distance between adjacent equivalent sources along the cortex, for two subjects, versus stimulus frequency.

bursts at a specific frequency, or clicks, or combinations of tones. Thus, it was possible to identify a "tonotopic" organization of the human auditory cortex by delivering the subject pure tones of increasing pitch, sinusoidally modulated at low frequency[5]. Fig.10 (top) shows the isofield maps obtained from one subject in response to tones with 200, 600, 2000, and 5000 Hz fundamental frequency. Beyond the clear dipolarity of all the maps, an increase of the distance between the two maxima with opposite polarity and a lateral translation of the midpoint were appreciable. The localization algorithm provided three-dimensional identification of the equivalent generators activated by different stimuli and permitted establishing that their location over the auditory cortex followed a clear logarithmic progression versus stimulation frequency - Fig.10, bottom. This was the first demonstration that in humans, as in animals, the auditory cortex is organized similarly to the cochlea, and that an equal number of neurons is devoted to each octave of sound.

The evoked field method seems particularly appropriate to study cortical organizations and functions, probably because it permits focusing investigation on a very specific phenomenon, and somewhat cancels all the other activity of the brain. Indeed, many results have been obtained in the last few years which definitively confirm this possibility[14,34,35]. The current state of the art permits the study of brain responses to different kinds of stimulation, including composite stimuli. The magnetic response can be analyzed in terms of isofield maps at steps of 1 ms, which seems appropriate to the speed of cerebral events. The localization algorithm can be sequentially carried out and provides the time-spatial evolution of the equivalent generators responsible for the studied activities. The lack of simultaneity still remains a problem, which will be overcome in the near future when large multichannel systems become available. Then experimentalists will have a unique tool to investigate, in almost real time, rapidly-occuring events in the brain, such as information processing.

5. CONCLUSIONS

The aim of this survey has been to provide at least a hint of current progress in SQUID-based biomagnetic research. Because of spatial limitations, no attempt has been made to be comprehensive. However, this is a rapidly growing area of interdisciplinary research, and biomagnetism embodies one of the most important application of SQUIDs. Still, some problems remain to be solved in order to achieve the goal of biomagnetic functional images. One such problem involves the need for liquid helium refilling. It should be noted that the cost of maintaining a class 100 system, i.e., a system with about 100 sensors, would be large enough to make the cost of helium a secondary contribution to the total cost. The development of closed cycle cryocoolers might also help to solve this problem. Finally, the impact of high T_c superconductors might dramatically modify the present situation. Even if the state of the art in this field is not yet adequately developed, it is conceivable that at least part of the superconducting circuitry may consist of high T_c materials, thus simplifying the cryogenic assembly and reducing the overall cost. How fast these improvements can be made is not easy to predict. The hope is that it will not take long.

6. ACKNOWLEDGEMENTS

The author is grateful to R.M. Chapman, C. Del Gratta, V. Pizzella and G. Torrioli for help provided during the preparation of this manuscript.

7. REFERENCES

1. Baule, G.M. and McFee, R.: Detection of the magnetic field of the heart. Am. Heart J. **66**, 95-96 (1963)
2. Cohen, D., Edelsack, E.A., and Zimmerman, J.E.: Magnetocardiograms taken inside a shielded room with a superconducting point contact magnetometer. Appl. Phys. Lett. **16**, 278-280 (1970)
3. Cohen, D.: Magnetoencephalography: detection of the brain's electrical activity with a superconducting magnetometer. Science **175**, 664-666 (1972)
4. Brenner, D., Williamson, S.J. and Kaufman, L.: Visually evoked magnetic fields of the human brain. Science, **190**, 480-482 (1975)
5. Romani, G.L., Williamson, S.J. and Kaufman, L.: Tonotopic organization of the human auditory cortex.Science, **212**, 1339-1340, (1982)

6. Barth, D.S., Sutherling, W.H., Engel, J. and Beatty, J.: Neuromagnetic localization of epileptiform spike activity in the human brain. Science, **218**, 891-894, (1982)
7. Chapman, R.M., Romani, G.L., Barbanera, S., Leoni, R., Modena, I., Ricci, G.B. and Campitelli, F.: SQUID instrumentation and the relative covariance method for magnetic 3-D localization of pathological cerebral sources. Lett. Nuovo Cimento, **38**, 549-554, (1983)
8. Barth, D.S., Sutherling, W.H., Engel, J. and Beatty, J.: Neuromagnetic evidence of spatially distributed sources underlying epileptiform spikes in the human brain. Science, **223**, 293-296, (1984)
9. Ricci, G.B., Leoni, R., Romani, G.L., Campitelli, F., Buonomo, S.,and Modena, I.: 3-D neuromagnetic localization of sources of interictal activity in cases of focal epilepsy. In: Biomagnetism: Applications and theory (H. Weinberg, G. Stroink and T. Katila eds.), pp. 304-310, New York-Toronto: Pergamon Press 1985
10. Williamson, S.J. and Kaufman, L.: Biomagnetism. J. Magn. Magn. Mat., **22**, 129-201 (1981)
11. Romani, G.L., Williamson, S.J. and Kaufman, L.: Biomagnetic instrumentation. Rev. Sci. Instrum., **53**, 1815-1845, (1982)
12. Williamson, S.J., Romani, G.L., Kaufman, L. and Modena, I. eds.: Biomagnetism: an interdiscilinary approach. New York-London: Plenum Press 1983
13. Proceedings of the 6th International Conference on Biomagnetism, Tokyo 1987, in press.
14. Romani, G.L. and Narici, L.: Principles and clinical validity of the biomagnetic method. Med. Progr. through Technol., **11**, 123-159 (1986)
15. Vacuumschmelze GMBH, Hanau, FRG
16. Williamson, S.J. and Kaufman, L.: Analysis of neuromagnetic signals. In: Handbook of Electroencephalography and Clinical Neurophysiology (A.Gevins and A. Remond eds), Revised Series, Vol.1, Amsterdam: Elsevier 1987
17. Hamalainen, M.S. and Sarvas, J.: Feasibility of the homogeneous head model in the interpretation of neuromagnetic fields. Phys. Med. Biol., **32**, 91-97 (1987)
18. Romani, G.L., Leoni, R and Salustri, C.: Multichannel instrumentation for biomagnetism. In: SQUID 85: Superconducting Quantum Interference Devices and their applications (H.D. Hahlbohm, H. Lubbig eds.), pp. 918-932, Berlin-New York: Walter De Gruyter (1985)
19. Lekkala, J.O. and Malmivuo, J.: Noise reduction using a matching input transformer. J. Phys. E: Sci. Instrum., **14**, 939-942 (1981)
20. Cohen, D.: Magnetic measurements and display of current generators in the brain: part II. Polarization of the alpha rhythm. Digest of the 12th Int. Conf. on Medical and Biological Engineering (Jerusalem) Beilinson Medical Center, Petah Tikva, Israel, 1979
21. Knuutila, J., PhD thesis, Helsinki University of Technology (1988)
22. Ketchen, M.B., Goubau W.M., Clarke, J. and Donaldson, G.B.: Superconducting thin film gradiometer. J. Appl. Phys., **49**, 4111-4116 (1978)
23. Carelli, P. and Foglietti, V.: A second-derivative gradiometer integrated with a dc superconducting interferometer. J. Appl. Phys., **54**, 6065-6067 (1983)
24. Ketchen, M.B.: Design of improved integrated thin-film planar dc SQUID gradiometers. J. Appl. Phys., **58**, 4322-4325 (1985)
25. Erne', S.N. and Romàni, G.L.: Performances of higher order planar gradiometers for biomagnetic source localization. In: SQUID 85: Superconducting Quantum Interference Devices and their applications (H.D. Hahlbohm, H. Lubbig eds.), pp. 951-961, Berlin-New York: Walter De Gruyter (1985)

26. Carelli, P. and Leoni, R.: Localization of biological sources with arrays of superconducting gradiometers. J. Appl. Phys., **59**, 645-650 (1986)
27. Bain, R.J.P., Jones, A.E. and Donaldson, G.B.: Design of high-order superconducting planar gradiometers with shaped asymmetric near-source response. IEEE Trans. Mag. **MAG-23**, 1146-1149 (1987)
28. Romani, G.L.: Biomagnetism: an application of SQUID sensors to medicine and physiology. Physica, **126B**, 70-81 (1984)
29. Williamson, S.J., Pelizzone, M., Okada, Y., Kaufman, L., Crum, D.B. and Marsden, J.R.: Magnetoencephalography with an array of SQUID sensors. In: Proc. of the 10th International Cryogenic Engineering Conference, Helsinki (H. Collan, P. Berglund and M. Krusius eds.), pp. 339-348, Westbury House: Butterworth, 1984
30. Ilmoniemi, R., Hari, R. and Reinikainen, K.: A four-channel SQUID magnetometer for brain research. Electroencephalogr. clin. Neurophysiol., **58**, 467-473 (1984).
31. Nowak, H., personal communication
32. Romani, G.L. and Leoni, R.: Localization of cerebral sources with neuromagnetic measurements. In: Biomagnetism: Applications and theory (H. Weinberg, G. Stroink and T. Katila eds.), pp. 205-220, New York-Toronto: Pergamon Press 1985
33. Erne', S.N., personal communication
34. Hari, R. and Kaurakanta, E.: Neuromagnetic studies of somatosensory system: principles and examples. Progr. in Neurobiol., **24**, 233-256, (1985)
35. Romani, G.L. and Rossini, P.: Neuromagnetic functional localization: principles, state of the art and perspectives. Brain Topography, **1**, (1988), in press
36. Modena, I., Ricci, G.B., Barbanera, S., Leoni, R., Romani, G.L. and Carelli, P.: Biomagnetic measurements of spontaneous brain activity in epileptic patients. Electroencephalogr. clin. Neurophysiol., **54**, 622-628 (1982)
37. Ricci, G.B.: Clinical magnetoencephalography. Nuovo Cimento, **2D**, 517-537 (1983)
38. Ricci, G.B., Romani, G.L., Pizzella, V., Torrioli, G., Buonomo, S., Peresson, M. and Modena, I.: Study of focal epilepsy by multichannel neuromagnetic measurements. Electroenceph. clin. Neurophysiol., **66**, 358-368 (1987)
39. Rose, D.F., Smith, P.D. and Sato, S.: Magnetoencephalography and epilepsy research. Science, **238**, 329-335 (1987)
40. Penfield, W. and Rasmussen, T. The cerebral cortex of man. New York-London: Hafner Publ. Co., 1968

SQUIDs for Everything Else

Gordon B Donaldson
Department of Physics and Applied Physics,
University of Strathclyde
Glasgow G4 0NG, Scotland

1. INTRODUCTION

The science of SQUIDs has been developing ever since 1964-5 when Jaklevic *et al* made a rudimentary DC SQUID [1], and Silver and Zimmerman discovered the RF device [2]. The growing understanding of how to operate SQUIDs, of their noise processes, and of their improvements is well documented in this volume and elsewhere [Refs 3,4,5,6,7].

As SQUIDs have developed, so have SQUID applications. Early DC SQUIDs were used as picovoltmeters [8,9], while RF SQUIDs were detecting cardiac magnetic fields within 5 years of their discovery [10]. In the last 20 years the number of applications has grown enormously: table I is not exhaustive, but gives an indication of the areas covered.

This chapter selectively indicates some ways in which SQUIDs have been applied. It does not cover biological applications or geophysics, nor the use of SQUIDs as DC voltmeters and small-signal RF amplifiers (particularly in nuclear resonance experiments), since these are dealt with elsewhere in this volume. Even with these exclusions, there is not enough space to discuss everything in Table I. An excellent review of non-standard applications of SQUIDS has been given by Odehnal [11].

We expect that the reader knows the basic principles of RF and DC SQUIDs, their flux-locked operation, and their noise processes. If not, it will be adequate to regard a SQUID and its input coil as a black-box, flux-to-voltage, converter, with typical flux noise spectral density $\Phi^2_N(f)$ where for $f > \approx 1$Hz up to about 2kHz (though extendible to 500MHz) the noise is white and given by

$$\text{RF SQUID}\ldots\Phi_N(f) = \Phi_W(f) = 1 \times 10^{-4}\Phi_0\text{-Hz}^{-1/2} \quad (1)$$

$$\text{DC SQUID}\ldots\Phi_N(f) = \Phi_W(f) = (0.5 - 20) \times 10^{-5}\Phi_0\text{-Hz}^{-1/2} \quad (2)$$

NATO ASI Series, Vol. F 59
Superconducting Electronics
Edited by H. Weinstock and M. Nisenoff

Pico-ammeter, pico-voltmeter pico-ohmeter	
MCG, foetal MCG MEG- epilepsy, cortical mapping, prenatal disorders Liver iron stores Magneto-pneumography	Magnetotellurics Schumann resonances Internal ocean waves Hydrocracks Tecto-,piezo-, and seismo-magnetism
NQR, NMR Static NMR (δM_z) ^{3}He gyroscope	Susceptometry Thermometry: 10^{-2}–10^{-9}K Rock magnetism Paleomagnetism
Monopoles Quarks Neutrinos h/m_e Ether drag	Non-destructive evaluation: Flaws Stress Corrosion
London moment gyroscope Baroelectric effect	Gravimeters Gravity gradiometers: Mineral surveying,
Submarine detection (MAD) ELF communication	Inertial navigation, Deviation from $1/r^2$ law
Charged beam current detectors	Gravitational wave detectors General relativity gravity probe B
A-D converters	

Table I: Some applications of SQUIDs

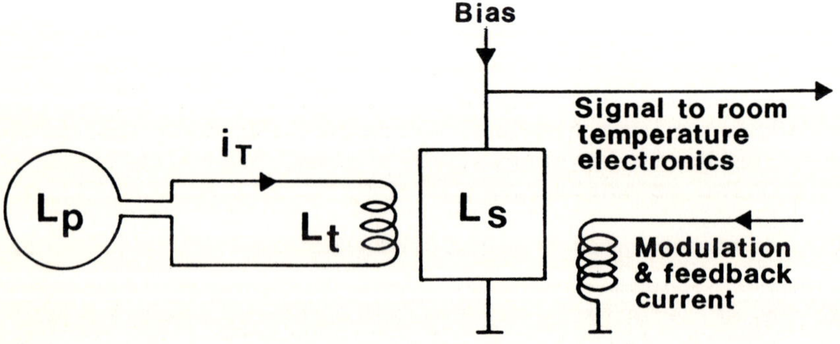

Figure 1: Schematic of typical SQUID operation; the input mutual inductance $M = \alpha(L_t L_s)^{1/2}$ where $\alpha \approx 0.2$–0.8

Below about 0.2-1Hz the noise is given approximately by a '1/f' formula:

$$\text{RF SQUID}\ldots\Phi_N(f) = \Phi_{1/f}(f) = (10^{-10}/f)^{1/2}\ \Phi_0\text{-Hz}^{-1/2} \quad (3)$$

$$\text{DC SQUID}\ldots\ \Phi_N(f) = \Phi_{1/f}(f) = (0.3 - 1) \times (10^{-10}/f)^{1/2}\ \Phi_0\text{-Hz}^{-1/2} \quad (4)$$

Considerable efforts are directed at understanding the origins of 1/f noise in various types and designs of SQUID, and to reducing it, because many SQUID applications are directed to frequencies far below 1Hz, where $\Phi_{1/f}$ dominates Φ_w.

Flux is usually applied (Figure 1) by a current i_t flowing in a superconducting input coil (inductance $L_t \approx 1\mu H$) tightly coupled (coefficient α) to the SQUID loop inductance L_s (typically 0.5-1nH). (Since the SQUID signal is periodic in net applied flux, it is usual to linearise the device by feeding back a current proportional to the output to null the flux that L_t applies to L_s; occasionally current feedback may null i_t itself, especially in SQUID voltmeters.) The noise can then be expressed in terms of an energy resolution, e_N, given by the energy $(1/2)L_t i^2_N$ associated with the current in L_t which can be detected by the SQUID with an S:N ratio of 1. This is

$$e_N = \Phi^2_N/2\alpha L_s \quad (5)$$

With special designs e_N can reach the quantum limit (of order $h = 6 \times 10^{-34}$J-Hz^{-1}), but for general applications designs, typical limits are

$$\text{RF SQUID}\ldots\quad e_N = 4 \times 10^{-29}\ \text{J-Hz}^{-1} \approx 10^5 h \quad (6)$$

$$\text{DC SQUID}\ldots\ e_N = (6 - 30) \times 10^{-32}\ \text{J-Hz}^{-1} \approx 100 - 500h \quad (7)$$

in the white-noise region.

2. Basic Concepts

2.1 Attractions of SQUIDs

SQUIDs have unparalleled electromagnetic energy sensitivity, at least at low frequency. But in most applications, it is unusual for SQUID noise (energy resolution) to be the limit on performance- hence the survival of the less sensitive RF SQUID.

Any one of the following (far from independent) properties can motivate a particular SQUID application:

(a) Extreme energy sensitivity - to 10^{-32} J-Hz^{-1} or better
(b) Extreme field sensitivity - to $\leq 10^{-15}$T-Hz$^{-1/2}$
(c) Use with superconducting circuitry, including persistent currents
(d) Use with gradiometric input
(e) Digital property (output periodic in Φ_o applied to L_s)
(f) Very large dynamic range (over 10^7), based ultimately on (e)
(g) Maintenance of sensitivity in high DC-bias fields, again related to (e).

Examples of each of these will be given.

2.2 Flux transformers; magnetometers and gradiometers

Figure 1 shows the input coil connected to a superconducting pick up loop to form a vector magnetometer. If flux Φ_A is applied to the pick-up coil (inductance L_p), the current induced is

$$i_t = \Phi_A/(L_p + L_t) \qquad (8)$$

causing

$$\Phi_S = Mi_t = \Phi_A \frac{\alpha(L_t L_s)^{1/2}}{(L_p + L_t)} \qquad (9)$$

to be applied to the SQUID. Maximum flux transfer is obtained for $L_p = L_t$, so that pick-up inductances are usually of order 1μH, and (setting $\alpha = 1$ for simplicity)

$$\Phi_S \approx \Phi_A \ (L_s/4L_t)^{1/2} \approx \Phi_A/60 \qquad (10)$$

If the pick-up coil has area A, and the applied field is B, then Φ_A = B x A: the field sensitivity (S:N ratio = 1) is then B_N, where in terms of the SQUID flux noise Φ_N,

$$B_N = \frac{\Phi_N}{A} \times \frac{(L_p + L_t)}{(L_t L_s)^{1/2}} \qquad (11)$$

For an n-turn coil of area A, $L_p = n^2K\mu_oA^{1/2}$ approximately, where K ($\approx$ 3) is a constant, so that

$$B_N = (\Phi_N/A)\frac{(n^2KA^{1/2} + L_t)}{(L_tL_s)^{1/2}} \qquad (12)$$

Equation 12 shows that increased pick-up area could improve field sensitivity, even though flux transfer efficiency might be degraded. In practice cryostat dimensions and other considerations limit the scope for this, so that a reasonable limit for B_N is indeed given from Eq 10 as

$$B_N \approx 60(\Phi_N/A) \qquad (13)$$

For a 4 turn 4cm² pick-up coil ($L_p \approx 1.2\mu H$) and a 10^{-6}(or 10^{-4})Φ_o-$Hz^{-1/2}$ DC (or RF) SQUID, Eq 13 yields

$$B_N = 10^{-15}(\text{or } 10^{-13})\ T\text{-}Hz^{-1/2} \qquad (14)$$

as a rule-of-thumb for SQUID-limited field sensitivity. In practice, figures up to 5-10 times better are possible.

Such sensitivities can only be utilised directly inside a superconducting magnetic shield or a magnetically shielded room [12]. Elsewhere, fluctuations in the earth's field, local mains fields and disturbances due to strong sources such as rotating machinery will dominate the applied fields.

Gradiometers (Figure 2) solve these problems. For a simple first-order structure (Figure 2B), uniform fields produce equal and opposite fluxes in the two equal-area opposed coils (P and Q); the induced current i_t is then zero, and there is no SQUID signal. But for a non-uniform field,

$$i_t \propto (\Phi_P - \Phi_Q) \qquad (15)$$

implying, <u>to first order in the appropriate Taylor series</u>,

$$i_t \propto dB_z/dz \qquad (16)$$

Thus, for an axial magnetic dipole m_z at distance h below the gradiometer face coil (h >> D, the coil spacing),

$$\Phi_S \propto D(dB_z/dz) \propto m_zD/h^4. \qquad (17)$$

The $1/h^4$ term shows that the gradiometer senses weak, but local, sources at the expense of strong, but distant, ones.

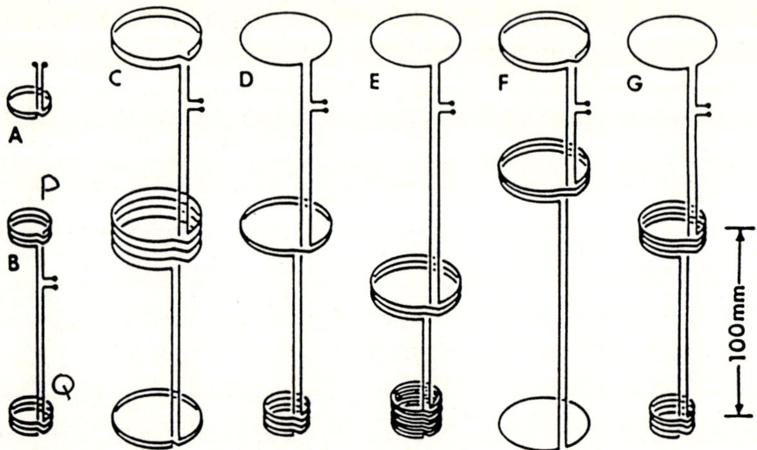

Figure 2: Pick-up structures (after Wikswo-Ref 13): A - Magnetometer; B - First order gradiometer; C - Basic second order (n = 2) gradiometer; D to G - Second order gradiometers with optimised flux transfer factors (F_L): for h << D the inductance should be predominantly in the face (bottom) coils.

Higher order gradiometers, sensitive to d^nB_z/dz^n but not to lower orders, can be produced with more complex windings. The n=2 case (Figure 2C), for example, with a turns ratio 1:-2:1, ignores uniform fields and their first differentials, and discriminates even more strongly in favour of nearby sources. For arbitrary n, equal-area coils, with successive turns ratios ${}_nC_r = (-1)^r\, n!/(n-r)!r!$ for r = 0,1,..n, are required.

However, there are drawbacks.

First, for a total gradiometer length D, Eq. 17 becomes:

$$\Phi_S \;\alpha\; \frac{1}{n!} \times \frac{d^nB_z}{dz^n} \sum_{r=0}^{n} (-1)^r \left(\frac{rD}{n}\right)^n {}_nC_r \qquad (18)$$

$$\alpha \left(\frac{m_z}{h}\right)^3 \left(\frac{D}{h}\right)^n \left(\frac{1}{n}\right)^n \frac{(n+2)!}{2n!} \sum_{r=0}^{n} \{(-1)^r\, r^n\, {}_nC_r\} \qquad (19)$$

$$\alpha\; (2\text{-}5) \times \left(\frac{m_z}{h}\right)^3 \left(\frac{D}{h}\right)^n \qquad (20)$$

so that each increase in order decreases the effective flux picked up by (D/h). Note that when $h \ll D$ and the Taylor series approximation is invalid, it is better to regard the system, not as a gradiometer, but as a single magnetometer coil sensing the field B_z due to m_z, connected to a set of bucking coils which cancel uniform fields and gradients due to distant sources up to order (n-1). In fact, it is important always to regard these systems primarily as flux differencing devices and only by approximation as gradiometers.

Secondly, the electromagnetic energy picked up in any coil has to be shared among all the coils and not just with the SQUID input inductance L_t. Flux transfer is decreased by a factor $F_L = L_t/(L_1 + L_2 + \ldots + L_t)$ which is given (very approximately) by $F_L = 1/\Sigma({}_nC_r)^2 \approx 2^{-2n}/3n$ for simple equal-area coil structures.

Finally, coil "balancing", to make their effective areas exactly equal in the first-order case, and to position them precisely in higher-order cases, is very difficult. Balancing usually involves distorting the applied fields by adjusting small tabs of superconducting material; balance factors of 10^{-6} can be achieved (defined as the signal produced by a uniform field relative to the signal produced by a single coil in the same field). But for $n > 2$, or, in the case of gradiometer arrays, for even lower order, the procedure is at best tedious and often impossible.

Flux sharing problems can be reduced by assymetric structures in which coil turns-areas are matched, but most inductance is associated with the face coil (see Figure 2D-G). Full discussions have been given by Wikswo [13] and others. There are also assymetric structures which give shaped and directional responses, peaking or nulling, for example, at a point outside the cryostat [14,15].

In thin-film technology, mechanical balancing can be eliminated with planar gradiometers produced by precise photolithography. Slight residual imbalance could be dealt with by "active balancing" - the use of a lower-order device on the same chip to provide a small correcting term.

Higher order thin-film gradiometers can be complex, because multiturn coils with superimposed loops are not possible. However, it is always possible to design [14, 16] an optimal n'th order gradiometer with only n crossovers (n+1 loops) by locating the crossovers at distances proportional to $\cos(r\pi/n+1)$, where $r = 1,2,\ldots.n$, along the baseline (where the ends of the baseline correspond to $r = 0$ and $n+1$). Shaped responses can be obtained by increasing the number of loops and formulae are available for the location of (m+n) crossovers (m in arbitrary positions) to produce n'th order gradiometers, for all $m \geq 0$. Extensions to two dimensions, for example to sense $d^{(p+q)}B/dx^p dy^q$ have been described [16,17].

2.3 Dipole mapping with coils and gradiometers

We now discuss (as rules-of-thumb, rather than precise algebra) the approximate response of various magnetometer/gradiometer structures to the passage of a magnetic dipole $\underline{m}$ beneath them. With the notation of Figure 3,

$$\underline{B}(\underline{r}) = (\underline{m}.\underline{r})\underline{r}/r^5 \quad . \tag{21}$$

If the dipole passes a distance h below a single magnetometer coil of diameter 2a, the peak field at the coil is (for h >> a)

$$B_{coil} \approx m/h^3 \tag{22}$$

and the peak flux

$$\Phi_{coil} = \pi a^2 m/h^3 \tag{23}$$

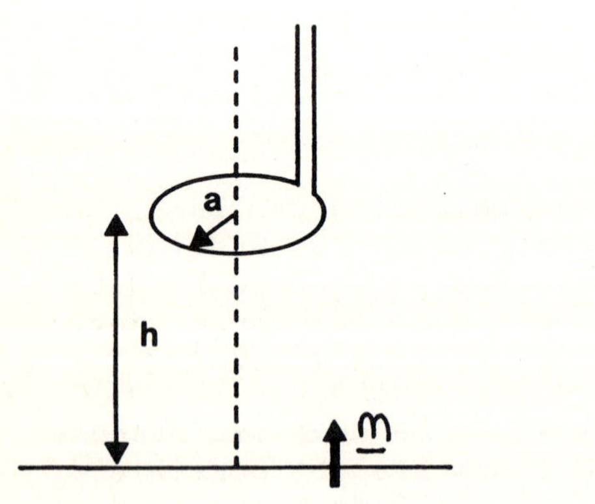

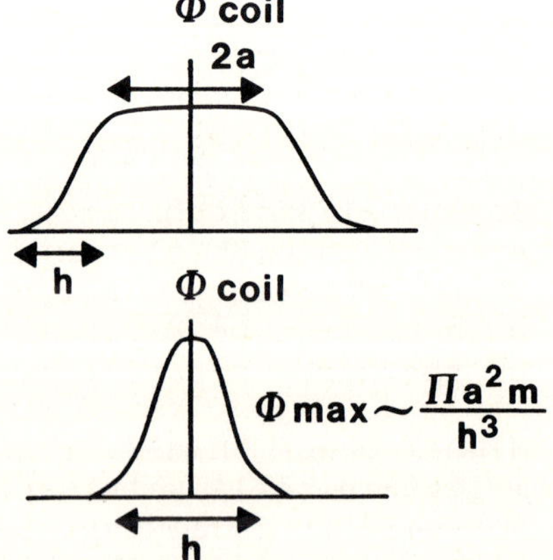

Figure 3: SQUID magnetometer signal when a dipole $\underline{m}$ tracks below a coil. The horizontal resolution is the lesser of the coil radius (a) and stand-off (h). For a gradiometer, use Eq. 25.

Similarly if the dipole is produced by polarising a volume V, with susceptibility χ, with a field B_{pol}:

$$\Phi_{coil} = \pi a^2(\chi V/h^3) \times B_{pol} \qquad (24)$$

For h << a, it is necessary to integrate $\underline{B}(\underline{r}).\underline{n}$ across the coil, where $\underline{n}$ is the unit vector normal to the coil. We see that the spatial resolution is the smaller of h or a.

When a gradiometer is used instead, it may be convenient to continue to think of the system as a single magnetometer consisting just of the face coil (area A), but linked by a reduced field B^*_{coil}. We estimate B^*_{coil} by using Eq. 20 together with the flux transfer factor $F_L = L_s/(L_1 + L_2 + ... + L_s)$, which decreases approximately as $2^{-2n}/3n$ for a simple case, though less steeply for special designs (see Wikswo [13]).

For an axial n'th order gradiometer of baseline D, we have

$$B^*_{coil} \approx mA/h^3 \times (D/h)^n \times F_L \qquad (25)$$

for h >> a, D, where the system is acting as a true gradiometer and

$$B^*_{coil} \approx mA/h^3 \times F_L \qquad (26)$$

for the case of h << D, when flux effectively links only the face coil.

These formulae, although approximate, allow design comparisons with single-coil magnetometers and comparable SQUIDs and their noise limits. Thus one may see a system quoted as capable of $B_N = 10^{-15}T\text{-}Hz^{-1/2}$, say. A comparison of B^* with B_N and with the magnitudes of the unwanted fields and gradients will indicate whether it is worth adopting n'th order gradiometers to measure a particular $\underline{m}$.

2.4 Susceptometers

The most basic use of magnetometers and gradiometers is to measure the magnetic moment of samples placed within their coils or passed from one coil to another. A polarising field, often produced by a superconducting magnet in the persistent mode, is usually involved. The technique was first used in low temperature thermometry, but was quickly extended to magnetochemistry on small samples. Sensitivities better than 10^{-10}emu for a $1cm^3$ volume of material in a 10mT field were achieved in the 1970's [17]; with integrated thin-film SQUID structures it is now possible to measure microgram specimens and detect as few as 2-3,000 magnetic spins-$Hz^{-1/2}$ [18,19]. Commercial models are available and are in routine use in, for example, biochemical [20] and archaeological [21] research.

The rest of the chapter discusses some recent applications which illustrate the principles outlined in this section.

3. Magnetic Monopole Detectors; other particles

Dirac's isolated magnetic pole [22] was predicted to carry a magnetic charge g = h/2e or an integral multiple n thereof. The flux through a surface surrounding it would be

$$\Phi = \int \underline{B}.d\underline{S} = 4\pi ng = 2n\Phi_o. \qquad (27)$$

The monopole is required by grand unified theories of fundamental interactions, and there are many predictions about its mass (≈ 10^{15} AMU), mean free path in matter (many earth diameters), lifetime (infinite), etc [23]. The monopole density is important in relation to the Parker [24] limits on their flux (of order 10^{-15}–10^{-12} particles $s^{-1}cm^{-2}sr^{-1}$ depending on the actual monopole mass). The limits are set by the possible contribution of monopoles to the mass of the universe, and by the size of the intergalactic magnetic field (≈ 10^{-10}T) which otherwise would have been quenched by energy transfer to monopoles if they were too numerous.

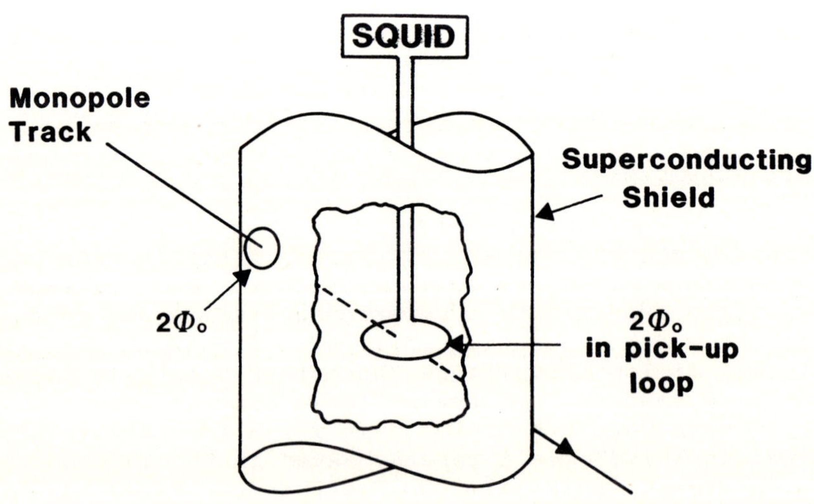

Figure 4: Detection of monopole by induction method.

SQUID monopole detection (Figure 4) relies upon the principles (c) and (d) given in Section 2.1, and not on high sensitivity. As a monopole passes through a superconducting loop which forms part of a flux transformer, it leaves flux $2\Phi_o$ (take n = 1 in Eq 27) trapped as a circulating current. Cabrera's "candidate" event of 1982 [25] at Stanford showed this. The output of a SQUID coupled to a 4-turn, $10cm^2$ coil, and mounted within a screened enclosure with residual field less than 10^{-12}T, suddenly jumped by $8\Phi_o$ referred to the pick-up, after the slight correction required if a monopole had left $2\Phi_o$ in the entry and exit walls of the surrounding superconducting shield.

Many similar experiments since 1982 [26, 17, 27 and others] have been characterised by (i) coil areas up to $0.2m^2$ to increase the likely rate of events; (ii) multiple SQUID channels with high-order gradiometers to reduce effects due to events in the walls, such as flux creep and mechanical vibration; and (iii) temperature, pressure and seismic detectors to correlate with spurious events. Appropriate coincidences and anti-coincidences would confirm the passage (and perhaps velocity) of a monopole through the apparatus. For example, a monopole should leave a persistent signal on two, and only two, of six pick-up coils arranged to form the faces of a rectangular box.

There has been at most one further event - an "unexplained event" in 1985 [28], but this scarcity is not surprising: in fact the rate so far (1 or 2 candidates in six years in a few m^2 of detector) is well <u>above</u> the Parker limit (which gives $\approx 10^{-4}$ events per year so far). Any conclusions must await much better statistics.

SQUIDs have also been used in searches for quarks using a method analogous to Millikan's method for the electron [28]. Some evidence for fractionally-charged particles was found, but there has been no further confirmation. There have also been proposals to use SQUIDs to detect neutrinos and other exotic particles [29]

4. Gravity Devices

Our next applications, by contrast with monopole detection, *do* rely on the extreme sensitivity of the SQUID. Indeed one of them - the problem of detecting gravitational radiation -

has prompted the development of SQUIDs with e_N close to the quantum limit [30,31,32].

4.1 Gravity Wave Detectors

A body emits gravitational waves when its mass distribution varies non-spherically, as in the rotation of a binary star, or the collapse of a star with the emission of 10^{45}–10^{48}J of gravitational radiation (over a time $\approx 10^{-3}$s). At the earth, this radiation causes instantaneous longitudinal strains. For the nearby supernova of April 1988, the strain at the earth ($\delta L/L$) will have been $\approx 10^{-15}$, but such events are very rare. To see a rate of a few events per decade requires looking at the galaxy as a whole; the typical signal is then much smaller ($\delta L/L \approx 10^{-18}$–$10^{-19}$ for an event at the centre of our galaxy at a range of 10kparsec.) For ten events per year a range of 3Mparsec ($\delta L/L \lesssim 10^{-22}$) is needed.

Several groups measure $\delta L/L$ in a large mechanical block using a SQUID sensor [29, 33, 34, 35]. The Stanford aluminium bar [29], for example, has mass $M \approx 5000$kg and length $L \approx 3$m, and operates at $T = 4.2$K. It resonates with $Q \approx 5 \times 10^6$ at $f_{res} = \omega_{res}/2\pi \approx 0.8$kHz (a period matching the expected length of a pulse due to stellar collapse). The thermal fluctuations

$$\langle \delta L \rangle_T \approx (4k_BTL/M\omega^2)^{1/2} \approx 10^{-17}\text{m-Hz}^{-1/2} \qquad (28)$$

are well above those induced by a gravity wave with $\delta L/L \approx 10^{-19}$. However, near the resonant frequency, the effective noise temperature of the bar is reduced to T/Q, because the bar retains a memory of energy deposited in it for approximately Q cycles. Induced amplitudes much smaller than the thermal fluctuations may then be detectable.

Readout involves the mechanical transformer illustrated in Figure 5. A thin niobium diaphragm of mass m resonating close to f_{res} is mounted on the end of the bar where it oscillates with an amplitude $(M/m)^{1/2} \approx 10^3$ times that of the bar, though with a response time $(M/m)^{1/2} \times 1/f_{res}$. It is mounted in a superconducting magnetic enclosure between superconducting pancake coils connected in parallel to a SQUID. These coils support a persistent current i_T of a few amps, whose field controls the effective stiffness of the diaphragm. Thus, by adjusting i_T (using a heat switch and small resistor), the diaphragm resonant frequency can be tuned to f_{res}. As the diaphragm moves, it modulates the inductances L_1 and L_2, and drives a small fraction of i_T into the SQUID transformer.

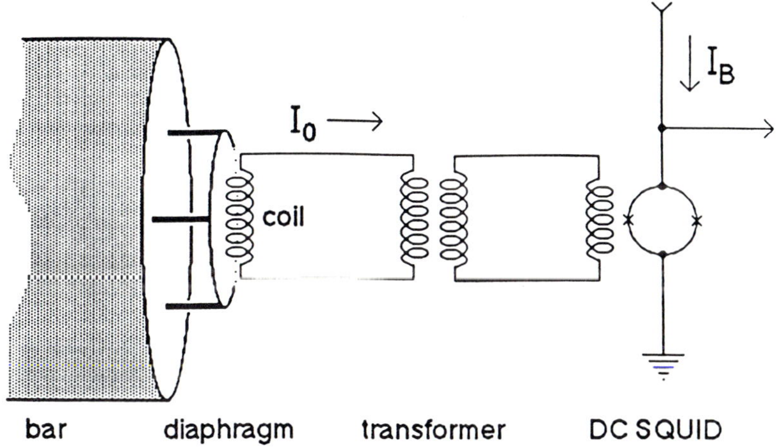

Figure 5: Mechanical transformer readout for gravity wave detector

With a commercial DC SQUID ($e_N \approx 10^{-30}$J-Hz^{-1}), the Stanford bar is sensitive to $\delta L/L \approx 10^{-18}$–$10^{-19}$; SQUID noise and thermal fluctuations are roughly equal. It is planned to cool to 30mK, giving an antenna noise temperature 30mK/Q, at which only zero-point motional strain ($\delta L/L \approx 3 \times 10^{-21}$) will be significant; the detector will be a DC SQUID with $e_N \approx 20h$.

To detect the more frequent events with $\delta L/L \leq 10^{-22}$ would depend on avoiding this apparent zero-point limit, which is set by the Heisenberg formula $\delta x.xp \approx h$. Naively, for a simple oscillator, one estimates δp as $M_{res}.\delta x$. "Quantum non-demolition" measurement procedures [36], in which δx is measured with arbitrarily fine precision while δp is in effect not measured at all, may be required.

4.2 Gravimeters and gravity gradiometers

SQUID instruments for measuring local gravitational effects include both gravimeters and tensor gradiometers. They are usually intended for mineral surveying or inertial navigation, though fundamental problems such as tests of the Universal Law of Gravitation have also been addressed.

A gravimeter (measuring g), based on a SQUID magnetometer sensing the equilibrium position of a superconducting ball levitated in a non-uniform magnetic field, was produced by Goodkind [37]. It had a sensitivity of 10^{-8}m-s^{-2}-Hz$^{-1/2}$ at low frequencies, which was usable for air-borne mineral surveys. However, for gravity anomalies, gradiometers, which we now discuss, are more attractive.

At h above the surface of a sphere of uniform density and radius R,

$$g(h) = \frac{4\pi\rho GR^3}{3(R+h)^2} \quad \text{and} \quad \frac{dg(h)}{dh} = -\frac{8\pi\rho GR^3}{3(R+h)^3} \qquad (29)$$

For $h/R \ll 1$, $g = 4\pi\rho GR/3(1-2h/R) \approx (4\pi\rho G/3)R$. But $dg/dh = 8\pi\rho G/3(1-3h/R) \approx (8\pi\rho G/3)$, so that its magnitude is virtually independent of R: in other words, the gravity gradient at the surface of an isolated small stone sphere equals that at the surface of the earth (approximately 3000 Eotvos, where 1 Eotvos = $10^{-9}s^{-2}$, or about $10^{-10}g_o$-m^{-1} where g_o refers to the earth's surface). Gravity gradiometers are therefore better than gravimeters for detecting local gravity anomalies, such as ore bodies. For mineral surveying and navigation one needs sensitivities of 1-10 Eotvos from DC to a few Hz.

Figure 6 shows one of a number [38,39] of gravity gradiometers derived from the gravity wave detector discussed in Section 4.1. It uses two niobium diaphragms mounted 10cm apart in an all-niobium enclosure. Each diaphragm has a pair of pancake coils mounted a distance $d \approx 100\mu m$ from either surface, and all the coils are connected in parallel to a SQUID input coil.

Like any spring-mass (M) combination with a natural oscillation frequency $\omega/2\pi$, a diaphragm will be displaced by x relative to its support, either by a low frequency acceleration a_1 or by a gravitational force Mg_1, where $x = a_1/\omega^2$ or $x = g_1/\omega^2$. For Figure 6, $\omega/2\pi$ = 1100Hz, giving a sensitivity to a_1 or g_1 of $(1100 \times 2\pi)^{-2}s^{-2}$, that is, about 20nm/ms^{-2}. The displacement towards a pancake coil produces a change in its effective inductance L given by $\delta L = -\beta L(x/d)$ where β is a geometrical factor ≈ 1. Thus $\delta L = a_1L/\omega^2d$ (or g_1L/ω^2d). So if diaphragm 1 experiences acceleration a and gravity g_1, while diaphragm 2 experiences a and g_2, then the current transferred to the SQUID input coil is, with obvious notation ($\tau \approx 0.7$):

$$I_3 - (2/\tau)\left|\frac{L}{L+4L_t}\right|\left|\frac{I_1(a+g_1)}{\omega_1^2d_1} - \frac{I_2(a+g_2)}{\omega_2^2d_2}\right| \qquad (30)$$

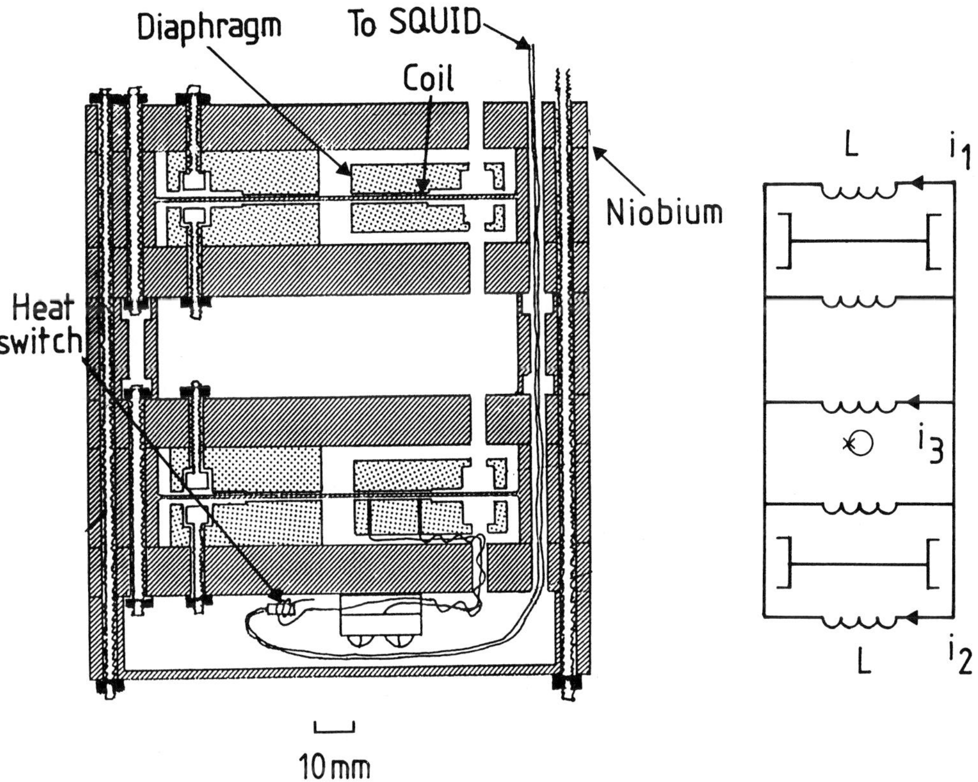

Figure 6: Double diaphragm gravity gradiometer (Ref 39)

This can be rewritten in common mode and difference terms:

$$I_3 = (2/\tau)\left|\frac{L}{L + 4L_t}\right| \times \left[\left|\frac{I_1}{\omega_1^2 d_1} - \frac{I_2}{\omega_2^2 d_2}\right|(a+\bar{g}) + \left|\frac{I_1}{\omega_1^2 d_1} - \frac{I_2}{\omega_2^2 d_2}\right| \frac{dg}{dh}\, b\right] \tag{31}$$

where b is the gradiometer baseline.

The gradiometer is balanced by making $I_1/\omega_1^2 d_1 = I_2/\omega_2^2 d_2$: a superconducting switch is used to insert a small series resistor (10^{-11}ohm) in one circuit, and the current is reduced slowly until the response to a uniform acceleration (produced by shaking) vanishes. To sense a dg/dh of 1 Eotvos in the earth's environment ($g = 10\text{ms}^{-2}$) with a 10cm baseline, a balance of 1 part in 10^9 is required: this is easily achieved and maintained because of the drift-free cryogenic and persistent-mode environment (Section 2.1c).

The apparatus in Figure 6 achieved a SQUID-noise-limited performance of 70 Eotvos-$Hz^{-1/2}$, corresponding to the detection of $(x_1 - x_2) \approx 10^{-16}$m-$Hz^{-1/2}$. The gravity gradient due to a 100kG mass pushed beneath the cryostat was easily seen.

Improved sensitivities have been achieved [40] by replacing the diaphragms by proof masses and cantilever springs with high compliance along the sensitive axis and high stiffness orthogonal to it. This reduces $\omega/2\pi$ to below 100Hz, decreasing the denominators in Eq 31.

Basic applications include gravity surveying for mineral prospecting and also geodesy, for which a 10^{-3}Eotvos satellite-borne instrument has been proposed. Interest has also been expressed for inertial navigation systems which use 3-axis accelerometers which record $(\underline{a}+\underline{g})$ from which tables are used to subtract $\underline{g}$ before a double integration of $\underline{a}$ with respect to time yields position. The inconvenience of tables could be removed by recording all terms dg_i/dx_j of the gravity gradient-tensor, and integrating with respect to displacement $\{x_1, x_2, x_3\}$ to give a running record of $\underline{g}(\underline{x})$. Designs for measuring the necessary off-diagonal terms in the gravity-gradient tensor have been proposed [41].

A gradiometer has been used to test for corrections to the Newton's law of gravity ($F = Gm_1m_2/r^2$), in which G is not constant but varies with distance. A possible form is

$$G(r) = G(1 + \alpha \exp(-\beta r)). \tag{32}$$

In such a case, the tensor trace

$$\Gamma = dg_x/dx + dg_y/dy + dg_z/dz \tag{33}$$

does not have the zero value which applies to a strict inverse square law. Chan et al. [42] have made laboratory measurements of (r ≈ 10m) with a 0.2 Eotvos gravity gradiometer, and find $\alpha = 0.024 \pm 0.036$ at $\beta^{-1} = 1$m, so that no significant corrections to G are established at this level. Further measurements, using tidal lakes and eccentric satellite orbits, have been proposed to test the law on larger length scales.

5. SQUIDs for Non-Destructive Evaluation (Non-Destructive Testing)

A growing use of SQUIDs is in the detection of defects in a variety of materials and structures, including electronic circuits. Frequent features from the list in Section 2.1 are (i) use of gradiometers and (ii) retention of pick-up sensitivity in large (e.g. 50mT) polarising fields. (The SQUID itself must be screened from such a field, but its niobium pick-up coils need not be.) This explains why systems as sophisticated as SQUIDs are called on when the resolutions needed are in the flux-gate magnetometer range. Only SQUIDs can be used as drift-free self-differencing gradiometers, and flux gate magnetometers do not maintain DC resolutions of 10^{-4} gauss in the presence of 500 gauss.

5.1 Remote magnetometry

5.1.1 NDE of steel plates

This application originates from biomagnetic work, originally by Wikswo *et al.* [43], involving the use of a SQUID gradiometer rigidly fixed with respect to a coil producing a persistent mode field. Because the SQUID senses only changes in the flux linking it, there is no signal due to this field itself, even though the gradiometer may not be perfectly balanced nor the field very uniform.

However, the gradiometer *does* sense (as in Section 2.3) the magnetic dipole created by the polarising field when a patient with a highly paramagnetic liver (due to excess iron stores) is placed under the cryostat. The resulting signal is a good measure of the total iron level, and the method has proved valuable in diagnosis, since the alternative to this "non-destructive" technique is a very painful, and possibly dangerous, biopsy.

These principles have been applied [44] to the remote detection of surface breaking cracks in ferromagnetic steel. Figure 7 shows apparatus derived from Wikswo's, in which a steel plate can be pushed below the cryostat. The magnetic permeability μ is very large ($\approx$ 500-700): field boundary conditions at a crack of roughly square cross section then imply that the polarising field is essentially excluded from the volume (V) of such a crack. The distortion of $\underline{B}_{pol}$

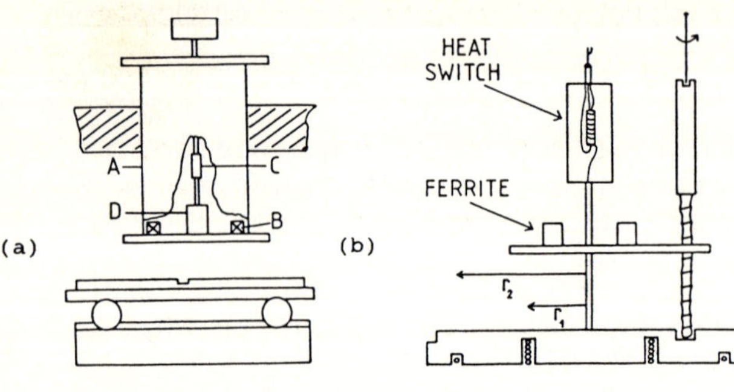

Figure 7:
Non-destructive
crack detection:

(a) cryostat (A)
magnet coil (B)
SQUID (C)
gradiometer (D)

(b) detail of
gradiometer

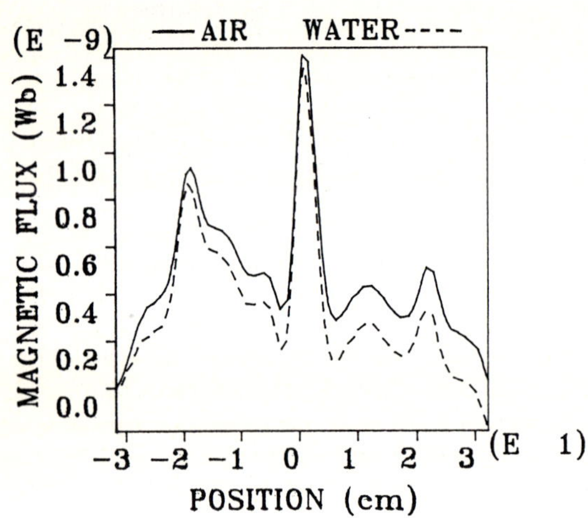

Figure 8:
Crack detection in steel

(a) scan over three slots
at 5cm stand-off in
air and salt water

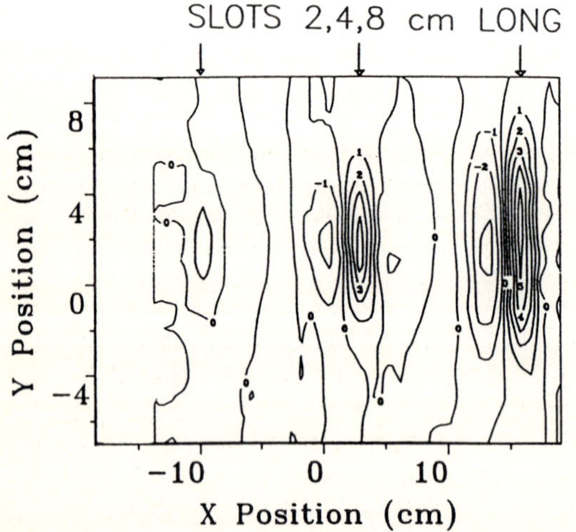

(b) x-y contour map of signal
above three slots

which this produces is equivalent to generating a dipole

$$\underline{m} = -\underline{B}_{pol}V \tag{34}$$

at the crack, which is then detected as it passes beneath the gradiometer (Figure 8). Standoff distances H (between the plate and the cryostat) of up to 10 cm are practicable, and immersion in salt water or covering with any non-magnetic material does not affect the signal.

The gradiometer design presents a problem, because of possible variations of the stand-off distance as the specimen is scanned. Boundary conditions at the uncracked plate show that $\underline{B}_{pol}$ makes a negligible angle (at most $1/\mu$) to the surface normal. This can be modelled by an image coil equal to the polarising coil and at a distance 2H below it. As H varies, say due to distortions of the plate, so too will the image field and its gradient at the sensor. With an axial gradiometer, these variations completely mask any crack signals. In fact, it is not usually even possible to "lock" the SQUID in such a case.

The problem is solved with a planar gradiometer - "balanced magnetometer" would be a better name. It has (Figure 7) two coaxial coils with N_1 and N_2 turns respectively, of radius r_1 and r_2 such that $N_1r_1^2 = N_2r_2^2$, connected in series opposition. Magnetic balancing (zero output in a uniform AC field) is performed with a movable ferrite ring. Now the transverse gradient of the image coil field is zero on the gradiometer axis, and is everywhere much less than the axial gradient. Thus, <u>changes</u> in the image coil field (as H varies) do not produce large changes in the net flux linked to the gradiometer. The SQUID therefore remains in lock at all times [45].

If SQUID noise were the limit, crack volumes V of $35\mu m^3$ would be detectable for H = 5cm and B_{pol} = 0.01T. However, large plate-dependent background signals, due to residual stand-off effects and spatial permeability fluctuations, limit the open-crack detectability to about $1mm^2$ cross-section, equivalent to distortions of the field of about 10^{-8}T at the gradiometer.

The spatial resolution is (see Section 2.3) the larger of H (stand-off) or d (gradiometer diameter): it is not possible to distinguish a crack in a weld from the magnetic anomaly of a weld itself (typically a 1cm heat affected zone) with a gradiometer which itself has dimensions ≈ 1-3cm . Nor is it possible to use the system within about H (or d) of the edges of plates.

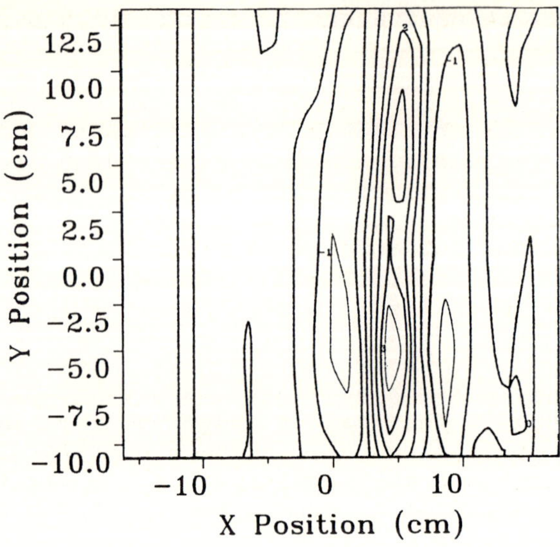

Figure 9: Contour map of signal above fatigued cracked plate. The crack itself is only 5cm long: the residual signal is due to μ variations produced by plastic flow in fatigue regions

If instead of an open crack the plate contains a region where μ has changed by $\delta\mu$, the effective dipole becomes

$$\underline{m} = (\delta\mu/\mu)\underline{B}V \tag{35}$$

Such variations (with μ decreasing by as much as 50%) occur in plastically deformed regions, close to fatigue cracks and along lines where cracks may yet develop. and can be seen by this method (Figure 9) [44].

5.1.2 Duplex stainless steels

Recently SQUID magnetometry has been applied to stainless steels which have been plastically deformed or heated above 500°C using the apparatus of Figure 7 [46]. Under these circumstances, small amounts of ferrite precipitate out and can be detected easily in a field of 10mT. Since the virgin steel is non-magnetic, there are essentially no background fields or effects due to specimen edges, and the field sensitivity is higher.

This method might be important in monitoring the aging of special steel vessels.

5.1.3 Plastic deformation and Barkhausen jumps in magnetic steels

A SQUID gradiometer has been used to monitor the magnetic state of a steel bar under stress [47]. No polarisation field was used. Under a reversible stress-strain regime, changing flux (presumably due to realignment of domains) could be sensed at distances of 20 cm. This flux too was reversible but did not change monotonically; instead, it passed through an extreme and declined again with increasing stress and strain. For plastic (non-reversible) strain, the flux changes also became irreversible.

In all cases, it was observed that the magnetoelastic coefficient changed sign at a stress value just below the elastic limit. A magnetomechanical method of strain monitoring based on the relative phase of the cyclic magnetic response to a cyclic stress was proposed.

With an increasing field, the usual Barkhausen jumps due to domain rotation produce detectable SQUID signals in unstressed polycrystalline iron samples.
Weinstock et al [48] have studied both Barkhausen emission and hysteresis onset in this way.

5.2 Remote Galvanometry

The sensing and locating of currents, perhaps flowing in faulty paths as a result of defects as been the object of several SQUID applications.

5.2.1 Pipelines

Weinstock and Nisenoff [47] showed that a metal pipe could be accurately located, in distance and direction, by using an axial SQUID gradiometer to sense the field due to a current in the pipe of about 1 amp at 4.6Hz. A scan across the line of the pipe (Figure 10) shows that when the pipe lies directly along the gradiometer axis there is a sharp zero in the detected signal which is then purely transverse. The horizontal resolution here is rather better than the h limit which we discussed earlier. Triangulation to determine the pipe position depends on a second scan with the gradiometer axis at about 30° to the vertical.

Distances up to 160cm were studied, but much greater values should be possible and the method should be applicable to buried pipelines.

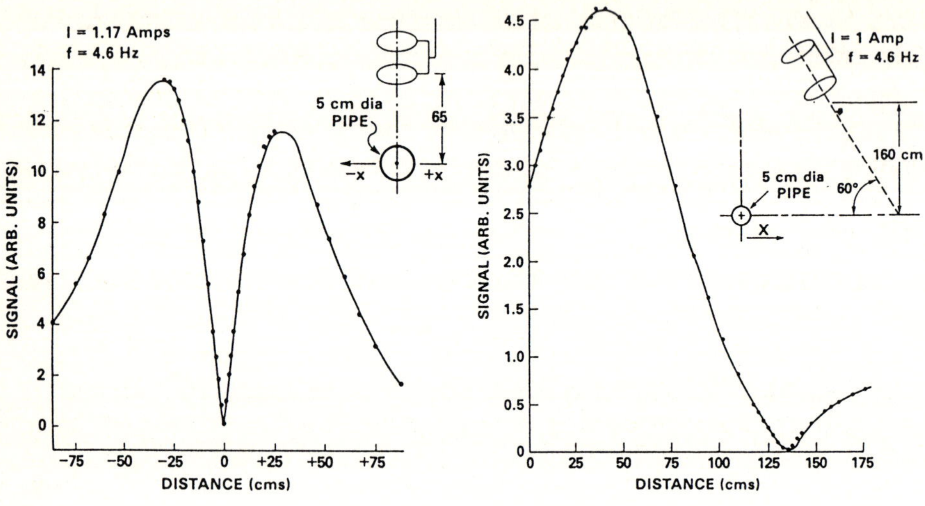

Figure 10: SQUID pipeline detection [48]

The method also detected current diverting flaws, such as holes or welds in the pipeline. They appear as anomalies in the signal as the SQUID is tracked along the line of the pipe.

Murphy [49] has extended these ideas to the detection of breaches in coatings on buried pipelines. A current passed along a pipeline (with earth/ground return) slightly leaks into the surrounding soil through the coating itself, but leaks strongly at a breach ("holiday"). The SQUID is tracked along the pipeline and measures the field due to the current distribution. A $1cm^2$ holiday can be observed as a sharp change in measured field: $1cm^2$ size breaks were detected at a depth of 1-2m.

5.2.2 Corrosion processes

Electrolytic corrosion processes in non-voltaic cells have been studied, using SQUIDs, by Bellingham et al [50,51]. In such cells both electrodes are made of the same metal and there is a single electrolyte: here Zn and HCl were used (Figure 11). A SQUID gradiometer mapped the field due either to an impressed either dc or ac current flowing in the cell. Comparisons are made with various electrochemical models of the electrode-electrolyte interface.

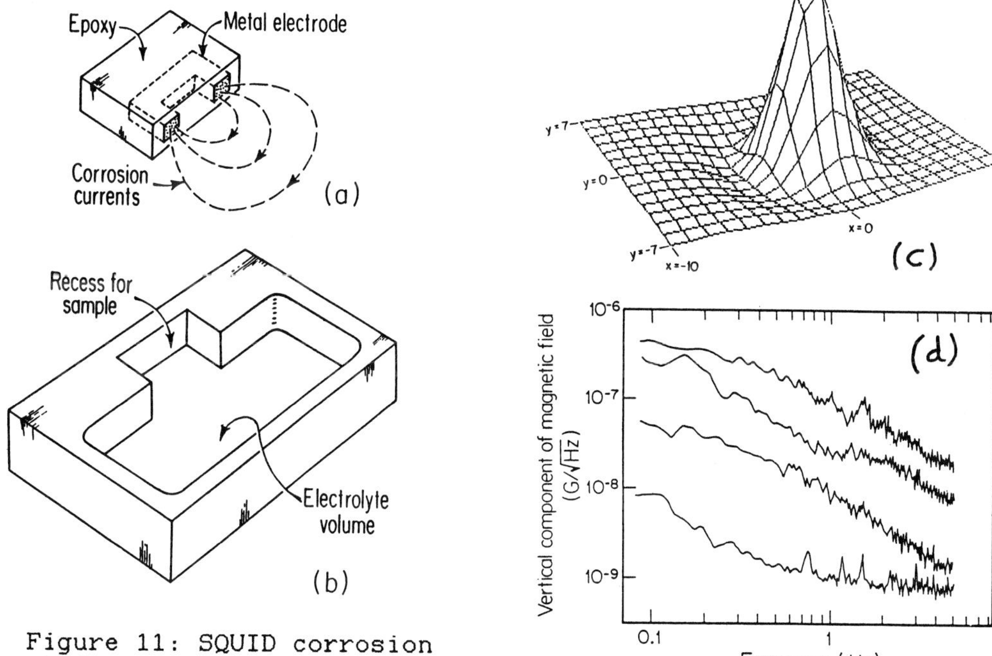

Figure 11: SQUID corrosion studies:
(a) and (b) Zn-HCl non-voltaic cell; (c) SQUID gradiometer signal; (d) Field spectra for HCl (10,3 and 1Molar strength); lowest curve is environmental noise)

Studies were made of the spectral density, up to a few Hz, of current fluctuations in the cell caused by different parts of each electrode operating at various times as a local anode or cathode. Diffusion, hydrogen bubbling, crack propagation, etch pit deepening and other processes appear to contribute to these fluctuations with characteristic signatures. Work is in hand to obtain greater spatial detail of the corrosion processes and current flow.

This is not likely to lead to a simple field-usable instrument, but rather to give laboratory information on corrosion processes for various metal/alloy-electrolyte combinations, and to help in evaluating them for various structural uses.

5.2.3 Integrated circuits.

Remote SQUID galvanometry could possibly map current distributions in integrated circuits for fault finding. Little practical progress has been made, for reasons we discuss in the next section, though Fagaly [52] has measured the audiofrequency fields above a timing device.

THE SPATIAL FILTERING APPROACH

$$\vec{B}(\vec{r}) = \frac{\mu_o}{4\pi}\int_v \frac{\vec{J}(\vec{r}')\times(\vec{r}-\vec{r}')}{|\vec{r}-\vec{r}'|^3}d^3r'$$

$$B_z(x,y,z) = \frac{\mu_o d}{4\pi}z\int_{-\infty}^{\infty}\int_{-\infty}^{\infty}\frac{J_y(x',y')}{[(x-x')^2+(y-y')^2+z]^{3/2}}dx'dy'$$

$$b_z(k_x,k_y,z) = FFT(B_z(x,y,z))$$

$$b_z(k_x,k_y,z) = g(k_x,k_y,z)j_y(k_x,k_y)$$

$$G(v,w,z) = \frac{\mu_o d}{4\pi}z\frac{1}{[v^2+w^2+z^2]^{3/2}}$$

$$g(k_x,k_y,z) = \frac{\mu_o d}{2}e^{-\sqrt{k_x^2+k_y^2}\,z}$$

$$j_y(k_x,k_y) = \frac{b_z(k_x,k_y,z)}{g(k_x,k_y,z)}$$

$$J_y(x,y) = FFT^{-1}(j_y(k_x,k_y))$$

Figure 12-I

The 2-D inverse solution [54]: Definition of functions

5.3 The 2-D inverse problem; improved spatial resolution

Developed SQUID NDE will require the ability to use a mapping of the magnetic field $\underline{B}(\underline{r})$ to derive details of the currents $\underline{J}(\underline{r}')$ or magnetic moments $\underline{m}(\underline{r}')$ flowing in the specimen under test. This inverse problem does not have a unique solution, so that in biomagnetism, for example, it is necessary to use heavily constrained models for the sources within the torso or brain [53]. However, if $\underline{J}$ or $\underline{m}$ are confined to 2 dimensions or less, the inverse problem does have a unique solution. A model calculation for a two dimensional current distribution has been given recently [54] and is demonstrated in Figure 12 (I-III) for two current dipoles: it involves two Fourier transformations with an intermediate Gaussian weighting operation.

Fig. 12-III (a)-(d) shows how the B field is calculated from the ideal current distribution I(z); Fig. 12-III(e) models a practical measurement of B by adding ("SQUID") noise. To reconstruct the sources (Fig. 12-III(f)-(h)),the signal is Fourier transformed (f), and filtered (g) to remove high spatial frequency noise components. This produces loss of detail in the final reconstruction (Fig. 12-III(h)). Some

THE SPATIAL FILTER INVERSE PROBLEM:
A PAIR OF CURRENT DIPOLES

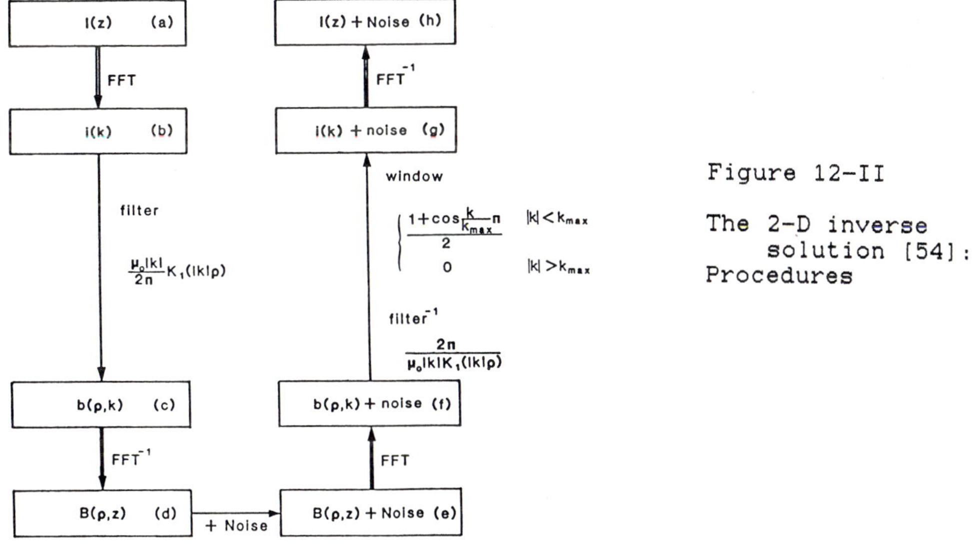

Figure 12-II

The 2-D inverse solution [54]: Procedures

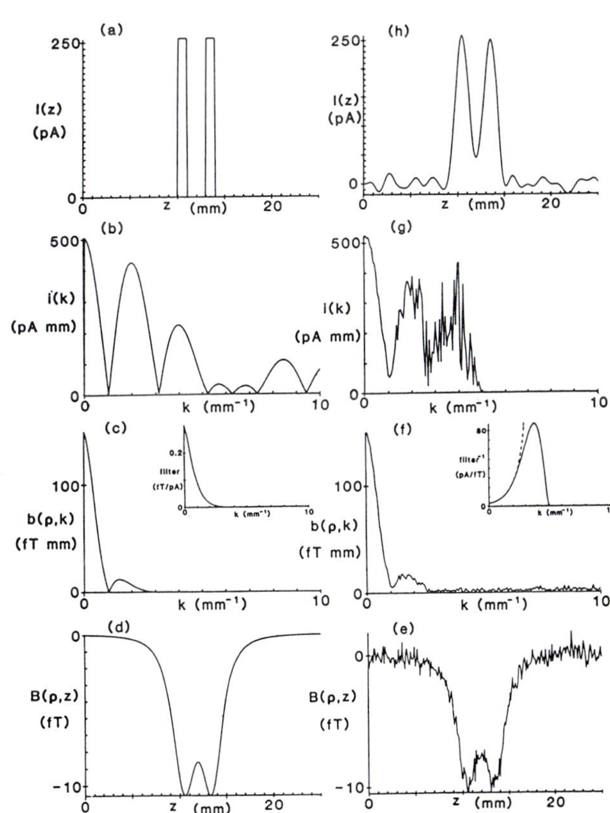

Figure 12-III

The 2-D inverse solution [54]: Implementation of procedures:

(a)-(d) - derivation of field due to current dipoles
(e) - addition of "SQUID" noise
(f)-(g) - spatial filtering
(h) - reconstructed dipole pair

improvements in spatial resolution can be obtained by improving the magnetometer signal-to-noise ratio, and by using a technique known as inward continuation. However the greatest improvements depend on applying the principles of Fig 3 : for optimum quality (i) coil-to-source spacing should be as small as possible, and (ii) coil diameter should equal coil-to-source spacing.

To be useful for integrated circuit mapping, spatial resolutions of order <<1mm, and preferably less than 10μm, will be necessary. Since the spatial resolution can not be much smaller than the coil-to-specimen distance, this calls for novel cryostat design! Solutions have been proposed by Wikswo - the first will indeed have a cryostat wall thickness (at least in the coil region) of only 1mm, and is under construction; the second would call for room temperature specimens to be inside the vacuum space itself and within a few microns of the superconducting coil.

5.4 High T_c (HTS) SQUIDs in NDE.

As we have seen, many NDE applications do not require high SQUID sensitivity, but do need reduced stand-off to improve spatial resolution. This points to NDE as an important field for the first applications of HTS SQUIDs: the simplicity of liquid nitrogen cooling means decreased wall thicknesses, while the white noise flux sensitivity of HTS SQUIDs (RF SQUIDs 10 x $10^{-4}\Phi_0$-$Hz^{-1/2}$; DC SQUIDs 5 x $10^{-5}\Phi_0$-$Hz^{-1/2}$) is already more than good enough. However, the large 1/f noise experienced so far at 77K may limit the performance at NDE frequencies. Moreover, good persistent mode HTS wires are needed before gradiometers can be made.

5.5 Summary

NDE is generating new SQUID applications, in which the ability to detect small changes remotely (perhaps in the presence of large polarising fields) is more important than fundamental sensitivity. It has even been suggested that there is a use for a fast digital SQUID (counting directly in Φ_0's) for such applications.

NDE may well be one of the first areas for application of HTS SQUIDs.

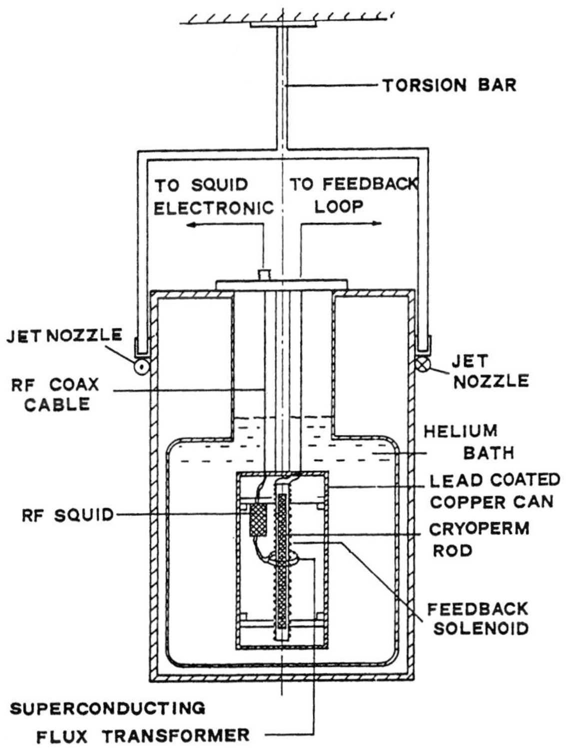

Figure 13: Magnetogyroscope [55]

6. Novel SQUID applications - Magnetogyroscope

There is a steady flow of novel SQUID uses. As an example, we choose the latest of a series directed at navigation (see also Section 4.2 on gravity gradiometers). It involves detecting absolute rotation by measuring the difference between the orbital magnetomechanical ratio (2m/e) of the superconducting electrons in a SQUID fixed inside a superconducting shield and the spin magnetomechanical ratio (m/e) of the electrons in a ferromagnetic rod fixed on the same axis [55].

A bulk superconductor (such as a SQUID and its coils) rotating at Ω generates the London field $\underline{B}^* = -2m\Omega/e$. The field is so small, however, (5.7×10^{-12} T/(rad-s^{-1})) that the SQUID could measure it only if a superconducting shield were used to screen external noise fields; but as that rotated with the SQUID, it would screen out B^*!

However, if a ferromagnetic rod rotates inside the superconducting shield, a magnetisation field arises of the form $2m\Omega/ge$, where g is the gyromagnetic ratio (closely equal to 2). If a SQUID embraces a ferromagnetic rod and a superconducting shield, and rotates rigidly with both, it will sense [56] a flux

$$\Phi = -(2m/e)\ S(1-1/g)(1-s/S)(\chi s/S)/(1+(\chi s/S)\Omega \quad , \qquad (36)$$

where s and S are the cross sections of the rod and the shield, and χ is the susceptibility of the ferromagnetic rod.

Vitale et al [56] have used a ferromagnetic rod (cryoperm10) mounted inside a lead box (Figure 13) to demonstrate the validity of Eq 36 within a few percent, and with high linearity over a range 0-0.8rad-sec^{-1}. Currently, the resolution is 0.01rad-sec^{-1}.

For navigation, the gyroscope must ultimately measure rotation rates Ω which are small fractions of the earth's rotation rate ($\approx 10^{-5}$Hz). The limiting sensitivity appears to be set by thermal fluctuations of the rod magnetisation

$$\langle \delta\Omega \rangle_T = (e/m)(\mu_o k_B T/V\nu)^{1/2} \ (\text{rad-s}^{-1}\text{-Hz}^{-1/2}) \qquad (37)$$

where V is the apparatus volume. and ν is a cut-off frequency for χ . Quoted ν values are 10^{12}Hz, so that at 4.2K, with V = 1 litre

$$\delta\Omega_{min} = 4.7 \times 10^{-8} \ \text{rad-s}^{-1}\text{-Hz}^{-1/2} \quad , \qquad (38)$$

which might just be adequate for navigation.

7. Integrated Thin Film Structures

Once thin film SQUIDs became available it was a short step to producing integrated instruments based on coupling to thin film gradiometers deposited directly on the same chip. Ketchen et al [18], for example, produced an integrated thin film magnetic susceptometer with 15 m coils capable of dealing with microgramme specimens and performing ultrafast magnetic spectroscopy [19]. A review of the basic principles of integrated devices was given by Donaldson et al. [57].

Integrated structures have grown more important as SQUID neuromagnetism has developed. Many parallel sensor channels (up to 100 have been suggested) will be needed if rapid mapping over a surface, such as a skull, is to be possible, and with it the precise location of a functional focus within the brain. In Japan a gradiometer specifically configured for brain sensing was reported recently [58], while an example of a system of 30 thin-film SQUID gradiometers has been described by Knuutila et al [59].

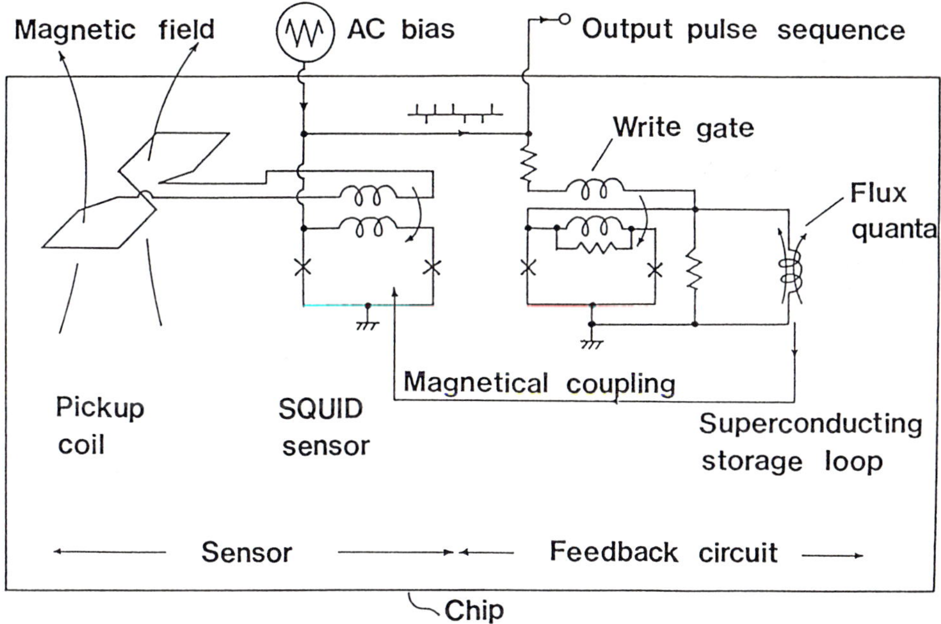

Figure 14: DC SQUID gradiometer with integrated feedback circuit [60]: "Chip" indicates 4.2K region

A problem with multi-channel DC SQUID systems (which is worse for RF SQUIDs) is the number of leads needed: at least 4 per channel because signals must pass from the SQUID to the room temperature electronics and back again for flux locking feed-back. A recent development may be very significant in this context.

It arises from the programmes which led to all-refractory tunnel junctions, and to entire logic circuits containing many thousand junctions. Only a few tens of such junctions might be needed to develop the necessary logic-based control electronics on the chip alongside a SQUID itself. The Fujitsu group [60] has recently used the $Nb/Al_2O_3/Nb$ process to make a DC SQUID with an integrated feedback circuit. A gradiometer (Figure 14) is coupled to an AC (600kHz) biased SQUID which produces output pulses whose polarity depends on the sign of the net flux input (including feedback). The pulses are counted algebraically outside the cryostat, but are also stored as $\pm\Phi_0$ in the write gate, from which a signal proportional to the flux stored is coupled back to the SQUID. The device thus seeks a zero net flux input at which the output pulses are alternately positive and negative, while the cumulative total of pulses counted outside the cryostat represents the *gradiometer* input flux.

At present, the slew rate is only $300\Phi_0 s^{-1}$, and the white noise $3 \times 10^{-3}\Phi_0\text{-Hz}^{-1/2}$. The dynamic range is also poor, but the device is already good enough for magnetocardiography (with gradient sensitivity $1.7 \times 10^{-9}\text{Tm}^{-1}\text{-Hz}^{-1/2}$). A high performance SQUID with digital, but room temperature, electronics has already been produced by Jutzi and Drung [61], and it may be possible to repeat this in Josephson technology. Multiplexing between different SQUIDs on the same chip may follow, pointing the way to multichannel magnetometers in which just a few data-highway lines connect the cryogenic environment with the outside world..

8. Conclusions

After twenty years the range of SQUID applications is still increasing. Not all will flourish, but some show signs of maturing and producing sophisticated multi-sensor instruments in which the distinction between analogue and digital Josephson electronics is blurred.

It is too soon to discern the role which HTS SQUIDs will play. White noise is already low, but 1/f corner frequencies are too high to permit most applications to move from 4.2K to 77K at present.

I thank many colleagues for useful discussions, and Marty Nisenoff and Harold Weinstock for their kind invitation to contribute to these proceedings.

REFERENCES

1. R C Jaklevic, J Lambe, A H Silver and J E Mercereau, Phys. Rev. Lett. 12, 274 (1964).
2. A H Silver and J E Zimmerman, Phys. Rev. Lett. 15, 888 (1965).

   For information on device physics and applications, see, for example:

3. J Clarke, Proc. IEEE 61, 8 (1973); Proc. NATO ASI'76, p67 (1976); IEEE Trans. Electron Devices, ED-27, 1896 (1980).
4. M B Ketchen, IEEE Trans Magnetics MAG-23, 1650 (1987).
5. A Barone and G Paterno, Physics and Applications of the Josephson Effect, John Wiley, London and New York, 1982.

6. Proceedings of successive Applied Superconductivity Conferences, published biennially in IEEE Trans Magnetics eg MAG 25 (to be published in 1989), MAG 23 (1987), MAG 21 (1985), MAG 19 (1983),.....
7. Proceedings of successive International Conferences on Superconducting Quantum Interference Devices in Berlin: SQUID '76, SQUID '80, SQUID '85, (H D Hahlbohm and H Lubbig, eds), de Gruyter, Berlin.

8. J W McWane, J E Neighbor and R S Newbower, Rev. Sci. Instrum., 37, 1602 (1966)
9. J Clarke, Proc. Roy. Soc., A308, 447 (1969).
10. D E Cohen, E A Edelsack and J E Zimmerman, Appl. Phys. Lett., 16, 278 (1970).
11. M Odehnal, Sov. J. Low Temp. Phys., 11, 1 (1985).
12. V O Kelha, pp33-50 and A Mager, S N Erne, H D Hahlbohm, H Scheer, Z Trontelj and J Palow, pp51-94, Biomagnetism, (S N Erne, H D Hahlbohm and H Lubbig, eds), de Gruyter, Berlin and New York, 1981.
13. J P Wikswo, AIP Conference Proceedings, 44, 145 (1978).
14. G B Donaldson and R J P Bain, Appl. Phys. Lett., 45, 360 (1984).
15. P Karp and D Duret, J. Appl. Phys., 51, 1267 1980).
16. C D Tesche, C C Chi, C C Tsuei and P Chaudhari, Appl. Phys. Lett., 43, 384 (1983).
17. E I Cukauskas, D A Vincent and B S Deaver, Rev. Sci Inst., 45, 1 (1974);.J S Philo and W M Fairbank, Rev. Sci Inst., 48, 1529 (1977).
18. M B Ketchen, T Kopley, and H Ling, Appl. Phys. Lett., 44, 1008 (1984).
19. D D Awschalom, M B Ketchen and W J Gallagher, to be published in IEEE Trans Magnetics MAG-25, £££ (1989).
20. See for example E P Day, T A Kent, P A Lindahl, E Munck, W H Orme Johnston, H Roder and A Roy, Biophys. J., 52, 837 (1987).
21. W S Goree and W L Goodman, Rev. Geophys. and Space Sci., 14, 59 (1976).
22. P A M Dirac, Proc. Roy. Soc., A133, 60 (1931); Phys. Rev., 74, 817 (1948).
23. An accessible review for the SQUID worker is D Fryberger, IEEE Trans Magnetics MAG-21, 84 (1985).
24. E N Parker, Astrophys. J., 166, 395 (1975); M S Turner, E N Parker, T J Bogdan, Phys. Rev., D26, 1296 (1982).
25. B Cabrera, Phys. Rev. Lett., 48, 1378 (1982).
26. M E Huber, B Cabrera, M Taber, R Gardner, IEEE Trans Magnetics MAG-23, 1134 (1987).
27. A D Caplin, M Koratzinos and J C Schouten, Nature, 321, 402 (1986).
28. W M Fairbank, Physica, 109-110, 1404 (1982).

29. B S Cabrera, private communication; A K Drukier and L Stodoslky, Phys. Rev. D, 30, 2295 (1984).
30. P Carelli and V Foglietti, J. Appl. Phys., 53, 7592 (1981).
31. M B Ketchen and J M Jaycox, Appl. Phys. Lett., 43, 736 (1982).
32. D J van Harlingen, R H Koch and J Clarke, Appl. Phys. Lett., 41, 197 (1982).
33. M Bassan, W M Fairbank, E Mapoles, M S McAshan, P F Michelson, B Moskowitz, K Ralls, R C Taber, Proc. 3rd Marcel Grossman Meeting on Gen. Relativity, H Ning ed, North Holland, New York (1983), p667.
34. E Amaldi, P Bonifazi, F Bordoni et al, Proc. 2nd Marcel Grossman Meeting on Gen. Relativity, R Ruffini ed, North Holland, Amsterdam (1982).
35. P J Veitch, D G Blair, M J Buckingham, C Edwards and F J van Kann, IEEE Trans Magnetics MAG-21, 415 (1985); J Kadlec and W O Hamilton, SQUID '80, (H D Hahlbohm and H Lubbig, eds), de Gruyter, Berlin 1980, p813.
36. C M Caves, K S Thorne, R W P Drever, V D Sandstrom and M Zimmerman, Rev. Mod. Phys., 52, 341 (1980).
37. W A Prothero and J Goodkind, Rev. Sci. Inst., 39, 1257 (1968); J Goodkind, J Geophys. Res., 91B, 9125 (1986).
38. H J Paik, E R Mapoles and K Y Wang, AIP Conference Proceedings, 44, 166 (1978); H J Paik, J. App. Phys., 47, 1168 (1976).
39. A B Colquhoun, N A Lockerbie and G B Donaldson, SQUID '85, (H D Hahlbohm and H Lubbig, eds), de Gruyter, Berlin 1985, p1191.
40. M V Moody, H A Chan and H J Paik, IEEE Trans Magnetics MAG-19, 461 (1983) and J. Appl. Phys, 60, 4380 (1986); F J van Kann, C Edwards, M J Buckingham and R D Penny, IEEE Trans Magnetics MAG-21, 610 (1985).
41. H J Paik, Phys. Rev. D, 35, 355 (1987).
42. H A Chan, M V Moody and H J Paik, Phys. Rev. Lett., 49, 1745 (1982).
43. J P Wikswo, J E Opfer and W M Fairbank, AIP Conf. Proc., 18, 1335 (1974) and Med. Phys. 7, 307 (1980); D E Farrell, J H Tripp, P Zanucchi, G M Brittenham, J W Harris, G M Brittenham and W A Muir, IEEE Trans Magnetics MAG-16, 818 (1980).
44. R J P Bain, G B Donaldson, S Evanson and G Hayward, SQUID '85, (H D Hahlbohm and H Lubbig, eds), de Gruyter, Berlin 1985, p841; R J P Bain, G B Donaldson and S Evanson, IEEE Trans Magnetics MAG-23, 473 (1987).
45. S Evanson, R J P Bain, G B Donaldson, G Stirling and G Hayward, to be published in IEEE Trans Magnetics MAG-25, (1989).

46. M Otaka, K Hasegawa, T Shimizu, K Takaku, S Evanson and G B Donaldson, submitted to Conf. Jap. Soc. Mech. Eng., 1989.
47. H Weinstock and M Nisenoff, *SQUID '85*, (H D Hahlbohm and H Lubbig, eds), de Gruyter, Berlin 1985, p843.
48. H Weinstock, T Erber and M Nisenoff, Phys. Rev. B, 31, 1535 (1987).
49. J C Murphy, R Srinivasan and R S Lillard, Workshop on SQUID NDE, Harpers Ferry, West Virginia, 1988.
50. J G Bellingham, M L A Macvicar, M Nisenoff and P C Searson, J Electrochem. Soc., 133, 1753 (1988).
51. J G Bellingham and M L A Macvicar, IEEE Trans Magnetics MAG-23, 477 (1987).
52. R L Fagaly, to be published in IEEE Trans Magnetics MAG-25, (1989).
53. See, for example, "The Biomagnetic Inverse Problem" S Swithenby ed, Physics in Medicine and Biology, 32, Number 1, (1987).
54. B J Roth, N G Sepulveda and J P Wikswo, J. Appl. Phys., 65, 361 (1989).
55. M Cerdonio, A Goller and S Vitale, *SQUID '85*, (H D Hahlbohm and H Lubbig, eds), de Gruyter, Berlin 1985, p1197.
56. S Vitale, European Physical Society SQUID Workshop, Rome 1988; S Vitale and M Cerdonio, to be published.
57. G B Donaldson, C M Pegrum, and R J P Bain, *SQUID '85*, (H D Hahlbohm and H Lubbig, eds), de Gruyter, Berlin 1985, p729.
58. M Koyanagi, N Kasai, K Chinone, M Nakanishi, S Kosaka, M Hoguchi and H Kado, to be published in IEEE Trans Magnetics MAG-25, (1989).
59. J Knuutila *et al.*, to be published in IEEE Trans Magnetics MAG-25, (1989).
60. N Fujimaki, H Tamura, T Imamura and S Hasuo, Proc. ICD-88, 33 (1988); S Hasuo, to be published in IEEE Trans Magnetics MAG-25, (1989).
61. W Jutzi and D Drung, to be published in IEEE Trans Magnetics MAG-25, (1989).

NONLINEAR PROPERTIES OF JOSEPHSON JUNCTIONS

N.F. Pedersen
Physics Laboratory I
The Technical University of Denmark
DK-2800 Lyngby, Denmark

Abstract

The basic equations for the Josephson junction are derived in a simple way. We discuss the properties of such junctions under various circumstances - including the effects of damping and capacitance - according to the shunted junction model. Also, the effects of an external rf bias current are discussed; this leads to a definition of the various characterizing frequencies and to the appearence of many new phenomena. rf-applications such as SIS mixers and parametric amplifiers will be mentioned. The introduction of spatial dimensions leads to problems concerning cavity excitations and solitons. Finally, conditions for the appearence of chaos in Josephson junctions will be discussed.

1. Introduction

The following description concentrates on the properties of superconducting Josephson junctions, in particular the unique nonlinear properties that have so much promise for useful applications. Although we largely think in terms of the "old" superconductors simply because the corresponding thin film Josephson junctions are well characterized, we expect most of the following to hold also for the new high-T_C ceramic superconductors, if and when high quality thin film Josephson junctions become available. Modifications will occur, of course, because of changed parameter values.

The basic physics of superconductivity ("old" and "new") has been discussed elsewhere in this volume [1, 2] and will not be dealt with in any detail here. However, section 2

NATO ASI Series, Vol. F 59
Superconducting Electronics
Edited by H. Weinstock and M. Nisenoff

gives a brief account of the properties of the (autonomous) Josephson junction. Section 3 deals with the Josephson junction in external circuits, in particular, a cavity. Section 4 discusses the properties of a long Josephson junction, i.e., a Josephson transmission line with solitons. Section 5 introduces very briefly the topic of external pumping with time varying signals, i.e., rf properties. Finally, section 6 discusses the chaos that appears in the special case of very large pumping signals. The paper is summarized in section 7.

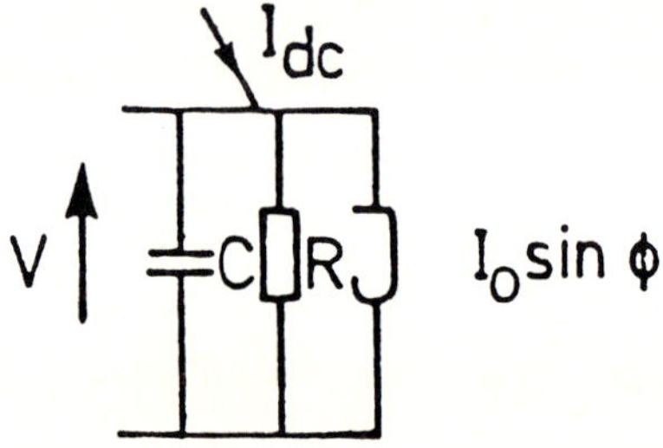

Fig. 1 Equivalent diagram for a small Josephson junction

2. The autonomous Josephson junction

The most widely investigated Josephson junction system is the current-driven Josephson junction, a description of which may be found in recent books on the subject [3,4].

The Josephson junction consists of two superconducting electrodes separated by a tunneling barrier. The equations with an external driving current I_{DC} may be written [3,4]

$$C dV/dt + V/R + I_0 \sin\Phi = I_{DC} \tag{1a}$$

$$d\Phi/dt = 2eV/\hbar, \tag{1b}$$

Equation (1b) is the famous Josephson frequency-to-voltage relation and Eq. (1a) is Kirchoff's law applied to the Josephson junction equivalent circuit shown in Fig. 1. The equation for the Josephson junction also describes other important physical systems, such as the synchronous motor, the phase-locked loop, pinned charge density waves, and the damped driven pendulum. In a Josephson junction the tunneling currents originate from two different kinds of charge carriers, Cooper pairs and normal electrons. The unique non-linear properties are due to the Cooper pair current, which may be expressed as $I_{DC}\sin\Phi$. Here Φ is the pair phase difference across the junction, and I_0 is the maximum pair current. In addition, a shunt resistance R carries a normal electron current (V/R), a capacitance C carries a capacitive current C(dV/dt), $\hbar$ is Planck's constant and e is the electron charge. With time normalized to the reciprocal plasma frequency $\omega_0^{-1} = (\hbar C/2eIo)^{\frac{1}{2}}$ and current normalized to the critical current I_0, these equations may be combined into a single dimensionless equation, [3,4]

$$\Phi_{tt} + \alpha\Phi_t + \sin\Phi = \eta \tag{2a}$$

$$v = \Phi_t \tag{2b}$$

The plasma frequency, ω_0, is the natural oscillation frequency for the Josephson junction, corresponding to the frequency in a pendulum. The McCumber parameter β_C is given by $\beta_C = 2eR^2I_0C/\hbar$, and the damping parameter α is given by $(1/\sqrt{\beta_C})$. η is the normalized (to I_0) dc bias current, and v is the voltage normalized to $\hbar\omega_0/2e$.

The dynamic behaviour of Eq. (2a, 2b) may be described in the following way. For (current) η below one, a time independent solution $\Phi = \arcsin\eta$ and voltage v = 0 is possible. For very large values of η the average voltage $\langle v \rangle$ is determined by the resistance (since the average of the capacitive current is zero, and the average value of the supercurrent is less than one), i.e., $\eta \approx \alpha\langle\Phi_t\rangle$. If we chose another normalizing frequency, the so-called characteristic frequency $\omega_C = 2eRI_0/\hbar$, and normalize the time to $1/\omega_C$ and the voltage to RI_0, we may obtain a convenient plot of a series of I-V curves with β_C as a parameter [3, 4]. These are shown in Fig. 2a. We note that for high damping, i.e., $\beta_C = 0$, the

I-<v> curve is single valued. For $\beta_C >> 0$ hysteresis occurs, i.e., for $\eta_C < \eta < 1$ a zero-voltage solution coexists with a solution at a finite voltage. The threshold bias value η_C is a function of the damping parameter, which is shown in Fig. 2b. The details of the dynamic behaviour of this so-called shunted junction model may be found in [3, 4]; it is the most important and most widely used Josephson junction model. It should be mentioned that no general analytical solution to it exists. However, the qualitative behaviour, approximation formulas, and numerical calculations are described in great detail in the literature [3,4].

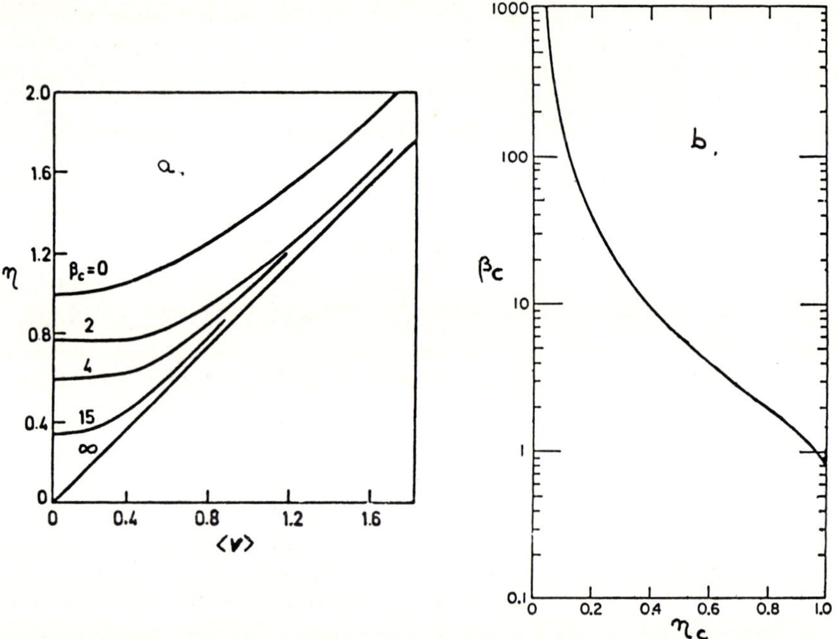

Fig. 2 (a) IV curves for the shunted junction model

(b) The McCumber curve

3. The Josephson junction with external circuits

In this section we write Eq. (2a,b) as a set of coupled first-order differential equations.

$$\dot{\phi} = v \tag{3a}$$

$$\dot{v} = \eta - \alpha v - \sin\phi - i_s \tag{3b}$$

where we have introduced i_S for the interaction with the external circuit - see Fig. 3. When $i_S = 0$,we have an isolated "bare" junction. For a Josephson junction interacting with external circuits, $i_S > 0$, additional equations describing those circuits are necessesary. An example is a Josephson junction coupled to a cavity. Its equivalent diagram is shown in Fig. 3a. With time normalized to the inverse plasma frequency, $\omega_0^{-1} = (\hbar C/2eI_0)^{\frac{1}{2}}$ it is described by Eqs. (3a,3b) plus

$$\dot{i}_S = (v - (\omega_C/\omega_0)(r'/R)i_S)/\beta_L \qquad (4)$$

for the series resonant circuit. Here r' is the cavity series resistance, R is the junction shunt resistance, and $\beta_L = 2eLI_0/\hbar$. Further, we have normalized to the characteristic frequency $\omega_C = (2eRIo/\hbar)$. For all practical purposes we have $r' << R$ (i.e., $r = r'/R << 1$) and may thus disregard the current through R. Accordingly, we will use r' instead of R in the definition of ω_C and β_C [5].

The Eqs. (3,4) then take the form of two equations for the Josephson junction, Eqs. (5a,5b) below and one for the external circuit, Eq. (5c), with a coupling between them in Eqs. (5b,5c) through i_S.

$$\dot{\Phi} = v \qquad (5a)$$

$$\dot{v} = (\eta - rv - \sin\Phi - i_S)/\beta_C \qquad (5b)$$

$$\dot{i}_S = (v - i_S)/\beta_L \qquad (5c)$$

The qualitative behaviour of a junction coupled to a cavity is the following. A Josephson junction with a dc voltage V_0 oscillates at a frequency which is given by $\omega = 2eV_0/\hbar$. When that frequency is in the vicinity of the cavity resonance frequency, a non-linear interaction with frequency locking may occur. For strong locking the oscillation frequency is determined by the cavity frequency. When the frequency is only sligthly detuned from resonance, the system may oscillate at the cavity frequency for some time, with intermittent shifts to the "Josephson frequency" [4,5]. The IV curve of such a system is shown qualitatively in Fig. 3b. The detailed dynamical behaviour of the system, Eq. (6) is described in [5].

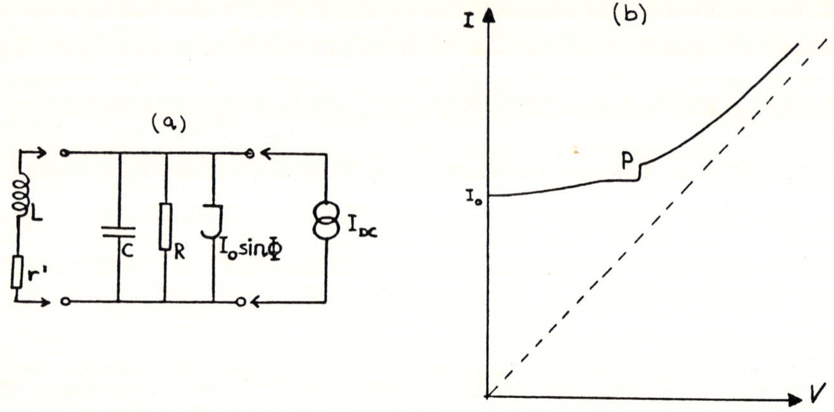

Fig. 3. Josephson junction coupled to a cavity.
(a) Equivalent diagram
(b) The IV curve with a cavity induced step near P.

4. Long Josephson junctions: spatial dependence

We have seen in the previous section that the cavity mode involves a spatial variation of the phase, although it is not explicitly visible in the lumped element equivalent circuit diagram.

As another extension of the simple Josephson junction let us consider what happens when a spatial variation of the pair phase is allowed. We may get a new type of excitation called a soliton. Indeed, the long Josephson junction, or the Josephson transmission line (JTL) is one of the physical systems where soliton propagation is accessible for direct experimental measurements [6,7]. For the purpose of this presentation it is sufficient to note that the physical manifestation of the soliton is a fluxon, i.e., a quantum of magnetic flux $\Phi o = h/2e = 2.064 \times 10^{-15}$ Vs. Moving fluxons in the Josephson transmission

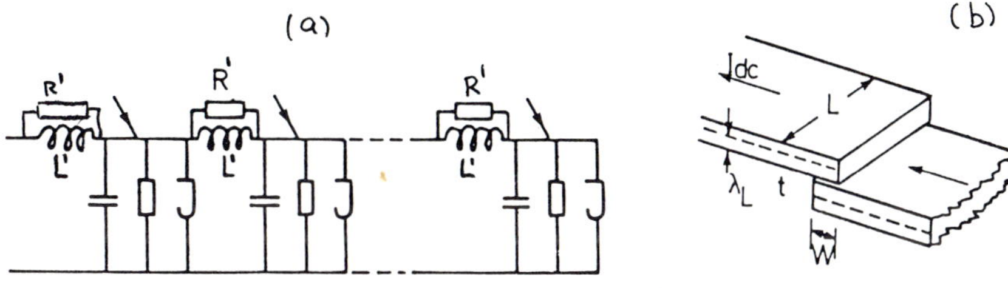

Fig. 4 (a) Equivalent diagram for a Josephson transmission line
(b) Schematic drawing of junction geometry

lines manifest themselves as the so-called zero field steps (ZFS) in the dc current-voltage characteristic of the Josephson junction - somewhat similar in appearence to the cavity step discussed in the previous section. Figure 4a shows the equivalent diagram for a JTL, and Fig. 4b shows the geometry of a long junction of the overlap type. The physical origin of the inductance L′ shown in Fig. 4a is Cooper pair currents within the London penetration layer of thickness λ_L (Fig. 4b). A possible resistance, R′ , due to a flow of normal electrons in the same layer is also shown. Taking all these circuit elements into account, the wave equation for the JTL may be written as an extension of Eq. (2)

$$-\phi_{xx} + \phi_{tt} + \sin\phi = \eta - \alpha\phi_t + \beta\phi_{xxt} \tag{6}$$

with (normalized) voltage $v = \phi_t$ (as in Eq. 2b) and (normalized) current $i = -\phi_x$. The additional normalizations used are as follows. Length is measured in units of the Josephson penetration depth $\lambda_J = \sqrt{(\hbar/2de\mu_0 J)}$ i.e., $l = L/\lambda_J$. The damping parameter β is given by $\beta = L'\omega_0/R'$, J is the current density, and d is the magnetic thickness of the junction, $d = 2\lambda_L + t$. The junction length, L, is assumed large, and the width, W, is assumed small compared to the Josephson penetration depth. Finally, velocities become normalized to the velocity of light in the barrier, $\bar{c}$, given by $\bar{c} = c\sqrt{t/d}$, where c is the velocity of light in vacuum. The expression for c reflects the fact that electric fields exist only across the tunneling barrier of thickness t, whereas magnetic fields exist in both the barrier and the penetration layers - see Fig. 4b. For typical experimental junctions (of the "old" superconductors), $\bar{c}$ is a few percent of c.

4a. Perturbation calculation for the infinite line

Soliton Dynamics

The methods in this section are based mainly on the work of McLaughlin and Scott [6]. With the right hand side equal to zero, Eq. (6) is the sine-Gordon equation. The loss and bias terms on the right hand side are considered as a perturbation to the sine-Gordon equation. The unpertubed sine-Gordon equation has the well known analytical single soliton solution [6],

$$\Phi = 4\tan^{-1} \exp\theta \qquad (7)$$

where $\theta = (x-ut)\gamma(u)$, and $(u) = 1/\sqrt{(1-u^2)}$ is the Lorentz factor. The solution gives rise to a 2π phase shift over a length of a few λ_J, and its derivative Φ_t represents a voltage pulse. Note that the form of the solution - a traveling wave in the parameter $\theta = \gamma(u)(x-ut)$ - is a consequence of the Lorentz invariance of the sine-Gordon equation. The sine-Gordon soliton behaves very much like a relativistic particle with energy H, and momentum P, given by

$$H = 8\gamma(u) \quad , \quad P = 8u\gamma(u) \qquad (8)$$

With the normalizations used here, the rest mass of the soliton is 8. In Eq. (7) the velocity u is a free parameter. As shown in [6] the perturbation terms are included by assuming a solution of the same form as that in Eq. (7), but with u to be determined by a power balance equation. Requiring either the Hamiltonian or the momentum to be independent of time, one finds the velocity (momentum) to be determined by [6,7,8]

$$\pi\eta/4 = u\gamma(u)(\alpha + \beta/(3(1 - u^2))) \qquad (9)$$

Equation (9) tells us that the velocity is determined by a balance between the losses represented by α and β and the energy input represented by the bias term . In most of what follows we will assume $\beta = 0$ and take only shunt losses into account. In that case the velocity may be found explicitly as

$$u = 1/\sqrt{(1 + (4\alpha/\pi\eta)^2)} \quad , \qquad (10)$$

Because Eqs. (9,10) are derived by a perturbational approach, they are not expected to be valid if the perturbing terms are large, i.e., if the bias term approaches one and/or the system is heavily damped. Various corrections have been considered in [9].

Another solution to the sine-Gordon equation that may be perturbed under the influence of bias and losses is the soliton-antisoliton solution that may be written [6,7,10]

$$\phi = 4\tan^{-1}(\sinh T/u\cosh X) \qquad (11)$$

where $T = u\gamma(u)t$ and $X = \gamma(u)x$. Assuming only shunt-losses - for $\beta > 0$ a calculation was done in [11] - it is possible in a manner similar to that for the single soliton case to perform a power balance calculation by requiring the time rate of change of the energy H to be zero, i.e., calculating the integral [7,10]

$$dH/dt = \int(\eta\phi_t - \alpha\phi_t^2)dx \qquad (12)$$

with ϕ inserted from Eq. (11). For this case a qualitatively new phenomenon occurs. For high incident energies the soliton and antisoliton will pass through each other with a phaseshift δ (spatial advance) given by [6]

$$= -2(1-u^2)\ \ln u \qquad (13)$$

For bias below a certain threshold, η_{TH}, to be calculated below, the soliton and antisoliton will annihilate each other, create a breather mode, and eventually die out as small amplitude damped plasma oscillations. In evaluating Eq. (12) one finds after rather lengthy calculations [10] that the collision gives rise to an energy loss ΔH, given by

$$\Delta H \cong 4\pi^2 \alpha \tag{14}$$

Part of this energy is dissipated in propagating, but decaying, oscillations of the line. The annihilation threshold [10] may be found by requiring that the total energy of the soliton and

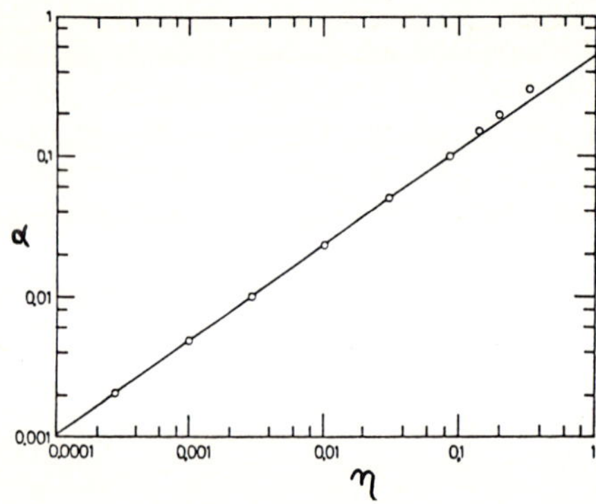

Fig. 5 Soliton-antisoliton annihilation curve on the infinite line. Full curve: Eq. 16. Circles: Numerical simulation

antisoliton before the collision, $H = 16\gamma(u)$, is equal to the energy loss plus the rest energy of a stationary soliton and antisoliton, i.e.,

$$16\gamma(u) \approx 4\pi^2\alpha + 16\gamma(0) \tag{15}$$

Eq. (15) together with Eq.(10) leads to [10]

$$\eta_{TH} \approx (2\alpha)^{3/2} \tag{16}$$

Figure 5 shows Eq. (16) together with a numerical simulation. The agreement is excellent except for α larger than approximately 0.2. Fig. 6 shows a numerical calculation of a soliton-antisoliton collision where both of the above phenomena - the energy loss and the phase shift - are easily observed.

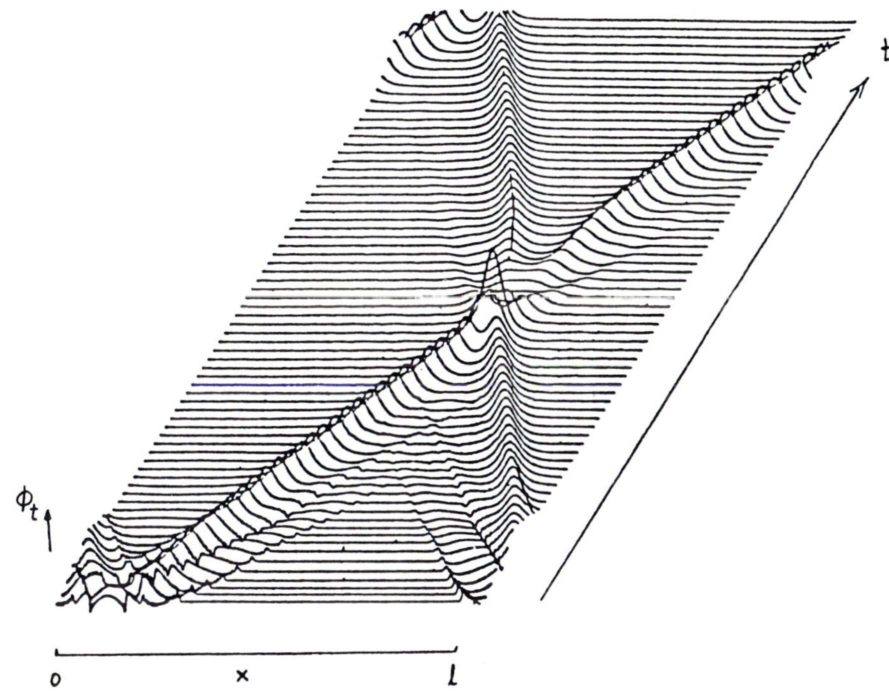

Fig. 6. Collision between a soliton and an antisoliton with α = 0.2. η = 0.22, normalized length l = 40. The wiggles for small time are transients that have not yet been damped out. (A. Davidson and N.F. Pedersen, unpublished)

4b. Soliton experiments: dc IV-curves

The overlap JTL

In the overlap junction (Fig. 4b) the bias current is uniformly distributed over the junction length, and $\eta = I_0/JWL$ may be assumed in Eq. (6). Due to the moving fluxon, a phase shift of 2π takes place in a time interval l/u, where l is the (normalized) length of the junction. This, in turn, gives rise to a (normalized) dc voltage v, given by

$$v = (2\pi/l)u \tag{17}$$

The overlap junction has boundary conditions requiring that no currents flow out at the ends, i.e.,

$$\phi_X(0,t) = \phi_X(l,t) = 0 \tag{18}$$

It may be shown that this boundary condition is mathematically equivalent to a soliton-antisoliton collision, which was treated in the previous section.

In the I-V curve the moving soliton gives rise to the so-called zero field steps (ZFS). The mechanism for the first ZFS, $n = 1$, is that a fluxon moves along the junction and is reflected at the boundary as an antifluxon. Since the reflection at $x = l$ is equivalent to a collision with a virtual antifluxon at $x = l$, the problem may be treated in the framework of Eq. (13) for the phase shift and Eq. (15) for the energy loss. If the junction length l is very large, the details at the boundaries play only a minor role, and the voltage of the first step is given by Eq. (17). For example, Pedersen and Welner [8] were able to neglect completely the effects of collisions in a comparison between experimental soliton ZFS on a very long overlap junction ($l = 45$) and perturbation theory. Fig. 7 shows an example of one of their experimental curves. If the junction length is smaller (for example of the order 5 - 10), the energy loss and the phase shift will give rise to corrections [10].

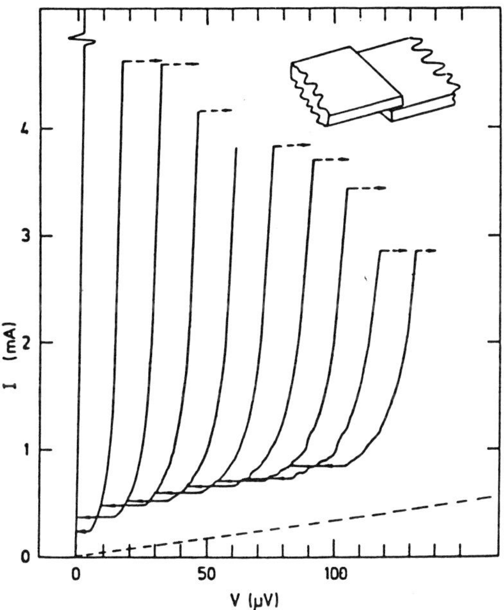

Fig. 7 Experimental zero field steps of a long overlap junction. (From ref. 8).

The annular JTL

This circular geometry, which looks mostly like an overlap junction that is folded back into itself, has the the simple periodic boundary conditions

$$\phi_X(0,t) = \phi_X(l,t) + 2p\pi \qquad (19)$$

used by many authors [12-13]. Here p gives the number of full phase rotations along the line. For topological reasons p is a conserved number; for example in experiments it may only be changed by taking the junction through the transition temperature, thus changing the superconducting wavefunction completely.

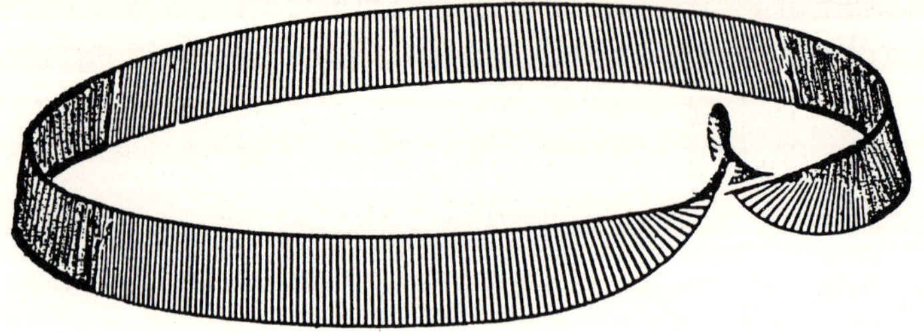

Fig. 8 Computer-generated equivalent pendulum array with a 2π kink. $\alpha = 0.02$, $\beta = 0.01$, $\eta = 0.4$, and $l = 8$. (From ref. 14).

The simplest case to consider is p = 1, i.e. a single fluxon on the circular line. This case is shown in Fig. 8 [14]. In that case there is no supercurrent, since as soon as a uniform bias current is applied, the fluxon starts moving with a velocity u determined by Eq. (10) and a voltage (Eq. (17)) develops. Hence for the annular junction with one trapped fluxon no supercurrent exists and the dc voltage is a direct measurement of the velocity of the single soliton [15]. The IV curve is shown qualitatively in Fig. 9. In addition to the single soliton, further solitons may be created only by introducing soliton-antisoliton pairs (for topological reasons), in which case the effects of collisions must be taken into account. Disregarding the collisions for simplicity, the voltages of these different configurations are given by multiplying the voltage in Eq. (17) with the total number n of fluxons and antifluxons, i.e., voltage steps are to be expected at voltages v_1, $v_3 = 3v_1$, $v_5 = 5v_1$, etc. (shown as the dashed curves in Fig. 9). For the higher order branches, v_3, v_5, ... the qualitative effect of the collisions is to lower the average voltage somewhat compared to nv_1, where n = 3, 5, 7,... Also a lower bias threshold η_{TH}, where a fluxon and an antifluxon annihilate each other (c.f. Eq. (16)), is to be expected. Figure 9 shows qualitatively the higher order steps (full curves) based on these arguments. Reference 13 shows for p=2 a numerically simulated IV curve which contains all the essential features of Fig. 9.

Experimental measurements of solitons on the annular junction have been reported [14,15]. In the experiment the p value could be changed only by taking the junction through the transition temperature. p = 1 appeared qualitatively as discussed above. p = 0 (zero fluxons trapped) showed the full supercurrent and fluxon-antifluxon steps at voltages $v_2 \approx 2v_1$, $v_4 \approx 4v_1$ ... etc.

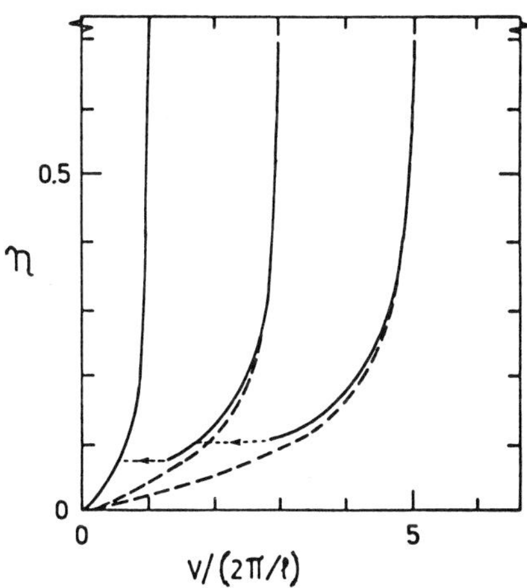

Fig. 9 Qualitative IV curve of the annular JTL with one trapped soliton. Details of the curves are discussed in the text. (From ref. 7).

The long annular junction experiment demonstrates in a very clean way the existence of a topological sine-Gordon soliton with a phase change of 2π. The experiment and the interpretation are elegantly connected with fundamental theory which requires the phase to change only in multiples of 2π around along a superconducting ring [1-4].

4c. Direct sampling measurement of the soliton waveform

The small magnitude of the flux quantum, $\Phi_0 \approx 2 \cdot 10^{-15}$ Vs, presents some severe limitations on the bandwidth and sensitivity of the amplifiers to be used. Most of the experimental work has been performed in Japan. The first direct measurement of a single fluxon on a JTL, was done in 1982 [16] by using conventional room temperature electronics and minicomputer signal processing.

Some details of the soliton-antisoliton collision [10,11], as discussed in section 4a, and threshold properties [17] have been confirmed by a direct measurement on the JTL with room temperature electronics. For example, the collision time delay due to dissipation (arising from the energy loss in Eq. (15)) was measured and found to agree with simulations.

Almost simultaneously a different experimental approach was taken by another Japanese group [18]. They fabricated a Josephson pulse generator and a Josephson sampling circuit directly on the same substrate as the JTL. They obtained a time resolution of less than 10 ps and a current sensitivity of a few microamps. They also investigated fluxon dynamics [19] by evaporating a resistor (of few micrometers length) on top of the JTL. By means of a control current supplied to the resistor it was found that a fluxon could be accelerated, made to pause, or even reverse its direction of propagation. Details of the reflection at an open boundary (section 4b) [20] and in particular of the fluxon-antifluxon collision [21] has been revealed by direct measurement with on-chip Josephson electronics. In that way most of the dynamical behaviour discussed in the previous sections now has been confirmed in direct measurements. Thus, Josephson soliton electronics can become a reality.

4d. Applications of solitons on the JTL

Solitons on the JTL may very well have technical applications. We will briefly discuss three possible applications that have emerged. These are (i) microwave oscillators and amplifiers, (ii) digital information processing, and (iii) analog amplifiers.

Microwave oscillators and amplifiers

Several designs of microwave oscillators based on the properties of fluxons in overlap or annular JTL's exist [6,22,23]. A particularly promising scheme is the flux flow oscillator [24] with a demonstrated performance of as much as 10^{-6} watt available on the substrate at frequencies tunable between 100 and 400 GHz. This is far superior to the results obtained for other technologies results and is sufficient for a pump source in an integrated Josephson junction millimeter wave receiver.

Digital information processing

The basic idea behind digital applications is the use of the soliton as the basic bit of information. In the early seventies the so-called flux shuttle was proposed by Fulton et al. [25]. In the flux shuttle, fluxons are situated in potential wells created by perturbing the geometry of the JTL at desired positions. The fluxons may be moved around and manipulated by applying currents and magnetic fields. Results described in [19] demonstrate that this is possible.

Analog amplifiers

The Josephson junction has been demonstrated to have superior properties in almost all areas of electronics. It is therefore remarkable that the fundamental element - a Josephson transistor - does not directly exist. However, Likharev et. al. [26] suggested that an overlap JTL with current injection at many points in parallel is an almost complete analog of a semiconductor transistor, where the role of electric charge carriers is being played by fluxons. In that scheme the control current is being applied to a film on top of the upper electrode, but is isolated from it. Somewhat similar concepts have been investigated experimentally [27-29]. Current gain of order 2-5, very fast response, and low power dissipation have been found.

5. rf properties of Josephson junctions

A very important application of Josephson junctions involves rf-devices. Since this topic is covered separately by Lukens [30] for millimeter wave sources and by Gundlach [31] for SIS mixers and detectors, I will mention only, that, for the latter devices, the situation is the same as in the next section, which deals with chaos. The main difference is that the pump power levels are below the threshold for chaos, though often not very far below for best performance. In the next section on chaos we will consider the situation when the pump strength is above the threshold.

6. Chaos in Josephson junctions

Another non-linear signature appearing quite often in Josephson junctions is chaos together with its accompagnying bifurcations. In fact, quite often the Josephson junction is used as a model system for chaos in numerical simulations. A particular feature of chaos in Josephson junctions is that both the effect of thermal noise and the effect of deterministic noise (chaos) are very important for experiments. The interplay between those two sources of noise is at best very complicated, and at worst makes it impossible to interpret experiments. This has led to new theoretical and numerical work on the non-linear interaction between thermal and deterministic noise.

6a. Deterministic chaos in the Josephson junction

The most widely investigated Josephson junction system is the rf-driven Josephson junction for which the equation may be obtained by adding a term $I_{RF}\sin\omega t$ to Eq. (2a), i. e. in normalized units ($\Omega = \omega/\omega_0$)

$$\Phi_{tt} + \alpha\Phi_t + \sin\Phi = \eta_0 + \eta_1 \sin\Omega t \tag{20}$$

Since analytical solutions do not exist, one has to do numerical simulations in the four dimensional parameter space of α, η_o, η_1, and Ω [32-37]. A particularly thorough investigation of the parameter space was done in Ref. [37]. Typically, the system has been investigated numerically in the η_1 - Ω plane for a fixed damping parameter α, a plot which more or less has become a standard for such systems [32-35]. Figure 10 shows such a plot

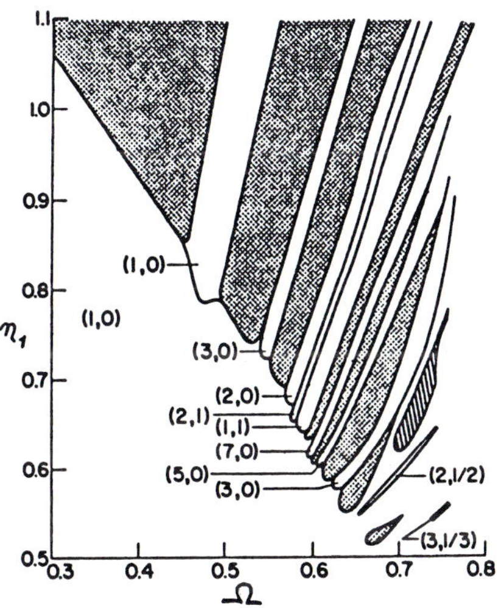

Fig. 10 Characterization of solutions in the Ω-η_1 plane for β_c = 25, η_o = 0. Crosshatched region: chaos. Hatched region: complicated periodic. Indexing (p,q) corresponds to the pth subharmonic on the qth rf-induced step. (From ref. 33).

with its complicated mapping of different dynamical behaviour. Note that for $\omega > \omega_o$, i.e., $\Omega > 1$, the threshold rises dramatically because the capacitor shorts out the applied rf current. For $\omega < 1/RC$, i.e., $\Omega < 1/\sqrt{\beta_c}$, the system is able to follow adiabatically the rf current, and chaos occurs only if $\eta_1 > 1$. For $\omega \approx \omega_0$, the threshold for chaos is lowest.

Another method of a somewhat computational nature is to use electronic analogs simulating the Josephson equation. Such systems have the advantage of being very fast, and Poincaré sections and bifurcation diagrams may be readily displayed [34,38]. The disadvantage is the limited precision and resolution, and the drift of analog electronic circuits.

6b. Thermally affected chaos in the Josephson junction

In ref. [39] a Josephson junction system with parameters such that two solutions existed, was investigated. The authors found that the basin boundaries between the two solutions were fractal, and thus the solutions could come infinitessimally close to each other in the phase plane. Under such circumstances a small amount of thermal noise may take the system back and forth between the solutions. The authors found that this mechanism gave rise to approximately 1/f noise for some parameter regions.

In another extensive numerical simulation including a thermal noise term in Eq. (2), Kautz [40] was able to obtain the very high noise temperatures ($\simeq 10^6$ K) that have been observed experimentally [41]. For a situation with overlapping rf-induced steps, the origin of the very high noise temperatures [40] was hopping between phase locked and metastable chaotic states induced by thermal fluctuations. This may even lead to the surprising result that the low frequency noise power increases as the temperature is reduced.

In the absence of thermal noise, numerical calculations of chaotic regions in the IV curve typically contain a wealth of complicated structure displaying bifurcations, chaos, periodic solutions, etc. This may be seen in Fig. 11, which shows a numerical calculation of an rf-induced step with loss of phase lock [42]. In experiments that have been performed, such interesting and complicated structure have been typically washed out because of thermal noise, and only a smooth curve which does not in a simple and convincing way demonstrate chaos, is obtained. Thermal smoothing due to a temperature of less than 100mK is sufficient to remove most of the traces of complicated dynamical behaviour. By comparing experiments with a calculation that includes thermal noise, however, the existence of chaos may be shown indirectly [42].

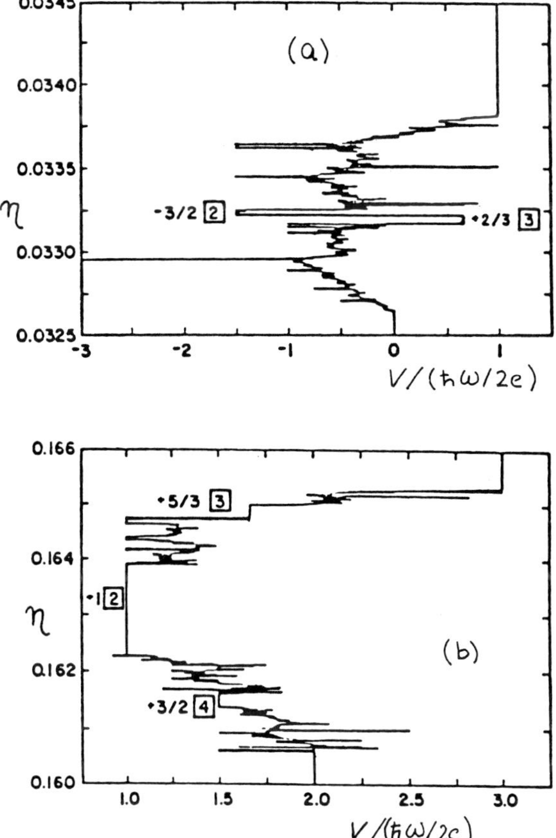

Fig. 11 Typical results of a numerical calculation for zero temperature, Ω = 0.16, β_c = 4, η_1 = 1.05: (a) the zeroth rf-induced step and (b) the second step. The numbers inside the squares denote the periodicity on the substep. (From ref. 42).

6c. Experiments on real Josephson junction

Common to all the experimental results is that they are not nearly as spectacular as the numerical simulations. The main reason is that thermal noise, which is most often not taken into account in simulations, has a major effect on the outcome of the experiments. This is because the energy levels in the thermal oscillations may very well be of the same order of magnitude as the intrinsic energy levels in the Josephson junction, and complicated non-linear interactions occur.

Thermal effects may produce not only quantitative changes, but also qualitative and quite dramatic changes as we shall see below.

(i) dc observation of chaos

Before the term chaos was connected to Josephson junctions, researchers sometimes noted very irregular and erratic IV curves in samples subject to strong applied rf signals. In many cases such junctions were discarded because of assumed defects during fabrication.

It is now known that such irregular IV curves may be a signature of chaos. Examples of such irregular behaviour, in particular the loss of phaselock on an rf-induced step, may be found in ref. [42-44].

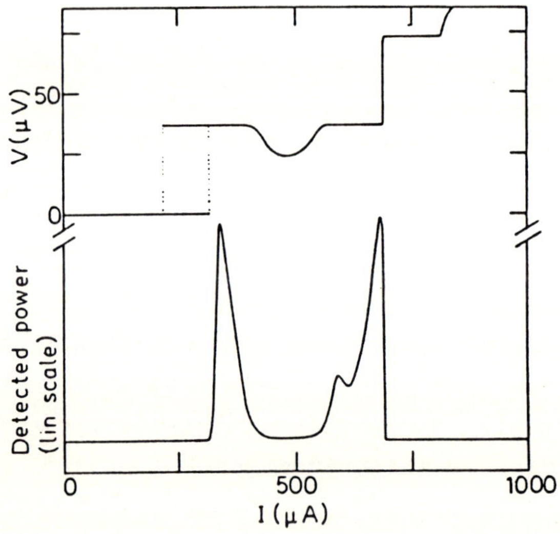

Fig. 12. Experimental microwave irradiated IV curve and corresponding half-harmonic generation. f = 17.6 GHz, T = 3.54K. (From ref. 44).

By comparing such experimentally obtained, irregular IV curves to numerically obtained ones, one has in principle the simplest experiment on chaos [42]. An example of such a dc experiment is illustrated in Fig. 12, which shows an experimentally obtained IV curve [44] with loss of phaselock on the rf-induced step. Also shown in the figure is the spectrum of half-harmonic generation as measured with a sensitive microwave spectrometer. Note that these experimental curves, which contain two period-doubling bifurcations and a chaotic region on an rf-induced step, can be considered as a standard example of the period-doubling route to chaos. These experimental results are very similar to the numerical results shown in Fig. 1 of ref. 36 and to analog results [38].

(ii) Chaos and parametric amplification

For Josephson junction parametric amplifiers, low noise temperatures were found in some cases, however, more often experiments showed considerable excess noise [45]. For experimentalists the observed noise rise has been a major puzzle. A large number of theoretical and numerical papers [46] have dealt with the problem. The conclusion is that the very large noise temperatures cannot be explained by traditional noise sources such as Johnson noise, shot noise, or quantum noise. Hubermann et al. [32] first suggested that chaos was the origin of the excess noise. This suggestion was further substantiated in ref. 33.

More recently a slightly different type of Josephson junction parametric amplifier was investigated [47] by another group, which drew the conclusion that noise in this amplifier cannot arise from deterministic chaos alone. The observed noise increase required the presence of thermal noise. They suggested that the noise increase was due to thermally induced hopping between a bias point that would be stable in the absence of thermal noise and an unstable point. This observation demonstrates the importance of thermal noise in modeling chaos in Josephson junction systems. A noise temperature of as much as 10^6K may be obtained.

(iii) Other Josephson junction systems

One of the first experiments to demonstrate chaos in Josephson junctions was done on a system different from that used to yield Eq. (16). Miracky et al. [5,48] used a junction shunted with a resistor having a substantial self inductance - see Fig. 3a,3b - i.e., a junction coupled to a cavity as described in section 3. By varying the bias current they found (experimentally) very large increases in the low frequency voltage noise, with noise temperatures as high as 10^6K or beyond. The excess noise arose from switching between subharmonic Josephson relaxation modes. More moderate noise increases (10^3K) could be characterized as noise affected chaos. The experiment was done in a 1 GHz bandwidth where low noise amplifiers and frequency-independent coupling is available.

Simulations of such systems indicated that for certain bias points the addition of thermal noise gave rise to an approximately 1/f noise spectrum by creating hopping between subharmonic modes.

7. Conclusion

This paper has discussed mainly the non-linear properties of Josephson junctions, which have so much promise for both applications and continued research on fundamental problems. The problems we have dealt with have all been defined on the basis of the "old" superconductors. Future work involving the new high T_c superconductors will most likely be required to deal not only with the same type of problems for different parameters, but also with completely new non-linear phenomena due, for example, to anisotropy. It may be safely predicted that much of interesting non-linear physics lies ahead.

References

1. J.R. Clem, this volume.
2. Y. Bruynseraede, this volume.
3. A. Barone and G. Paterno, Physics and Applications of the Josephson Effect, New York: Wiley-Interscience, 1982.
4. K.K. Likharev, Dynamics of Josephson junctions and Circuits. Gordon and Breach, New York (1986).
5. R.F. Miracki, M.H. Devoret, and J. Clarke, Phys. Rev. A31, 2509 (1985)
6. D.W.McLaughlin, and A.C.Scott, Phys.Rev. A18, 1652 (1978)
7. N.F. Pedersen in " Solitons in Action" MPCMS, Eds. A.A. Maraduddin and V.H. Agranovich, North Holland, 1986. p 469.
8. N.F. Pedersen and D. Welner, Phys.Rev. B29, 2551 (1984)
9. M. Buttiker and H. Thomas, Phys. Rev. A37, 235 (1988)
   M. Buttiker and R. Landauer, Nonlinear Phenomena at Phase Transitions and Instabilities, (Plenum Publishing Corp.), 1982, p.111
10. N.F. Pedersen, M.R. Samuelsen, and D. Welner, Phys.Rev.B30, 4057 (1984)
11. A. Matsuda, Phys. Rev. B34, 3127 (1986)
12. P.M.Marcus and Y.Imry, Sol.Stat.Comm. 33, 345 (1980)
    S.E. Burkov and A.E. Lifsic, Wave Motion 5, 197 (1983).
13. A. Davidson and N.F. Pedersen, Appl. Phys. Lett. 44, 465 (1984)
14. A. Davidson, B. Dueholm, and N.F.Pedersen, J. Appl. Phys. 60, 1447 (1986)
15. A. Davidson, B. Dueholm, B.Kryger, and N.F.Pedersen Phys. Rev. Lett. 55, 2059 (1985)
16. A. Matsuda and S. Uehara, Appl. Phys. Lett. 41, 770 (1982).
17. J. Nitta and A. Matsuda, Phys. Rev. B35, 4764 (1987)
18. S. Sakai, H. Akoh, and H. Hayakawa, Jap. Journ. of App. Phys. 22, L479 (1983).
19. H. Akoh, S. Sakai, A. Yagi, and H. Haykawa, IEEE Trans. Magn. **MAG-21**, 737 (1985).
20. H. Akoh, S. Sakai, and S. Takada, Phys. Rev. B35, 5357 (1987)
21. A. Fujimaki, K. Nakajima and Y. Sawada, Phys. Rev. Lett. 59, 2895 (1987)
22. S. Sakai, H. Akoh, and H. Hayakawa, Jap. Journ. of App. Phys. 23, L610 (1984).
23. B. Dueholm, O.A. Levring, J. Mygind, N.F. Pedersen, O.H. Soerensen, and M. Cirillo, Phys. Rev. Lett. 46, 1299 (1981).
24. T. Nagatsuma, K.Enpuku, K.Yoshida, and F.Irie, J. App. Phys. 56, 3284 (1984).
    T. Nagatsuma, K.Enpuku, K. Sueoka, K.Yoshida, and F.Irie, J. App. Phys. 58, 441 (1985).
25. T.A. Fulton,, R.C. Dynes, and P.W. Anderson, Proc. IEEE 61, 28 (1973).
26. K.K. Likharev, V.K. Semenov, O.V. Snigirev, and B.M. Todorov, IEEE Trans. Magn. **MAG-15**, 420 (1979).
27. T.V. Rajeevakumar, App. Phys. Lett. 39, 439 (1981).
28. B.J. van Zeghbroeck, IEEE Trans. Magn. **MAG-21**, 916 (1985).

29. D.P. McGinnis, J.E. Nordman and J.B. Beyer, IEEE Trans. Magn. **MAG-23**, 699 (1987)
30. J. Lukens, this volume.
31. K. H. Gundlach, this volume.
32. B.A. Huberman, J.P. Crutchfield, and N.H. Packard, Appl. Phys. Lett. 37, 750 (1980)
33. N.F. Pedersen and A. Davidson, Appl. Phys. Lett. 39, 830 (1981)
34. D. D'Humieres, M.R. Beasley, B.A. Huberman, and A. Libchaber, Phys. Rev. A26, 3483 (1982)
35. R.L. Kautz, J. Appl. Phys. 52, 3528 (1981)
M. Octavio, Phys. Rev. B29, 1231 (1984)
Kazuo Sakai and Yoshihiro Yamaguchi, Phys. Rev. B30, 1219 (1984)
36. R.L. Kautz, J. Appl. Phys. 52, 6241 (1981)
37. R.L. Kautz and R. Monaco, J. Appl. Phys. 57, 875 (1985)
38. M. Cirillo and N.F. Pedersen, Phys.Lett. 90A, 150 (1982)
W.J. Yeh and Y.H. Kao, Appl. Phys. Lett. 42, 299 (1983)
H. Seifert, Phys. Lett. 98A, 213 (1983)
Da-Ren He, W.J. Yeh and Y.H. Kao, Phys. Rev. B31, 1359 (1985)
V.K. Kornev, K.Yu. Platov and K.K. Likharev, IEEE Trans. Magn. MAG-21, 586 (1985)
39. M. Iansiti, Quing Hu, R.M. Westervelt, and M. Tinkham, Phys. Rev. Lett. 55, 746 (1985)
40. R.L. Kautz, J. Appl. Phys. 58, 424 (1985)
41. M. Octavio and C. Readi Nasser, Phys. Rev. B30, 1586 (1984)
42. D.C. Cronemeyer, C.C. Chi, A. Davidson, and N.F. Pedersen Phys. Rev. B31, 2667 (1985)
43. K. Okuyama, H.J. Hartfuss, and K.H. Gundlach, J. Low Temp. Phys. 44, 283 (1981)
44. N.F. Pedersen, O.H. S¢rensen, B. Dueholm, and J. Mygind, J. Low Temp. Phys. 38, 1 (1980)
45. R.Y.Chiao, M.J. Feldman, D.W. Peterson, B.A. Tucker, and M.T. Levinsen, Future trends in superconductive electronics, AIP Conference Proceedings 44, 259 (1978)
46. For references on this see e.g. N. F. Pedersen in SQUID 80, edited by H.D. Hahlbohm and H. Lubbig (de Gruyter, Berlin, 1980), p739.
47. R.F. Miracky and J. Clarke, Appl. Phys. Lett. 43, 508 (1983)
48. R.F. Miracky, J. Clarke, and R.H. Koch, Phys. Rev. Lett. 50, 856 (1983)

APPLICATION OF JOSEPHSON EFFECT ARRAYS FOR SUBMILLIMETER SOURCES

J.E. Lukens, A.K. Jain, K.L. Wan
Department of Physics
State University of New York at Stony Brook
Stony Brook, New York 11794

I. Introduction

As the most fundamental of the Josephson equations (Eq. 1) immediately makes clear, Josephson junctions are intrinsically voltage controlled oscillators. The average voltage $\bar{V}$ across the junction is related to the frequency of supercurrent oscillation ν in the junction by this Josephson equation,

$$\bar{V} = \frac{h}{2e}\nu \ , \quad (1)$$

through the fundamental constants e (the electron charge) and h (Planck's constant). Indeed this remarkable result has been shown[1] to be independent of the materials or structure of the junctions being used with a precision of better than 1 part in 10^{15}. One would thus hope that Josephson junctions would be useful tunable sources operating up to the superconducting gap frequency – a few THz for conventional superconductors and perhaps tens of THz for the new high T_c materials.

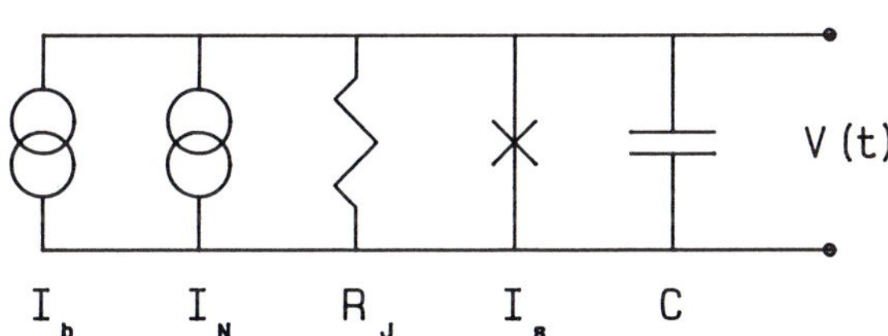

Fig. 1. Equivalent circuit model for resistively shunted junctions (RSJ) with bias current I_b, noise current I_N, shunt resistance R_J and capacitance C and supercurrent $I_s = I_c \sin\phi$.

Although a great deal of work in many laboratories has gone into the development of practical sources based on the Josephson effect, the serious problems of the very

NATO ASI Series, Vol. F 59
Superconducting Electronics
Edited by H. Weinstock and M. Nisenoff

low power and source impedance of these junctions have been difficult to surmount. One technique for overcoming these problems, which will be the focus of this lecture, is to use arrays of junctions in place of single junction sources. First, the properties of single junction sources along with the results of the initial attempts to use arrays for sources will be reviewed in order to get a perspective on the problems to be solved. Next, perturbative techniques for analyzing arrays will be developed based on the RSJ model without capacitance. While this model is clearly only approximately correct for real tunnel junctions, it can produce analytic results useful for achieving an insight into the design of coherent arrays. In addition, these results have provided provide a very good description of many of the experiments to date. The final part of this chapter will explore a number of questions which are important for practical arrays; among these are effects of junction capacitance and effect of having the junctions distributed over many wavelengths.

II. Single junction sources

The properties of single junction sources will be described first in order to see where arrays may be useful. The RSJ model (Fig. 1) with $C = 0$ will be used to describe

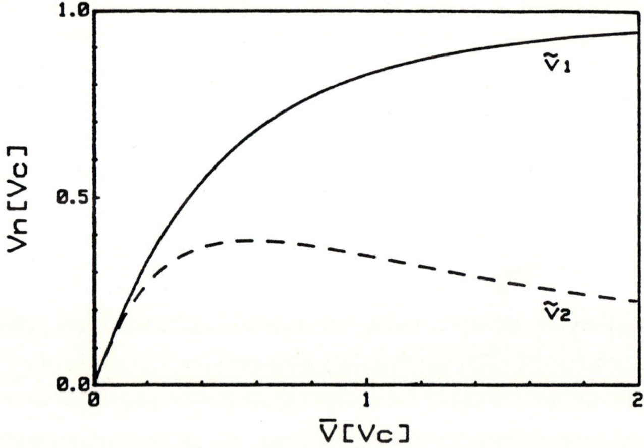

Fig. 2. Fundamental and second harmonic of junction voltage (RSJ, $C = 0$) vs. dc voltage across junction.

the junction's behavior since, for this case, analytic solutions exist[2] which will provide a useful guide to understanding. When the bias current I_b is increased above I_c, the junction's phase ϕ begins to increase with time, producing an oscillating supercurrent (since $I_s = I_c \sin(\phi)$) and consequently an oscillating voltage. For I_b near I_c the voltage waveform is nearly a spike, having a large harmonic content. As I_c is increased the higher harmonic content of the waveform decreases, giving a nearly sinusoidal wave for $\bar{V} \geq V_c \equiv I_c R_J$. The amplitudes of the harmonics are given by

$$\tilde{V}_n \equiv V_c \tilde{v}_n = V_c \frac{2\bar{v}}{[(1+\bar{v}^2)^{\frac{1}{2}}+\bar{v}]^n} \qquad (2)$$

where, for the RSJ model, $\bar{v} \equiv \bar{V}/V_c = (i^2 - 1)^{\frac{1}{2}}$, $i \equiv I_b/I_c$. Also, $\omega_c \equiv 2\pi V_c/\Phi_0$. As can be seen in Fig. 2, which shows the first two harmonics, the waveform has become nearly sinusoidal by $\bar{v} = 1$. Since for most applications one would like a reasonably sinusoidal source, we will impose the constraint that $\bar{v} \geq 1$ at the desired operating frequency.

Since the junction impedance at the Josephson frequency is $Z \simeq R_J$ for $\bar{v} \geq 1$, at high frequencies the junction can be viewed as an oscillator of amplitude $\tilde{V}_1$ at the Josephson frequency in series with R_J as shown in Fig. 3a. Thus the power available to a matched load from a single junction is

$$P_1(\nu) = \frac{1}{8}\frac{\tilde{V}_1^2}{R_J} \to \frac{1}{8} I_c^2 R_J, \quad \bar{v} \geq 1 \; . \qquad (3)$$

There are, however, limits on both I_c and R_J which limit the maximum power obtainable at a given frequency. The discussion so far has assumed that the junction is one dimensional, that is, the phase difference across the junction is independent of the position in the electrode transverse to the direction of current flow. This in general means that the junction size is limited to several times the Josephson penetration depth, which decreases with increasing current density. This limits the useful critical current to several milliamps, above which phenomena such as flux flow appear, and much of the power is dissipated in the junction rather than being coupled to the load.

Having fixed I_c, R_J should be adjusted using a shunt resistor depending on the desired operating frequency and limited, of course, by the material-dependent intrinsic $I_c R_J$ product of the junction. The condition $\bar{v} \geq 1$ for a sinusoidal waveform limits the maximum resistance for a given I_c and frequency ν. Further, if R_J is reduced below the value for which $\bar{v} = 1$, Eq. 2 indicates that the power will decrease. Thus, one should select the shunt resistance such that

$$R_J \simeq \nu\Phi_0 / I_c \; , \qquad (4a)$$

giving $$P_1 \simeq 1\text{nW/GHz} \; , \tag{4b}$$

where the maximum for I_c has been taken as 4ma. For example, one might expect a maximum power of about 1μW at 1THz. While this is already enough for some applications, it is about the limit, with proportionally less power available at lower frequencies. Also, the source impedance for this 1 μW source would be less than an ohm. This would require a substantial transformer ratio for typical loads and could be a problem if wide tuning were desired.

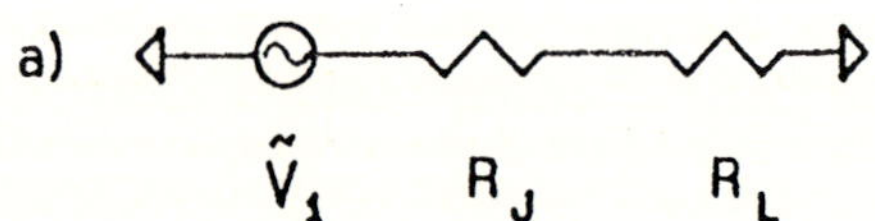

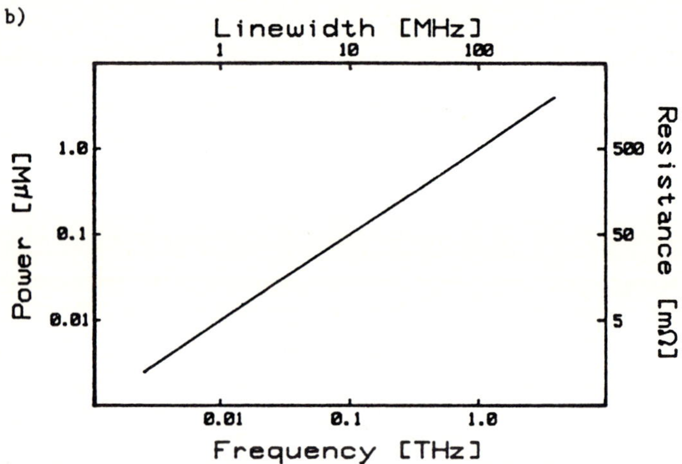

Fig. 3. a) Equivalent circuit for a junction near the Josephson frequency ($\bar{v}\geq 1$) with voltage source $\tilde{V}_1$ and rf impedance R_J driving a load R_L. b) Power, linewidth and source impedance of the junction for a matched load using the model in a) with parameters described in the text.

A final consideration related to single junction sources is their linewidth. The linewidth of the Josephson radiation is determined by frequency modulation due to low frequency voltage noise across the junction,[3] with frequencies up to about the linewidth $\Delta\nu$ being important, where

$$\Delta\nu = \frac{1}{2}\left(\frac{2\pi}{\Phi_0}\right)^2 S_I(0)\, R_d^2 \; . \tag{5}$$

Here $S_I(0)$ is the low frequency current noise spectral density, and R_d is the differential resistance at the operating voltage. The lower limit for $\Delta\nu$, when the current noise is just the Johnson noise of the junction resistance, is shown in Fig. 3 along with the maximum power and junction impedance as a function of operating frequency. This is only a rough guide, since S_I at low frequencies will in general be increased due to such things as 1/f noise prevalent in high J_c junctions, as well as to down-converted quantum noise from near the Josephson frequency. On the positive side, since only low frequency noise is important in determining $\Delta\nu$, one can in principle make the linewidth arbitrarily small by shunting the junction at low frequencies without reducing its high frequency impedance. This technique has been successfully used, although it can have drawbacks.

One can thus summarize the properties of single junction sources, as shown in Fig. 3, by saying that in general such sources have too little power, too low an impedance and too broad a linewidth, although the power and impedance begin to become useful for some applications as terahertz frequencies are approached. Next we will take a brief rather elementary look at arrays to see to what extent the replacement of single junctions by arrays of junctions might solve these problems.

III. Arrays

Interest in Josephson arrays was sparked in the late 1960's by a paper by Tilley[4] who predicted superradiance in such arrays, much as in a collection of atoms in a cavity. One signature of this superradiance was a prediction that the output power would scale as the square of the number of junctions. This led to a hope – rather naive in retrospect – that significant power levels could be obtained from Josephson junctions simply by connecting a large number of junctions together without worrying in detail about just how they were coupled. The initial experiments were done by Clark[5] on arrays of superconducting balls, Josephson-coupled through their oxide coating. These indeed showed evidence of interactions among the junctions. Experiments of this type have, however, never produced significant levels of power. The development of Josephson-effect arrays, involving hundreds of workers, has been covered in detail in two review papers by Jain, Likharev, Lukens and Sauvageau (JLLS)[6] and by Lindelof and Hansen,[7] and more recently in a book by Likharev.[8] Readers are referred to these sources for a comprehensive review.

In more recent work, including successful attempts to obtain increased power from arrays of junctions, the junctions are simply treated as classical oscillators, i.e., the junction's current and phase are classical variables as in the discussion of single-junction sources above. This is the approach taken throughout this chapter. It is worth emphasizing the distinction between the present work based on the classical picture and the initial discussion of Josephson arrays in terms of superradiance, since much confusion has been caused over the years by not fully appreciating this distinction. This confusion has been compounded by the fact that there have been observations of the power from arrays increasing as N^2, as predicted by superradiance. As far as we know all these observations can be explained in terms of purely classical circuit analysis, as we shall see, for example, below.

To begin the study of classical arrays, imagine stringing a number of junctions together in series to drive a load as shown in Fig. 4. The immediate effect of this, assuming all of the junctions oscillate in phase, is to increase the source impedance to

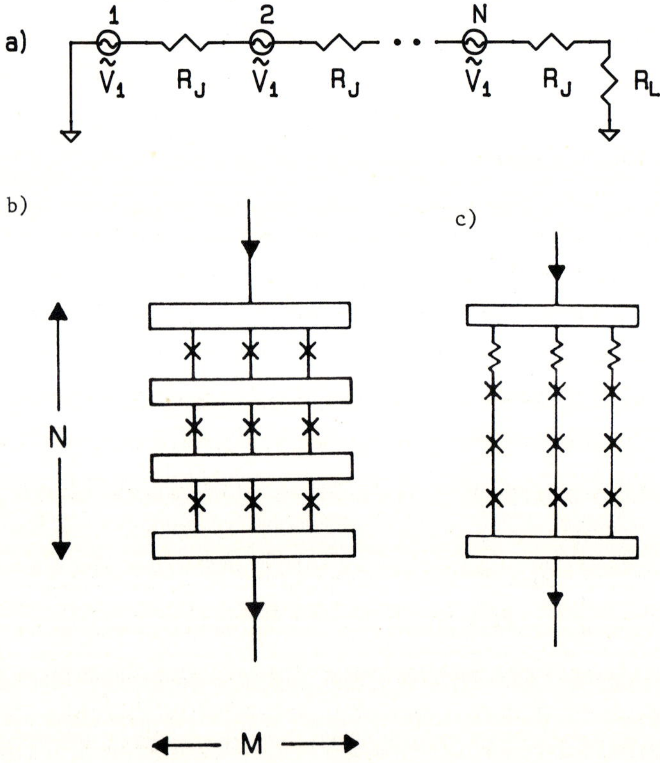

Fig. 4. a) High frequency equivalent circuit for a one-dimensional array. b,c) Two possible junction layouts for a 2D linear array.

more reasonable levels. If one chooses the number of junctions N equal to R_L/R_J then an impedance match is achieved without the use of transformers. Also, since the available power from a junction is not really changed by placing it in an array, the total power delivered to the load increases by a factor of N (as can be verified easily by replacing each junction by its equivalent circuit shown in Fig. 3a). As can be seen from this model, one example where the power for an array of classical Josephson junctions increasing as N^2 occurs for the case with $NR_J \ll R_L$. To estimate what one might expect from a *matched* array, consider an array of junctions matched to a 50 Ω load. Then

$$N = \frac{R_L}{R_J} \approx \frac{10^5}{\nu[\mathrm{GHz}]} , \tag{6a}$$

and the power to the load is

$$P_N = NP_1 \approx \frac{1}{8} I_c^2 R_L \approx 0.1\mathrm{mW} . \tag{6b}$$

Thus, the number of junctions needed to achieve a match varies inversely with frequency, but the power to be expected from such an array, optimized for a given frequency, would be about 0.1 mW, independent of frequency. The numerical estimates have been obtained using the constraints and maximum I_c from the discussion of single junctions above. A discussion of the linewidth expected from arrays will be deferred until we have discussed phase-locking. However, as noted above, it is not an intrinsic property of the array since it can be affected by low-frequency shunting.

Next, we ask whether anything is gained by replacing the one-dimensional array by a two-dimensional array, as in Figs. 4b,c. We imagine making such an array by replacing each junction in Fig. 4a by a parallel string of junctions, M junctions wide amd tramsverse to the rf current. If all of these junctions were identical and all oscillated in phase, this would be equivalent to replacing each junction in Fig. 4a with a junction having $I_c' = MI_c$ and $R_J' = R_J/M$. One would then need M times as many of these series junctions to match the load; thus the power delivered to the load would be increased by a factor M^2.

As an example, at 1 THz approximately 100 junctions would be required in a one-dimensional array to match a 50 Ω load producing in a power of 0.1 mW. If a two-dimensional array were used with M = 100, then the matched array would have 10^6 junctions and deliver 1 watt of power. It is not difficult to fabricate a million-junction array with modern lithographic techniques. The real question is whether all of the junctions could be made to oscillate in phase as assumed above, particularly since the motivation for thinking about a two-dimensional array is that the useful critical current of a single junction is limited due to phase instabilities that arise at larger values of I_c.

The arrays discussed above should be considered as linear arrays, since for proper operation the phase should vary only in one direction, even in the two-dimensional array. It is important to distinguish this situation from truly two-dimensional arrays where the phase varies along both dimensions. There has been a great deal of very interesting work, primarily to study phase transitions, in these latter arrays. This work will be completely ignored here, since it really does not address the problems related to using Josephson arrays as radiation sources. A second important topic which will be neglected here is that of long junctions. A number of groups have observed radiation from such junctions, which can be thought of as a limiting case of a parallel array. These junctions are discussed in a separate chapter of this volume.

IV. Phase-locking

It should be clear from the brief discussion above that the real key to the usefulness of arrays is how, or if, the junctions phase-lock. Even if all of the junctions are identical, one must still ask if the "uniform phase" condition (in which all junctions have the same phase relative to the locking current) is a solution, and if so is it a stable solution. If there is such a stable solution, the next problem is to find out what happens if all of the junctions are not identical. In real arrays there is always some degree of scatter in the junction parameters, e.g., the critical current, as well as random noise, which tend to make the junctions of the array oscillate at different frequencies.

It is possible to get much insight into both the stability and strength of phase-locking in arrays by considering the well known phenomenon of a single junction phase-locking to external radiation. We will start by using perturbation theory to study the effects of external radiation on an RSJ for which analytic solutions are available. Later, the effects of junction's capacitance in the low β_c ($\beta_c \equiv \omega_c R_J C \leq 1$) limit will be included. The perturbation techniques used are standard and have been applied to Josephson junctions by several authors.[9,10] Here the key ideas of the theory will be reviewed briefly and then applied to the phase-locking problem. To begin, a quantity related to the junction phase ϕ, called the "linearized phase", is defined by

$$\Theta = \hat{\omega} t \ , \tag{7}$$

where $\hat{\omega}$ is the junction's frequency averaged over a time long compared to a period of a Josephson oscillation, yet short enough to respond to the low frequency noise and modulation. The ˆ symbol is used in general to indicate averaging over the time scale which is long compared to $1/\omega$. The success of the perturbation theory depends on the wide separation of the Josephson frequency from the low frequency currents which are important in fixing the linewidth and oscillation frequency.

The essential results of the perturbation theory are shown schematically in Fig. 5. The junction is represented by an equivalent circuit with two parts, one for the high frequency (HF) (near ω) behavior, and the second modeling the low frequency (LF) response. The high frequency circuit consists of the Josephson oscillator (with amplitude $\tilde{V}_1$ given by Eq. 2 and frequency ω) and the source impedance R_J. This HF section is coupled to the LF section through ω, which is determined by the LF voltage through the Josephson equation (Eq. 1).

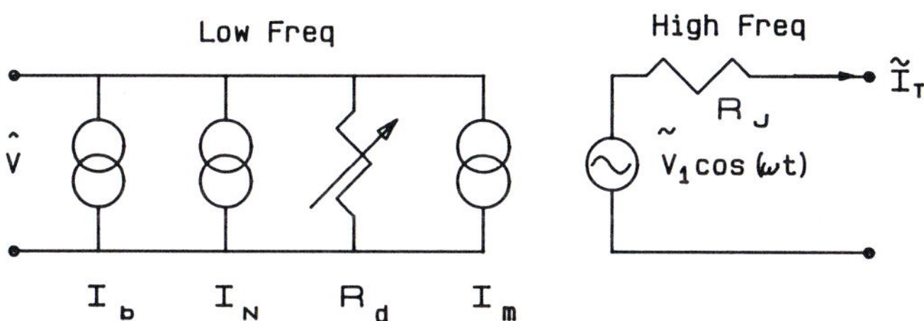

Fig. 5. Equivalent circuit for the RSJ model from perturbation theory. Low frequency (LF) section models response for $\omega \ll \omega_J$ ($\omega_J = 2\pi\hat{V}/\Phi_0$) and is coupled to the high frequency (HF) section by the mixing current I_m – see text. HF section contains the Josephson oscillator $\tilde{V}_1 \cos\omega_J t$ and source impedance R_J. The perturbation consists of an rf current $\tilde{I}_T$ with $\omega \sim \omega_J$ flowing through the HF terminals.

The perturbations which we wish to consider are caused by an rf current $\tilde{I}_T$ with a frequency near ω flowing through the HF terminals. This in turn affects the LF voltage (and thus ω) through the presence of a "mixing current,"

$$I_m = \widehat{\alpha\,(2\,\tilde{I}_T \cos\Theta)}\ , \tag{8}$$

in parallel with the bias and noise currents on the LF side. Here α is the conversion coefficient, which is given in the RSJ model as

$$\alpha = \frac{1}{2(1+\bar{v}^2)^{\frac{1}{2}}}\ . \tag{9}$$

The cause of this perturbation might, for example, be either an external rf current source or a load placed across the HF terminals, or both. So

$$\hat{\omega} = \hat{\omega}_u\,(\hat{I} + I_m)\ , \tag{10}$$

where the subscript u refers to the value of the variable (ω) in the absence of the HF perturbation. In other words, in the presence of a HF perturbation the junction will oscillate at the same frequency as an unperturbed junction biased with a current equal to the sum of the bias and mixing currents in the perturbed junction.

In order to apply this technique to understand the phase-locking of a junction to external radiation, we take the perturbing rf current to be that due to an external current source with amplitude I_e and frequency ω_e near ω, so

$$\tilde{I}_T = I_e \cos(\omega_e t)\ . \tag{11}$$

This gives a mixing current

$$I_m = \alpha\, I_e \cos(\delta\Theta)\ . \tag{12}$$

where $\delta\Theta = \Theta - \omega_e t$.

Equation 10 is actually a differential equation for Θ which can be rewritten by expanding ω (I) about I_b using the differential resistance of the unperturbed junction and remembering (Eq. 7) that $\dot{\Theta} = \hat{\omega}$. Thus,

$$\frac{\Phi_0}{2\pi R_d}\,\dot{\Theta} - \alpha\, I_e \cos(\delta\Theta) = \frac{\Phi_0}{2\pi R_d}\,\omega_u\,(\hat{I})\ . \tag{13}$$

If a new variable, $\theta \equiv \Theta - \omega_e t - \pi/2$ is defined, then Eq. 13 becomes

$$\frac{\Phi_0}{2\pi R_d}\,\dot{\theta} + I_L \sin\theta = \delta I\ , \tag{14}$$

where I_L is

$$I_L = \alpha\, I_e \tag{15a}$$

and

$$\delta I = I_b - I_{be}\ . \tag{15b}$$

That is, δI is the difference between the actual bias current and the bias current I_{be} which would make the unperturbed junction oscillate at frequency ω_e. This equation is just the familiar equation for the phase of a RSJ with critical current I_L, resistance R_d and bias current δI; hence the solutions are well known.

The main result which we need is the locking strength, that is, the range of bias current over which $\dot{\theta} = 0$. Equation 14 clearly has a constant θ solution for $-I_L \le \delta I \le I_L$, with $0 \le (\Theta - \omega_e t) \le \pi$. Thus, as seen in Fig. 6a, $\bar{V}$ remains constant over a range of bias currents $2\,I_L$ about the bias current I_{be} for which the unperturbed junction would have frequency ω_e. Note that this locking strength could also be expressed in terms of the variation in I_c (since I_{be} is a function of I_c), which is possible at fixed bias without losing phase-lock. This latter view is more relevant for arrays where we may wish to bias a string of junctions with a common current and ask how large a scatter (e.g., in I_c) can be tolerated. In this sense a junction is most strongly locked when biased in the center of the current step where the difference between the phase of the junction's oscillation and that of the external radiation is $\pi/2$, i.e., $\delta\Theta = \pi/2$. For this bias, the

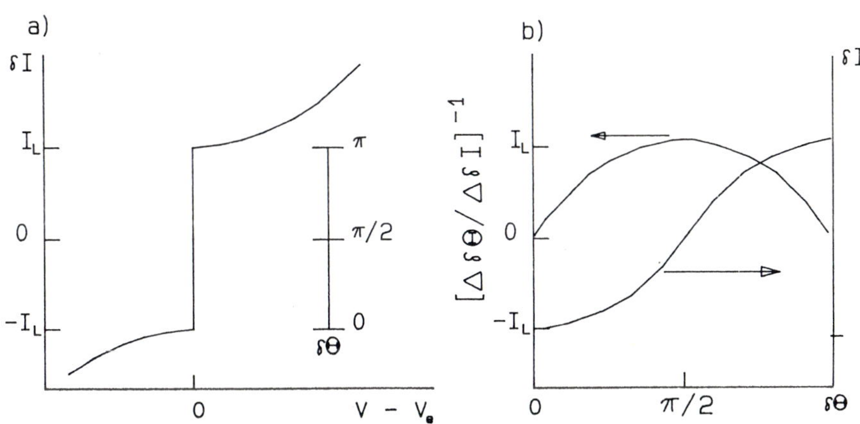

Fig. 6. a) Junction I-V curve in neighborhood of radiation induced step. b) Phase-locking stability along step, $[\partial\delta\Theta/\partial\delta I]^{-1}$. Note greatest stability is for $\delta\Theta = \pi/2$. The region $0>\delta\Theta>-\pi$ is unstable.

greatest deviation of I_c is possible in a random direction. The presence of this $\pi/2$ phase shift for strongest locking has, important implications for the design of arrays as we shall see below. Another measure of locking strength is the variation in $\delta\Theta$ with δI. This is shown in Fig. 6b. Again one sees that $\delta\Theta$ is most stable with respect to changes in δI for $\delta\Theta = \pi/2$ and becomes completely unstable at the edges of the step, $\delta\Theta = 0$ or π. Note that for the range of negative $\delta\Theta$, where the current leads the oscillator phase, $d\delta\dot{\Theta}/d\delta\Theta>0$; hence the phase-locked solution is unstable.

V. Phase-locking in arrays

A detailed analysis of phase-locking in arrays has been carried out in JLLS (Chap.6), as well as in Ref. 8 (Chap. 13). The discussion presented here in terms of phase-locking to external radiation will, it is hoped, be intuitive while minimizing the mathematical complications. For large arrays this approach gives nearly the same results as does the more exact analysis. To begin, consider a series array of identical junctions, modeled by their HF equivalent circuits and connected in a loop through a load Z_ℓ as shown in Fig. 7. If all of the oscillators have the same phase, then the rf current which flows in series through all of the junctions and the load is

$$\tilde{I}_\ell = \frac{\tilde{V}_1}{R_J z_c} , \qquad (16a)$$

where the coupling impedance per junction, z_c, is given in terms of the load impedance Z_ℓ and the junction impedance R_J as

$$z_c = \frac{1}{NR_J}(NR_J + Z_\ell) . \qquad (16b)$$

To calculate how much the critical current of one of the junctions can be varied without having it come unlocked from the array, we can just treat this current as external radiation assuming that the array is large enough that a variation of the phase of a single junction will have a negligible effect on $\tilde{I}_\ell$.

When all of the oscillators are running in phase, the relative phase of an oscillator and the locking current is fixed by the loop impedance. Since this impedance always

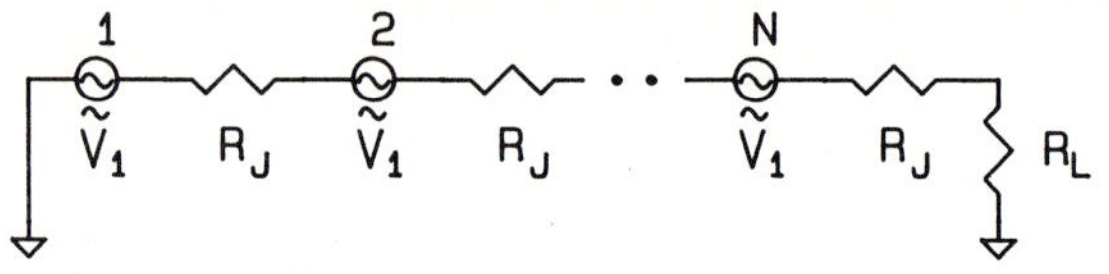

Fig. 7. Equivalent HF circuit for an array terminated in load Z_ℓ.

contains a real part equal to the sum of the junctions' resistances plus the load resistance, it is clear that the ideal situation of having the locking current lag the oscillator phase by $\pi/2$ cannot be achieved for the circuit shown in Fig. 7 unless $\mathrm{Im}(Z_\ell) \to \infty$. Thus, in optimizing $\mathrm{Im}(Z_\ell)$ for the maximum locking strength, there is a tradeoff between the amplitude of the locking current and its phase, the phase being given by

$$p_\ell = \tan^{-1}\left[\frac{-\mathrm{Im}(z_c)}{\mathrm{Re}(z_c)}\right] . \qquad (17)$$

We now wish to see how far the bias current (or critical current) of the kth

junction can be varied from the mean for the array without the junction coming unlocked. The mixing current for this junction, I_{mk}, is

$$I_{mk} = I_{k_L} \cos\left(p_\ell + \delta\Theta_k\right) , \tag{18a}$$

where $\delta\Theta_k \equiv \bar{\Theta} - \Theta_k$, $\bar{\Theta}$ being the mean phase of the oscillators, and

$$I_{k_L} = I_c \frac{\alpha\, \tilde{v}_1}{|z_c|} . \tag{18b}$$

Note that the product $\alpha\, \tilde{v}_1$ is a maximum near $\bar{v} = 1$, since $\tilde{v}_1 \propto \bar{v}$ for $\bar{v} \ll 1$ and $\alpha \propto 1/\bar{v}$ for $\bar{v} \gg 1$.

As for the case of a single junction locked to external radiation, phase-locking will be maintained for $-p_\ell \leq \delta\Theta_k \leq \pi - p_\ell$. Since $p_\ell \neq \pi/2$, it will be possible to shift I_{ck} farther in one direction than in the other. In a real array one would likely have a symmetric, roughly gaussian distribution of critical currents with the operating frequency of the array determined by the mean I_c of the distribution. In that case the maximum width which this distribution could have and still maintain complete locking would be set by the lesser of the two deviations, i.e., by

$$\delta I_k\,[\max] = I_L \min\{[1+\cos(p_\ell)],\, [1-\cos(p_\ell)]\} . \tag{19}$$

These phase relations are illustrated in Fig. 8.

Just as with a junction locking to external radiation, the stable situation is when the locking current lags the Josephson oscillator, i.e., the load must be inductive. The inductance which gives the largest locking strength can easily be determined by maximizing Eq. 9 with respect to L. If the load impedance is $Z_\ell = R_\ell + jL\omega$, then the locking strength is a maximum for

$$L\omega = \sqrt{3}\,(NR_J + R_\ell) . \tag{20}$$

For strong locking the coupling impedance must have a large reactive component with an inductive character. One can also see the importance of this reactance by calculating the variation in $\delta\Theta_k$ with changes in I_{bk} from Eq. 18a. Near equilibrium($\delta\Theta_k = 0$) this variation is

$$\left[\frac{\partial\delta\Theta_k}{\partial I_k}\right]^{-1} \propto |\tilde{I}_\ell| \sin(p_\ell) \propto \frac{\sin(p_\ell)}{|z_c|} , \tag{21}$$

i.e., the phase stability is proportional to the $\mathrm{Im}(y_c)$, where $y_c = 1/z_c$. Subject to this constraint the maximum power will be delivered to R_ℓ when $R_\ell = NR_J$. Thus $L\omega = 2\sqrt{3}\ NR_J$ gives $p_\ell = \pi/3$ and a value for δI_k [max] of

$$\delta I_k\,[\max]\,/I_c = \frac{\alpha\,\tilde{v}_1}{8}\ . \tag{22a}$$

Using the RSJ values for α and $\tilde{v}_1$ with $\bar{v} = 1$ gives

$$\delta I_k\,[\max]\,/\,I_c \simeq .04\ . \tag{22b}$$

Therefore, the total spread in I_{bk} for this type of array can be about 8% before junctions will start to unlock. Since $I_b\,/\,I_c$ is about 1.4 (in the RSJ model) for $\bar{v} = 1$, this implies

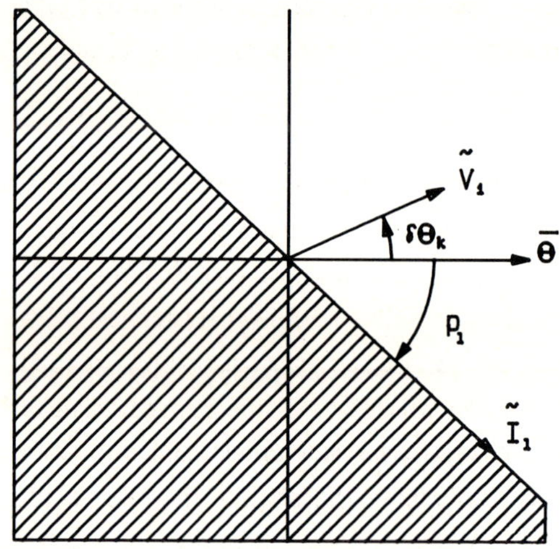

Fig. 8. Relative phase of the mean oscillator voltage (real axis), load current $\tilde{I}_\ell$ and k^{th} junction with different bias or critical current for array modeled in Fig. 7. Hatched region is unstable.

a permissible variation in I_c of about 11%. This number is a maximum since the spread of I_c for the junctions near the center of the distribution will produce some scatter in their phases with a consequent small reduction in the locking current $\tilde{I}_\ell$.

If we are concerned about the unlocking of the first few junctions in the tails of a large distribution, the estimate above is rather close, as can be shown from computer simulations.[11] It is worth noting that the interaction range of the junctions in this type of array is essentially infinite, i.e., the interaction of the kth and ℓ^{th} junctions does not depend on their separation. As a consequence, the unlocking of several junctions in a large array has a negligible effect on the phase locking among the remaining junctions. It may be undesirable to have even one junction unlocked, however, since if its frequency is close enough to that of the array, mixing will occur which will modulate the array frequency to some degree. As the width σ of the I_c distribution is increased and additional junctions unlock, $\tilde{I}_\ell$ will begin to decrease, causing yet more junctions to unlock and leading to a rapid uncoupling of the array with increasing σ. Computer simulations on a 40-junction array show that this "catastrophic" failure occurs for a value of σ about twice that at which the junctions in the tail of the distribution first unlock.

Understanding the effects of array size on radiation linewidth requires the more complete theory as presented in JLLS Chapter 6. One needs to calculate the change in the frequency of the array when the bias current through one junction is changed due to noise. Here we have treated the rest of the array as "external radiation" and assume it to be unaffected by the single junction. The result in JLLS is that, neglecting the effects of LF shunting, $\Delta\nu$ is reduced by a factor 1/N in an array. In practical arrays LF shunting will probably be important, leading to an even greater reduction in $\Delta\nu$.

We conclude this section on phase-locking with some brief comments on the prospects for 2D linear arrays. As discussed above, when RSJ's are connected in an inductive loop, their rf voltage tends to add inphase around the loop. For the 2D array shown in Fig. 4b, the lowest impedance path seen by a junction is the inductive path through the junction in parallel with it. For the configuration in Fig. 4b, the tendency is for the rf polarities to change along a parallel chain of junctions with the result that circulating currents are set up within the chain. Hence power is dissipated internally instead of being coupled to the load. For the 2D array in Fig. 4c, on the other hand, the lowest impedance path for all of the series chains is the (presumably inductive) path through the load. Consequently, one would expect a constant phase transverse to the current flow, as desired, for this array. Capacitive coupling between the chains might further stabilize this situation. These stability agreements are developed in much greater detail in JLLS.

VI. Effects of capacitance

So far only junctions with no capacitance have been considered. In this section we will examine what changes in the behavior of arrays one might expect if $\beta_c \neq 0$. First, point to be aware that capacitance is dangerous. A clue to this is seen in the locking of a single junction to external radiation where the situation corresponding to a capacitive load, i.e., the current phase leading the oscillator, is unstable. It has been shown that for an array like that in Fig. 7, the uniform phase solution is unstable if Z_ℓ is capacitive[6,8]. Instead, the rf voltage tends to sum to zero around the loop. Further, there are many examples of chaotic behavior in capacitive junctions subject to applied radiation or external loads.

There are several reasons to consider using capacitive junctions in arrays. First, by far the most advanced technology for making Josephson junctions is for tunnel junctions where capacitance is unavoidable. Next, as we shall see below, the presence of a small shunt capacitance can, under certain conditions, enhance the locking strength in the array. Also, there have been very successful examples of locking large arrays of high capacitance tunnel junctions to external radiation,[12] as well as demonstrations that even junctions with $\beta_c \gg 1$ can phase lock and generate radiation.[13-18] Furthermore, it has been shown that tunnel junctions generate significant power levels, at least up to the sum of the gap frequencies.[19] It is fortunate that a very general technique for analyzing the stability of the uniform solution in arrays with arbitrary β_c and Z_ℓ has been developed by Hadley, Beasley and Wiesenfeld (HBW) [20,21] and this should be of great help in designing arrays of capacitive junctions to avoid the "dangerous" regions of parameter space.

In this section the analysis above for arrays of junctions with $C = 0$ will be extended, using perturbation theory, to the case of small C, i.e., for $\beta_c \lesssim 1$, in order to develop some insight into the effects of capacitance. This low C region near $\bar{v} \gtrsim 1$, where one can still hope to obtain meaningful results from perturbation theory is also the

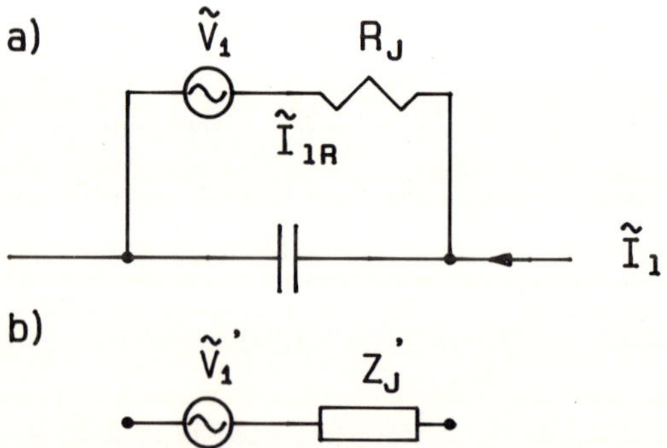

Fig. 9. a) HF equivalent circuit of a junction with a capacitor for perturbation. b) Equivalent circuit as seen by the rest of the array.

region in which the analysis of HBW indicates that the uniform solution should have the greatest stability. In order to include the effects of junction capacitance, a capacitor will be connected across the HF terminals of the junction and treated as an additional perturbation.

The effects of the various perturbations to the junction are additive, since the circuits are linear. The direct effect of the shunt capacitor on the junction will be to change the voltage $\bar{V}$ obtained for a given bias current. This will be ignored since it does not influence the locking behavior of the junction, but just means that a slightly different bias must be used to achieve the desired frequency. The most important effects of the capacitance are to change the effective impedance and rf voltage of the junction as seen by the rest of the circuit. This is illustrated in Fig. 9. The Thévenin equivalent for the junction with capacitance is shown in Fig. 9b, with the equivalent source voltage $\tilde{V}_1'$ being

$$\tilde{V}_1' = \frac{\tilde{V}_1}{1 + jR_J C\omega}, \tag{23}$$

and the equivalent impedance being

$$Z_J' = \frac{R_J}{1 + jR_J C\omega} \quad . \tag{24}$$

Placing an array of these junctions in series with a resistive load as shown in Fig. 7 gives for the rf loop current generated by the junctions

$$\tilde{I}_\ell = \frac{\tilde{V}_1}{R_J} \frac{1}{(r_\ell + 1 + j r_\ell \beta_c \bar{v})}, \tag{25a}$$

where r_ℓ is the load resistance expressed in units of NR_J; also we have expressed capacitance in terms of β_c and frequency in terms of the reduced voltage $\bar{v}$ through the relation $\beta_c \bar{v} = \omega R_J C$. The phase of this loop current relative to that of $\tilde{V}_1$ is

$$p_\ell = \tan^{-1}\left[\frac{-\beta_c \bar{v} r_\ell}{r_\ell + 1}\right] \quad . \tag{25b}$$

We see, not surprisingly, that the phase shift produced by this shunt capacitance has the same sign as that due to a series inductance in the load. Thus the phase relationship between the locking current and the junction is that required for stable locking.

To calculate the mixing current I_{mk} and hence the locking strength for the junction, it is important to remember that only that part of the loop current flowing through the HF terminals of the RSJ mixes with the junction's oscillations. This current $\tilde{I}_{\ell R}$ is

$$\tilde{I}_{\ell R} = \frac{\tilde{V}_1}{R_J} \frac{1}{(1+j\beta_c\bar{v})\ (r_\ell+1+jr_\ell\beta_c\bar{v})} \tag{26a}$$

and is phase shifted relative to $\tilde{I}_\ell$ by $\tan^{-1}(-\beta_c\bar{v})$, giving a total phase shift for $I_{\ell R}$ of

$$p_{\ell R} = \tan^{-1}\left[\frac{-\beta_c\bar{v}(1+2r_\ell)}{1+r_\ell[1-(\beta_c\bar{v})^2]}\right]. \tag{26b}$$

These phase relations are illustrated in Fig. 10 and show a potential advantage of a shunt capacitance over a series inductance. Recall from section V. that the locking

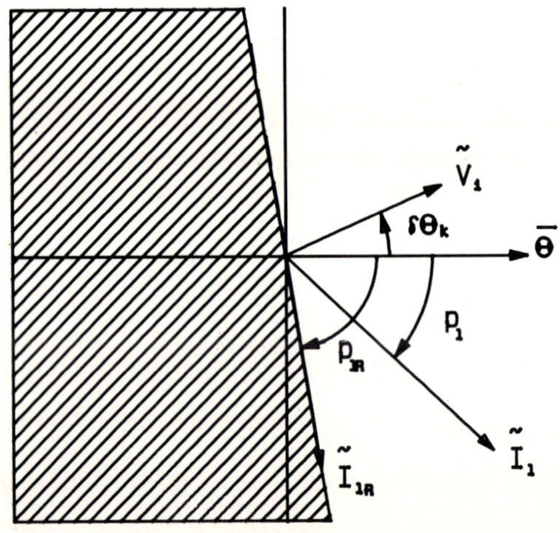

Fig. 10. Relative phases of the oscillators, loop current $\tilde{I}_\ell$ and locking current $\tilde{I}_{\ell R}$ for an array of capacitive junctions.

strength and hence the acceptable scatter in I_c was substantially reduced, since it was not possible to have a $\pi/2$ phase shift between the mean phase of oscillators and the locking current. With a shunt capacitance one can, as seen from Eq. 26b, achieve this optimum phase shift by choosing $\beta_c\bar{v} = \left[\frac{1 + r_\ell}{r_\ell}\right]^{\frac{1}{2}}$.

There is a large parameter space in $\bar{v}$, β_c and Z_ℓ that can be explored to optimize the power and locking strength for a given application. To get a feel for the performance of these arrays with $\beta_c > 0$, let us take the purely resistive load which maximizes the load power for given β_c and $\bar{v}$. This gives

$$r_\ell = [\,1 + (\beta_c\bar{v})^2]^{-\frac{1}{2}} \quad . \tag{27}$$

Substituting this load resistance into Eq. 26a, we see that the desired $\pi/2$ phase shift is obtained for $\beta_c\,\bar{v} = \sqrt{3}$. For these values the locking strength is

$$\delta I_k \,/\, I_c = \frac{\alpha\tilde{v}_1}{2\sqrt{3}} \;, \tag{28}$$

more than twice that for the RSJ array from Eq. 22a, indicating that complete locking should still be possible with a total spread in I_c of greater than 20%. For smaller values of $\bar{v}$ the limits of perturbation theory are being pushed, so the exact values need to be compared with computer simulations. We note that the estimate from perturbation theory is in line with the result of simulations done by HBW on 100-junction arrays with $\beta_c \simeq 0.75$ and $i \simeq 2.3$, where locking was still observed with greater than a 15% scatter in R_J, C and I_c.

VII. Distributed arrays

In all of the discussions above, it has been assumed that the dimensions of the arrays were much less than the wavelength λ. As a result, the lumped circuit approximation could be used. To see if this is realistic, note that if the entire array is to have a length less than $\lambda/8$ the junction spacings must be

$$s = \frac{1}{8}\frac{v_p}{\nu N} \simeq 0.1\ \mu m \;, \tag{29}$$

where $v_p \simeq 10^8$ m/s is the propagation velocity in the superconducting transmission line connecting the junctions, and the value of νN from Eq. 6a has been used. Unfortunately, 0.1 micron spacing is about two orders of magnitude closer than is practical to place the junctions in the array when such things as heating and the limits of lithography are considered. We conclude that in order to achieve maximum power, even from one-dimensional arrays, the junctions must be distributed over a wavelength or more.

The analysis of phase-locking above has shown that the phase of the junction's oscillations relative to the locking current flowing in the coupling circuit is crucial. In general, an oscillator in a transmission line will generate waves propagating in both directions. This makes it impossible to maintain the same phase relationship between all of the oscillators and the locking current when the junctions are placed at arbitrary positions along the transmission line. There have been several proposals[6,22,23] for

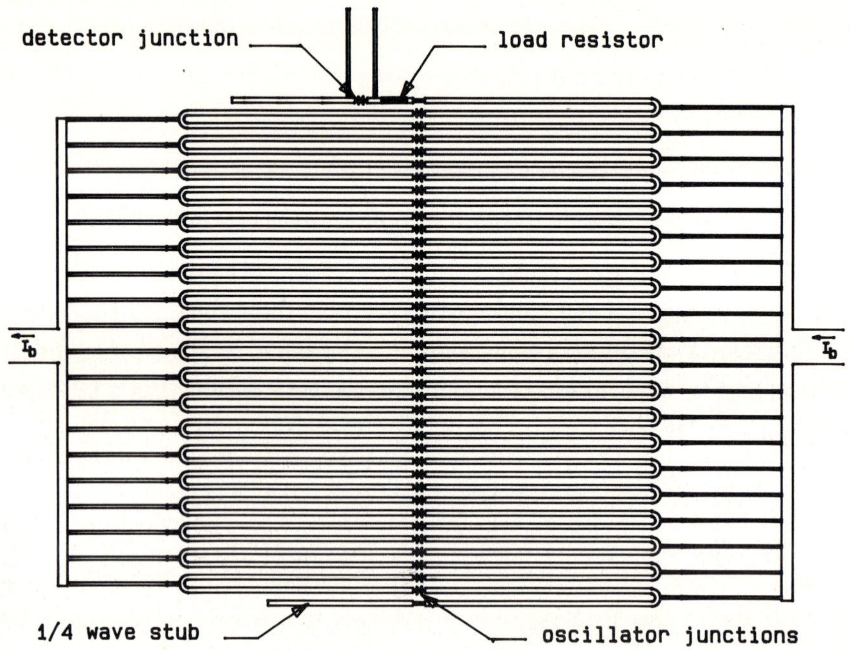

Fig. 11. Layout of distributed array. Oscillator junctions (×) are placed at wavelength intervals along the serpentine microstrip. Load resistor and detector junction to monitor the load current are shown at upper left.

placing junctions along a transmission line such that they will phase-lock. The approach taken in the experiments described in this section is probably the simplest: the junctions are placed at wavelength intervals along the transmission line. Hence, all junctions see the same impedance and the same relative phase. The analysis of this circuit[24] at the frequency ν_0, where the spacing is equal to λ , is identical to the lumped circuit analysis above. The disadvantage of this design is that the frequency of the array is not continuously tunable over a large range.

Figure 11 shows a schematic of such a distributed array. The junctions (indicated by ×) are placed at λ intervals along a serpentine microstrip transmission line. An independently biased detector junction is placed immediately after a load resistor in the line. By measuring the range of detector bias current over which the detector phase locks to the array-generated locking current flowing through the load resistor, the power to the load can be determined for each operating frequency of the array. The ends of the array are terminated with $\lambda/4$ stubs so that, to the junctions, the array appears grounded through the load resistor. Additional length can be added to the stub in order to achieve an inductive component to the load for increased locking strength.

A micrograph of a forty-junction array fabricated using this design is shown in Fig. 12. The junctions are resistively shunted lead-alloy tunnel junctions having an area of about 1.5 μm^2. These junctions have been described in detail elsewhere.[11] The copper shunt resistor ($\simeq$0.5 Ω) for the junction is placed directly under the junction to ensure a

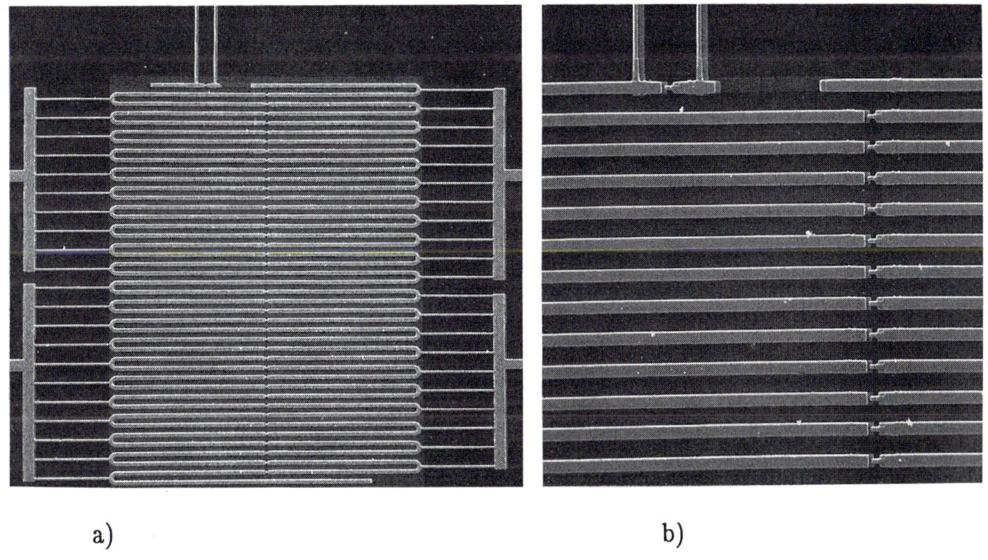

a) b)

Fig. 12. a) Micrograph of distributed array. b) Blowup of a), top left, showing detector junction, load resistor and several oscillator junctions on the right. The vertical separation of the oscillators is 10μm.

very low parasitic inductance. Critical currents are typically in the range 500 μA < I_c < 1mA giving a value for β_c of 0.2 to 0.4. The junctions are placed at 350 μm intervals along the lead-alloy microstrip, which has a width of 4 μm and is separated from the Nb ground plane by 0.7 μm of SiO dielectric, giving an impedance of 20 Ω and a propagation velocity of about 120 μm/ps. Thus ν_0 for this array is about 350 GHz. Since our fabrication process does not yield sufficiently uniform critical currents to insure that locking would be achieved if all junctions were biased with the same current, a "parallel" bias scheme has been used. The superconducting bias leads inject (remove) current at alternate bends in the microstrip, with the result that each junction is part of two interlocking dc SQUIDs. All junctions then have the same average voltage, which alternates in polarity along the microstrip. This forces the bias current to divide so as to compensate to first order for the variations in the junctions' critical currents. The rf locking current is still crucial however, since without it the phases of each junction would be essentially random due to random flux linking the SQUIDs. Further, noise currents would cause voltage (and frequency) fluctuations among the junctions. Phase-locking for this type of parallel biasing has been analyzed in detail in Ref. 6. It is primarily as discussed above for series-biased junctions except that the effective scatter in I_c approximately equals Φ_0/L, where L is the inductance of the dc SQUID. For this circuit $\delta I_c \simeq 5$ μA.

The power (as determined from the detector junction) vs. array frequency is shown in Fig. 13. As expected, the power peaks near ν_0 with a maximum value of

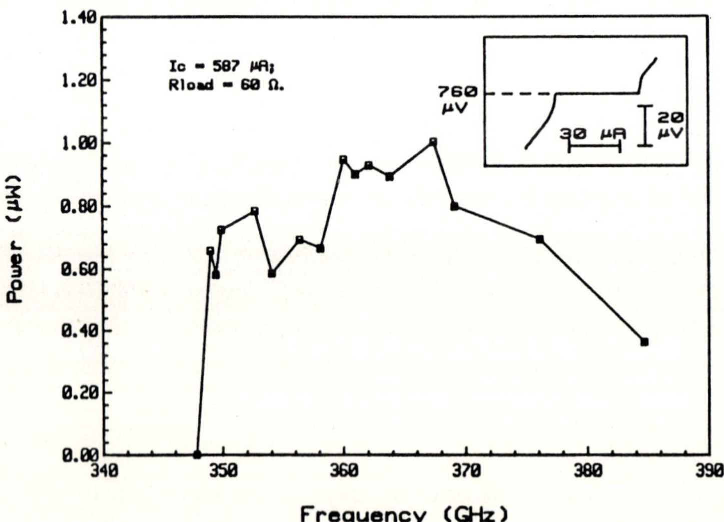

Fig. 13. Power delivered to a 60 Ω on-chip load (by a 40-junction array) vs. frequency. Inset: Detector response at maximum power.

1μW, consistent with all of the junctions being in phase, and is greater than $1/2\mu$W over a 10% band between 350GHz and 380GHz. The insert shows the response of the detector junction when the array is operating at 370GHz.

VIII. Conclusions

These results, along with those of other groups, make it likely that Josephson effect submillimeter sources with power levels approaching a milliwatt can be developed. Relatively straightforward extensions of the results presented here, using higher T_c conventional superconductors and larger arrays, should give a power level of about 0.1mW at $1-2$ THz.

Improved fabrication technology and possibly the use of capacitance to increase locking strength can make series biasing and a resultant simplification in array design possible. If ideas for two dimensional arrays prove correct, power levels of the order of a watt can be contemplated. Finally, if Josephson junctions can be successfully fabricated from the new high temperature superconductors (HTS), Josephson effect sources might well work to above 10 THz. It is worth noting that high quality tunnel junctions are not needed for sources, so this application may prove to be one of the least demanding for HTS fabrication.

IX. Acknowledgements

The preparation of this manuscript was supported in part by the Air Force Office of Scientific Research. Work on parallel distributed submillimeter arrays was supported in part by the Innovative Science and Techology office of the Strategic Defense Initiative Organization, and managed by the Electromagnetics Directorate of Rome Air Development Center (Contract # F19628-86-K-0039). We thank Baokang Bi for computer simulations of phase-locking in capacitive junctions.

References

1. J.S. Tsai, A.K. Jain and J.E. Lukens, Phys. Rev. Lett. 51, 316 (1983).
2. See for example Ref. 8, Chap. 4.
3. K.K. Likharev and V.K. Semenov, JETP Lett. 15, 442 (1972).
4. D.R. Tilley, Phys. Lett. A 29, 11 (1969).
5. T.D. Clark, Phys. Rev. B 8, 137 (1973).
6. A.K. Jain, K.K. Likharev, J.E. Lukens and J.E. Sauvageau, Phys. Rep. 109, 309 (1984).
7. P.E. Lindelof and J. Bindslev Hansen, Rev. Mod. Phys. 56, 431 (1984).
8. K.K. Likharev, Dynamics of Josephson Junctions and Circuits, Gordon and Breach Science Publishers, New York (1986), Chap. 13.
9. P.W. Forder, J. Phys. D 10, 1413 (1977).
10. L.S. Kuzmin, K.K. Likharev and G.A. Ovsyannikov, Radio Eng. and Electron. Phys. 26, No.5, 102 (1981).
11. J.E. Sauvageau, Ph.D. dissertation, State University of New York at Stony Brook (unpublished, 1987).
12. R.L. Kautz, C.A. Hamilton and Frances L. Lloyd, IEEE Trans. Magn., MAG-23, 883 (1987).
13. T.F. Finnegan and S. Wahlsten, Appl. Phys. Lett. 21, 541 (1972).
14. G.S. Lee and S.E. Schwarz, J. Appl. Phys. 55, 1035 (1984).
15. G.S. Lee and S.E. Schwarz, J. Appl. Phys. 60, 465 (1986).
16. L.S. Kuzmin, K.K. Likharev and E.S. Soldatov, IEEE Trans. Magn., MAG–23, 1051 (1987).
17. V.W. Krech and M. Reidel, Ann. Phys. (Leipzig) 44, 329 (1987).
18. A.D. Smith, R.D. Sandell, A.H. Silver and J.F. Burch, IEEE Trans. Magn., MAG-23, 1267 (1987).
19. R.P. Robertazzi, B.D. Hunt and R.A. Buhrman, IEEE Trans. Magn., MAG–23, 1271 (1987).
20. P. Hadley, M.R. Beasley and K. Wiesenfeld, Appl. Phys. Lett. 52, 1619 (1988).
21. P. Hadley, M.R. Beasley and K. Wiesenfeld, preprint 1988.
22. A. Davidson, IEEE Trans. Magn., MAG-17, 103 (1981).
23. J.E. Sauvageau, A.K. Jain, J.E. Lukens and R.H. Ono, IEEE Trans Magn., MAG-23, 1048 (1987).
24. J.E. Sauvageau, A.K. Jain and J.E. Lukens, Int. J. of Infrared & Millimeter Waves 8, 1281 (1987).

PRINCIPLES OF DIRECT AND HETERODYNE DETECTION WITH SIS JUNCTIONS

K.H. Gundlach
Institut de Radioastronomie Millimétrique (I.R.A.M.)
Domaine Universitaire
38406 St. Martin d'Hères, France

1. Radio Astronomical Observations

Radio astronomy has stimulated the development of low-noise receivers for millimetre and sub-millimetre electromagnetic radiation. A branch of radio astronomy is devoted to the study of continuum radiation and requires low-noise detectors of large instantaneous bandwidth, ideally of the order of 100 GHz. Continuum observations include the 3 K cosmic background radiation, thermal radiation from dust in interstellar clouds, synchrotron radiation and free-free emission from ionized regions.

Another important branch of radio astronomy deals with the study of molecules in comets, stellar atmospheres, circumstellar shells, proto-planetary nebulae and interstellar clouds. Most molecules are observed in rotational emission lines. The rotational excitation usually occurs through collision with H_2 molecules and He atoms having kinetic temperatures of 10 to 200 K.

Fig. 1.1 IRAM 15-m radio telescopes on the Plateau de Bure, near Grenoble, France, at an altitude of 2550 m

Fig. 1.2 shows the rotation lines of the (2–1) transition of carbon monoxide and the (7–6) transition of aluminium fluoride; a rotation line of an isotope of the ring molecule SiCC can just be resolved (Guélin et al., 1988). The brightness temperature in Fig. 1.2 is the Rayleigh-Jeans temperature of an equivalent black body that would give the same signal power as the source.

So far, more than 70 molecules are known in the interstellar space, including complex organic molecules, such as ethyl alcohol C_2H_5OH, acetic acid CH_3OOH, acetonc $(CH_3)_2CO$; exotic species like isohydrocyanic acid HNC or heavy radicals as C_5H and C_6H. Other examples are the ring molecules C_3H_2 and SiC_2 (Guélin, 1986). Water vapour H_2O and silicon monoxide SiO occur as maser lines (Downes, 1983).

NATO ASI Series, Vol. F 59
Superconducting Electronics
Edited by H. Weinstock and M. Nisenoff

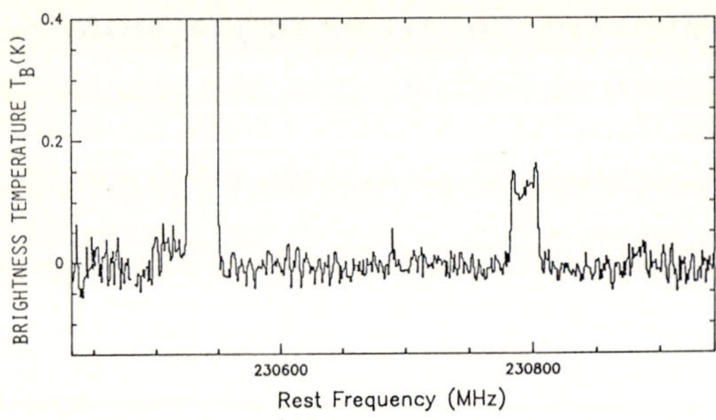

Fig. 1.2 Record of rotation lines from the outer shell of a carbon-rich star with an SIS receiver on the IRAM 30-m telescope in southern Spain. Observation time τ about 27 minutes (Guélin et al., 1988)

2. Direct Detection

The direct detector, also known as the video detector, is of interest for continuum observations. So far, mostly bolometers are used. The incoming electromagnetic radiation passes appropriate broadband frequency filters and is converted into heat which changes the temperature of the bolometric device. The information on the phase of the incoming radiation is lost at the output of the bolometer. It is an incoherent detector. The best bolometers use doped germanium, cooled to about 0.3 K by a ^{3}He cryostat. The sensitivity of a direct detector can be characterized by its noise equivalent power (NEP), which will be explained later. At the operation frequency of about 230 GHz, an NEP of $3{\cdot}10^{-16}$W has been achieved with a typical bandwidth of 50 GHz (Kreysa, 1984, 1987).

The superconductor-insulator-superconductor (SIS) tunnel junction has been investigated for direct detection using the Josephson effect (Hartfuss et al., 1980) and the quasiparticle tunneling characteristic (Richards et al., 1980, Hartfuss and Gundlach, 1980, 1981).

We discuss now the quasiparticle direct detector. The quasiparticle current of the SIS junction has a sharp onset at the gap voltage $V_g = 2\Delta/e$. Weak electromagnetic radiation slightly rounds off the onset due to photon-assisted tunneling as shown in Fig. 2.1.

2.1 The current responsivity

For a direct detector, the signal power P_s should induce a large change ΔI of the junction current. In the classical limit, the current-to-power responsivity

$$\eta_c = \frac{\Delta I}{P_s} \tag{2.1}$$

can be derived for small signal power as follows:

$$I(t) = I(V_o + V_{rf} \cdot cos\omega t) \tag{2.2}$$

$$\simeq I(V_o) + V_{rf} \cdot cos\omega t \cdot \frac{dI}{dV} + \frac{1}{2} V_{rf}^2 \cdot cos^2\omega t \cdot \frac{d^2 I}{dt^2} + ... \tag{2.3}$$

$$\overline{I(t)} \simeq I(V_o) + \frac{1}{4} V_{rf}^2 \frac{d^2 I}{dV^2} \tag{2.4}$$

$$\Delta I = \overline{I(t)} - I(V_o) \simeq \frac{1}{4} V_{rf}^2 \frac{d^2 I}{dV^2} \tag{2.5}$$

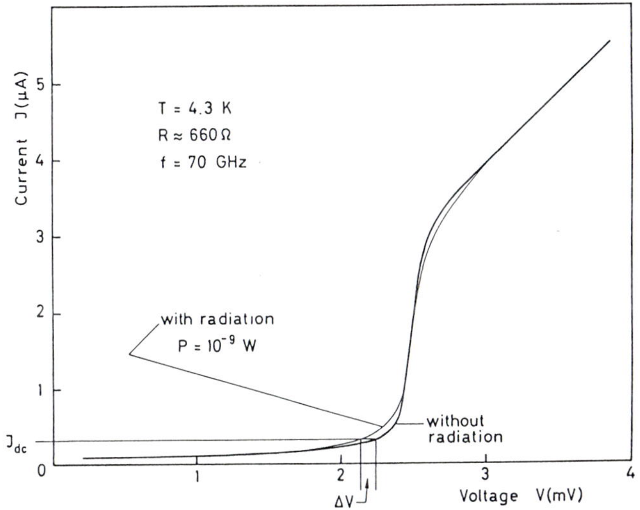

Fig. 2.1 The quasiparticle current (I)-voltage (V) characteristic of an SIS tunnel junction without and with weak 70 GHz radiation. Impressing a current I_{dc}, a voltage jump ΔV occurs when the radiation is switched on. The voltage $\Delta V = \Delta I \cdot R_d$ can be measured by a lock-in amplifier chopping the incoming radiation (Hartfuss and Gundlach, 1981).

The absorbed signal power reads

$$KP_s = \frac{V_{rf}^2}{2R_d} \tag{2.6}$$

From eqs. (2.1), (2.5) and (2.6) the current-to-power responsivity (η_c) becomes

$$\eta_c = \frac{K}{2} \cdot \frac{d^2I/dV^2}{dI/dV} \tag{2.7}$$

In eqs. (2.2) to (2.7) V_{rf} is the *rf* voltage across the junction, $\omega/2\pi$ the signal frequency, $R_d = dV/dI$ the dynamic resistance at the d.c. bias voltage V_o, $\overline{I(t)}$ is the time average of the *rf* current I(t), and K is the ratio of absorbed to incident signal power.

At the operating point V_o, just below the gap voltage $V_g = 2\Delta/e$ of a high quality SIS junction, the curvature d^2I/dV^2, and thus η_c given by eq. (2.7), can become arbitrarily large. Here the classical theory is no longer valid, and eq. (2.7) has to be replaced by the corresponding quantum mechanical expression. Using quantum mechanics Tucker (1979) derived for the time dependent quasiparticle current the expression

$$I(t) = \sum_{m,\ell=-\infty}^{+\infty} J_m(\alpha)J_{m+\ell}(\alpha)[I(V_o + m\hbar\omega/e) \cdot cos(\ell\omega t) + I_{KK}(V_o + m\hbar\omega/e) sin(\ell\omega t)] \tag{2.8}$$

where $J_m(\alpha)$ and $J_{m+\ell}(\alpha)$ are the Bessel functions with indices m and m+ℓ and the argument $\alpha = eV_{rf}/\hbar\omega$. The term $I(V_o + m\hbar\omega/e)$ represents the *dc* I-V characteristic without radiation and shifted in voltage by integral multiples of the photon voltage $\hbar\omega/e$. Finally, $I_{KK}(V)$ is the Kramers-Kronig transform of the *dc* I-V characteristic without radiation.

The *dc* I-V characteristic under irradiation follows from eq. (2.8) setting $\ell = 0$, with the result

$$I = \sum_{m=-\infty}^{+\infty} J_m^2(\alpha) I(V_o + m\hbar\omega/e) \tag{2.9}$$

For small signal power (α <<1) one obtains

$$I = I(V_o) + \frac{V_{rf}^2}{4} \cdot \frac{I(V_o + \hbar\omega/e) - 2I(V_o) + I(V_o - \hbar\omega/e)}{(\hbar\omega/e)^2} \tag{2.10}$$

The current change ΔI due to an *rf* signal becomes

$$\Delta I = I - I(V_o) = \frac{V_{rf}^2}{4} \frac{I(V_o + \hbar\omega/e) - 2I(V_o) + I(V_o - \hbar\omega/e)}{(\hbar\omega/e)^2} \tag{2.11}$$

In a similar way, the dynamic resistance R_d at frequency ω can be derived from eq. (2.8) retaining only the terms with $\ell = 1$ and $\ell = -1$. This gives

$$R_d^{-1} \simeq \frac{I(V_o + \hbar\omega/e) - I(V_o - \hbar\omega/e)}{2\hbar\omega} \tag{2.12}$$

From eqs. (2.1), (2.6), (2.11) and (2.12) the current-to-power responsivity becomes

$$\eta_c = K \frac{e}{\hbar\omega} \frac{I(V_o + \hbar\omega/e) - 2I(V_o) + 2I(V_o - \hbar\omega/e)}{I(V_o + \hbar\omega/e) - I(V_o - \hbar\omega/e)} \tag{2.13}$$

It is seen that the derivatives in the classical expression (2.7) are replaced by finite differences. At low frequency ($\hbar\omega/e \to 0$), or for an I-V characteristic without a sharp current onset at the gap voltage, expression (2.13) approaches the classical expression (2.7). On the other hand, biasing a junction with a sharp I-V curve just below the gap voltage so that $I(V_o - \hbar\omega/e) \simeq I(V_o) \simeq 0$, the current responsivity from eq. (2.13) approaches the quantum limit

$$\eta_c = e/\hbar\omega \tag{2.14}$$

for a matched junction: one additional quasiparticle per incident signal photon traverses the tunnel barrier. This gives e.g. $e/\hbar\omega$ = 3450 A/W for $\omega/2\pi$ = 70 GHz. One disadvantage of the SIS direct detector is that its responsivity decreases with increasing frequency.

2.2 The noise equivalent power

The sensitivity of an SIS direct detector can be expected to be limited by the noise due to the bias and the signal current. The signal power which gives a signal-to-noise ratio of one for a postdetection bandwidth Δf = 1 Hz is called the noise equivalent power, henceforth NEP. It may be written as (Richards et al., 1980, Hartfuss and Gundlach, 1980)

$$NEP = \frac{\sqrt{\langle i^2 \rangle}}{\Delta I / P_s} \qquad (2.15)$$

Here $\langle i^2 \rangle$ is the mean-square noise current. At T$\simeq$ 4 K and the operating point V_o just below the gap voltage V_g

$$eV_o > 2kT \qquad (2.16)$$

In this case, shot noise dominates (Tucker, 1979), and we obtain

$$\langle i^2 \rangle = 2e(I_o + I_p) \cdot \Delta f \qquad (2.17)$$

where I_o is the "dark" current, that is, the bias current without signal power, and I_p is the signal current. Inserting eq. (2.17) into eq. (2.15) and taking the maximum current responsivity $\Delta I / P_s = e/\hbar\omega$, yields

$$NEP = 1.4\hbar\omega \cdot \Delta f \sqrt{\frac{I_o + I_p}{e \cdot \Delta f}} \qquad (2.18)$$

The expression under the square root can be understood as the number N of electrons passing the tunnel barrier in the resolution time $\Delta t - 1/\Delta f$. For an ideal junction at T = 0, the dark current I_o = 0. This gives the noise equivalent power

$$NEP = 1.4 \cdot \hbar\omega\Delta f \sqrt{N} \qquad (2.19)$$

where N is the number of photons absorbed in the time Δt. The ultimate noise limit is reached when one signal electron passes the barrier in the resolution time Δt. At 70 GHz one obtains

$$NEP = 1.4\hbar\omega\Delta f \simeq 10^{-23} W \qquad (2.20)$$

for Δf = 1 Hz. In real junctions, I_o may be about 1μA yielding NEP $\simeq 2 \cdot 10^{-16}$W at 70 GHz; about $4 \cdot 10^6$ signal photons per resolution time Δt are required to obtain a signal-to-noise ratio of one. So one is far from counting single photons (Hartfuss and Gundlach, 1981).

The first direct detection experiments using the quasiparticle current in SIS junctions were made by Richards et al. (1980) at 36 GHz. The best current responsivity was within a factor of 2 of the quantum limit $e/\hbar\omega$ and with NEP = $2.6{\cdot}10^{-16}$W. Hartfuss and Gundlach (1980, 1981) performed similar experiments at 70 GHz. The best current responsivity also was about $0.5{\cdot}(e/\hbar\omega)$, but the NEP was only about 10^{-15}W. It appears that the dark current I_o and 1/f noise from the junction have prevented better results so far. Using heterodyne frequency mixers, the dark current can effectively be suppressed (cf. Chapter 3.). The 1/f noise is circumvented because the IF-amplifiers operate at a frequency of 1 GHz or higher. The direct detection experiments were performed using lead-alloy SIS junctions. It may be interesting to repeat the experiments using Nb-Al oxide-Nb (e.g. Morohashi et al., 1985; Imamura et al., 1987) or Nb-Ta oxide-Pb/Bi junctions (Face et al., 1986) which can have a considerably better quality than lead alloy junctions. Up to now, the SIS direct detector is hardly competitive with ^{3}He-cooled bolometers. For space and other applications for which ^{3}He-cooling poses problems, the SIS direct detector could be of interest, even though its NEP is proportional to the signal frequency ω and the bandwidth of bolometers is intrinsically very large.

3. Heterodyne Detection

3.1 Principle of operation

For the observation of molecular lines below about 50 GHz, a maser is used as the input stage amplifier followed by a heterodyne system. Maser amplifiers have noise temperatures as low as 10 K in the K band, but require large pump power. The instantaneous bandwidth can be about 200 MHz. Above 50 GHz amplifiers with sufficiently low noise temperature are not available. The signal is therefore coupled to a low-noise heterodyne system. Fig. 3.1 shows a block diagram of a heterodyne receiver followed by a back-end spectrometer. The essential component for us is the mixer diode. Its non-linear current-voltage curve is used to mix the signal frequency f_S with the frequency f_{LO} of a local oscillator to generate the intermediate frequency $f_{IF} = |f_S - f_{LO}|$. The intermediate frequency is generally at about 1.5 or 4 GHz, where amplifiers with noise temperatures less than 10 K are available. For a double sideband (DSB) mixer, the output at the intermediate frequency (IF) is equally sensitive to input signals at either side of f_{LO}, that is, at $f_S = f_{LO} + f_{IF}$ and $f_{-S} = f_{LO} - f_{IF}$. The output of a single side band (SSB) mixer is only sensitive to the signal frequency f_S or its image frequency f_{-S}. The IF amplifier has typically a bandwidth of 400 to 600 MHz. Its output signal is passed to a back-end spectrometer. Each channel of it has a much smaller bandwidth $\Delta\nu$, for instance, 1 MHz or even 100 kHz.

The signal-to-noise ratio achieved in the observation time τ in a channel of bandwidth $\Delta\nu$ of the back-end spectrometer is

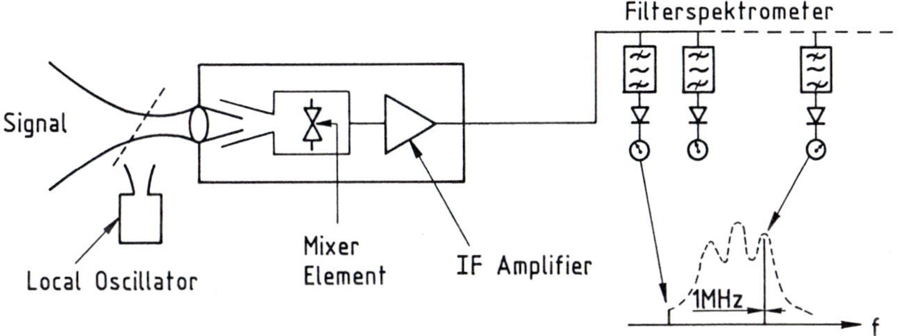

Fig. 3.1 Block diagram of a heterodyne receiver with back-end spectrometer

$$\frac{S}{N} \simeq \frac{1}{2} \frac{T_B \cdot \sqrt{\tau \cdot \Delta\nu}}{T_{rec} + T_{atm} + T_{gr}} \tag{3.1}$$

where T_{rec} is the noise temperature of the entire receiver, T_{atm} is the noise from the atmosphere, and T_{gr} is the noise from the ground. Under good conditions, $T_{atm} + T_{gr}$ is 40 to 50 K around 100 GHz. If $T_{rec} >> T_{atm} + T_{gr}$, an improvement of T_{rec} by a factor of 2 reduces the observation time by a factor of 4 or corresponds to a doubling of the surface of the dish. Reduction of T_{rec} below $T_{atm} + T_{gr}$ is less rewarding. Ground-based observations are restricted to atmospheric frequency windows. Therefore, one also uses airborne and balloon telescopes; space telescopes are planned. In these cases, it would be interesting to reduce T_{rec} down to about 10 K. Referring to the schematic block diagram in Fig. 3.1, the receiver noise temperature may be written as

$$T_{rec} = T_{in} + L_{in} \cdot T_M + L_{in}\, L_M \cdot T_{IF} \tag{3.2}$$

Here L_{in} and T_{in} are the loss and noise temperature, respectively, associated with the receiver input; T_M and T_{IF} are the noise temperatures of the mixer and IF amplifier. The mixer conversion loss $L_M = P_s/P_{IF}$ is the ratio of the available signal power at the mixer input to the power P_{IF} coupled to the IF amplifier. It is seen from eq. (3.2) that a mixer with conversion loss ($L_M > 1$) increases the importance of the IF amplifier noise, but a mixer with conversion gain ($L_M < 1$) would reduce the contribution of the IF amplifier noise to the receiver noise temperature. We come back to this point later.

3.2 Classical mixing with the Schottky diode

For a classical mixer, the diode is considered as a non-linear resistor. It is a passive element which responds without time delay to external *rf* signals. Quantum effects of the photon field are negligible. The *dc* I-V characteristic of the diode is sensitive to the signal power, but not to the signal frequency. A diode operating as a non-linear resistor cannot produce conversion gain unless there is a region of negative dynamic resistance on the *dc* I-V curve. If the signal power P_s is equally coupled into the sidebands above and below f_{LO}, the lowest DSB value is $L_M = 1$. If the signal power is coupled only into one of the two sidebands, the signal power is split between the two sidebands. If they are equally terminated, the best possible SSB value is $L_M = 2$. If the unused sideband is strongly mismatched, then the lower limit is $L_M = 1$ (Richards and Shen, 1980).

It has been shown that the ideal classical mixer is a switch (Barber, 1967). The physics appears to be similar to that of a stroboscope. If the pulse duty ratio of the switch $t/t_o \rightarrow 0$, the conversion loss $L_M \rightarrow 1$. The mixer noise temperature in this model may be written as

$$T_M = T_D(L_M - 1) \tag{3.3}$$

The diode noise temperature T_D will be evaluated below. Eq. (3.3) reminds one of the noise temperature of a lens or of an attenuator with the physical temperature T_D and an absorption of (L-1). It follows from eq. (3.3) that in the classical theory $T_M \rightarrow 0$ for $L_M \rightarrow 1$. This obviously means that the ideal classical mixer can completely suppress shot noise (cf. also Tucker and Feldman, 1985).

The question is how can a switch be realised? We consider a Schottky diode (Phillips and Woody, 1982, Gundlach et al., 1985). Its *dc* I-V curve at a temperature $T \simeq 15$ K may be approximated by the expression

$$I_{dc} = I_s[exp(eV/E_o) - 1] \tag{3.4}$$

One can argue that the local oscillator voltage periodically switches the diode between high and low conductance states (Barber, 1967). To approach the ideal behaviour of a switch, the *dc* I-V curve should be as non-linear as possible. Hence, the tunneling parameter E_o should have the smallest possible value. A value of $E_o \simeq 8$ meV has been achieved. Smaller values of E_o also yield a lower diode noise temperature T_D. The diode current I_{dc} at the bias point V_o gives rise to shot noise. Its mean-square noise current is

$$\langle i_s^2 \rangle = 2e\, I_{dc}(V_o) \cdot \Delta f \tag{3.5}$$

where Δf is the bandwidth. One can associate a noise temperature to the shot noise current by presuming that it arises from Nyquist noise:

$$\langle i_N^2 \rangle = \frac{4kT_D}{R_d} \cdot \Delta f \tag{3.6}$$

Taking $R_d = dV/dI_{dc}$ from eq. (3.4), and setting $\langle i_s^2 \rangle = \langle i_N^2 \rangle$, yields

$$T_D = \frac{E_o}{2k} \tag{3.7}$$

For $E_o = 8$ meV one obtains $T_D \simeq 40$ K; a value which is not particularly low. It also has been shown that the exponential *dc* I-V curve of the Schottky diode would require an unrealistically large local oscillator voltage to approach the ideal behaviour of a switch (Barber, 1967).

After the discovery of quasiparticle tunneling between two superconductors by Giaever (1960), it was suggested that the SIS junction should be more appropriate to realise the switch for frequency mixing, because it has a considerably stronger non-linear current-voltage characteristic than does the Schottky diode. It was, however, also recognised that the classical picture of frequency mixing is too simple for the SIS junction. To illustrate this, Ibrügger (1987) has calculated the effect of microwave irradiation on the *dc* I-V curve of the SIS junction using classical theory. This result, shown in Fig. (3.2), is inconsistent with the experiment. Microwave irradiation of frequency $\omega/2\pi$ induces steps of width $\hbar\omega/e$ in the quasiparticle *dc* I-V curve as shown in Fig. 3.3 These steps, first observed by Dayem and Martin (1962), are explained in terms of photon-assisted tunneling (Tien and Gordon, 1963). The absorption of n photons by the SIS junction provides the energy $n\hbar\omega$ to open a path for quasiparticle tunneling at the bias voltages $V = (2\Delta - n\hbar\omega)/e$. Eq. (2.9) accounts for the step structure in the d.c. I-V curve.

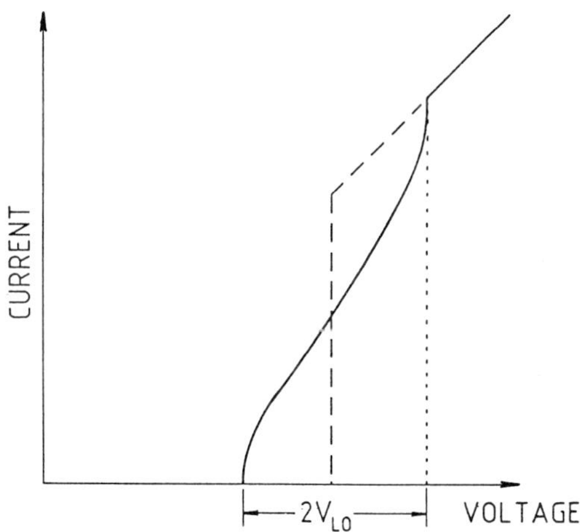

Fig. 3.2 The effect of microwave irradiation on the quasiparticle characteristic of an SIS junction using a classical calculation (Ibrügger, 1987).

3.3 Quantum mixing with the SIS junction

Tucker (1979, 1980), Shen (1981), Hartfuss and Tutter (1983, 1984), Tucker and Feldman (1985), Devyatov et al. (1986), Feldman (1987, 1987a), Winkler (1987) and others have analysed the performance of quasiparticle mixing with SIS junction using quantum theory. We discuss the results of these studies in some detail.

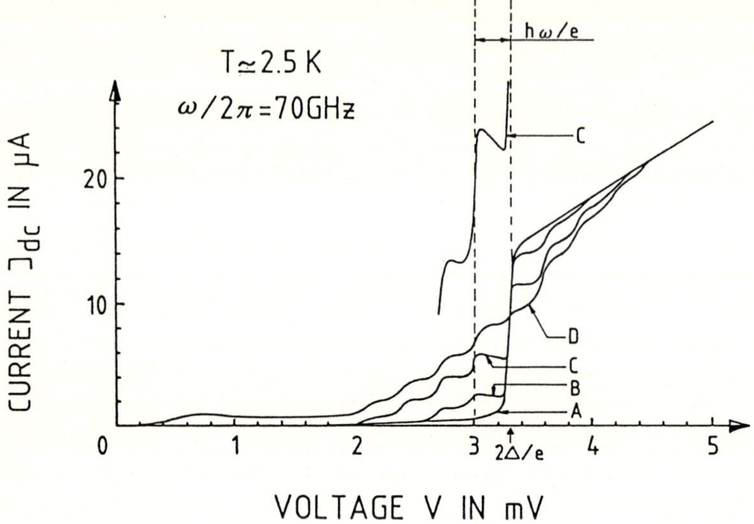

Fig. 3.3 The effect of 70 GHz microwave irradiation on the quasiparticle characteristic of a Pb/Bi/In-oxide-Pb/Bi junction. Curve (A) without, curves (B) to (D) with microwave irradiation of increasing power (Gundlach et al., 1985).

(i) Mixer conversion gain

Contrary to the classical resistive mixer, the quasiparticle mixer can have conversion gain: L_M can be smaller than one. We mentioned already that conversion gain reduces, according to eq. (3.2), the contribution of the IF amplifier noise to the receiver noise temperature. The conversion gain was first observed by Shen et al. (1980) and Kerr et al. (1981). It is not easy to understand and explain in simple terms why the SIS mixer can produce conversion gain.

The first thing to note is that, from eq. (2.8), an *ac* voltage $V_{rf}\cdot\cos\omega t$ across the SIS junction not only generates a dissipative (in-phase) current, but also a reactive (out-of-phase) current. This implies that an SIS junction reacts with time delay to an external *ac* voltage. Thus, one can suggest that the reactances represented by the SIS junction give rise to parametric effects including parametric amplification.

A quite different approach to discuss the conversion gain is to analyse the mixer performance in the limit of small local oscillator power. In this case, one does not rely on computer calculation and can derive an analytical expression for L_M. We do not repeat here the detailed derivation of L_M, which has been reported by Gundlach et al. (1981), but present only the essential points. We consider the simpliest case, a two-port mixer. It has only a signal and IF port and its minimum conversion gain may be written as

$$L_M = \frac{G_{10}}{G_{01}} \cdot \frac{1+\sqrt{1-\eta}}{1-\sqrt{1-\eta}} \tag{3.8}$$

where

$$\eta = \frac{G_{10} \cdot G_{01}}{G_{11} \cdot G_{00}} \tag{3.9}$$

Subject to the conditions

$$\alpha_{LO} = \frac{eV_{LO}}{\hbar\omega_{LO}} << 1 \qquad \alpha_s = \frac{eV_s}{\hbar\omega_s} << 1 \qquad \alpha_{IF} = \frac{eV_{IF}}{\hbar\omega_{IF}} << 1 \tag{3.10}$$

the mixer matrix elements

$$G_{00} = \frac{I_{IF}}{V_{IF}}, \qquad G_{01} = \frac{I_{IF}}{V_s}, \qquad G_{10} = \frac{I_s}{V_{IF}}, \qquad G_{11} = \frac{I_s}{V_s} \tag{3.11}$$

can be shown to take the simple forms in the intervall $2\Delta - \hbar\omega_{LO} < eV_o < 2\Delta$:

$$G_{00} \simeq \frac{\alpha_{LO}^2}{4R_N} \qquad G_{01} \simeq \frac{\Delta \cdot \alpha_{LO}}{\hbar\omega_{LO} \cdot R_N} \tag{3.12}$$

$$G_{10} \simeq \frac{\alpha_{LO}}{4R_N} \qquad G_{11} \simeq \frac{\Delta}{\hbar\omega_{LO} \cdot R_N}$$

Here R_N is the normal state resistance of the junction. It follows from eqs. (3.9) and (3.12) that $\eta = 1$, and thus from eqs. (3.8) and (3.12),

$$L_M = \frac{G_{10}}{G_{01}} \simeq \frac{\hbar\omega_{LO}}{4\Delta} \tag{3.13}$$

It is seen that $L_M < 1$ for $\hbar\omega_{LO} < 4\Delta$. The conversion gain arises because (cf. eqs. (3.11))

$$\frac{I_{IF}}{V_s} > \frac{I_s}{V_{IF}}, \tag{3.14}$$

which means that the signal voltage V_s creates more efficiently a current I_{IF} at the intermediate frequency than does the voltage V_{IF} for the signal current I_s. As a consequence, the mixer can convert power more efficiently power from the signal to the IF frequency than from the IF to the signal frequency. The quasiparticle mixer is non-reciprocal ($G_{10} \neq G_{01}$). A classical resistive mixer is reciprocal ($G_{10} = G_{01}$). The non-reciprocity of the quasiparticle mixer is related to the fact that the signal photons induce photon-assisted tunneling, but the IF photons do not do likewise because the IF frequency is too low. The usual bias point

$$V_o \simeq (2\Delta - \frac{\hbar\omega_{LO}}{2})/e \tag{3.15}$$

prevents photon-assisted tunneling at the intermediate frequency.

If this conclusion is correct, the gain should disappear for bias voltage V_o in the interval

$$(2\Delta - \frac{\hbar\omega_{IF}}{2})/e < V_o < 2\Delta/e \tag{3.16}$$

The results presented by eqs. (3.12) hold only in the limit $\omega_{IF} \to 0$. We have evaluated the mixer matrx elements for finite intermediate frequency (Gundlach et al., 1981) and discuss the results for

$$\omega_{IF} = \frac{\omega_{LO}}{3} = \frac{\omega_s}{2} \tag{3.17}$$

In this case, the mixer matrix elements for $2\Delta - \hbar\omega_{IF} < eV_o < 2\Delta$ are

$$G_{00} \simeq \frac{\Delta}{\hbar\omega_{IF} \cdot R_N} \qquad G_{01} \simeq \frac{\Delta \cdot \alpha_{LO}}{\hbar\omega_s \cdot R_N} \tag{3.18}$$

$$G_{10} = \frac{-\Delta \cdot \alpha_{LO}}{2\hbar\omega_{IF} R_N} \qquad G_{11} \simeq \frac{\Delta}{\hbar\omega_s R_N}$$

Inserting these results into eq. (3.8) yields the conversion loss

$$L_M \to \infty \qquad \text{for} \qquad \alpha_{LO} = \frac{eV_{LO}}{\hbar\omega_{LO}} \to O \tag{3.19}$$

It is seen that the conversion gain not only disappears in the intervall $2\Delta - \hbar\omega_{IF} < eV_o < 2\Delta$, but that the conversion loss becomes infinitely large. This result is analogous to that of the classical mixer for which the conversion loss approaches infinity as the local oscillator power (α_{LO}) approaches zero. It should be noted that conversion gain still exists for $2\Delta - \hbar\omega_{LO} + \hbar\omega_{IF} < eV_o < 2\Delta - \hbar\omega_{IF}$. Table 1 shows the dependence of L_M on the *dc* voltage bias V_o as has previously been derived (Gundlach et al., 1981).

In conclusion, the mixer matrix elements given by eq. (3.12) and (3.18) and the calculation of L_M from eq. (3.8) reveal that the conversion gain predicted for small local oscillator power occurs if the intermediate frequency voltage V_{IF} creates only a small current I_{IF} at the intermediate frequency (cf. G_{00}), as well as only small current at the signal frequency (cf. G_{10}). This is possible because the *dc* bias voltage $V_o < (2\Delta - \hbar\omega_{IF})/e$ prevents (low-order) photon-assisted tunneling current at the intermediate frequency, whereas signal photons create photon-assisted tunneling and thus not only large current I_s at the signal frequency (cf. G_{11}), but also large current I_{IF} (cf. G_{01}).

The above discussion on the performance of the quasiparticle mixer for small local oscillator voltage was introduced to provide some insight into the physics of the conversion gain. We now turn to the more realistic case of moderate local oscillator voltage ($eV_{LO}/\hbar\omega_{LO} \simeq 2$). It has been argued (Richards and Shen, 1980) and there is also experimental evidence (Shen et al., 1980, Kerr et al., 1981) that the conversion gain is proportional to the dynamic resistance R_d at the photon steps shown in Fig. 3.3. The mixer can be considered as a current source for I_{IF} shunted by R_d. The value of R_d

depends on the signal source impedance, on R_N and the local oscillator power. Even regions with negative values of R_d can be obtained (McGrath et al., 1981; Kerr et al., 1981). The change of R_d by varying the local oscillator power is exemplified in Fig. 3.3. Negative resistance implies arbitrary large conversion gain. Large values of $|R_d|$ involve the problem of impedance matching to the IF amplifier which is usually optimised for a load resistance of about 50Ω. A transformer between the SIS junction and the IF amplifier can be used for impedance matching. So far, such transformers considerably reduce the desired IF-bandwidth from 500 MHz or more to about 50 MHz (Räisänen et al., 1986). Negative values of R_d involve the problem of oscillation. Until now, conversion gain has not been utilised in receivers being used in radio telescopes. In practise, though quasiparticle mixers have less conversion loss than cooled Schottky mixers.

Voltage Range	L_M
$(2\Delta - \hbar\omega_{IF})/e < V_o < 2\Delta/e$	∞
$(2\Delta - \hbar\omega_{LO} + \hbar\omega_{IF})/e < V_o < (2\Delta - \hbar\omega_{IF})/e$	$\simeq \hbar\omega_s/4\Delta$
$(2\Delta - \hbar\omega_{LO})/e < V_o < (2\Delta - \hbar\omega_{LO} + \hbar\omega_{IF})/e$	$> \hbar\omega_s/4\Delta$
$(2\Delta - \hbar\omega_s)/e < V_o < (2\Delta - \hbar\omega_{LO})/e$	$\simeq \omega_s/\omega_{IF}$

Table 1 Dependence of the mixer conversion loss $L_M = P_s/P_{IF}$ on the *dc* bias voltage V_o for $eV_{LO}/\hbar\omega_{LO} << 1$ and large IF frequency ($\omega_{IF} = \omega_{LO}/3 = \omega_s/4$).

(ii) Mixer noise temperature

Tucker (1979, 1980) realized that his quantum theory of quasiparticle mixing yielded very low mixer noise temperatures T_M for certain sets of parameters. It was soon reported that T_M can reach even the quantum limit, which was expected to be of the order of $T_M \simeq \hbar\omega/k$. This can be seen as follows. The quantum limit of T_M is a consequence of the Heisenberg uncertainty principle. According to Tucker and Feldman (1985) the uncertainty in the measurement of a flow of photons is at least one photon in the observation time Δt, that is, $\hbar\omega/\Delta t$. As the observation time is related to the bandwidth Δf such that $\Delta t = 1/\Delta f$, the observation of a single photon corresponds to an input power of $\hbar\omega\Delta f$. Expressing this power as a temperature by equating it to $kT \cdot \Delta f$, yields a minimum noise temperature of $\hbar\omega/k$.

Caves (1982) has derived the noise power, referred to the input of a linear, high-gain amplifier with the result

$$P_{in} = [A + \frac{1}{2}coth(\frac{\hbar\omega}{2kT_N})]\hbar\omega \cdot \Delta f \tag{3.20}$$

Here, A is the photon number equivalent noise added by the amplifier to the input signal. The minimum value for A is 1/2. Thus, the minimum noise power added to the signal is $(\hbar\omega/2)\cdot\Delta f$. Equating this to $kT\cdot\Delta f$, the minimum noise temperature of the amplifier is $\hbar\omega/2$k. A high-gain amplifier produces a large multiplication of the number of photons. The mixer is a high-gain amplifier in the sense that it produces a large number of IF photons per signal photon and preserves the phase information of the signal input. It has been suggested (Feldman and Rudner, 1983) and later theoretically shown (Feldman, 1987) that the minimum noise temperature obtainable for a mixer is also $T_M = \hbar\omega/2k$. It may preferable to say: the minimum noise power the mixer adds to the signal power is $(\hbar\omega/2)\cdot\Delta f$. There are several ways to convert this noise power into a noise temperature. To have a linear relationship between noise power and noise temperature, one may write

$$\frac{\hbar\omega}{2}\Delta f = kT_M\Delta f \tag{3.21}$$

and thus $T_M = \hbar\omega/2k$. This is only a convention which has the advantage that one can add noise temperatures.

We will now discuss why the SIS quasiparticle mixer can reach the quantum limit of noise. The main reason is probably the shape of the I_{dc}-V characteristic of the SIS junction. For the ideal junction at T = 0, the *dc* current in the operating point is zero. Thus, the unpumped junction produces no shot noise. As we assumed T = 0, there is also no thermal noise. One can expect that at finite temperature a high quality junction produces only little noise (contrary to the Schottky diode). In additon, the strong non-linear *dc* I-V characteristic requires only little local oscillator power P_{LO}. In a classical picture, the local oscillator voltage V_{LO} switches the SIS junction on and off. Hence, at the operating point

$$V_o = \frac{2\Delta}{e} - \frac{\hbar\omega_{LO}}{2e} \tag{3.22}$$

the local oscillator voltage should be larger than $\hbar\omega_{LO}/2$e. To estimate roughly the shot noise generated by the local oscillator, we take

$$V_{LO} = \frac{\hbar\omega_{LO}}{e} \tag{3.23}$$

The local oscillator power may be written as

$$P_{LO} \simeq \frac{V_{LO}^2}{2R_N} = \frac{(\hbar\omega_{LO})^2}{2e^2R_N} \tag{3.24}$$

The current responsivity reads

$$\eta_c = \frac{I_{LO}}{P_{LO}} \tag{3.25}$$

If each local oscillator photon creates one electron passing the tunnel barrier, η_c also may be written as

$$\eta_c = \frac{e}{\hbar\omega_{LO}} \tag{3.26}$$

From eqs. (3.24-3.26) we obtain

$$I_{LO} = \frac{\hbar\omega_{LO}}{2eR_N} \tag{3.27}$$

Thus the mean square of the shot noise current

$$\langle i_s^2 \rangle = 2eI_{LO} \cdot \Delta f \tag{3.28}$$

becomes

$$\langle i_s^2 \rangle = \frac{\hbar\omega_{LO}}{eR_N}\Delta f \tag{3.29}$$

One can associate a noise temperature to this mean square noise current by presuming that it arises from Nyquist noise:

$$\langle i_N^2 \rangle \simeq \frac{4kT_D}{R_N} \cdot \Delta f \tag{3.30}$$

From

$$\langle i_N^2 \rangle = \langle i_s^2 \rangle \tag{3.31}$$

one finds

$$T_D = \frac{\hbar\omega}{4k} \tag{3.32}$$

It is seen that the resulting noise temperature referred to as diode noise temperature, is smaller than the above discussed quantum limit $T_M = \hbar\omega/2k$, indicating that the shot noise created by the local oscillator is so low that one can expect to approach the quantum limit of the mixer noise temperature.

In the evaluation of the mixer noise generated by the local oscillator, one must consider that all mixer sidebands contribute to the noise. Taking a three-port mixer and assuming equal conversion efficiency from the signal and image frequency to the intermediate frequency, the expression for the mean square noise current, referred to the mixer input port, becomes (Hartfuss and Tutter, 1984)

$$\langle i^2 \rangle = [H_{00} + 2(\lambda_{01} + \lambda_{01}^*)H_{10} + (\lambda_{01}^2 + \lambda_{01}^{*2})H_{1-1} + 2|\lambda_{01}|^2 \cdot H_{11}] \cdot \Delta f \tag{3.33}$$

The term λ_{01} is related to mixer conversion loss: $|\lambda_{01}|^2 = G_L/(G_S\, L_M)$ where G_L is the matched output conductance, and G_S is the conductance of the signal source. The matrix element H_{00} describes the mean square noise current due to the *dc* current generated by the local oscillator. $H_{11} = H_{-1-1}$ are the components of H_{00} at signal and image frequency. H_{10} and H_{1-1} are the corresponding cross correlations of the mean square noise current. The matrix element H_{00} has the form

$$H_{00} = 2e \sum_{n=-\infty}^{+\infty} J_n(\alpha) I_n \, coth(\frac{eV_n}{2kT}) \tag{3.34}$$

The current I_n represents the *dc* I-V characteristic for zero local oscillator power evaluated at voltages $V_n = V_o + n\hbar\omega/e$. If $\alpha = eV_{LO}/\hbar\omega_{LO}$ is small enough, large values of the index n do not contribute to H_{00}. Furthermore, if the physical temperature T of the junction is sufficiently low, so that

$$eV_n > 2kT, \tag{3.35}$$

we obtain

$$H_{00} \cdot \Delta f = 2e \sum I_n(\alpha) \cdot J_n \cdot \Delta f \tag{3.36}$$

This expression describes the shot noise. For zero local oscillator power ($\alpha = 0$) we obtain

$$H_{00} \cdot \Delta f = 2e\, I_o \cdot \Delta f, \tag{3.37}$$

the "classical" shot-noise formula.

The other current correlation matrix elements H_{10}, H_{1-1}, and H_{11} have forms similar to that for H_{00}. They can be found, e.g., in the paper of Feldman (1987).

An interesting observation was noted by Hartfuss and Tutter (1984). The cross correlation matrix elements H_{10} and H_{1-1} are **not** always positive as can be seen in Fig. 3.4. The noise currents from the various mixer ports are correlated and can interfere with each other in such a way that the mean square noise current given by eq. (3.33) can be

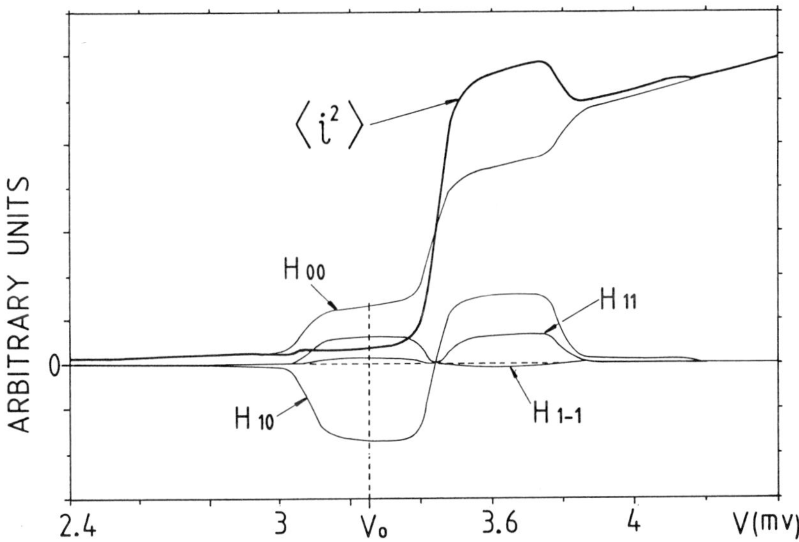

Fig. 3.4 The quantities H_{00}, H_{11}, H_{10}, H_{1-1} and their sum $\langle i^2 \rangle$ according to eq. (3.33) (Hartfuß and Tucker, 1984).

smaller than $H_{00} \cdot \Delta f$, the mean square noise current due to the *dc* current generated by the local oscillator.

We have seen that the local oscillator produces only small shot noise currents in and between the various sidebands of the mixer. The interference of these noise currents can reduce the noise further. The detailed evaluation of the minimum mixer noise temperature performed by Feldman (1987) has yielded the following results:

SSB three-poort mixer

If the image port is reactively terminated so that none of the image field noise is matched to the junction, the minimum mixer noise temperature is

$$T_{min} = \hbar\omega/2k$$

In this case, all the quantum noise results from shot-noise current created by the local oscillator. Any noise entering the signal port is counted as a part of the signal; this includes even the zero-point fluctuations. T_{min} represents the minimum noise power the mixer adds to the total radiation coming into the signal port of the mixer.

DSB three-port mixer

(1) The interesting input enters only the signal port.

$$T_{min} = \hbar\omega/2k$$

In this case, the shot-noise power is zero. The entire quantum noise results from zero-point fluctuations in the external electromagnetic field at the image frequency.

(2) The interesting input enters the signal as well as the image port, and the radiation fields at the signal and image port are assumed to be correlated. In this case

$$T_{min} << \hbar\omega/2k$$

This surprising result for the DSB mixer has been derived by Likharev and Zorin (1984), Zorin(1985) and Devayatov et al. (1986). Some explanations may be useful. As outlined under (1), for a DSB mixer of which the interesting input enters only the signal port, the entire quantum noise results from zero-point fluctuations at the image port. If the interesting input also enters the image port, zero-point fluctuations at the image port (as for the signal port) are counted as part of the signal and not as noise.

Does the statement that the DSB mixer has no fundamental bound to the noise it can add to the signal not conflict with Heisenberg's uncertainty principle? The answer is no because the DSB mixer is only sensitive to one quadrature component of the incoming radiation. There is an uncertainty relationship between the two quadrature components of the incoming radiation. The one which is in phase with the local oscillator wave can (in principle) be measured exactly with the DSB mixer, whereas the information in the other quadrature component is completely lost. A SSB mixer is sensitive to both quadrature components of the incoming radiation. The DSB mixer is of interest for the measurement of states for which one quadrature component has reduced quantum noise and the other component has increased quantum noise. Such states of the electromagnetic field are known as squeezed states (Leuchs, 1987).

(iii) Local Oscillator Power

A further advantage of the SIS quasiparticle mixer is the small amount of local oscillator power required for optimum operation. This also arises from the strong non-linear current-voltage characteristic. The local oscillator power of the quasiparticle mixer is only approximately $5{\cdot}10^{-8}$W at 240 GHz; about two to three order of magnitude lower than for cooled Schottky mixers. This is of great practical importance as high frequency oscillators, such as klystrons and carcinotrons, are expensive, voluminous, and have short lifetimes. For the quasiparticle mixer, frequency multipliers driven by a lower frequency source, such as Gunn oscillator or klystron, can provide sufficient local oscillator power. In the future it may be possible to use the Josephson oscillation in a series array of small SIS junctions as a local oscillator source (I.E. Lukens et al., 1988). The oscillatory motion of magnetic flux quanta in a series array of long SIS junctions could eventually also be exploited as a local oscillator source (Pederson, 1986, Monaco et al., 1988).

4. Practical SIS Heterodyne Receivers

4.1 Millimetre wavelengths

Currently about 12 SIS heterodyne receivers are in routine use in radio telescopes. The lowest frequency is about 50 GHz, and the highest is about 250 GHz. There is little doubt that SIS receivers, although they are cryogenically more complicated, will eventually replace Schottky receivers. The receiver at about 50 GHz uses a series array of Nb–AlO_x–Nb junctions. It is installed at the Nobeyama Radio Observatory in Japan, and has an SSB noise temperature of about 100 K (Inatani et al., 1987). Most SIS receivers work in the 100 GHz frequency range. The IRAM 100 GHz receiver uses a Pb/Bi/In-oxide–Pb/Bi junction (Gundlach et al., 1982). This receiver has been in operation for almost three years, and has a DSB noise temperature of about 100 K (Blundell et al., 1983). So far, the lowest receiver noise temperatures in the 100 GHz frequency range are about 80 K SSB (Pan et al., 1983, 1987). It is, of course, difficult to compare receivers. They generally have different input and/or IF bandwidths, and differ in the variation of the receiver noise temperature with local oscillator frequency. Some receivers have a single, movable, *rf* tuning element, while other receivers have two.

In the 150 GHz frequency range, DSB receiver noise temperatures around 100 K have been achieved (Ibrügger et al., 1984, 1987, Hilberath et al., 1985). The noise temperature of a receiver covering the frequency range 215–250 GHz is plotted in Fig. 4.1 (Blundell et al., 1988). At 220 and 230 GHz, the DSB noise temperature is only 80 K. We now describe this receiver in greater detail.

The SIS mixer incorporates a single Pb/Bi/In-oxide–Pb/Bi tunnel junction as the mixing element. The junction has an area of 1.5 μm^2 and a normal state resistance of 50 Ω. It was evaporated together with integrated low-pass filter structures onto a fused quartz substrate 5 mm long, 0.5 mm wide and 0.1 mm thick. The substrate is held in a waveguide mixer block in such a way that the junction is at the centre of the waveguide cross section (0.13 x 1.1 mm), as can be seen in Figure 4.2. One side of the junction is electrically connected, via a low-pass filter structure and a spring-loaded contact, to the mixer block. The other side is connected, again via a low-pass filter structure, to a modified SMA connector through which *dc* bias is applied to the junction. The IF output is via the same SMA connector to an amplifier chain of centre frequency 3.95 GHz and bandwidth 650 MHz. The first stage of the IF amplifier chain incorporates a low-noise HEMT cooled to 15 K. The average IF noise temperature is 11–12 K. Input to the mixer is made via a corrugated feed horn and rf tuning of the mixer is achieved using a sliding short circuit of the contacting type.

The low temperature required to operate the SIS mixer is obtained with a hybrid cryostat (Blum 1986). A commercially available closed-cycle He refrigerator (CTI–350 CP) is used to cool radiation shields to 70 and 15 K. These shields surround a liquid

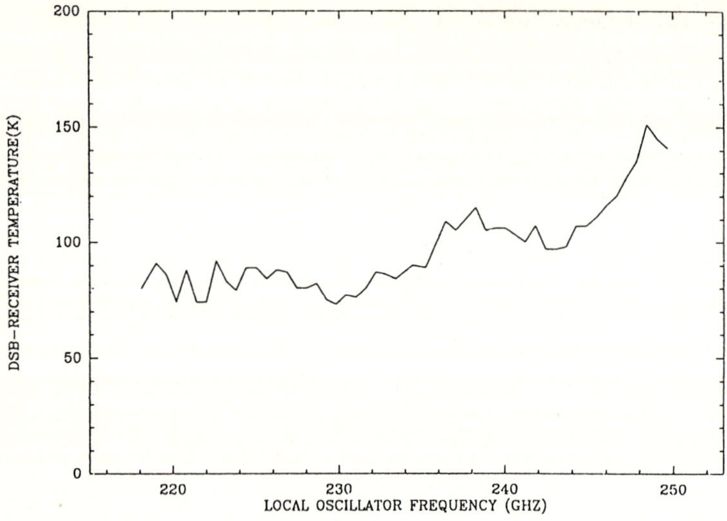

Fig. 4.1 DSB receiver noise temperature as a function of LO frequency (Blundell et al. 1988).

helium reservoir whose temperature can be lowered to about 2.5 K by reducing the pressure above the boiling liquid. The helium boil-off gas is used to cool the mixer, which is mounted in a separate chamber, indicated in Figure 4.2. The signal and LO power, derived from a klystron–frequency tripler combination, are coupled to the mixer feed horn via a quasi-optical diplexer and a room temperature, high-density polyethylene lens. Low loss windows on the 70 and 15 K radiation shields stop a large fraction of the incoming infrared radiation, thus reducing the heat load on the cryostat. Finally, a fused quartz window, approximately half a wavelength thick at 230 GHz, forms a helium-tight seal on the mixer chamber. This receiver has been in routine use at the IRAM 30-m telescope since September 1987.

4.2 Submillimetre wavelengths

In the past 3 to 4 years, SIS mixers for submillimetre wavelengths (Wengler et al., 1985, Büttgenbach et al., 1988) have been under development. Low-noise quasiparticle mixing can be expected for frequencies f_l up to twice the gap frequency $4\Delta/h$. For all Pb/Bi(In) junctions, $\Delta \simeq 1.7$ meV, yielding $f_l \simeq 1.6$ THz. However, pair tunneling in SIS junctions can set a considerably lower frequency limit. The pair current causes the so-called drop-back voltage

$$V_d = K\sqrt{\frac{\hbar\omega \cdot 2\Delta}{e^2\omega R_N \cdot C}} \tag{4.1}$$

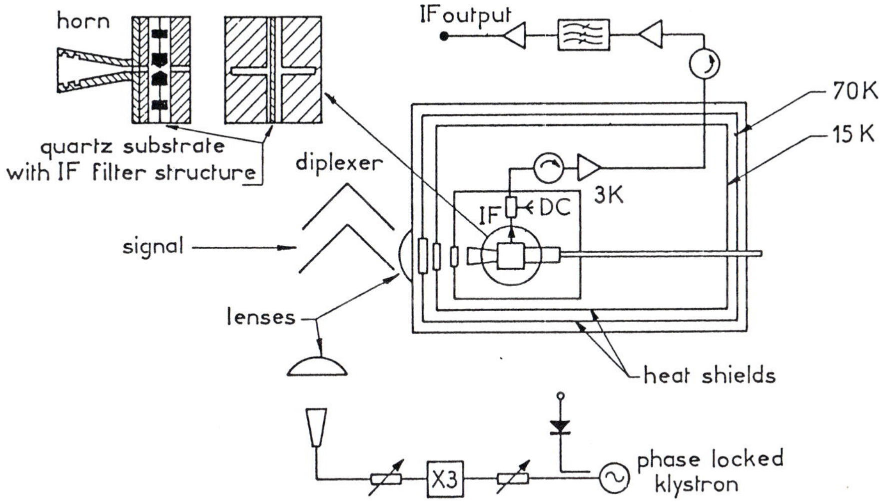

Fig. 4.2 Block diagram of the 1.3 mm receiver. The mixer assembly is shown in detail at the upper left (Blundell et al. 1988).

where K is a contant (Tucker and Feldman, 1985). The drop-back voltage creates a sharp rise of the mixer's output noise for *dc* bias voltage below the threshold value

$$V_N = V_{LO} + V_d \tag{4.2}$$

In principle, the zero-voltage pair current in the *dc* I–V characteristic, and thus the drop-back voltage, can be quenched by applying a magnetic field so that one flux quantum is within the junction cross-sectional area. Quasiparticle mixers for submillimetre wavelengths require high current density junctions ($\geq$ 1000 A/cm^2) with areas below 1 μm^2. In such junctions, the pair current, and thus V_d, can often not be suppresed completely, so the upper frequency limit will be below $2\Delta/e$, depending on the remaining value of the threshold voltage V_N.

There are other Josephson phenomena which can conflict with optimum quasiparticle mixing. For junctions with a gap voltage of about 3.4 mV, the first quasiparticle conversion curve is smooth for frequencies up to about 250 GHz, even though no magnetic field is applied. This can be seen, for example, in Fig. 1 of the paper of Ibrügger et al. (1985) and in Fig. 1 of the paper by Blundell and Gundlach (1985). With increasing frequency, Josephson steps of lower order, and hence larger amplitude, enter the region of the photon step used for quasiparticle mixing, and cause distortion of the smooth conversion curve. Feldman (1987a) pointed out that this effect reduces the saturation power of the SIS mixer. The Josephson interference can be so strong that the conversion curve is drastically deformed, even in the presence of a magnetic field which minimizes the *dc* pair current (Wengler et al. 1985). As a consequence, not only the saturation power is reduced, but the mixer also operates in a combination of Josephson and quasiparticle modes. It is still an open question if and how seriously Josephson mixing in SIS junctions generates additional noise. In point contacts, Josephson mixing was not successful. It was suggested that down conversion of noise from higher sidebands has hindered good performance (Tucker and Feldman, 1985).

In spite of these problems, SIS receivers for submillimetre wavelengths have been built. For the coupling of the *rf* signal to the junction, one can use either standard waveguide techniques or quasi-optical approaches. With increasing frequency, waveguide mixers become more difficult to fabricate and to handle. On the other hand, quasi-optical systems with planar antenna structures are well adapted for SIS junctions. Wengler et al. (1985) used quasi-optical *rf* coupling via a bow-tie antenna on a quartz lens. The DSB receiver noise temperature was 205 K at 116 GHz, 375 K at 350 GHz, and 815 K at 466 GHz. The polar diagram of a bow-tie antenna deposited on a substrate has large side lobes which prevent efficient performance at the telescope. Better in this respect is the planar two-arm logarithmic spiral antenna. It has been used by Büttgenbach et al. (1988) for an SIS receiver, and provides a very large input bandwidth. The receiver was reported to operate at 115 GHz with a DSB noise temperature of only 33 K, and at 762 GHz with a DSB receiver noise temperature of 1100 K.

A general feature of SIS receivers above 300 GHz is that their noise temperatures are considerably higher than theoretically expected for quasiparticle mixing. This can be due partly to the fact that it is more difficult to build a receiver for submillimetre than for millimetre wavelengths, and partly due to the interference from Josephson effects. To reduce the interference from Josephson effects, one could use junctions with higher gap voltages, such as all Pb/Bi (Gundlach et al., 1982), or even better, all NbN junctions (Shoji, 1987). Alternative approaches involve series arrays of SIS junctions or SIN (superconductor–insulator–normal metal) junctions. For a series array, the individual junction can be larger in size, so the *dc* Josephson current can be suppresed with a smaller magnetic field than would be needed for a single junction. The SIN junction has no Josephson current, but its current–voltage characteristic is less sharp than that of an SIS junction. Thus, the lower frequency limit of quantum mixing with an SIN junction is higher than with the corresponding SIS junction. The highest frequency used so far in mixing with SIN junctions is about 230 GHz (Blundell and Gundlach, 1987). The lowest DSB receiver noise measured was 230 K at 218 GHz. At submillimetre wavelengths, lower noise temperatures can be expected because the current–voltage characteristic will be sharper on the photon voltage scale $\hbar\omega/e$. Consequently, the SIN junction can be an interesting alternative for high-frequency heterodyne receivers where pair tunneling in SIS junctions could give rise to interference from Josephson effects.

References

Barber, M.R. (1967), "Noise figure and conversion loss of the Schottky barrier mixer diode", *IEEE Transactions on Microwave Theory and Techniques*, **MTT-15**, 629

Blum, E.J. (1986), "A cryostat for radio astronomy receivers using superconducting mixers at millimetre wavelengths", *Advances in Cryogenic Engineering*, **31**, 551

Blundell, R., Gundlach, K.H., Blum, E.J. (1983), "Practical low-noise quasiparticle receiver for 80–100 GHz", *Electronics Lett.*, **19**, 498

Blundell, R., and Gundlach, K.H. (1985), "SIS junction response from the millimetre into the submillimetre wave region", *SPIE Vol.* **598** Instrumentation for Submillimeter Spectroscopy, pp. 16

Blundell, R. and Gundlach, K.H. (1987), "A quasiparticle SIN mixer for the 230 GHz frequency range", *Int. J. of Infrared and Millimeter Waves*, **8**, 1573

Blundell, R., Carter, M., Gundlach, K.H. (1988), "A low-noise SIS receiver covering the frequency range 215–250 GHz", *Int. J. of Infrared and Millimeter Waves*, **8**, 361

Büttgenbach, T.H., Miller, R.E., Wengler, M.J., Watson, D.G., Phillips, T.G. (1988), "A broadband low noise SIS receiver for submillimeter astronomy", to be published

Caves, C.M. (1982), "Quantum limit of noise in linear amplifiers", *Phys. Rev.*, **D26**, 1817

Dayem, A.H. and Martin, R.J. (1962), "Quantum interaction of microwave radiation with tunneling between superconductors", *Phys. Rev. Lett.*, **8**, 246

Devyatov, I.A., Kuzmin, L.S., Likharev, K.K., Migulin, V.V., Zorin, A.B. (1986), "Quantum – statistical theory of microwave detection using superconducting tunnel junctions", *J. Appl. Phys.*, **60**, 1808

Downes, D. (1983), "H_2O masers in star-forming regions", in *Birth and Infancy of Stars*, edited by R. Lucas, A. Omont and R. Stora, North-Holland, Amsterdam, pp. 557

Face, D.W., Prober, D.E., McGrath, W.R., Richards, P.L. (1986), "High quality tantalum superconducting tunnel junctions for microwave mixing in the quantum limit", *Appl. Phys. Lett.*, **48**, 1098

Feldman, M.J. and Rudner, S. (1983), "Mixing with SIS arrays", in *Reviews of Infrared and Millimeter Waves*, edited by K.J. Button (Plenum Publishing Corporation), Vol. **1**, pp. 47

Feldman, M.J. (1987), "Quantum noise in the quantum theory of mixing", *IEEE Transactions on Magnetics*, **MAG-23**, 1054

Feldman, M.J. (1987a), "Saturation of the SIS mixer", Extended Abstracts of the *International Superconductivity Electronics Conference* (ISEC '87), August 28–29, Tokyo, Japan, pp. 290

Feldman, M.J. (1988), "Theoretical considerations for THz SIS mixers", *Int. J. of Infrared and Millimeter Waves*, **8**, 1287

Giaever, I. (1960), "Energy gap in superconductors measured by electron tunneling", *Phys. Rev. Lett.*, **5**, 147

Guélin, M. (1988), "Organic and exotic molecules in space", in *Molecules in Physics, Chemistry and Biology*, edited by D. Reidel, in press

Guélin, M., Cernicharo, J., Penalver, J. (1988), unpublished data

Gundlach, K.H., Hartfuss, H.J., Takada, S. (1981), "Photon-assisted tunneling and frequency mixing in SIS junctions", Max-Planck-Institut für Physik und Astrophysik, Institut für Astrophysik, *International Report MPA 21*

Gundlach, K.H., Takada, S.,Zahn, M., Hartfuss, H.J. (1982), "A new lead alloy tunnel junction for quasiparticle mixing and other application", *Appl. Phys. Lett.*, **41**, 294

Gundlach, K.H., Blundell, R., Blum, E.J. (1985), "Eine Neuentwicklung für die Radioastronomie: der SIS-Empfänger", *Mikrowellen Magazin*, **11**, 32

Hartfuss, H.J., Gundlach, K.H., Kadlec, J. (1980), "Video detection of mm-wave radiation using SIS Josephson junction", *SQUID '80*, Walter de Gruyter and Co., Berlin – New York, pp. 841

Hartfuss, H.J. and Gundlach, K.H. (1980), "MM-wave detection using quasiparticle tunneling", *SQUID '80*, Walter de Gruyter and Co., Berlin – New York, p. 851

Hartfuss, H.J. and Gundlach, K.H. (1981), "Video detection of mm-waves via photon-assisted tunneling between two superconductors", *Int. J. of Infrared and Millimeter Waves*, **2**, 809

Hartfuss, H.J. and Tutter, M. (1983), "Numerical design calculation of mm-wave mixer with SIS tunnel junction", *Int. J. of Infrared and Millimeter Waves*, **4**, 993

Hartfuss, H.J. and Tutter, M. (1984), "Minimum noise temperature of a practical SIS quantum mixer", *Int. J. of Infrared and Millimeter Waves*, **5**, 717

Hilberath, W., Vowinkel, B., Gundlach, K.H. (1985), "145–GHz front end with SIS mixer", *SPIE Vol.* **598**, Instrumentation for Submillimeter Spectroscopy, pp. 20

Ibrügger, J., Okuyama, K., Blundell, R., Gundlach, K.H., Blum, E.J. (1984), "Quasiparticle 150–GHz mixer", *Proceedings of the 17th International Conference on Low Temperature Physics*, edited by U. Eckern, A. Schmid, W. Weber, and H. Wühl (North-Holland, Amsterdam), pp. 937

Ibrügger, J. (1987), unpublished result

Imamura, T., Hoko, H., Hasno, S. (1987), "Integration process for Josephson LSI based on Nb-Aloxide-Nb junctions", Extended Abstracts of the *International Superconductivity Conference* (ISEC '87), August 28–29, Tokyo, Japan, pp. 57

Inatani, J., Sakamoto, A., Tsuboi, M. (1987), "Nb-Aloxide-Nb junctions for mm-wave mixers", Extended Abstracts of the *International Superconductivity Conference*, August 28–29, Tokyo, Japan, pp. 103

Kerr, A.R., Pan, S.K., Feldman, M.J., Davidson, A. (1981), "Infinite available gain in a 115–GHz SIS mixer", *Physica*, **108 B**, 1369

Kreysa, E. (1984), "Bolometer systems developed at the MPIfR", Int. Symp. on *Millimeter and Submillimeter Wave Radio Astronomy U.R.S.I.*, Granada, Spain, pp. 153

Kreysa, E. (1987), private communication

Leuchs, G. (1986), "Photon statistics, antibunching and squeezed states in non-equilibrium quantum statistical physics", *Frontiers of Non-Equilibrium Physics*, edited by G.T. Moore and M.O. Scully, Plenum Press, New York, London, pp. 329

Likharev, K.K. and Zorin, A.B. (1984), "Quantum limitation of superconducting mixers", *Proceedings of the 17th International Conference on Low Temperature Physics*, edited by U. Eckern, A. Schmid, W. Weber and H. Wühl (North Holland, Amsterdam), pp. 1153

Lukens, J.E., Jain, A.K., Wan, K.L. (1988), "Application of Josephson effect arrays for submm sources", this volume

McGrath, W.R., Richards, P.L., Smith, A.D., van Kempen, H., Batchelor, R.A., Prober, D.E., Santhanan, P. (1981), "Large gain, negative resistance, and oscillations in superconducting quasiparticle heterodyne mixers", *Appl. Phys. Lett.*, **39**, 655

Monaco, R., Pagano, S., Costabile, G. (1988), "Superradiant emission from an array of long Josephson junctions", to appear in *Phys. Lett. A*

Morohashi, Skinoki, S.F., Shoji, A., Aoyagi, M., Hayakawa, H. (1985), "High quality Nb/Al-Aloxide-Nb Josephson junction", *Appl. Phys. Lett.*, **46**, 1179

Pan, S.-K., Feldman, M.J., Kerr, A.R., Timbie, P. (1983), "A low-noise 115–GHz receiver using superconducting tunnel junctions", *Appl. Phys. Lett.*, **43**, 786

Pan, S.-K., Kerr, A.R., Lamb, J.W., Feldman, M.J. (1987), "SIS mixers at 115 GHz using Nb/Al–Al_2O_3–Nb junctions", National Radio Astronomy Observatory, Charlottesville, Virginia, *Electronics Division International Report No. 268*

Pedersen, N.F. (1986), "Long Josephson junctions", in *Josephson Effect – Achievements and Trends*, edited by A. Barone (World Scientific, Singapore)

Phillips, T.G. and Woody, D.P. (1982), "Millimeter and submillimeter wave receivers", *Ann. Rev. Astron. Astrophysics*, **20**, 285

Richards, P.L. and Shen, T.M. (1980), "Superconductive devices for millimeter wave detection, mixing and amplification", *IEEE Trans. Electron. Devices*, **ED–27**, 1909

Richards, P.L., Shen, T.M., Harris, R.E., Lloyd, F.L. (1980), "Superconductor–insulator–superconductor quasiparticle junctions as microwave photon detectors", *Appl. Phys. Lett.*, **36**, 4802

Räisänen, A.V., Crété, D.G., Richards, P.L., Lloyd, F.L. (1986), "Wide band low noise mm-wave SIS mixers with a single tuning element", *Int. J. of Infrared and Millimeter Waves*, **7**, 1835

Shen, T.M., Richards, P.L., Harris, R.E., Lloyd, F.L. (1980), "Conversion gain in mm-wave quasiparticle heterodyne mixers", *Appl. Phys. Lett.*, **36**, 777

Shen, T.M. (1981), "Conversion gain in millimeter wave quasiparticle heterodyne mixers", *IEEE J., Quantum Electron.*, **QE–17**, 1151

Shoji, A., Aoyagi, M., Kosaka, S., Shinoki, F. (1987), "NbN–MgO–NbN Josephson tunnel junctions", *Superconductivity Electronics*, edited by K.O. Hara, Ohmsha, Kanda, Tokyo 101, pp. 182

Tien, P.K. and Gordon, J.P. (1963), "Multiphoton process observed in the interaction of microwave fields with the tunneling between superconductor films", *Phys. Rev.*, **129**, 647

Tucker, J. (1979), "Quantum limited detection in tunnel junctin mixers", *IEEE J., Quantum Electron.*, **QE–15**, 1234

Tucker, J. (1980), "Predicted conversion gain in superconductor– insulator–superconductor quasiparticle mixers", *Appl. Phys. Lett.*, **36**, 477

Tucker, J. and Feldman, M.J. (1985), "Quantum detection at millimeter wavelengths", *Rev. Mod. Phys.*, **57**, 1055

Wengler, M.J., Woody, D.P., Miller, R.E., Phillips, T.G. (1985), "A low-noise receiver for millimeter and submillimeter wavelengths", *Int. J. of Infrared and Millimeter Waves*, **6**, 697

Winkler, D. (1987), "Properties of quasiparticle mixers at frequencies corresponding to the superconducting energy gap", Department of Physics, Göteborg, ISBN 91–7032–307–0, Chalmers Bibliotheks Tryckeri

Zorin, A.B. (1985), "Quantum noise in SIS mixers", *IEEE Trans. Magn.*, **MAG–21**, 939

SIGNAL PROCESSING

T. Van Duzer
Department of Electrical Engineering and Computer Sciences
and the Electronics Research Laboratory
University of California, Berkeley, California 94720

INTRODUCTION

Signal processing places ever-increasing demands on circuit performance and is, therefore, an application that can take advantage of the extensive capabilities of superconductor circuits. In this chapter we will discuss both analog and digital superconductive devices and the roles they can play in signal processing. Some of the applications areas include high-resolution imaging radars, signal-source identification, spread-spectrum communications, and infrared imaging. These requirements press for the highest possible performance.

In a typical system for sensing and signal processing, the front end could be a wideband millimeter-wave heterodyne receiver containing antenna, mixer, local oscillator, and intermediate-frequency (and possibly radio-frequency) amplifier, or it might be an array of cooled infrared detectors in an imaging system. A millimeter-wave system for radar or spread-spectrum communications might contain a filter matched to one in the transmitter ("matched filter") for chirp radar or secure communications, a Fourier transformer to do spectrum analysis, or, perhaps, a correlator for signal pattern identification. These devices are coupled to analog-to-digital converters or comparators to digitize the signals in preparation for further manipulation in a digital processor. Infrared detector arrays for imaging might require analog-to-digital conversion or comparators to prepare the data for digital processing. In the absence of a superconductive processor, the data require multiplexing for coupling out of the cryogenic environment.

It is important to compare superconductive analog signal processors with room-temperature devices that perform the same functions and also to review the relative advantages of analog and digital processors. A comparison of digital and analog processors can be made using the matched filtering as an important example. [1] Digital programmable matched filtering can employ fast Fourier transforms (FFT) to minimize the number of operations. With that assumption, the required computation rate is at least $20B \log_2 TB$, where T is the length of the signal and B is its bandwidth. The product TB is called the "processing gain" and equals the improvement in signal-to-noise ratio that can be achieved with the processor. For most signal processing functions,

NATO ASI Series, Vol. F 59
Superconducting Electronics
Edited by H. Weinstock and M. Nisenoff

TB > 100 is required, and TB > 1000 satisfies almost all of the present requirements. Reible has plotted the useful graph in Fig. 1 which shows the bandwidth and processing

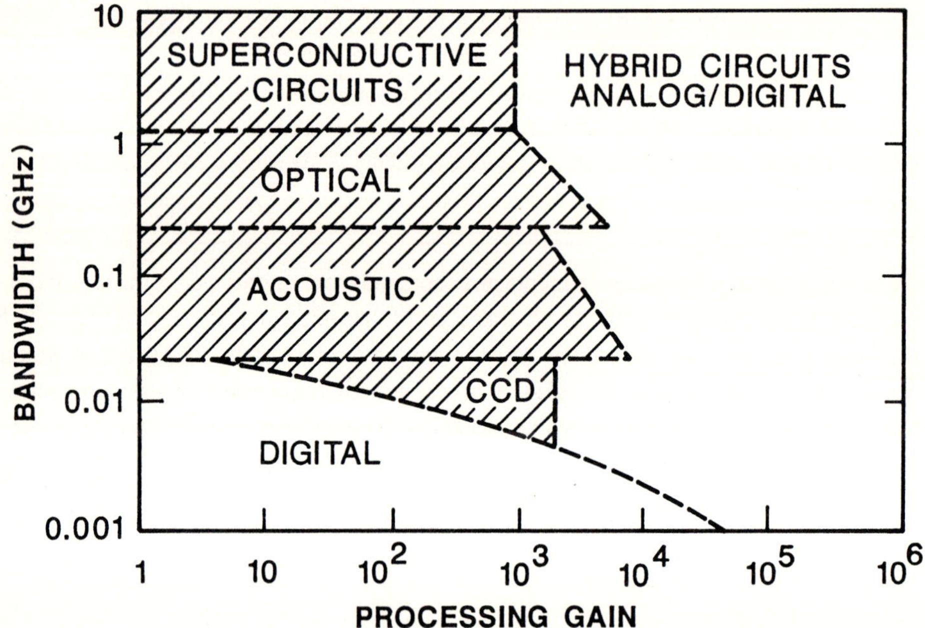

Fig. 1 Expected bandwidth and processing gains for several types of signal-processing devices. Extension of processing gain of analog devices (shown cross hatched) can be achieved by combinations with digital devices in hybrid systems, as suggested by the white area.

gain that can be achieved with various analog approaches (shown cross-hatched) and compares that with a dedicated semiconductor digital processor having a computation rate of 2×10^9 operation/second. [2] (It is possible that superconductive digital processors could provide higher operation rates.) Surface acoustic wave (SAW) devices are highly developed and provide up to about 300 MHz of bandwidth. Charge-coupled devices (CCDs) are somewhat slower but are more flexible. Optical and superconductive devices are experimental; the widest bandwidths (up to 10 GHz) are afforded by use of superconductive delay-line processors. Hybrid processors combining analog and digital components offer the possibility of the wide bandwidth of the analog device along with extended processing gain, as suggested by the white area marked "hybrid" in Fig. 1. It would be natural to do both the analog and digital parts with superconductive circuits. Superconductive analog-to-digital converters and shift registers promise order of magnitude improvement over semiconductor devices. The advantage in data processors may be less than that but very significant advances should be possible. For example, there is a recent report of a 4-bit microprocessor made in niobium superconductor technology that is more than ten times faster than its semiconductor counterpart, as will be discussed later in this chapter.

Another degree of flexibility may be offered by the new high-temperature superconductors. At convenient temperatures, such as 30-50 K, one might have sufficiently low RF surface losses to permit the use of superconductive delay lines but with enough cooling capacity to allow semiconductor circuits for the required auxiliary functions and for a digital processing component. Alternatively, if Josephson or other suitable devices are developed using the new superconductors, the entire processor could be superconductive.

In this chapter we first discuss superconductive transmission lines as developed at the MIT Lincoln Laboratory for the analog signal-processing applications and review the basic ideas behind several processing techniques. In the second part of the chapter, we explain the basic ideas of superconductive digital circuits and the achievements made to date. Finally, we present various accomplishments and proposals for analog-to-digital (A/D) converters and shift registers.

ANALOG SIGNAL PROCESSORS

The basic component of any analog signal processor is a tapped delay structure (Fig. 2) which must have low dispersion and loss, be sufficiently compact, and have

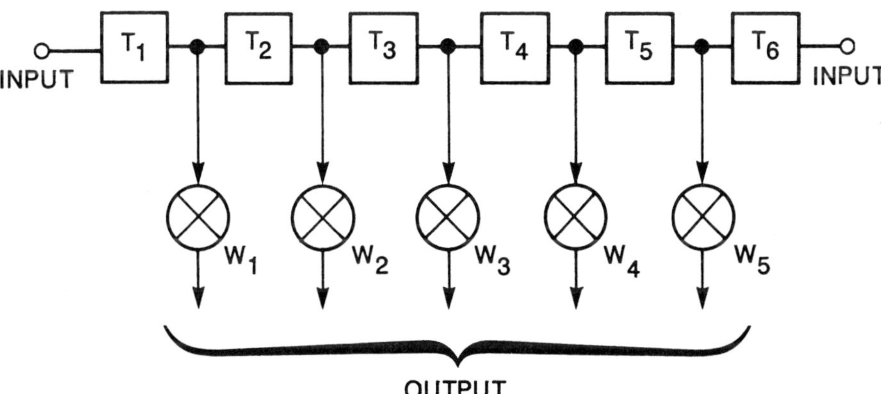

Fig. 2 A tapped delay structure, which constitutes the fundamental component of an analog signal processor. Depending on the application, inputs may or may not be fed into both ends of the delay structure. The delay structure under consideration here is a superconductive strip line.

taps with accurate weights. Required auxiliary functions are multiplication with sufficient dynamic range, circuits for spatial summation or temporal integration, and readout logic circuits. The processors to be considered here employ superconductive transmission lines with superconductive auxiliary circuits. The work has been done exclusively at the MIT Lincoln Laboratory.

Superconductive Tapped Delay Lines

Thin-film niobium delay lines have been developed with delays as long as about 100 ns and sufficiently low loss and dispersion for operation at 4.2 K with a center frequency of 4 GHz and a bandwidth of 2.7 GHz. [3] To minimize the area required, the lines are either wound in a re-entrant spiral form or as a meander line. The former gives the longest line for a given area. The length is limited by the need for sufficient spacing between adjacent lines to have adequate isolation. A 40 ns reentrant spiral of two parallel lines will be discussed below in the Matched Filter section. It is a stripline structure with 125-μm-thick silicon insulators and a total line length of about 3 m. The line spacing is 200 μm to avoid cross talk.

Evaluation of a proposed transmission-line structure can be done by testing a resonator with the same structure. The Q of a transmission-line resonator can be shown to have the value π/α where the attenuation α is in nepers/wavelength. [4] Or, for a line of length L, $Q=(\pi/\alpha_T)(L/\lambda)$ where α_T is total attenuation and λ is wavelength. Also, the product of storage time and frequency (corresponding to λ) is $Tf=L/\lambda$, the number of information cycles. If we assume the $f \simeq 1.5B$, where B is bandwidth, then

$$Q=\frac{13\pi}{\alpha_T}\,TB \tag{1}$$

where α_T is the total line attenuation in decibels. If the total attenuation is required to be a less than about 4 dB, the required Q for a given TB is 10 TB. Measurements on striplines with 125-μm-thick sapphire dielectrics (sapphire losses are about the same as silicon) for Nb and NbN are shown in Fig. 3 for various temperatures and frequencies. [5] It is seen that the Qs of the Nb lines at frequencies up to 8 GHz are adequate for TB values near 1000 at 4.2 K and for TB > 100 even up to about 8 K. The NbN lines can achieve TB > 1000 at f = 4.5 GHz up temperature of 12 K. The line losses increase somewhat faster than the f^2 of the two-fluid model, which predicts one order of magnitude difference between 4.5 GHz and 14.5 GHz. Time-bandwidth products in excess of 100 can be achieved with the NbN line at 14.5 GHz at 10 K.

In order to achieve processing times greater than about 40 ns on a 5-cm diameter wafer, it is necessary to reduce line spacing. This requires making the dielectrics

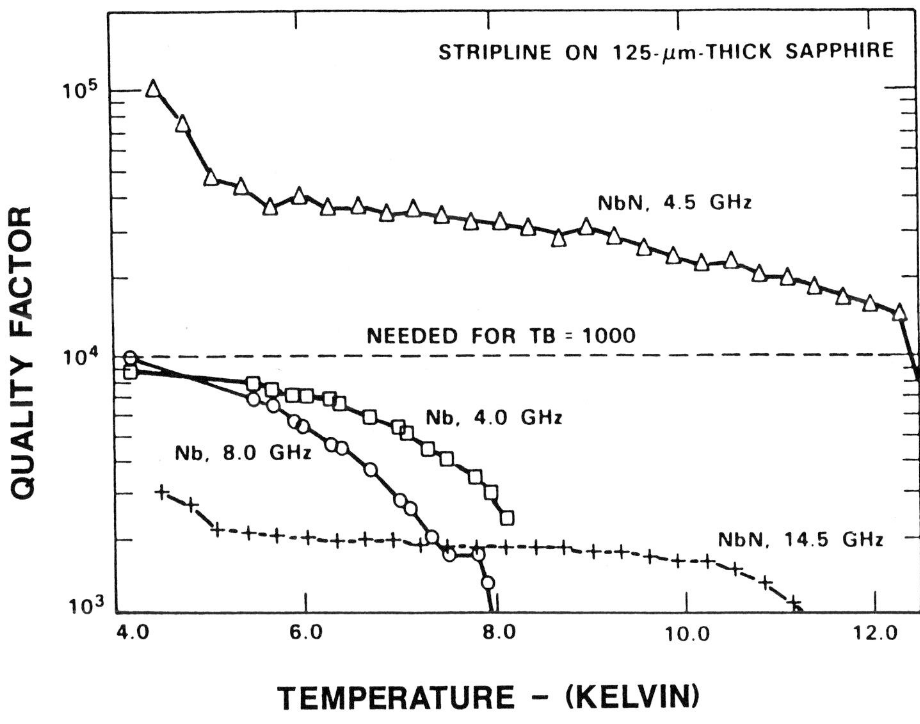

Fig. 3 Measurement of quality factor Q for strip lines of niobium and niobium nitride with sapphire insulators, as a function of temperature. Quality factor is directly related to the time-bandwidth product.

thinner since the minimum tolerable line-to-line crosstalk was taken into account in the 40-ns lines. A technique has been developed in which 15-μm-thick silicon epitaxial layers are used as the dielectrics. [6] The fabrication procedure is shown in Fig. 4. A high-resistivity epitaxial layer is grown on a low-resistivity substrate (Fig. 4a) and a niobium film is deposited on the epitaxial layer. A pyrex sheet is then bonded to the niobium (Fig. 4b) and the low-resistivity silicon wafer is etched away. One half of the stripline structure is completed by deposition and patterning of the delay line on the remaining epitaxial layer. (Fig. 4c) The other half is formed by the same procedure except that no delay line is formed. The two halves are pressed mechanically face-to-face on each other to complete the stripline (Fig. 4d). Test results on resonant lines of this form have given $Q=2.9\times10^5$ for a frequency of 562 MHz. [7] This would be adequate for a processor with TB $\simeq$ 1600, center frequency of 10 GHz, and processing

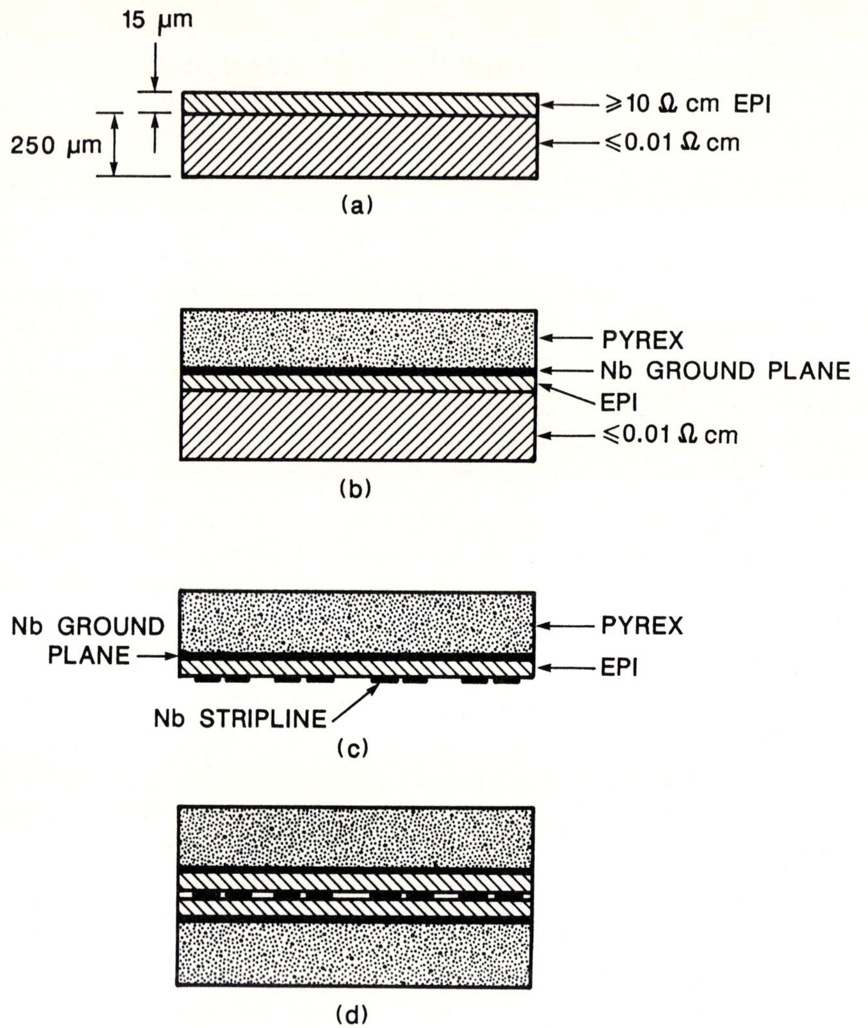

Fig. 4 Fabrication steps to produce a strip line with 15-μm-thick epitaxial silicon insulators. (a) The epitaxial layer on a silicon wafer. (b) A pyrex sheet is bonded to a niobium film deposited on the epitaxial layer. (c) The low-resistivity silicon wafer is etched away from the high-resistivity epitaxial layer and the strip line is deposited. (d) The structure is completed by pressing to the structure in (c) a similar one without the strip lines.

time of about 240 ns. With the thin dielectrics and T $\simeq$ 240 ns, the lines can be spaced enough to avoid significant crosstalk on a 5-cm-diameter wafer.

Fabrication of these devices using the new perovskite superconductors to take advantage of more convenient operating temperatures is attractive from a system point of view. It is not clear yet that the losses in the new superconductors will be low enough. The losses also increase strongly with temperature, so the optimum operating temperature will be a trade-off between tolerable losses and system convenience. So far, the formation of new superconductors with the best properties has required substrates such as $SrTiO_3$, which has unsatisfactory dielectric properties. The choice of substrate for signal processor delay lines will require considerations of dielectric constant (higher is better in order to shorten the line), dielectric losses (lower temperature also helps here), and feasibility of use to construct stripline structures. Operation at temperatures higher than 4.2 K permits the use of cryogens with greater heat capacity; therefore, it should be feasible to use semiconductor transistors for the auxiliary circuits if suitable superconductive junction devices cannot be produced.

Chirp Filter

A so-called "chirp filter" has the characteristic that the frequency components of the input are spread out in time in the filter's output. For example, an impulse, with its broad range of component frequencies fed at the input will produce an output as shown in Fig. 5. The frequencies in the output can be either increasing in time ("up-chirped")

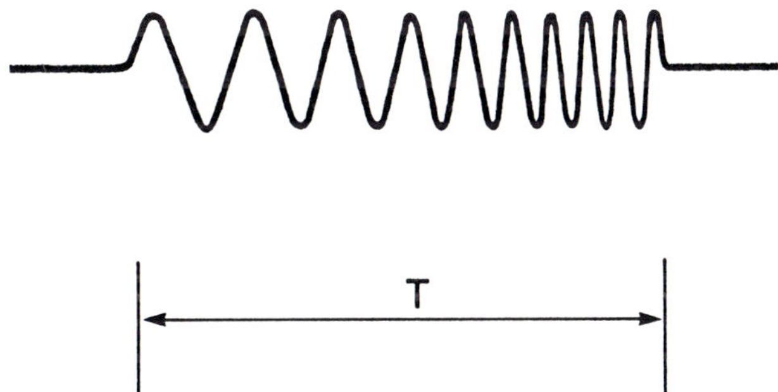

Fig. 5 RF pulse with up-chirped carrier.

or decreasing ("down-chirped"). These devices find use in various signal processing applications, including chirp radar and Fourier analysis. The reason for using a chirp filter in radar applications is to spread out the transmitted pulse to lower the peak power

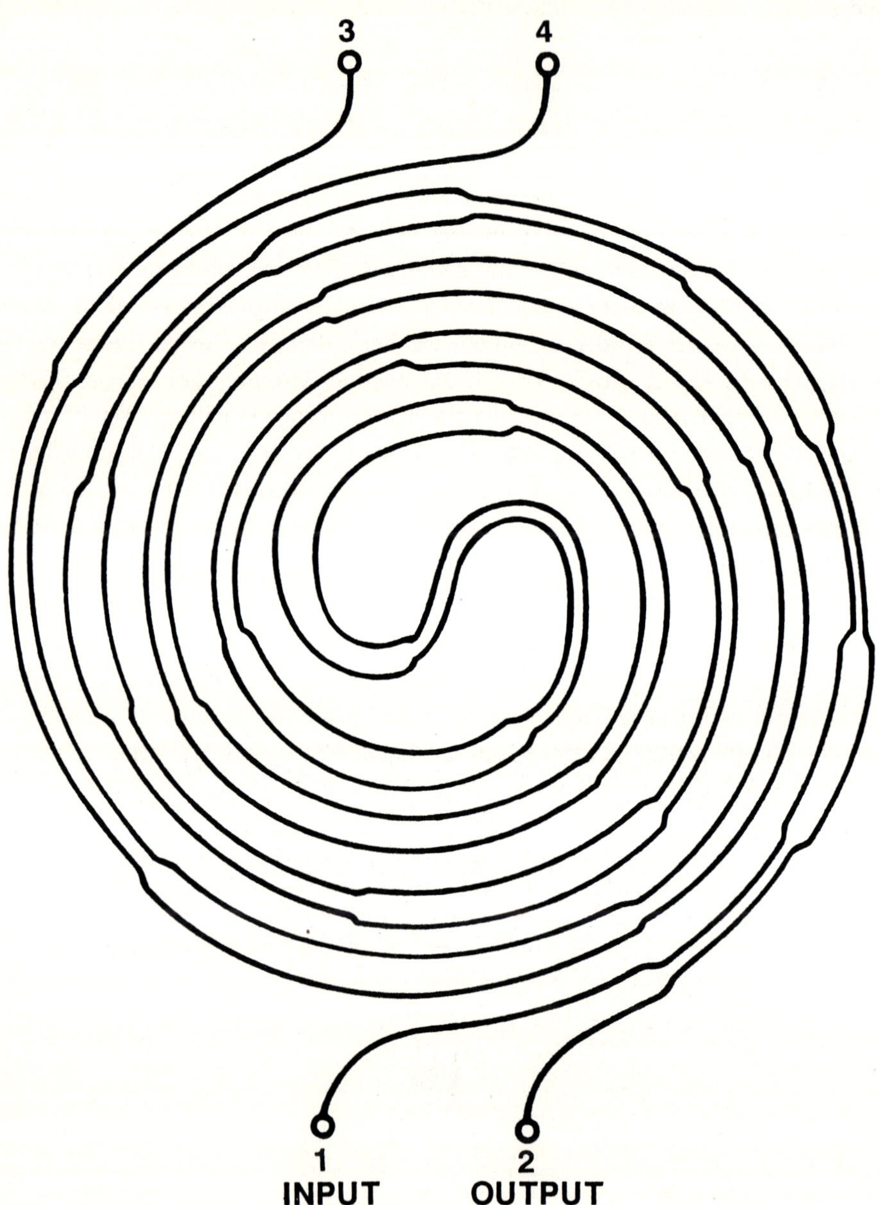

Fig. 6 Schematic of a tapped spiral stripline pair designed as a chirp filter. Sections that are more closely spaced are quarter-wave backward-wave couplers from the input line (1) to the output line (2).

requirement on the transmitter. This must be done in such a way as to avoid loss of range accuracy; by producing a chirped output and then filtering the return signal through a matched filter with the opposite chirp, signal time compression is achieved and the time resolution of the return pulse is not degraded relative to the short pre-chirped transmitter pulse. The use in a Fourier analysis system will be discussed below. Systems of this kind are presently employed in radar and spread-spectrum communication systems in which the chirp filter is a surface-acoustic-wave (SAW) device.

Superconductor delay lines can be used to produce a chirp filter as shown in Fig. 6. [8] Two transmission lines are run parallel to each other with spacing adequate to avoid coupling between them except in certain regions where coupling is desired. There the lines are brought close to each other for a distance equal to one-quarter wavelength of a chosen frequency. The ideal quarter-wavelength coupler produces a power flow backwards in the output line relative to the direction of flow in the input line. It is clear in Fig. 6 that the highest frequency components will be coupled closest to the input and will therefore arrive at the output first. The lower frequency components will travel farther down the line and will arrive later at the output. The coupler lengths and distance between the couplers is increased linearly along the line to achieve a linear down-chirp. Terminals 3 and 4 are matched to avoid reflections.

The superconductive realization of the chirp filter is shown in Fig. 7. Two strip-

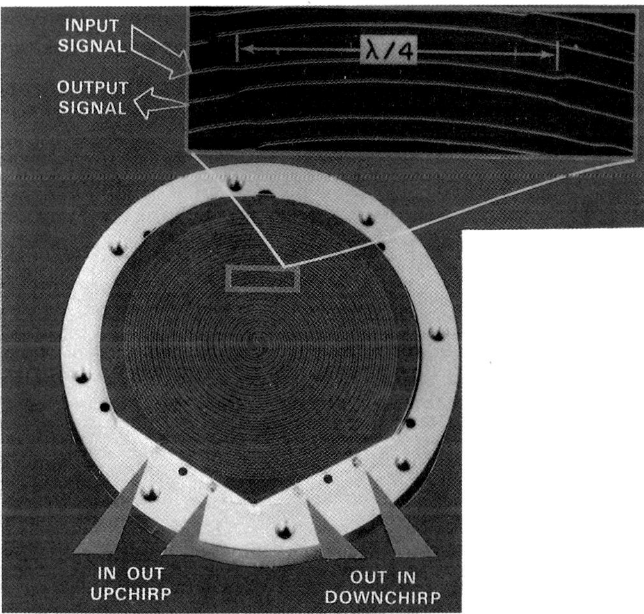

Fig. 7 Photograph of the realization on a 5-mm-diameter wafer of the filter shown in Fig. 6.

lines, formed as discussed in the preceding section, are run parallel to each other in a double spiral that first has radius decreasing to the center and then increasing to the output terminals. Depending on whether the input is at the end with the shortest couplers or at the opposite end, the output is either down-chirped or up-chirped, respectively. This two-inch silicon wafer carries three meters of lines. The change of spacing to achieve coupling is shown in the inset. Measurements made on this structure show excellent agreement with design values for the frequency dependence of insertion loss and deviation of phase from the quadratic dependence that is expected in a linearly chirped output. In this realization of a Hamming-weighted filter, the mid-band insertion loss is about 6 dB. Time-bandwidth products of about 100 are achieved typically with lines on a two-inch wafer.

Spectral Analysis

The chirp filters described above can be combined in systems that perform spectral analysis. [9] The system shown in Fig. 8 performs a spectral analysis of the input sig-

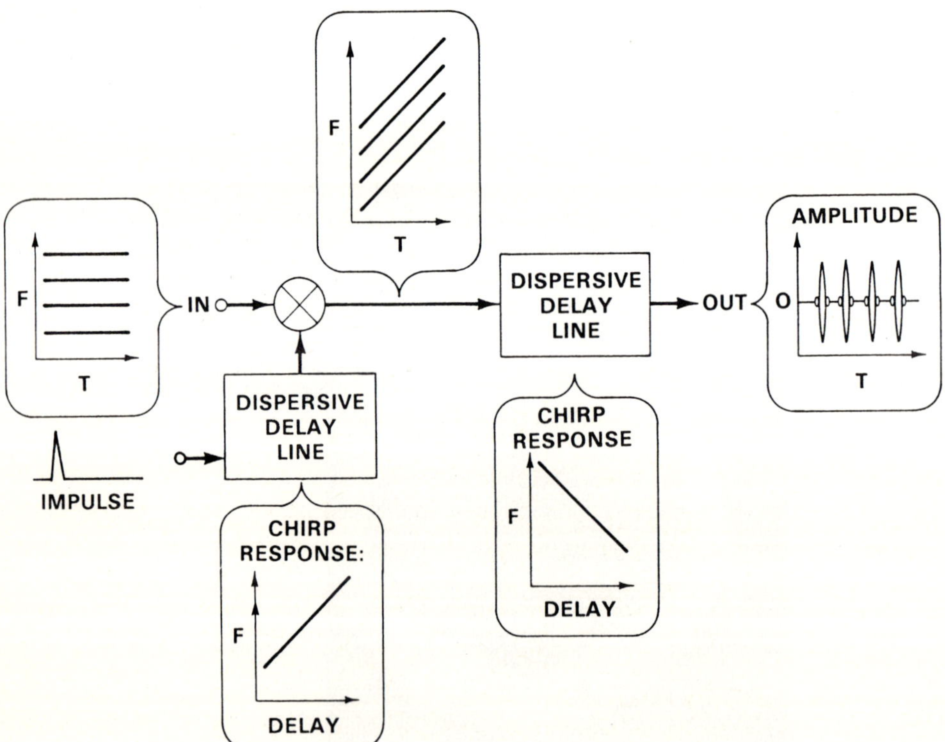

Fig. 8 System for spectral analysis by means of the chirp Fourier transform. The spectrum of the input signal is converted to a function of time at the output.

nal with the f(ω) of the input converted into f(t); the spectrum of the input signal appears at the output as a function of time. The frequency components of the input are multiplied by the up-chirped output of the dispersive delay line "A" that results from an impulse input. The output of the multiplier is a signal in which all frequency components of the input are up-chirped. When this signal is passed through the down-chirp dispersive delay line "B", the output is an f(t) which has the same functional dependence on time as the system input had on frequency.

The spectral representation signal f(t) from the transform system shown in Fig. 8 can be digitized by the scheme shown in Fig. 9. [10] The envelope of f(t) is fed to a set

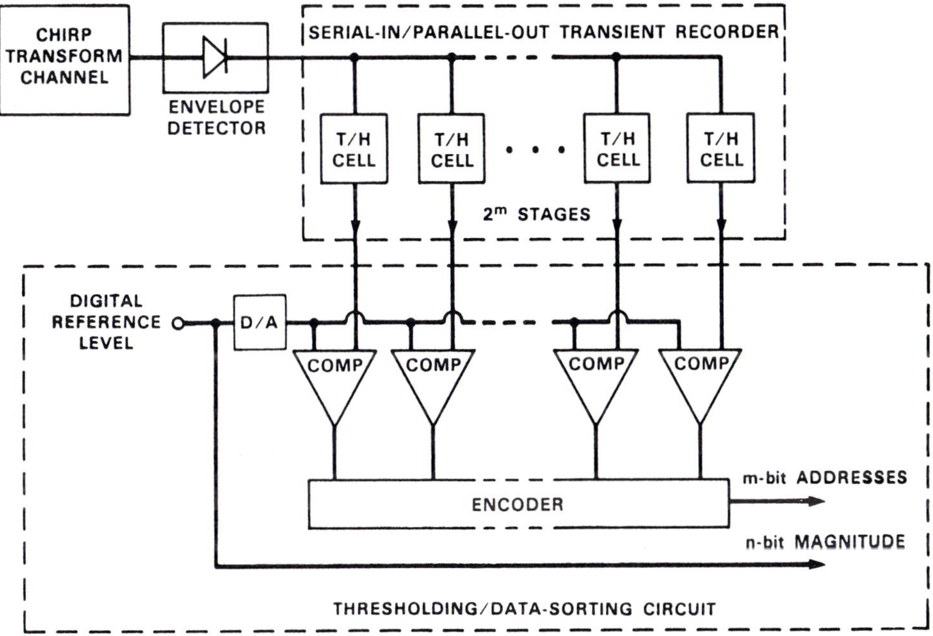

Fig. 9 A proposed system for digitizing the output of the chirp transform system shown in Fig. 7.

of track-and-hold cells in the transient recorder. Each cell tracks the signal until its clock switches and the value of the signal at that instant is held. The clocks are delayed relative to each other so that the group of cells holds a set of values of the signal that constitutes a discrete analog representation. In the Thresholding/Data-Sorting Circuit, the values held in the T/H cells are compared in the comparators (COMP) with a succession of digital levels produced from a binary input by the digital-to-analog

(D/A) converter. For each level at which one of the comparators switches, the address of that comparator is recorded along with the magnitude at which the switching occurred. The address is equivalent to the value of frequency in the spectrum of the signal f(ω) at the input to the transform circuit. Thus, the spectrum is recorded digitally. To obtain a spectrum defined at, say, 100 points, would require 100 track-and-hold circuits and comparators and a corresponding encoder.

An alternative to the track-and-hold scheme would be to directly feed the output of the envelope detector to a high-speed analog-to-digital converter, the output of which could be stored in a shift register. The currently made lines typically produce a 40 ns long signal out of the transform circuit. If this could be sampled at 2.5 gigasamples per second, the spectrum would be represented by 100 samples.

Convolver

The tapped superconducting delay line can be used to perform convolution of two signals, each having a time duration as long as the propagation time through the line. [11,12] Such a device can serve as a programmable matched filter for extremely wide-band spectral analysis or spread-spectrum communications systems.

If s(t) and r(t) are the two functions to be convolved, the convolution is defined by

$$c(t) = \int_{-\infty}^{\infty} s(\tau) r(t-\tau) d\tau \tag{2}$$

In the analog convolver, the integration is replaced by a summation

$$c(t) = \tau_i \sum_n s(n\tau_i) r(t - n\tau_i) \tag{3}$$

in which the products are summed at taps spaced by a delay τ_i along the line at each instant during the time that the signals overlap. The realization of the convolver is shown in Fig. 10. The signal s(t) is fed to the input at the left end of a long meander line and the reference signal r(t), with which s(t) is to be convolved, is fed in the opposite end of the line. Taps are made at the bends of the meander line to sample the sums of the functions counter-propagating on the line. The sums are coupled to superconductive mixers, the outputs of which are proportional to the products of the functions. Self-products $r^2(t)$ and $s^2(t)$ are eliminated by using the balanced mixers. The total length of the line in which the mixer outputs are added is less than 2 cm so that the products are summed with time skew on the order of 0.01 of the signal lengths. This

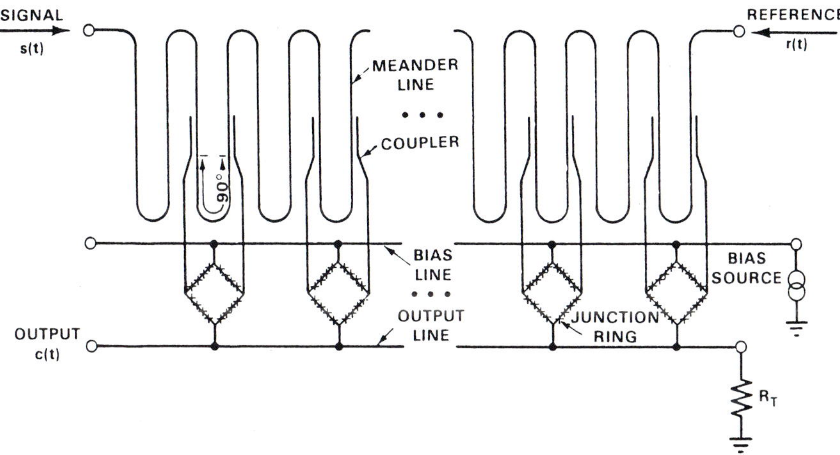

Fig. 10 Tapped-delay-line convolver. The output line is very short compared with the tapped delay line and sums the products $s(t)r(t-n\tau_i)$ taken from the set of balanced mixers.

skew can be compensated by off-setting the center frequency of the signal from that of the reference.

The convolution operation is illustrated in Fig. 11. The time delay τ is the vari able of integration in (2); it is plotted leftward in Fig. 10c to correspond to the direction on the delay-line schematic. An example signal is shown as a function of τ by the solid line and the reference $r(t-\tau)$ is shown by broken lines for several different values of time t. The time base is selected so that $t = 0$ at the time that the signals just begin to overlap at the center of the line and the output convolution begins to appear. The overlap of the signal and reference is finished when $t=T$, the propagation time through the meander line (as well as the signal length). The position of the delay line relative to τ changes with time as shown by the bars under the graph in Fig. 11. The correct duration of the convolution is 2T. In this delay-line convolver, time is compressed by a factor of two. The center frequency and bandwidth are correspondingly increased by a factor of two.

The mixers can employ series arrays of SIS junctions in order to achieve sufficient dynamic range, which should exceed the signal processing gain. The meander line should be designed to avoid excessive scattering from the bends and to otherwise have negligible loss and dispersion caused either by the superconductors or the dielectric.

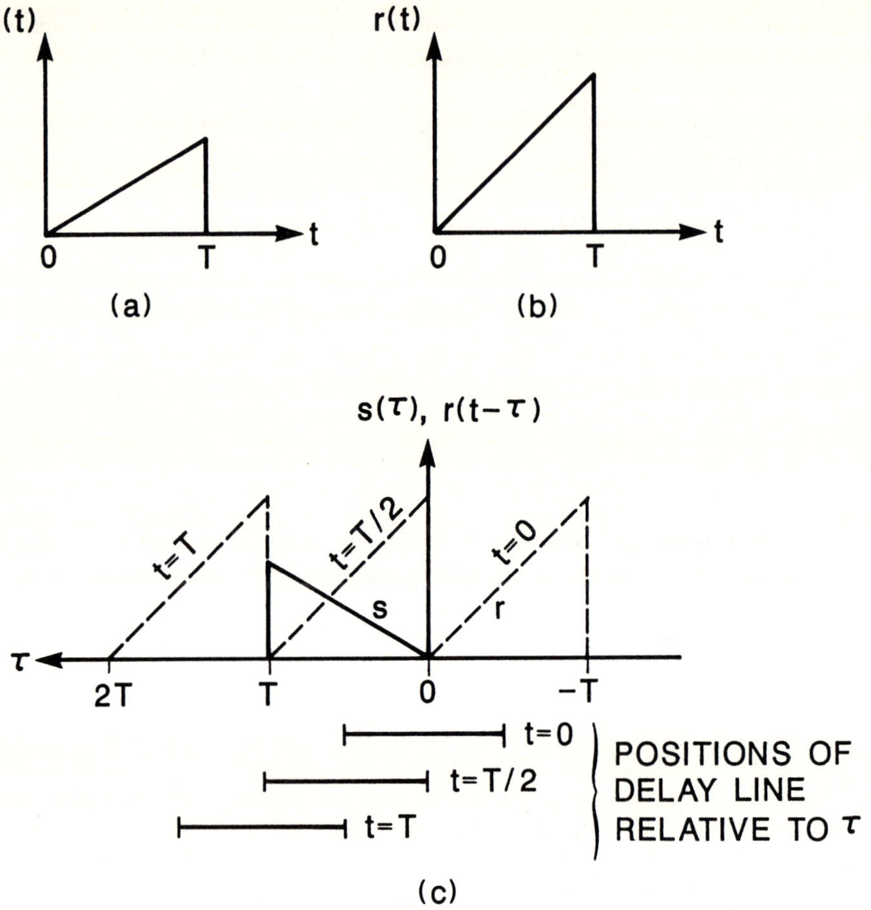

Fig. 11 The waveforms in (a) and (b) are representative signal s(t) and reference r(t) for illustration. Figure (c) shows the convolution operation in Equation (1). The solid line is $s(\tau)$ and the broken lines show $r(t-\tau)$ for various times t. The delay τ is measured from a zero that moves along the line with time, as shown by the bars below the graph.

Proximity taps with a coupling factor of -25 dB are satisfactory. Other design improvements have been suggested. [12]

Time-Integrating Correlator

Another important function that can be performed with the tapped delay line is the cross-correlation between two signals s(t) and r(t), which is defined by

$$\phi_{rs}(\tau_i) = \int_{-\infty}^{\infty} s(t)r(t-\tau)dt \qquad (4)$$

where τ is the time shift between the two signals. [13,14] The system shown in Fig. 12

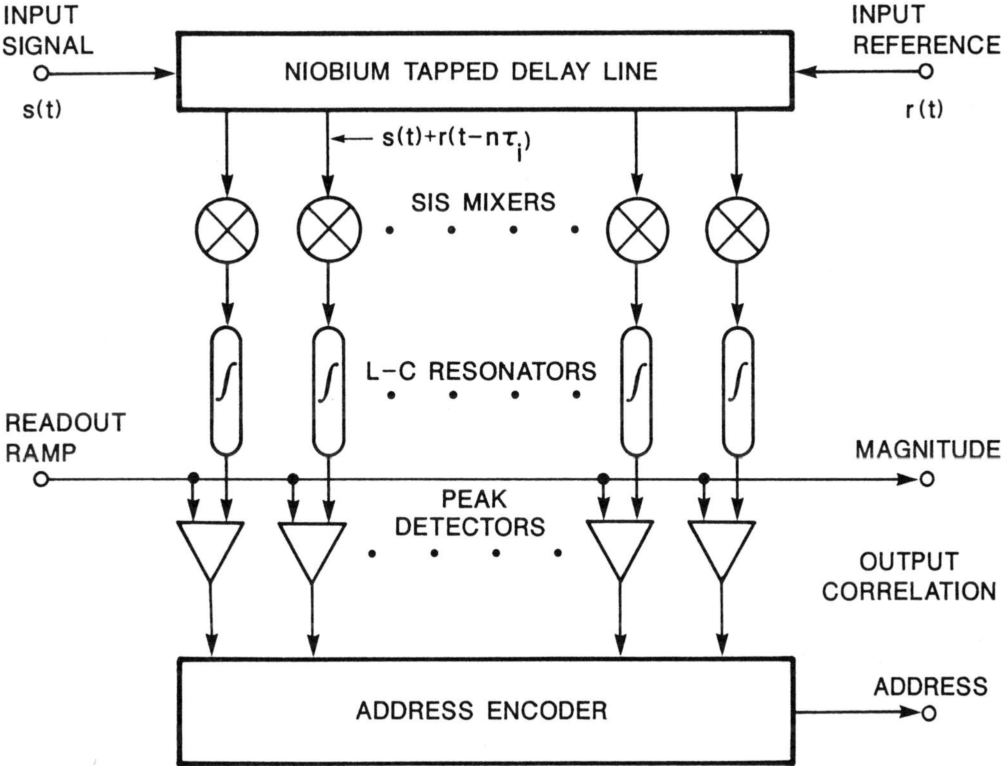

Fig. 12 Proposed correlator system based on a tapped strip line. The products $s(t)r(t-\tau_i)$ formed in the mixers are integrated in time in the resonators for each of a set of τ_is. The integrals are read by the method shown in Fig. 13. The magnitude and address (τ_i) from each integrator are read out as shown.

has been devised to permit formation of the correlation function for signals of duration up to several microseconds with a set of time delays $n\tau_i$. The line is long enough to contain a time delay equal to the total required range. The tap at the center of the line will be reached by s(t) and r(t) at the same time so that $n=0$ there; $n > 0$ to the left of center and vice versa. The line is configured as a meander-line, as in Fig. 10 to facilitate tapping. The multiplications are performed for each $n\tau_i$ in an SIS mixer diode and the integration is done with a high-Q L-C resonator. In order to avoid contributions to the integrand by the self-products $s^2(t)$ and $r^2(t)$, the signal s(t) is mixed with a sinusoidal carrier at the integrator's resonator frequency f_0 before it is fed into the tapped transmission line. The details of the mixer-integrator circuit are shown in Fig. 13a. A series array of junctions is used in the mixer to achieve a greater dynamic range than would be achieved with single junctions. The product of the $r(t-n\tau_i)$ and the signal s(t) (which has been pre-mixed with a 23 MHz signal) feeds the L-C resonator which has a Q of 600. The time over which the signal amplitude in the tank circuit increases linearly with the input is approximately 2 μs; this sets the limit on the length of signals that can be processed. The amplitude of the oscillations in the tank circuit should be proportional to the integral of $s(t)r(t-n\tau_i)$, so that an evaluation of the amplitudes in the various tank circuits will give $\phi_{rs}(n\tau_i)$.

The technique under investigation for evaluation of the oscillation amplitudes is indicated in Fig. 13b. A threshold read-out ramp current is fed into the circuit at the point shown in Fig. 13a and adds to the oscillation current. When the sum reaches the critical current of the peak-detector Josephson junction, the junction voltage is detected on the line feeding the read-out ramp current; it triggers a reading of the address of the circuit that switched. The time at which it switched is a direct measure of the amplitude of the oscillation. The amplitude-and-address pairs are thus read into the external circuitry, giving a recording of $\phi_{rs}(n\tau_i)$.

SWITCHING CIRCUITS FOR SIGNAL PROCESSING

Before discussing analog-to-digital converters and shift registers that are being developed for signal-processing applications, we will present some background on superconductive digital circuits. The motivation for using Josephson-junction circuits rather than transistors is the ability to achieve faster switching with much reduced power dissipation. The low power dissipation allows packing the circuits closer together to minimize propagation delay. Speeds of the Josephson circuits exceed those of equivalent semiconductor devices typically by a factor of about five for multipliers, with power dissipation typically two orders of magnitude lower. Some new types of logic circuits are under study in which the devices only enter the voltage state long enough to transfer a flux quantum. These have still lower power dissipation, about

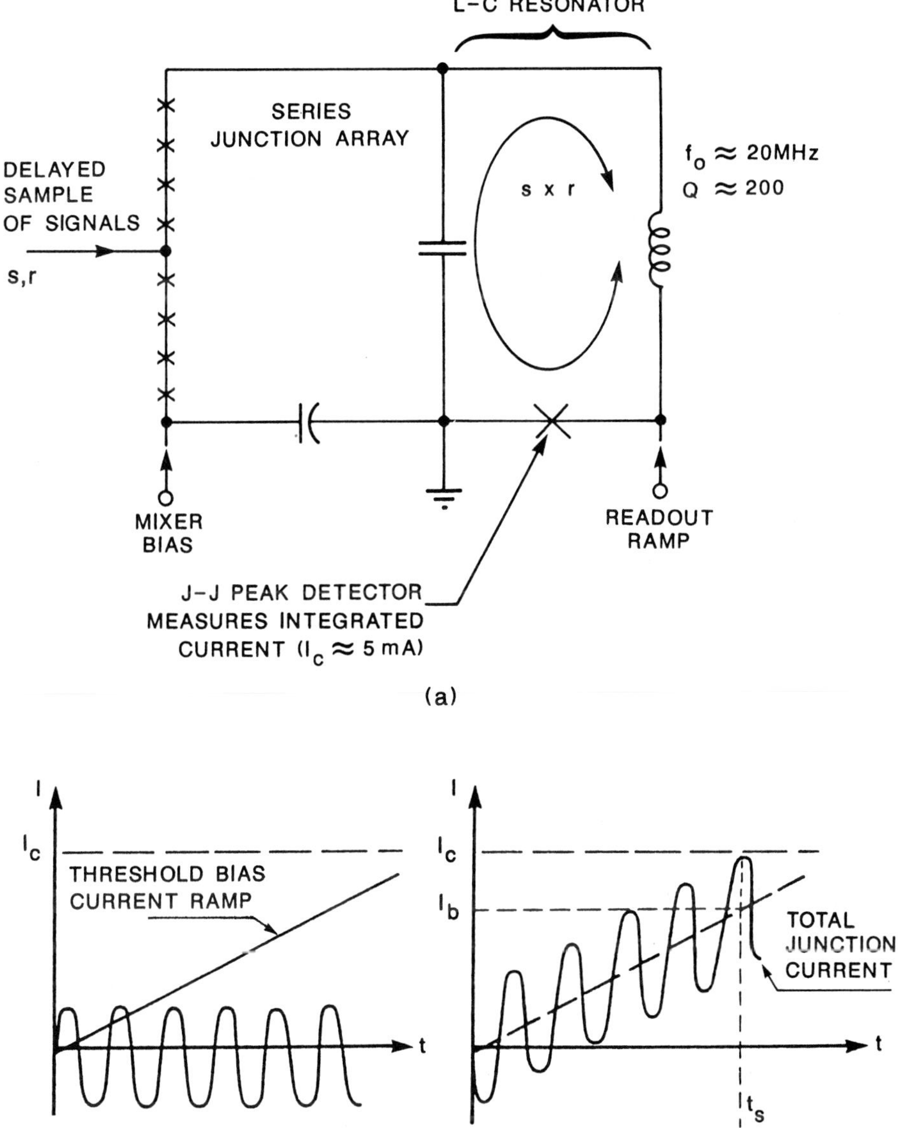

Fig. 13 (a) Mixer-integrator-peak detector circuit for one of the correlator delay channels. The level of oscillation in the resonator is the integral of the product $s(t)r(t-\tau_i)$. (b) A ramp current is added to the oscillating current. The time at which the sum reaches I_c of the peak detector is a measure of the amplitude of the oscillating current.

three orders of magnitude below the currently used Josephson circuits. We will concentrate below on the voltage-state logic circuits since they are more highly developed.

Logic and Memory Circuits [15]

Voltage-state logic circuits can be divided into two categories, as shown in Fig. 14.

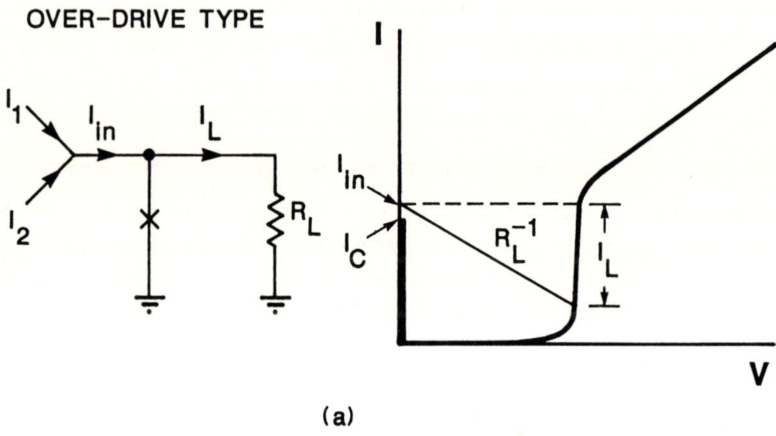

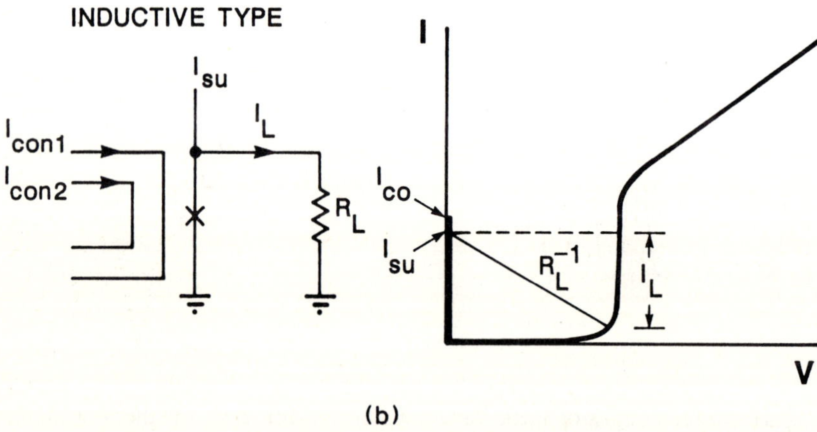

Fig. 14 General classification of Josephson logic circuits.

The devices represented by junction symbols could be either Josephson junctions or SQUIDs. The load resistance should approximately equal the junction R_n. The family of circuits in which switching takes place because the input currents overdrive the maximum zero-voltage current I_c are represented by the circuit in Fig. 14a. If the circuit were an OR gate, either I_1 or I_2 would be sufficient to produce an input greater than I_c. If it were an AND gate both I_1 and I_2 would be required. The current passes through the junction until it reaches I_c. After switching, the junction becomes a high resistance (subgap resistance is typically about $20\,R_n$) and the current is diverted to the load R_L, which would typically be a matched transmission line with characteristic impedance of R_L. The current transferred to the load is the difference between I_{in} and the small current remaining in the junction.

The other family of logic, represented by Fig. 14b, employs inductive coupling to suppress the critical current below the supply current I_{su}. The design is such that either one or both of the inputs I_1 and I_2 are required to reduce I_c below the supply current, depending on whether it is an OR or an AND gate, respectively. The amount of current transferred to the load is comparable to that of the overdrive family in Fig. 14a.

Figure 15 shows a typical switching event for a Josephson device. If the supply

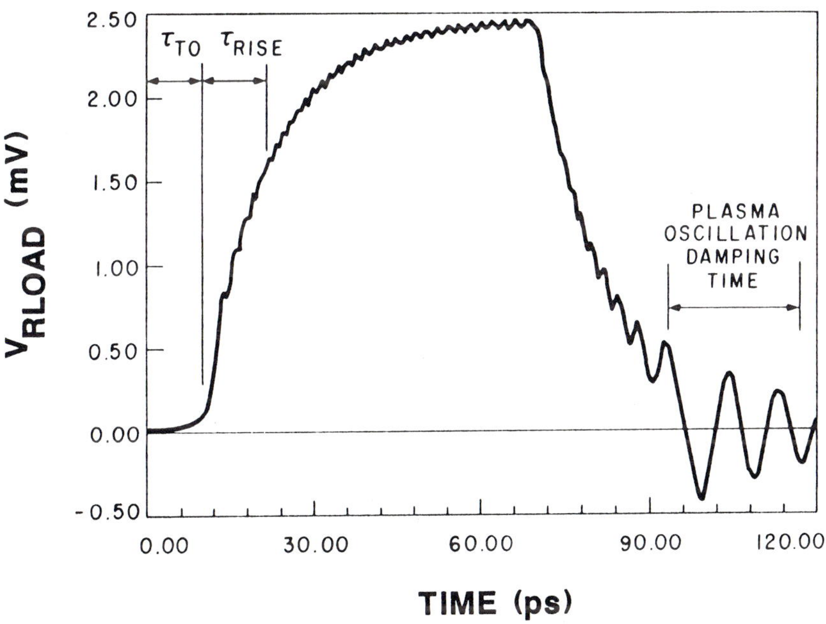

Fig. 15 Typical switching of a Josephson junction from the $V = 0$ state of the $V \neq 0$ state and back, with the switching produced by stepping the supply current.

current to a junction is raised above the critical current I_c at $t=0$, the voltage does not rise immediately but has a latency period called the "turn-on delay" τ_{TO}. This is followed by a period of rising voltage, the "rise time" τ_{rise}, which is usually determined by the RC time constant of the junction and its load resistor. The oscillations visible on the rising and falling edges and on top are Josephson oscillations at frequency $f_J=2ev(t)/h$. After the junction voltage reaches zero, there are oscillations at approximately the so-called "plasma frequency" $f=(1/2\pi)(2eI_c\hbar C)^{1/2}$, where C is the capacitance of the junction. The switching speed of a junction increases with the critical tunneling current density J_c and switching delays as low as 2.5 ps have been achieved.

We will introduce here briefly some examples of different types of logic circuit. The logic family used in the IBM project at the time of its termination in 1983 was the so-called "current-injection logic" (CIL). [16] The AND gate is a two-junction SQUID (Fig. 16a) with directly injected currents which represent the logic states of the gates

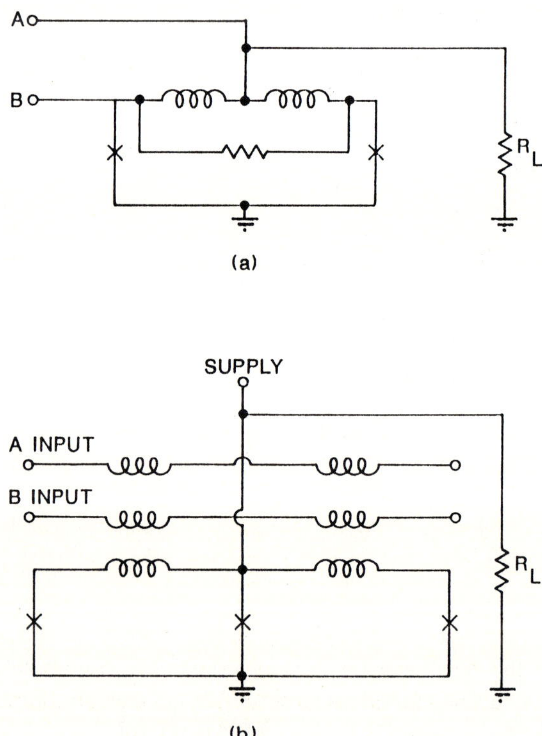

Fig. 16 Gates for the IBM current-injection logic (CIL) family. (a) AND (b) OR.

that feed the one under consideration. The AND gate provides no isolation so it is always preceded by inductively coupled SQUID OR gates (Fig. 16b) which do have output-input isolation. The isolation results from the fact that transients in the output enter the OR SQUID symmetrically, with equal currents passing through the two inductive arms so that the signals coupled into the input lines cancel. The complete family requires also a timed inverter and a latch.

One of several logic families that employ combinations of resistors and Josephson junctions is the Resistor Coupled Josephson Logic (RCJL) family shown in Fig. 17.

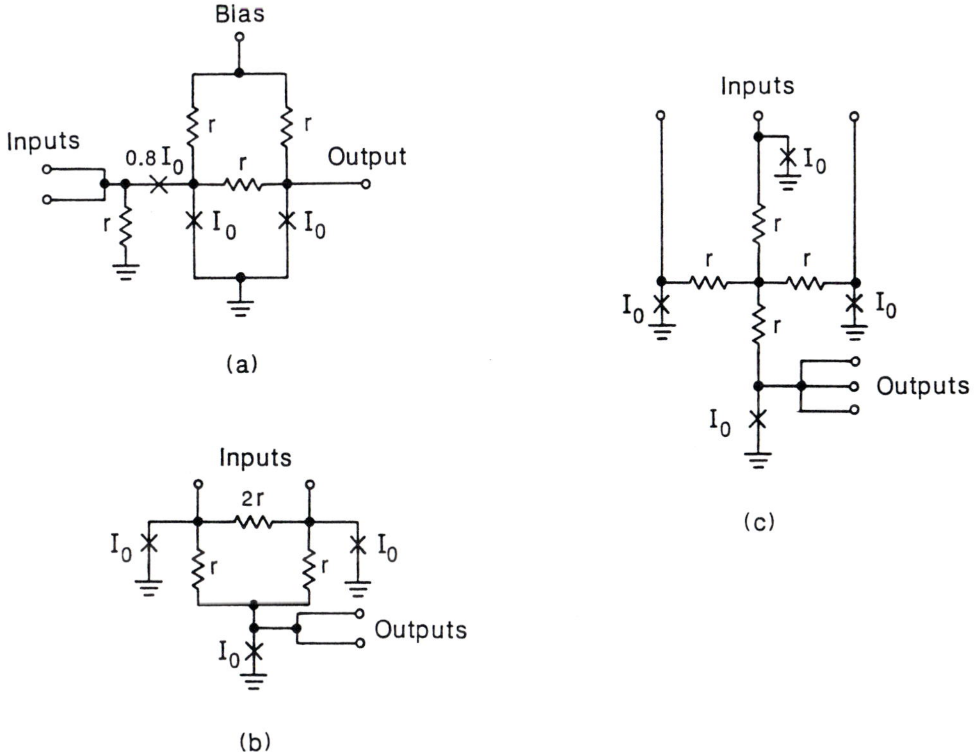

Fig. 17 Gates for the NEC resistor-coupled Josephson logic (RCJL) family. In these gates there are no closed superconducting loops so there is less problem with trapped flux.

[17] The three gates constitute a complete logic family. The AND and 2/3 MAJORITY gates have no output-input isolation and must be used in combination with the OR

gate. The isolation is achieved in the OR gate by switching of the junction marked "0.8 I_0" into the high-resistance state.

The circuit shown in Fig. 18 is called the "4JL gate." [18] A dot-OR input can be

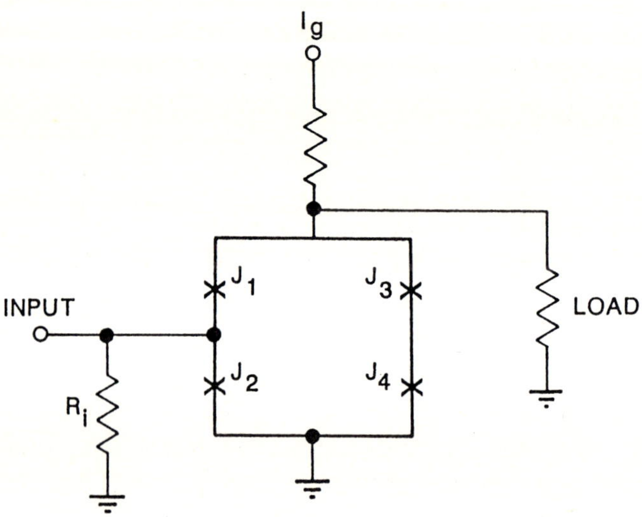

Fig. 18 The four-Josephson-junction-logic (4JL) gate developed at the Electrotechnical Laboratory.

used. No intentional inductance is needed in the loop since all phase shifts occur in the junctions. Thus, the gate is very compact. When a sufficiently large input is provided, junction J_2 switches to the voltage state. The input current is then diverted around the loop and adds to the gate current I_g in junctions J_3 and J_4, which then also switch to the voltage state. A redistribution of the current then causes J_1 to switch; the gate current is diverted to the load, and the input goes to ground through R_i. A complete family has been built around the 4JL gate and a number of moderately large logic circuits have been demonstrated.

The family called "modified variable threshold logic" (MVTL) employs a combination of inductive coupling and direct injection. [19] The OR gate is shown in Fig. 19a

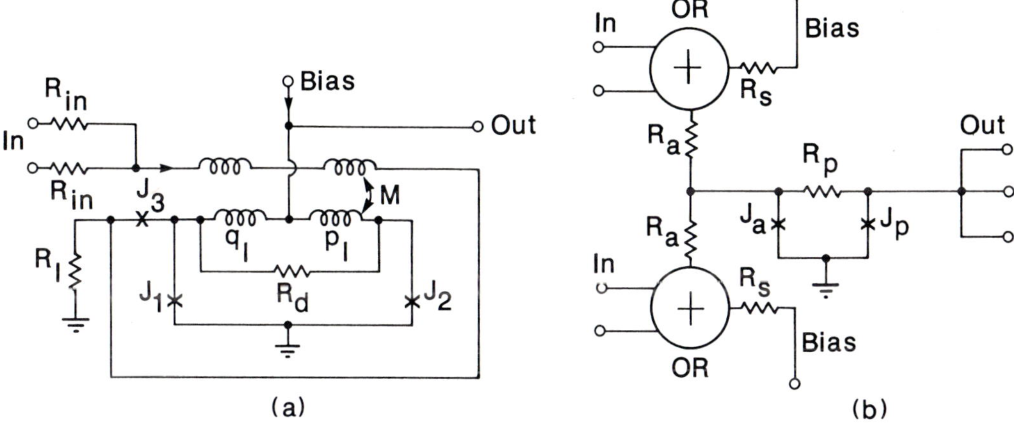

Fig. 19 Modified-variable-threshold-logic (MVTL) gates developed at Fujitsu Laboratories. (a) OR gate. (b) AND gate comprising J_a, J_p, and R_p driven by two OR gates.

and a unit cell comprising two OR gates and an AND gate is in Fig. 19b. The AND gate clearly has no output-input isolation and that must be provided by the OR gates again in this case. These OR gates have been shown to have lower delays (2.5 ps) than any other gate [20], superconductor or semiconductor, and a number of large logic circuits have been made.

Since the Josephson device is symmetrical, it can do logic functions with gate currents of either polarity. Furthermore, the gates latch into the voltage state when switched and must be reset to the zero-voltage state after each logic operation. It is therefore convenient to use an ac power supply so that the gates are reset during the changes of polarity. One problem is that if the polarity change is too rapid, the "1" logic state punches through to a "1" of the other polarity instead of resetting to zero. The punchthrough error can be avoided in a single-phase powering system only at the expense of a significant slowing of the logic. In present logic circuits, either two-phase or three-phase power is being used to avoid the punchthrough delays.

Random-access memory (RAM) has been the greatest bottleneck in the development of superconductive computers. Most memory cells are variants of the cell in Fig. 20, which was developed by IBM, in the sense that data are stored as circulating

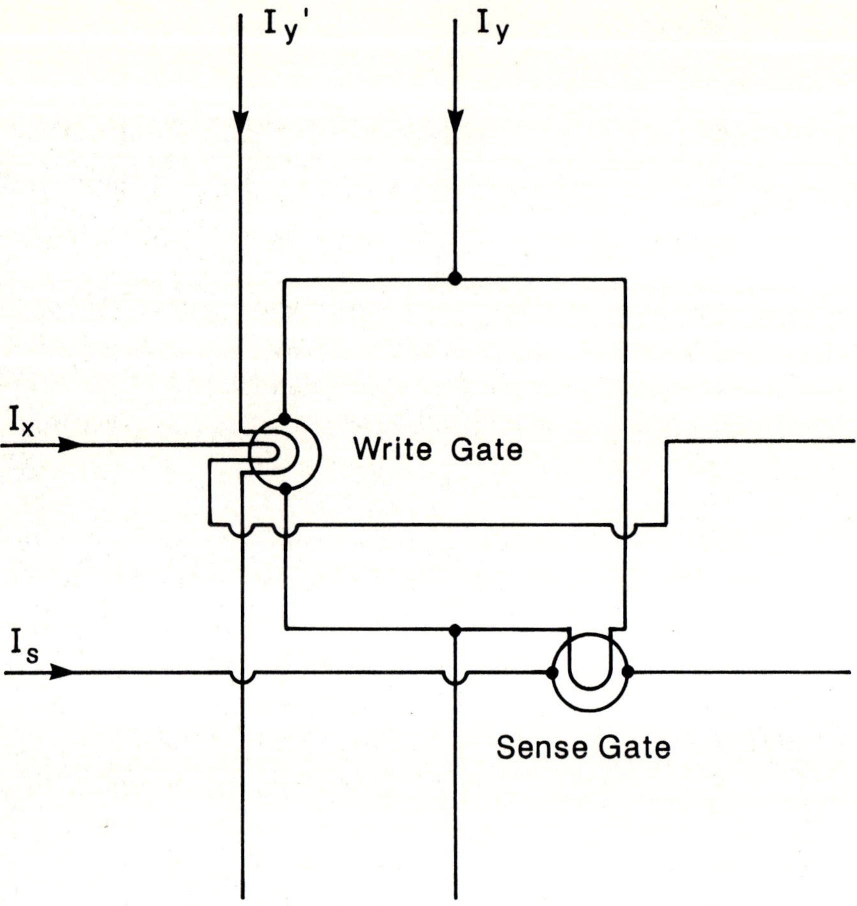

Fig. 20 Circulating-current memory cell developed by IBM. Most other memory cells are variants of this one.

currents. Partially working memory units have been demonstrated. Recently, a 1 K-bit RAM was shown to have a minimum access time of 570 ps but was not completely functional. [21] Work is in progress on a 4 K-bit RAM by the same group at NEC.

The technology for Josephson circuits has improved tremendously since 1983, when IBM was using Nb-NbOx-Pb alloy junctions. Most circuits now incorporate Nb-AlOx-Nb Josephson junctions made with a process in which the entire junction structure

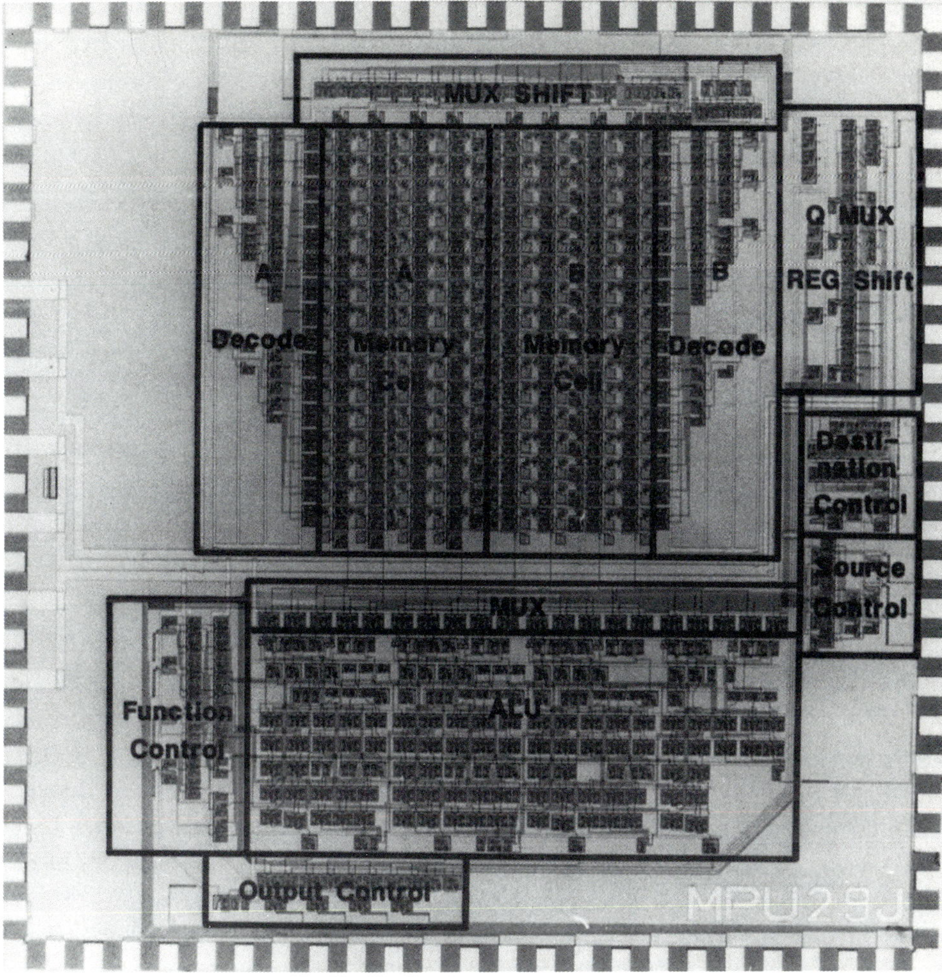

Fig. 21 Four-bit bit-slice microprocessor developed at Fujitsu Laboratories.

is made in a single vacuum system pump-down. This has led to barriers with high integrity. The spreads of critical currents can be attributed almost entirely to lithography imperfections. The junctions also can be stored without degradation and thermally cycled indefinitely without change. Large circuits have been made with reasonable yields by very small research groups. A 3000-cell gate array was demonstrated, and an array of 8000 memory cells (not connected as a memory) was made successfully as a test of the technology. It seems reasonable that with efforts in any way comparable with those devoted to Si or GaAs technology, the Nb Josephson technology should give high yields.

Several groups are developing microprocessors of various kinds to demonstrate the achievements of the technology. One such processor for four bits made at Fujitsu already has been reported. [22] It lacks a ROM so does not have an instruction set but does contain an arithmetic logic unit and a 64-bit memory. It performs functions similar to those performed by a device that was made in GaAs and was operated at 72 MHz with a 2.2 W power dissipation. The Josephson circuit was operated successfully at 770 MHz under worst-case conditions with a 5 mW power dissipation. Figure 21 shows a photograph of the Fujitsu Josephson circuit with the various components designated.

High-Accuracy Analog-to-Digital Converters

The work on Josephson A/D converters has been on devices in two different categories. In the first are the high-accuracy, lower speed devices that have as one important application, infrared detector arrays for which dynamic range is of great importance. Typically, 16-bit accuracy and 5 MHz bandwidth are desired. Conversion is achieved in one of two ways, both of which require counting. In one study [23] the analog signal is applied as a voltage to a Josephson junction. The oscillations of the ac Josephson current through the junction can be counted as a measure of the voltage across the junction since $f = (2e/h)V$ in the Josephson theory. A fundamental hurdle for this type of converter is the difficulty of achieving a linear relation between the analog signal and the voltage across the low-impedance junction. The other circuit under study is one that tracks the signal variations. [24] The signal is coupled inductively to an input interferometer (SQUID) as shown in Fig. 22. If the signal rises sufficiently to introduce an additional flux quantum into the SQUID, a voltage pulse appears across junction A. If it lowers by that amount, a voltage pulse appears across junction B. There follows a circuit that counts both sets of pulses and gives an output which is a binary measure of the signal level. In both types of converter, the counting period must be controlled to within one part in 2^n, where n is the number of bits (1/65536 for a 16-bit A/D converter). There are also design issues relating to the pulse width and the need to start counting immediately after the end of the sampling period. The required count rate is 1 THz for a 16-bit, 5 MHz converter. A counter with a rate in excess of 100 GHz already has been demonstrated. [25] Theoretically, at a critical current density of 2×10^4 A/cm^2, the counting rate should approach 1 THz.

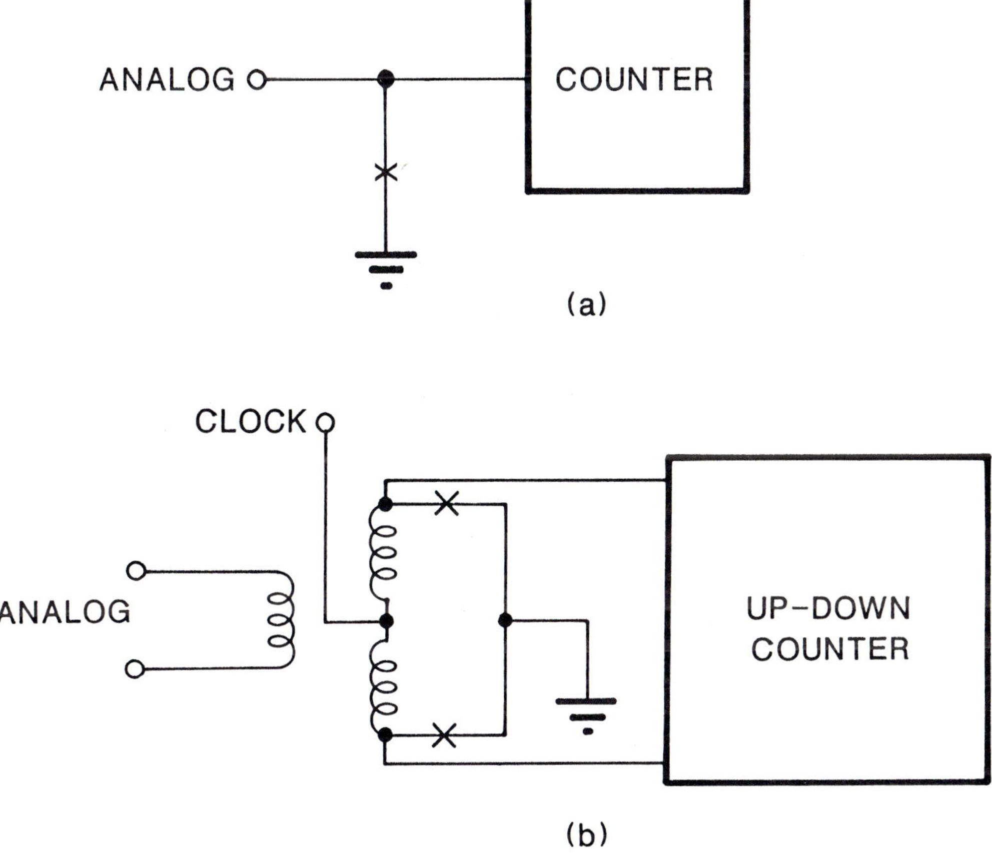

Fig. 22 Two types of counting A/D converter. (a) Analog voltage across a Josephson junction produces Josephson oscillations which are counted as a measure of the voltage. (b) Tracking-type A/D converter. Pulses are generated across one junction or the other, depending on whether the signal is increasing or decreasing, and are counted to track the signal.

Flash-Type A/D Converters

Our concentration in this section is the achievement of A/D conversion with the highest possible bandwidth. This is done at the sacrifice of dynamic range in a so-called flash-type A/D converter.

For many types of A/D converter, samples are taken at evenly spaced intervals T_s, and the time required to take the sample is called the aperture time. Each sample amplitude must be converted into a binary word during the time between samples. There are two major criteria that limit the bandwidth. The Nyquist criterion requires that samples be taken at twice the highest frequency f_{max} of the analog signal. Thus, $f_{max}=2/T_s$. Also, the aperture time must be short enough that the signal does not change by as much as one least significant bit while the sample is being taken. Using the maximum rate of change of a sinewave of frequency f_{max} it is easy to see that the aperture time $\tau=[\pi 2^n f_{max}]^{-1}$, where n is the number of bits. In what follows we will assume that it is always desired to have a bandwidth given by the Nyquist frequency so that the sampling frequency f_s is twice the bandwidth f_{max}. It should be noted that the aperture time requirement is quite severe for large bandwidths. As examples, for four bits $\tau=20\,\mathrm{ps}$ if $f_{max}=1.0\,\mathrm{GHz}$ and $\tau=2\,\mathrm{ps}$ if $f_{max}=10\,\mathrm{GHz}$. A switching time of 2 ps is pressing the limits of any kind of electronic circuit. Other factors that affect the accuracy of converter A/D performance include noise, aperture jitter, and parameter spread in fabrication.

Bit-parallel flash-type A/D converters

Zappe first suggested the use of the SQUID periodicity to make possible A/D conversion with one input comparator circuit for each bit. [26] The basic idea is illustrated in Fig. 23, which shows a set of four threshold characteristics for the SQUIDs of the comparators of a four-bit A/D converter. The periodicity differs by a factor of two between each successive comparator; this can be achieved in various ways as will be seen below. The most significant bit is on the top line in Fig. 23 and the least, on the bottom line. The "0"s and "1"s superimposed on the chart indicate the binary levels associated with the various positions along the horizontal axis (analog signal strength). The coding is Gray code in which only one of the bits changes value on crossing from one digital level to the next. In natural binary, if one of the comparators were to change at a different point from the others as a result of circuit imperfections, large errors in the binary word value would result. In Gray code, if that happens, the maximum error is one least significant bit. The broken lines on the threshold curves are set to give equal division of "1"s and "0"s. The method of setting this threshold depends on the choice of circuit realization.

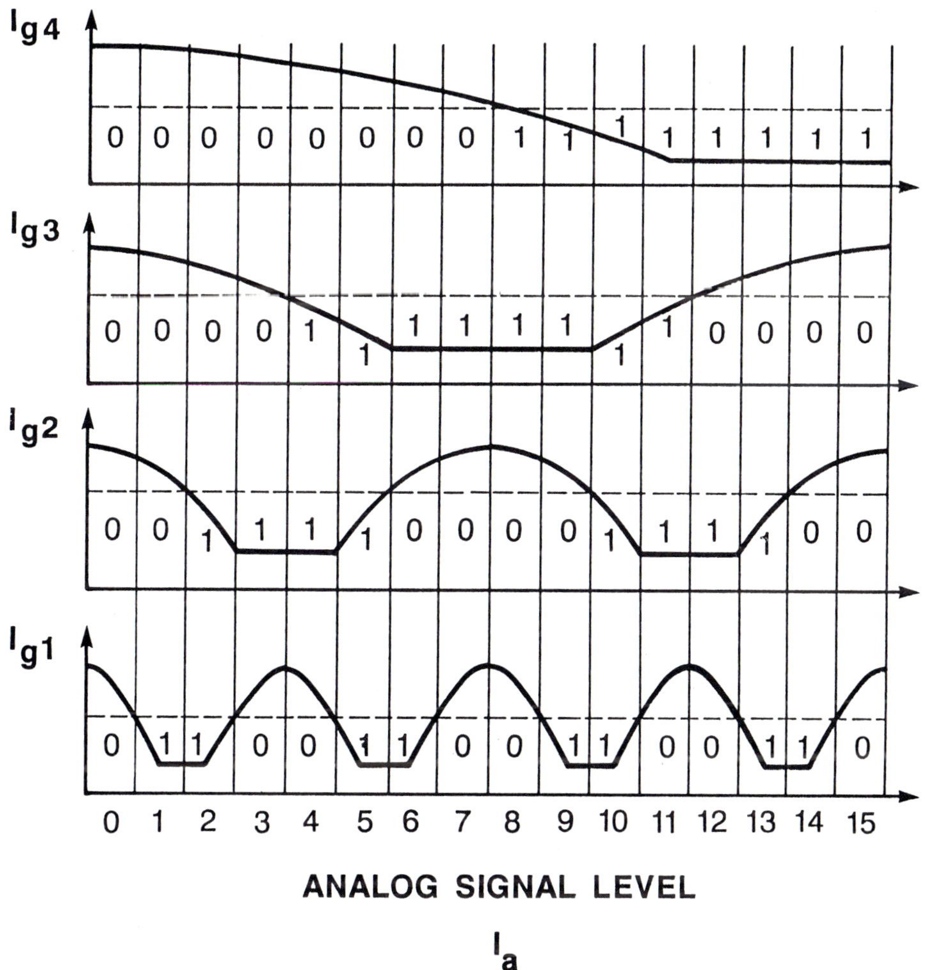

Fig. 23 The basic concept of bit-parallel A/D converters, which require devices with periodic dependence on analog signal level.

The first experiments on this type of A/D converter used one SQUID for each bit, with the various sensitivities to analog signal effected by changing the coupling to the SQUIDs for successive bits by a factor of two. This was achieved with accuracy sufficient also to make a 6-bit converter. [27] The comparator SQUIDs for two of the bits are shown in Fig. 24. An ac trapezoidal clock signal is applied to the gates of all

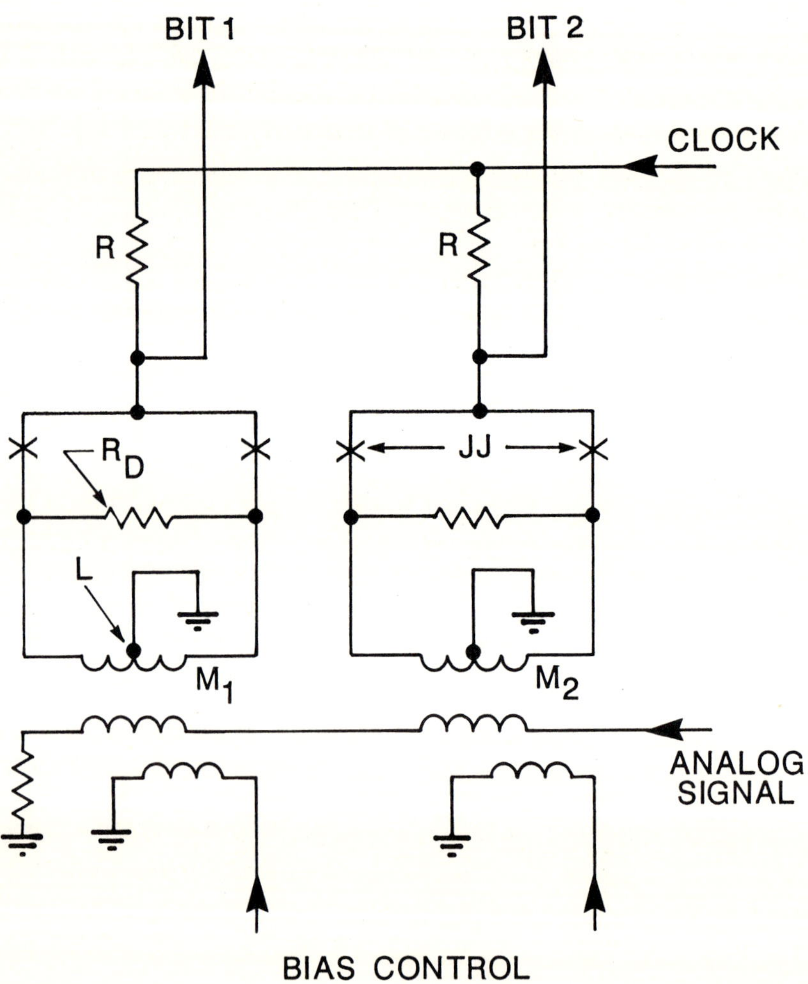

Fig. 24 Two comparator stages in a bit-parallel A/D converter, in which factor-of-two sensitivity difference is achieved by a difference of mutual coupling factors M_1 and M_2.

SQUIDs simultaneously and the analog signal is applied to the control lines; the analog signal determines the position on the horizontal axis in Fig. 23. Then the gate current rises to the level of the broken line. If located in a region marked "0" in the graph, the SQUID does not switch and if in a region marked "1", it crosses the threshold characteristic and switches to the voltage state. Thus, for example, for the four-bit converter depicted in Fig. 23, a digital level "8" would be represented by Gray code word of 1100. It was shown that this arrangement could convert a low-frequency analog signal into 6-bit Gray code words at a rate of four gigasamples per second. However, these simple SQUID comparators had no way to achieve the small aperture required for large analog bandwidth.

In a subsequent work, an A/D converter with the required short aperture time was studied. [28,29] The comparator circuit for each bit is an edge-triggered latch, the state of which is determined by a race in the input circuit that occurs during the rise of the clock. It was adapted from use as computer circuit where it was called a Self-Gating AND gate and is shown in Fig. 25. The input I_{bias} to G_2 is such that it switches at one

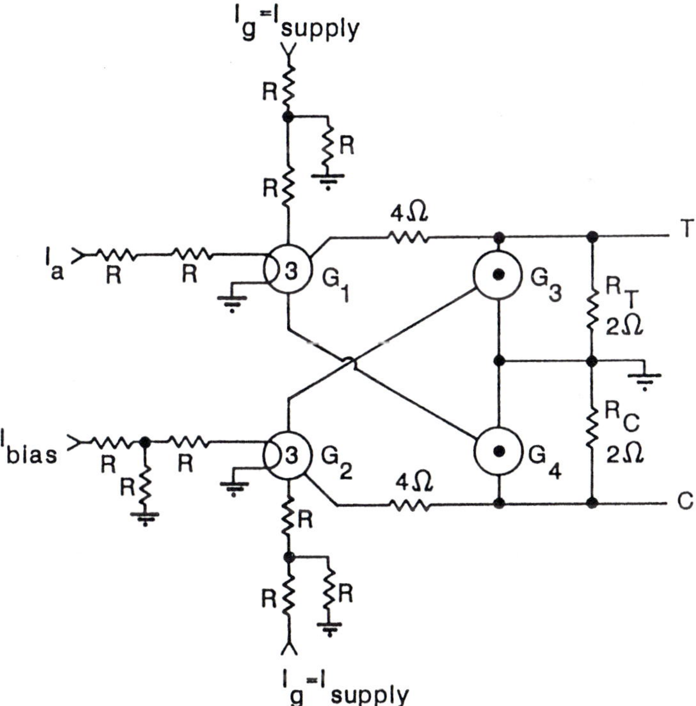

Fig. 25 Self-gating AND circuit used as an edge-triggered latch for an A/D comparator.

of the broken lines in Fig. 23. The SQUID G_1 switches at the threshold characteristic, which depends on the value of the analog current. As the clock rises, either G_1 or G_2 will switch first, depending on I_a. If I_a is in a "0" range G_2 switches first and the output stage latches into a state with C = 1 and T = 0. When I_a is in a "1" region, the opposite result obtains. By having a fast rising clock, the decision is based on the value of the analog signal in a very short aperture time.

This circuit has been demonstrated experimentally in a 4-bit A/D converter. See Fig. 26. The four comparators, which correspond to the four threshold characteristics in

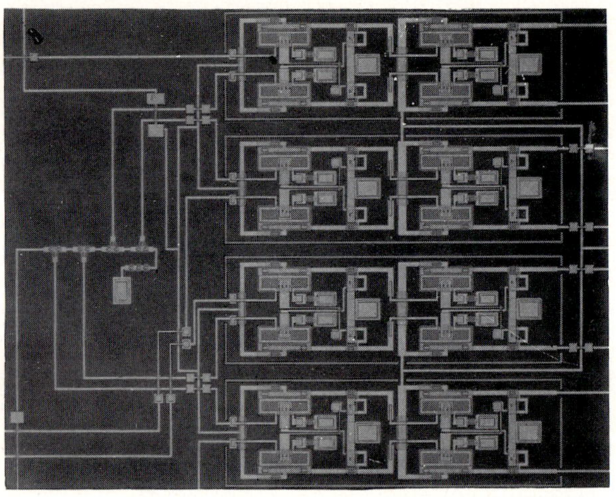

Fig. 26 Realization of a 4-bit A/D converter based on the edge-triggered latch in Fig. 25. The four latching comparators in the left column are fed by a binary resistor divider and are clocked at 1 GHz. The right column of latches is subharmonically clocked to select samples from the comparators for data-rate reduction.

Fig. 23, form the left column of circuits in Fig. 26. These were clocked at 1 gigasample/second so that the Gray code data coming out of the comparators were at too high a rate for the room-temperature test circuits. Therefore, a second set of latches (circuits identical to the comparators) was included on the chip to reduce the data rate. The second column of latches was clocked at 1/32 of the 1 gigasample/second clock rate so the data taken off the chip are at about 31 megasamples/second. By using a beat-frequency test scheme, in which the frequency of the analog signal to be evaluated is slightly different from a subharmonic of the clock frequency, the complete analog sinewave is mapped out. The data were fed into a minicomputer and the sinewave was reconstructed. The result was that a 500 MHz sinewave was converted with 3-bit accuracy.

This type of circuit has been evaluated theoretically and in simulation to determine the factors limiting its performance. [30] The circuit involves a transfer of current, as illustrated by the crossed lines in Fig. 25, when one of the input SQUIDs switches upon winning the race during the rise of the clock. This transfer of current is the limiting factor. The circuit can be improved within the same general structure, and simulations indicate that four bits at $f_{max} = 500$ MHz is about the best that can be done.

Other circuits have been proposed that eliminate the limitation in the above-described edge-triggered latch. In one of these, the comparator consists of two SQUIDs connected in series. [31] One is biased with a fixed control current and the other has the analog signal as control current. The operation is similar to the race discussed above except that here the circuits are in series. This comparator eliminates the delay involved in the current transfer in connection with the circuit in Fig. 25. The aperture time is a fraction of the clock rise time here also. In another circuit, the periodic threshold of a two-junction SQUID is still used, but a short access time is achieved by switching a one-junction SQUID on the rising edge of the clock; the aperture time is 0.5-1.0 of the rise time of the clock. [32] The dominant limitation on the two types of A/D converter discussed in this paragraph is the distortion of the threshold characteristics of the least-significant-bit circuit that results from the rapid changes of analog control currents in the SQUIDs. It has been estimated that, for any 4-bit flash-type A/D converter which employs the periodic threshold characteristic of multi-junctions SQUIDs, the analog bandwidth will be limited to about 1.5 GHz. [30]

A recently reported comparator circuit employs the periodic characteristic of a one-junction SQUID as the basis for a bit-parallel A/D converter. [33,34] The one-junction SQUID in Fig. 27a has a periodic relationship between the junction current I_j and the SQUID current I_a. The relationship is single-valued as in Fig. 27b if the product $\beta_L = 2\pi L I_c/\Phi_0 \leq 1.0$, where I_c is the junction critical current and Φ_0 is the flux quantum, but is multi-valued for larger β_L. The design of this comparator circuit depends on the I_j–I_a relationship being single-valued. The comparator circuit actually uses a quasi-one- junction SQUID in which the leg with the junction in Fig. 27a has two junctions in series, but with one having so much larger critical current than the other that the circuit behaves essentially as a one-junction SQUID. The comparator for one bit is shown in Fig. 28. The inductor L, along with J_0 and J_s form the quasi-one-junction SQUID. A set of four such comparators fed by a binary resistor divider (as used in the circuit of Fig. 26) will give a Gray code output, as introduced in connection with Fig. 23. The pulse generator at the top of the comparator circuit in Fig. 28 is shared among the four comparators in order to synchronize the sampling. The pulse width sets the aperture time. Simulations showed successful 4-bit conversion of a 10

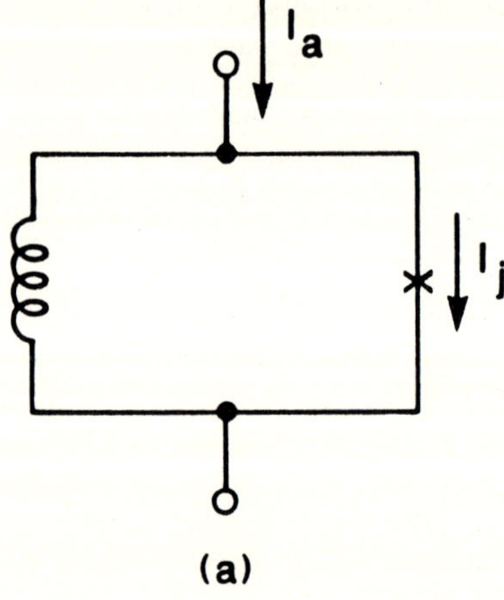

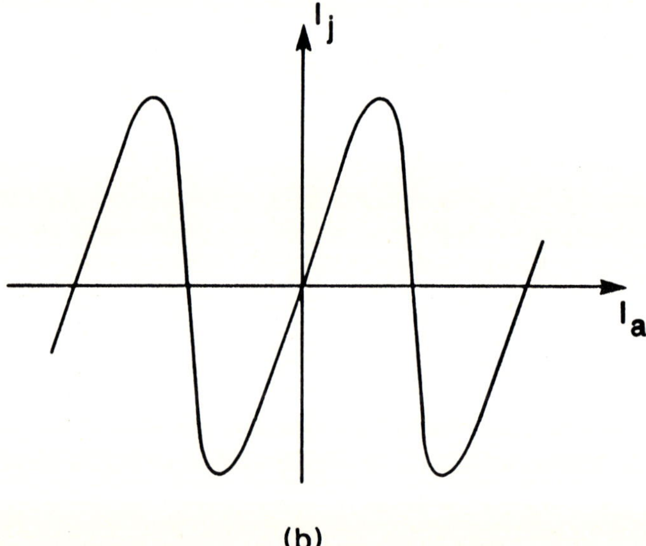

Fig. 27 (a) A one-junction SQUID. (b) Single-valued relation between junction current and total SQUID current that obtains for $\beta_L^2 = LI_c/\Phi_0 \leq 1.0$.

GHz analog sinewave using a 20 GHz clock [33] and ≈5 bit accuracy in a 6-bit converter clocked at 20 GHz [34]. The higher bandwidth compared with the circuits using multi-junction SQUIDs results from the better dynamic behavior of the one-junction SQUID.

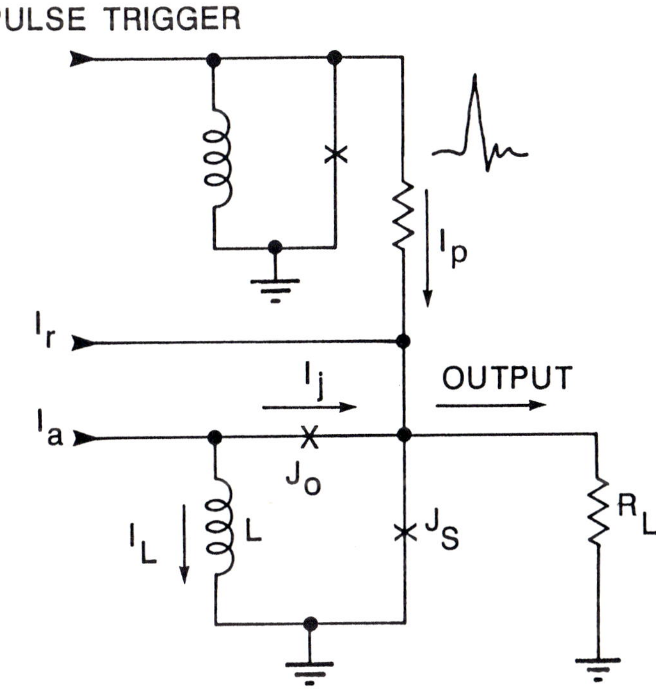

Fig. 28 A comparator based on the one-junction SQUID characteristic shown in Fig. 27b. The key element is the quasi-one-junction SQUID involving L, J_0, and J_s.

Fully parallel flash-type A/D converters

The architecture of the fully parallel flash-type A/D converter employs a set of comparators to act as a digital "thermometer" for the analog signal amplitude, which is fed to all of the comparators in parallel and is there compared with references that represent the 2^n-1 digital levels (15 for a 4-bit converter). The outputs of the comparators must then be combined in an encoder to give the binary word output. We examine in this section two different circuits that use this architecture.

The circuit shown in Fig. 29 is being studied as a comparator for a fully parallel

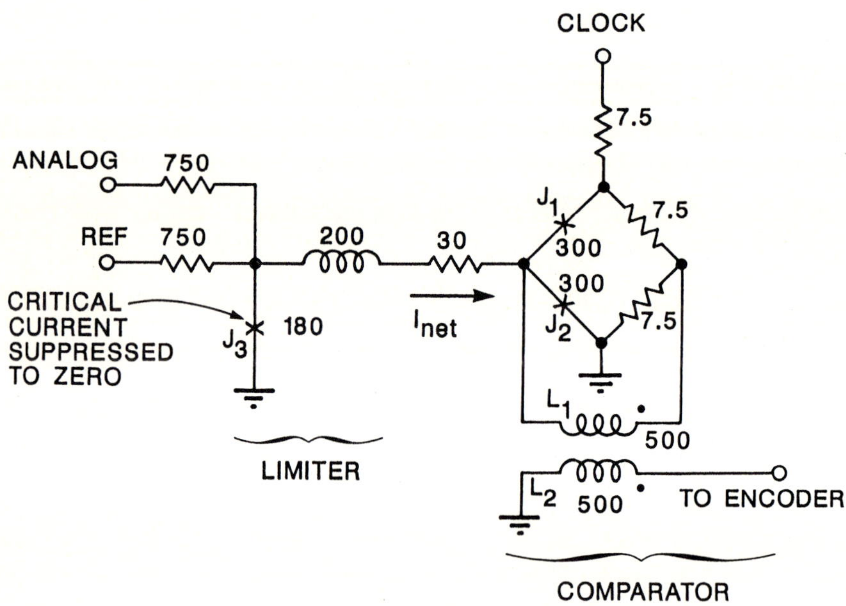

Fig. 29 Comparator with input limiter for an experimental fully parallel A/D converter. Resistor values are in ohms, inductances in picohenrys, and junction critical currents in microamperes.

A/D converter. [35] The central element of the comparator is the bridge containing two junctions and two resistors. The circuit is clocked with a fast rise-time (≈ 5 ps) supply. If no current I_{net} were present, the clock current would pass through the superconducting junctions J_1 and J_2 to ground. The current I_{net} is proportional to the difference between the analog and reference inputs. If I_{net} is positive, it adds to the clock current in J_2 causing it to switch and subtracts from the current in J_1, which does not switch. The result is that the clock current circulates through L_1 in a counterclockwise direction, passing through J_1 and the lower 7.5 Ω resistor. On the other hand, if I_{net} is negative, the clock current circulates clockwise through the top 7.5 Ω resistor in the bridge and through L_1 and J_2. The current in L_1 induces current in the secondary circuit to the encoder. Circuit simulations showed a 3.5 GHz bandwidth for a 4-bit A/D converter.

The limiter circuit preceding the bridge is included to increase the dynamic range to a value adequate for a 4-bit A/D converter, while maintaining sufficient design margins. [36] The critical current of the limiter junction is suppressed to zero and the quasi-particle part of the I-V characteristic is used to make a soft limiter. The junction capacitance and the inductor form a low-pass filter to remove high-frequency interference. Dynamics of the comparator with its limiter were verified experimentally.

The encoder is a set of four multiple-input two-junction SQUIDs. The fifteen lines are used in various combinations for the SQUIDs to convert the digital levels to natural binary code. In each SQUID, the various control lines cause transitions back and forth across one threshold and the problem of the threshold distortion that occurs in multi-junction SQUID bit-parallel A/D circuits is thereby avoided. Work is still in progress on this converter.

Another circuit that is designed for use in a fully parallel A/D converter is based on the current latching property of a one-junction SQUID. [32] The comparator circuit is shown in Fig. 30. The central element in the comparator is the one-junction SQUID

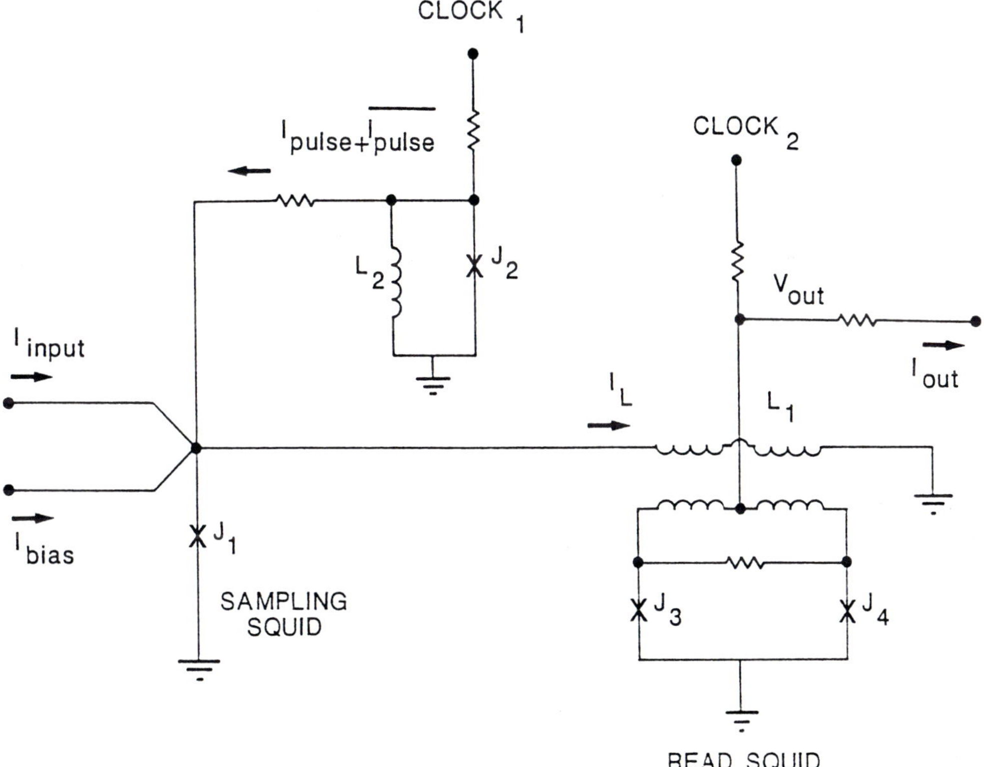

Fig. 30 Comparator based on a one-junction SQUID for a fully parallel A/D converter.

consisting of J_1 and L_1. It is fed the analog current input, a dc bias current, and pulses generated in the one-junction SQUID at the top of the drawing. The power supply $CLOCK_1$ has a trapezoidal form so that, when applied to the J_2–L_2 SQUID, a sharp positive pulse is produced on the leading edge and a negative pulse on the trailing edge. The multi-valued relation between current in the inductor L_1 and the current applied to the node above J_1 is shown in Fig. 31. Shown there is a bias current adjusted so that, if any positive analog signal were present, the positive pulses would drive past the critical point on the I_L vs. I_{ex} curve and the SQUID would switch up to the higher I_L. When the negative pulse arrives at the end of $CLOCK_1$, the circuit is reset. The two-junction SQUID is switched to the voltage state if the larger value of I_L passes through L_1 while $CLOCK_2$ is high. The aperture time of the comparator is less than the width of the positive pulses produced by J_2–L_2, which can be a few picoseconds. Thus for a 4-bit A/D converter, the bandwidth should be in excess of 5 GHz. A pipelined encoder can be made with circuits identical to the comparator since it can perform the AND-OR function. For a complete converter, four clocks with appropriate phase shifts are required.

Some other flash-type A/D converters, those with comparators based on the quantum flux parametron (OFP), have recently been reported. [37] Some of this work also suggests the possibility of multigaghertz operation.

SHIFT REGISTERS

A number of different circuits using Josephson junctions have been proposed as shift registers. A few circuits have been evaluated by simulation and low-clock rate experiment, and one has been reported with high-clock rate test results. We concentrate here on the most recent work.

The only high-speed results were reported for an 8-bit shift register powered by a three-phase sinusoidal clock with a frequency of 2.3 GHz. [38] The gates used in the circuit are modified variable threshold logic (MVTL) gates, circuits devised at Fujitsu Laboratories. The shift-register circuit diagram is shown in Fig. 32 for one bit. The output for the 1-bit shift register is given by the logical function $S{\cdot}DS+L{\cdot}DL+H{\cdot}T_{\phi 3}$ where S, L, and H represent the control signals for SHIFT, LOAD, and HOLD, respectively, DS and DL represent the data for SHIFT and LOAD operations, respectively, and $T_{\phi 3}$ is the output. The 8-bit shift register used a circuit area of $1.1\times 2.1\,mm^2$ and contained 328 Josephson junctions and 516 resistors. It was fabricated using niobium technology with Nb-AlOx-Nb junctions having critical current density of $1700\,A/cm^2$.

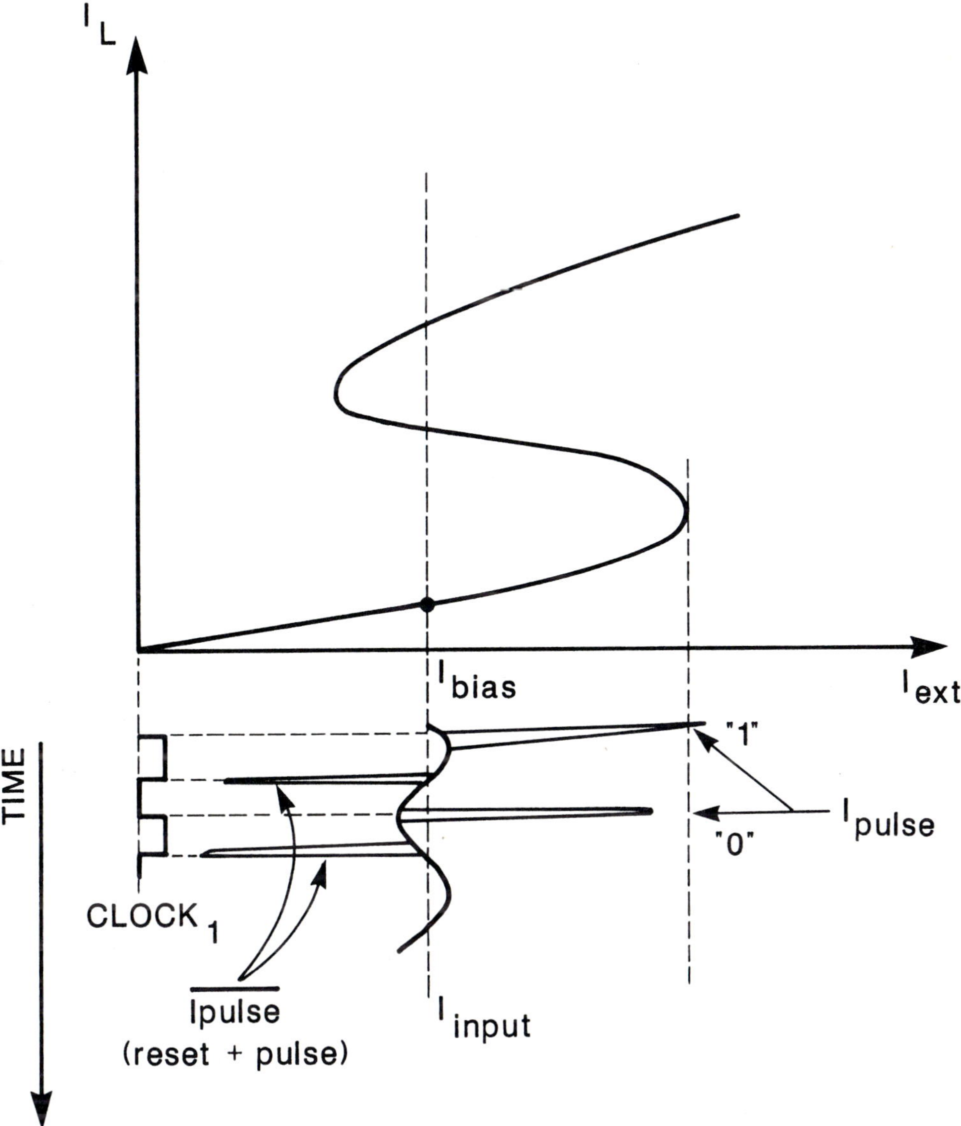

Fig. 31 Characteristic of the one-junction SQUID for the circuit of Fig. 30, shown to illustrate the principle of operation.

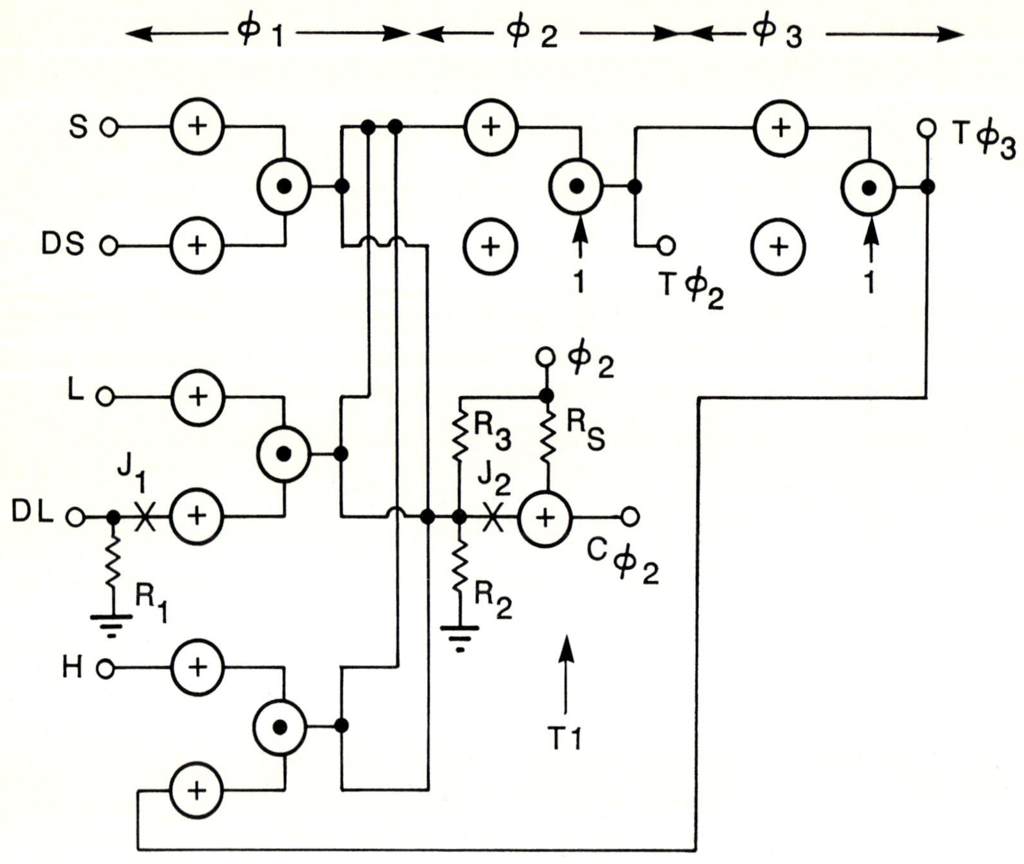

Fig. 32 One bit of a shift register with a clocking freqeuncy of 2.3 GHz. The symbols ⊕ and ⊙ are OR and AND gates, respectively.

Another circuit that has received some experimental evaluation, but at low clock rate, as well as extensive simulation, is shown in Fig. 33. [39] The clocking is three-phase; I_{c1}', I_{c2}', and I_{c3}' are currents with the three phases. A flux quantum is shifted from one loop to the next by application of the clocks. Simulations indicated that the clocking can be done successfully at 55 GHz. The circuits go into a voltage state only momentarily, in order to transfer the flux. To read out the data, a latching two-junction SQUID is connected in the third loop. The last loop on the right in Fig. 33 is the first stage of the next bit.

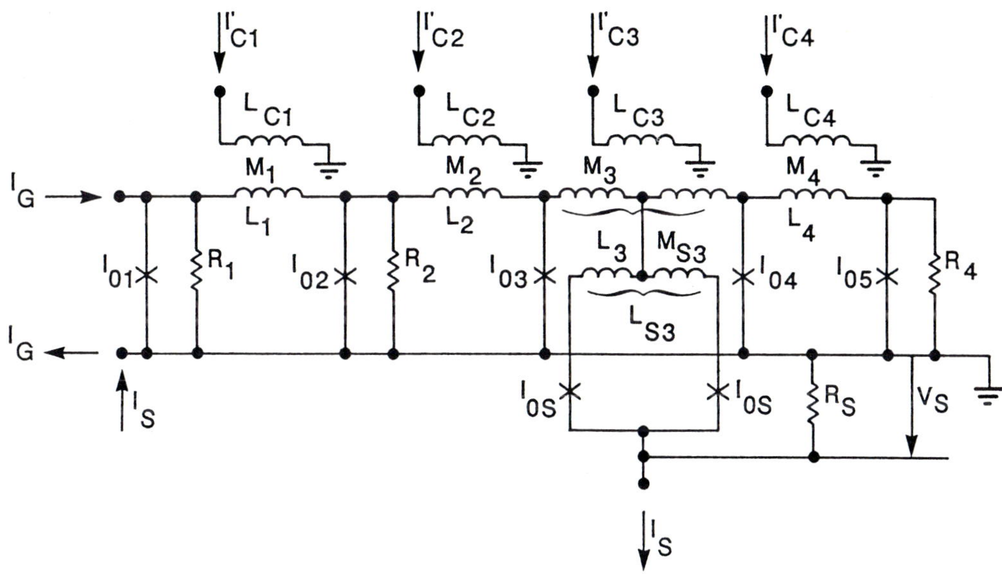

Fig. 33 A flux-transfer type of shift register. Loops 1-3 form one bit. Read-out is by means of the two-junction SQUID containing L_3 and M_{S3}. The fourth loop is the first one of the next bit.

Several other circuit designs that involve flux transfer and memory by circulating currents have appeared. Work is in progress on a circuit that involves shifting currents between arms of a superconducting loop to represent the logic states. [40] See Fig. 34. Every memory location consists of two superconducting loops, each of which contains two-junction SQUIDs for redirecting a dc current from one leg of a loop to the other. Logical "1" and logical "0" are represented by the paths taken by the currents. If the data signal D is high when CLOCK comes up, the SQUID S_L switches to the voltage state and diverts the portion of I_{dc} that was flowing through it to the right side, as shown by the curved arrow. The SQUID S_L then quickly resets to the zero-voltage state. When CLOCK falls and $\overline{\text{CLOCK}}$ rises, the current is forced to the left side of the second loop because of the switching of the SQUID on the right side of that loop. The current latched into the left side of the loop can be read by the READ GATE without interrupting the shifting operation. The power dissipation is extremely low, at 0.1 μW/bit, with worst-case switching at 62.5 GHz.

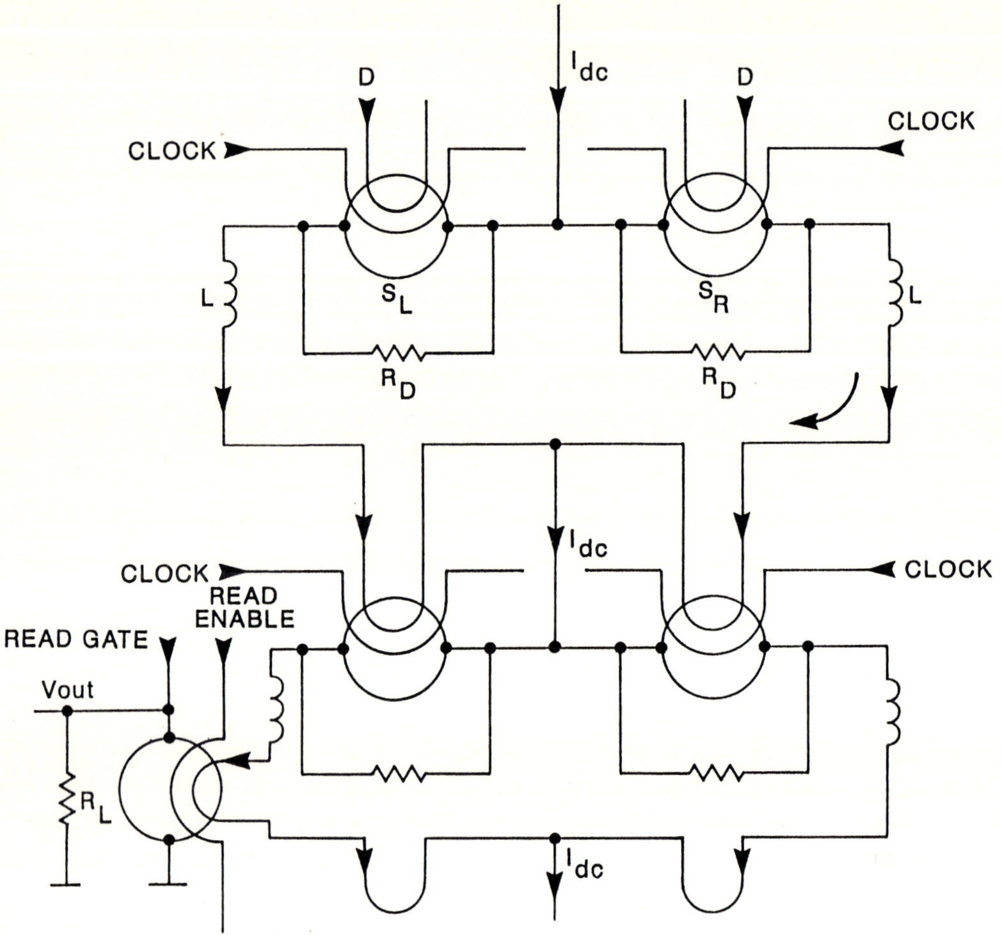

Fig. 34 One bit of a shift register based on current steering.

Other circuits that show promise for multi-gigahertz clocking have been discussed. Some of these employ resistor-junction AND gates; three such gates with three-phase clocking can form a one-bit storage circuit. The circuit in Fig. 35 showed correct operation at 35 GHz in simulations. [41] Experiments at low speed were reported recently. [42] Attention should also be brought to another single-flux-quantum shift register that has been announced. [43]

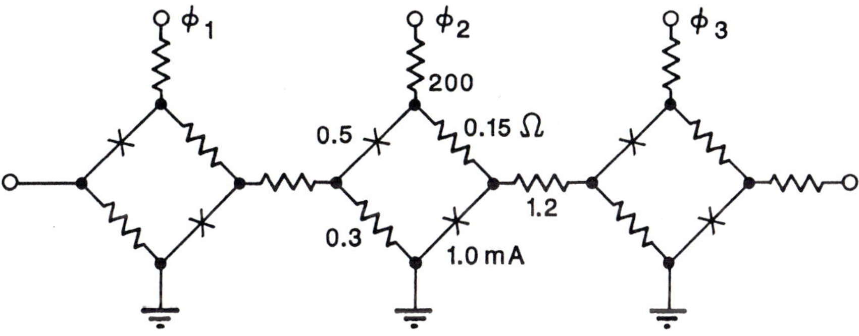

Fig. 35. One bit of a three-phase-powered shift register based on resistor-junction AND gates.

Digital circuits with high-temperature superconductors

The advantage of using high-temperature superconductors lies mostly in the convenience of refrigeration. More than an order-of-magnitude reduction of refrigerator weight is achieved by operating in the range 20-80 K, compared with operation at 4.2 K. On the other hand, almost everything electrical worsens. The noise increases with temperature and currents must be increased proportionally to avoid erroneous switching of digital circuits. Also, if the mode of operation is similar to that used in the low-temperature circuits, the voltage levels will increase in proportion to the transition temperature. With both the voltage and current increasing, significantly more power will be dissipated. There is the possibility of compensating for the higher power level by changing the choice of logic circuits to the flux-transfer type. For most scaling systems that have been considered, it seems difficult not to also lose on circuit speed.

ACKNOWLEDGEMENTS

The author would like to express appreciation for the help given by Richard Ralston with the sections on analog signal processing. The support of AF Contract F19628-86-K-0033 with RADC funded by SDIO/IST for work on A/D converters and shift registers is gratefully acknowledged.

REFERENCES

1. Cafarella, J.H.: Wideband signal processing for communication and radar. Proc. NTC'83 -- IEEE National Telesystems Conf., November 1983. CH1975-2/83/0000-0055.
2. Reible, S.A.: Future of cryogenic devices for signal processing applications. In: Research on Superconductive Signal-Processing Devices. MIT Lincoln Laboratory Report, 30 November 1984
3. Green, J.B.: Private communication.
4. Ramo, S., Whinnery, J.R., and Van Duzer, T.: Fields and waves in communication electronics. 2nd ed. p. 245, New York: Wiley 1984
5. Anderson, A.C.: Private communication
6. Withers, R.S., Anderson, A.C., Green, J.B., and Reible, S.A.: Superconductive delay-line technology and applications. IEEE Trans. Magn. **MAG-21**. 186-192 (1985)
7. Anderson, A.C., Marden, J.A., and Withers, R.S.: Thin stripline dielectrics with passivated superconductors. MIT Lincoln Laboratory Report. Solid-State Research 1985: 3
8. Withers, R.S., Anderson, A.C., Wright, P.V., and Reible, S.A.: Superconductive tapped delay lines for microwave analog signal processing, IEEE Trans. Magn. **MAG-19**. 480-484 (1983)
9. Withers, R.S. and Reible, S.A.: Superconductive chirp-transform spectrum analyzer. IEEE Electron Device Lett. **EDL-6**. 261-263 (1985)
10. Ralston,R.W.: Private communication
11. Reible, S.A., Anderson, A.C., Wright, P.V., Withers, R.S., and Ralston, R.W.: Superconductive convolver. IEEE Trans. Magn. **MAG-19**. 475-479 (1983)
12. Reible, S.A.: Superconductive convolver with junction ring mixers. IEEE Trans. Magn. **MAG-21**. 193-196 (1985)
13. Green, J.B., Smith, L.N., Anderson, A.C., Reible, S.R., and Withers, R.S.: Analog signal correlator using superconductive integrated components. IEEE Trans. Magn. **MAG-23**. 895-898 (1987)
14. Green, J.B., Anderson, A.C., and Withers, R.S.: Analog superconductive correlator for wideband signal processing. Extended Abstracts. 1987 International Superconductivity Electronics Conference (ISEC'87) pp. 49-52. Tokyo, August 28-29, 1987
15. For a more extensive review of logic and memory, see: Van Duzer, T.: Superconductive digital ICs, Chapter 16 in VLSI Handbook. J. Di Giacomo, ed. McGraw-Hill, 1989
16. Gheewala, T.R.: Josephson-logic devices and circuits. IEEE Trans. Electron Devices. **ED-27**. 1857-1869 (1980)

17. Sone, J., Yoshida, T., and Abe, H.: Resistor-coupled Josephson logic. Appl. Phys. Lett. **40**. 741-744 (1982)

18. Nakagawa, H., Sogawa, E., Kosaka, S., Takada, S., and Hayakawa, H.: Operating characteristics of Josephson four-junction logic (4JL) gate. Japanese J. App. Phys. **21**. L198-L200 (1982)

19. Kotani, S., Fujimaki, N., Imamura, T., and Hasuo, S.: Ultrahigh-speed logic gate family with Nb/Al-AlO_x/Nb Josephson junctions. IEEE Trans. Electron Devices. **ED-33**. 379-384 (1986)

20. Kotani, S., Imamura, T., and Hasuo, S.: A 2.5-ps Josephson OR gate. Tech. Digest of IEDM, Washington, D.C., Dec. 6-9, 1987. pp. 865-866

21. Wada, Y., Hidaka, M., Nagasawa, S., and Ishida, I.: AC and DC powered subnanosecond 1K-bit Josephson cache memory design. IEEE J. Solid-State Circuits. **SSC-23**. 923-932 (1988)

22. Kotani, S., Fujimaki, N., Imamura, T., and Hasuo, S.: A Josephson 4-bit microprocessor. Extended Abstracts. 1988 Int. Solid-State Circuits Conf. San Francisco, February 18, 1988. pp. 150-151

23. Hamilton, C.A.: Private communication

24. Hurrell, J.P., Pridmore-Brown, D.C., and Silver, A.H.: Analog-to-digital conversion with unlatched SQUID's. IEEE Trans. Electron Devices. **ED-27**. 1887-1896 (1980)

25. Hamilton, C.A. and Lloyd, F.L.: 100 GHz binary counter based on DC SQUID's. IEEE Electron Device Lett. **EDL-3**. 335-338 (1982)

26. Zappe, H.H.: Ultrasensitive analog-to-digital converter using Josephson junctions. IBM Tech. Disclosure Bull. **17**. 3053-3054 (1975)

27. Hamilton, C.A. and Lloyd, F.L.: A superconducting 6-bit analog-to-digital converter with operation to 2×10^9 samples/second. IEEE Electron Device Lett. **EDL-1**. 92-94 (1980)

28. Dhong, S.H., Jewett, R.E., and Van Duzer, T.: Josephson analog-to-digital converter using self-gating-AND circuits as comparators. IEEE Trans. Magn. **MAG-19**. 1282-1285 (1983)

29. Petersen, D.A., Ko, H., Jewett, R.E., Nakajima, K., Nandakumar, V., Spargo, J.W., and Van Duzer, T.: A high-speed analog-to-digital converter using Josephson self-gating-AND comparators. IEEE Trans. Magn. **MAG-21**. 200-203 (1985)

30. Fang, E.S. and Van Duzer, T.: Speed-limiting factors in flash-type Josephson A/D converters. To appear in IEEE Trans. Magn. **MAG-25**. (1989)

31. Hamilton, C.A., Lloyd, F.L., and Kautz, R.L.: Superconducting A/D converters using latching comparators. IEEE Trans. Magn. **MAG-21**. 197-199 (1985)

32. Fang, E., Nandakumar, V., Petersen, D.A., and Van Duzer, T.: High-speed A/D converters and shift registers. Extended Abstracts. 1987 International Superconductivity Electronics Conference (ISEC'87), August 28-29, 1987, Tokyo, pp. 325-328

33. Ko, H. and Van Duzer, T.: A new high-speed periodic-threshold comparator for use in a Josephson A/D converter. IEEE J. Solid-State Circuits. **23**. 1017-1021 (1988)

34. Ko, H.: A flash Josephson A/D converter constructed with one-junction SQUIDs. To be published in IEEE Trans. Magn. **25**. (1989)

35. Petersen, D.A., Ko, H., and Van Duzer, T.: Dynamic behavior of a Josephson latching comparator for use in a high-speed analog-to-digital converter. IEEE Trans. Magn. **MAG-23**. 891-894 (1987)

36. Petersen, D.A., Hebert, D., and Van Duzer, T.: A Josephson analog limiter circuit. To appear in IEEE Trans. Magn. **MAG-25**. (1989)

37. See Proc. 1988 Applied Superconductivity Conference. San Francisco, August 21-25, 1988. To appear in IEEE Trans. Magn. **MAG-25**. (1989)

38. Fujimaki, N., Kotani, S., Imamura, T., and Hasuo, S.: Josephson 8-bit shift register. IEEE Trans. Solid-State Circuits. **SC-22**. 886-891 (1987)

39. Jutzi, W., Crocoll, E., Herwig, R., Kratz, H., Neuhaus, M., Sadorf, H., and Wunsch, J.: Experimental SFQ interferometer shift register prototype with Josephson junctions. IEEE Electron Device Lett. **EDL-4**., 49-50 (1983)

40. Nandakumar, V. and Van Duzer, T.: Design of a fast variable-frequency shift register. IEEE Trans. Circuits and Systems. **CAS-35**. 1172-1174 (1988)

41. Przybysz, J.X. and Blaugher, R.D.: Josephson data latch for frequency agile shift registers. IEEE Trans. Magn. **MAG-23**. 777-780 (1987)

42. Przybysz, J.X. and Blaugher, R.D.: Josephson shift register design and layout. To appear in IEEE Trans. Magn, **MAG-25**. (1989)

43. Kuo, F., Whitely, S.R., and Faris, S.M.: A fast Josephson SFQ shift register. To appear in IEEE Trans. Magn. **MAG-25**. (1989)

JOSEPHSON LSI TECHNOLOGY AND CIRCUITS

Hisao Hayakawa

Department of Electronics, Nagoya University
Furo-cho, Chikusa-Ku
Nagoya 464-01
Japan

1 Introduction

A Josephson tunnel junction can change its state from a superconductive state to a resistive state by application of a weak magnetic field or injection of a current to the junction. This means the Josephson junction can be operated as a current switch. Josephson junctions can switch with high speed and very low power dissipation so that logic gates and memories based on Josephson junctions can be expected to be circuit elements for ultra-fast computer systems.

IBM initiated the development of Josephson computer technology in the mid 1960's soon after Josephson's prediction. They developed junction integration technologies based on Pb-alloy junctions, circuit technologies for logic and memory, and system concepts including power and packaging [1]. These achievements attracted worldwide attention and served as an incentive to other laboratories to start research on Josephson digital applications.

Although much progress in Josephson technology had been made, IBM announced a halt to their Josephson computer project in 1983, a decision based partially on the belief that the performance of semiconductor devices was becoming competitive with Josephson devices. Despite the IBM decision, efforts for developing Josephson computer technology were maintained in Japan by taking new approaches in material and device technologies.

The most important achievement made by Japanese efforts is the progress in Josephson integration technologies based on all -refractory Josephson junctions. The introduction of junctions made of refractory materials (such as Nb and NbN instead of Pb alloys) into Josephson circuits has improved the reliability, stability

NATO ASI Series, Vol. F 59
Superconducting Electronics
Edited by H. Weinstock and M. Nisenoff

and the uniformity of circuits, resulting in the realization of Josephson LSI circuits. Moreover, the performance of Josephson circuits also has been improved and experimentally confirmed to be superior to that of the most advanced semiconductor devices. Using these technologies, Japanese laboratories now are working on small-scale computer systems to test the feasibility of future large scale Josephson computer systems.

This chapter treats the recent advances in Josephson digital technologies which have been developed mainly in Japan. The chapter consists of two major sections: one concerned with all-refractory Josephson integration technology, and the other with Josephson circuitry.

2 Josephson LSI Technology

2-1 Overview of Junction Technology

The first Josephson tunnel junctions available for circuit applications were Pb-alloy junctions developed by IBM [2]. Junctions made of pure Pb films exhibit good tunneling characteristics. However, short circuits were often found after only one thermal cycle between room temperature and liquid helium temperature. This is due to the formation of hillocks in the films. Thermal strains in films induced by the difference in thermal expansion coefficients between substrate and film cause dislocation slide at crystal grain boundaries, forming hillocks. These hillocks easily produce short circuits in thin tunnel barriers.

The Pb-alloy junction was developed to improve the reliability of Pb junctions for thermal cycling. The base electrode of the junction is made by a sequential evaporation of Au, Pb and In (typically, Au:Pb:In = 4 wt% : 84 wt% : 12 wt%). Alloying of these materials is easily completed by heating the film to 80°C for one hour. The alloy films exhibit high resistance to the formation of hillocks by thermal cycling.

The ε phase Pb-Bi-alloy film is used for the counter electrode. Before the counter electrode is evaporated, the base electrode is oxidized with a plasma oxidation to form a tunnel barrier. A novel method called sputter-oxidation was developed to control the thickness of the oxide tunnel barrier [3]. The sample is placed on the cathode in an RF-oxygen plasma, where sputter-etching and oxidation occur simultaneously. These processes finally reach a condition where the etching rate and the oxidation rate are equal. After reaching this condition, the thickness of the oxide layer no longer changes with time, which is useful for controlling the barrier thickness precisely.

The stability of Pb-alloy junctions with respect to thermal cycling also was improved by crystallographic research on the Pb-alloy films. It was found that films with smaller grain size exhibit higher resistance to hillock formation [4]. Junctions whose base electrodes were evaporated at a low substrate temperature, hence producing smaller grain size, exhibited excellent stability after more than 1000 thermal cycles [5].

Using these Pb-alloy junctions, a large number of Josephson devices were fabricated and successfully operated. Although the Pb-alloy junctions were useful for demonstrating device operation in the early stages of development, it became clear that they have problems which make it difficult to fabricate circuits with LSI-level complexities. Two major problems are noted: one is a large initial failure rate; the other is degradation in tunneling characteristics during storage.

Initial failure rates for various types of junctions measured by the Fujitsu group are listed in Table 1 [6]. As is seen from this table, more than 1% of the Pb-alloy junctions are shorted just after fabrication, almost two orders of magnitude larger than for all-Nb junctions. This means that complete operation of a circuit based on Pb-alloy junctions becomes more difficult as the scale of the circuit increases.

Table 1 Initial failure rate for various junctions [6].

Junction	No. of Test Cells	No. of Shorted Junctions	Failure Rate(%)
Pb-In-Au/Pb-Bi	24,576	260	1.06
Nb/Pb-Bi	65,536	83	0.13
Nb/Al0x/Nb	49,152	8	0.016

Another problem with Pb-alloy junctions is that a junction's tunneling characteristics may change during storage at room temperature. In order to avoid this problem, junctions must be stored in a low-temperature environment, i.e., in liquid nitrogen, which is quite inconvenient for practical use.

Since 1980, intensive efforts have been made to develop more reliable junctions based on refractory superconductors such as Nb and NbN. In the process of developing all-refractory junctions, semi-hard junctions in which only base electrodes are replaced by Nb or NbN were considered. Nb/oxide/Pb and NbN/oxide/Pb junctions exhibited good tunneling characteristics, good stability to thermal cycling and smaller initial failure rates than for Pb-alloy junctions. NbN/oxide/Pb showed excellent tunneling characteristics with very low leakage currents in the subgap region [7,8].

A disadvantage of a semi-hard junction relates to the larger junction capacitance caused by the relatively large dielectric constant of the niobium oxide used as the tunnel barrier. The junction capacitance per unit area of Nb/oxide/Pb and NbN/oxide/Pb junctions (~12 μF) is about three times larger than that of Pb-alloy junctions. The large junction capacitance makes switching speeds slow. In order to use these junctions for high-speed digital devices, the junction sizes must be small to reduce the junction capacitance. Edge junctions, making use of the edge of the niobium base electrode to define the junction area, were developed to meet this requirement [9].

Although some attempts had been made to fabricate LSI circuits using semi-hard junctions, i.e., NbN/oxide/Pb junctions [10], junctions for circuit applications have been replaced by newly developed all-hard (or all-refractory) junctions since the mid 1980's. All-refractory junctions in which both base and counter electrodes are made of refractory superconductors such as Nb or NbN also have completely changed Josephson integration processes, allowing for the operation of LSI logic and memory circuits with sufficient stability.

In the following sections, the all-refractory junctions and their integration processes will be described.

2-2 All-Refractory Junction

It is natural to use an oxide layer obtained by oxidizing the surface of a base electrode as a tunnel barrier. A number of attempts were made to fabricate all-refractory junctions using native-oxide tunnel barriers formed on the Nb base electrodes, i.e., Nb/oxide/Nb junctions [11]. However, these attempts were not successful. It was difficult to make junctions with good quality. Usually the junctions exhibited large leakage currents in the subgap region; the V_m value was less than 10 mV (where the V_m value is a measure of subgap leakage current and is defined as the product of the Josephson critical current times the subgap resistance). The reason for this is that the niobium-oxide tunnel barrier easily reacts with Nb during the evaporation of the counter electrode, forming low-T_c materials at the interface. Contrary to the leakage problems of Nb junctions, NbN junctions with oxide barriers were found to have fairly good tunneling characteristics (V_m > 20 mV) [12]. This is due to the fact that NbN is less chemically active than Nb. The ETL group made Josephson LSI circuits using these NbN/oxide/NbN junctions for the first time in 1985 [13]. This effort showed NbN to be a good candidate for junction electrodes. NbN is attractive as a junction material because it has a T_c of about 15 K, which is much higher than that of Nb, and NbN films are easily obtained by reactive sputtering of Nb at a relatively low substrate temperature.

Although the NbN/oxide/NbN junctions exhibited good tunneling characteristics, the measured gap voltage (V_g = 3.8 mV) was found to be much less than expected (5 mV) from the value of T_c. This indicates that the reaction between the oxide barrier and the counter electrode reduces the T_c of the NbN counter electrode just above the barrier, resulting in the lower gap voltage. Thus, it is important to prevent the counter electrode from reacting with the barrier in order to make ideal all-refractory junctions.

The development of artificial barriers to replace native-oxide barriers has solved this problem. Figure 1 shows various junction structures which have been fabricated successfully to produce high quality junctions.

The first success in fabricating an all-Nb junction using an artificial barrier was made by the Sperry group [14]. They used amorphous Si (a-Si) as a tunnel

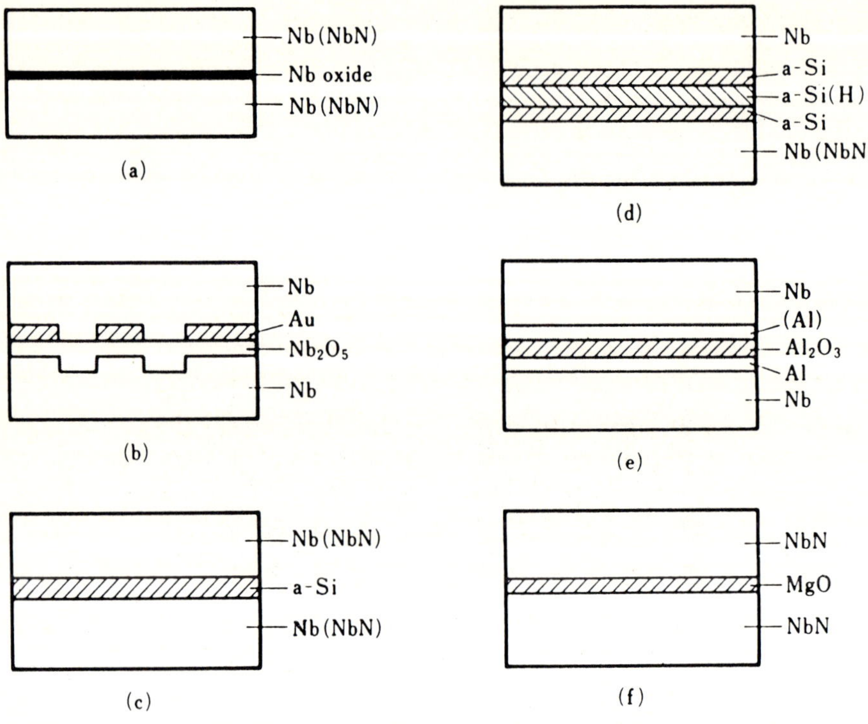

Fig. 1 Junction structures of various all-refractory Josephson junctions.

barrier. Junctions with a single-layered a-Si barrier exhibited a V_m of 11 mV. The V_m value was improved by using a multilayered barrier with a hydrogenated a-Si layer sandwich with very thin unhydrogenated a-Si layers (<10 A) as shown in Fig. 1(d) [15]. The hydrogenation of a-Si drastically reduces the number of states in the barrier, which suppresses the tunneling current, resulting in the improvement of V_m. A V_m value of 28 mV was reported for a Nb/a-Si·a-Si(H)·a-Si/Nb junction.

Improvement of the junction quality of the Nb/oxide/Nb junction also was made by introducing a gold (Au) film on the oxide barrier as shown in Fig. 1(b) [16]. The Au layer was used as a buffer layer, preventing the Nb counter electrode from reacting with the oxide barrier. Pin holes in the Au film were filled by strong oxidation after evaporation of the Au layer. A V_m of 26 mV was obtained for this Nb/oxide·Au/Nb junction.

Important progress in the fabrication of all-Nb junctions was made by the development of a junction with an aluminum-oxide barrier. The AT&T group developed an all-Nb junction with an Al-oxide barrier thermally grown on a thin Al overlayer on the Nb base electrode as shown in Fig. 1(e) [17]. The junction had excellent tunneling characteristics: a very low leakage current in the subgap region and a sharp current rise at the gap voltage. A V_m value of 40 mV was obtained. A typical I-V curve of the Nb/Al·Al_2O_3/Nb junction is shown in Fig. 2.

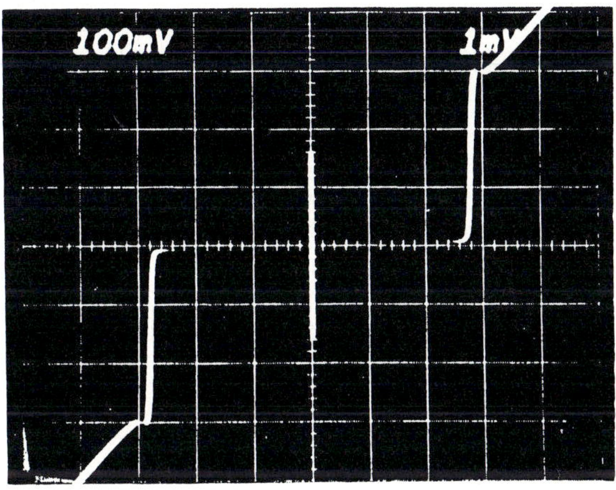

Fig. 2 A typical current-voltage curve for a Nb/Al·Al_2O_3/Nb junction. Junction area is 5 μm x 5μm. Vertical scale: 1 mA/cm. Horizontal scale: 1 mV/cm.

The excellent tunneling characteristics of the Nb/Al·Al_2O_3/Nb junction have attracted the attention of many researchers, and improvements have been made in tunneling characteristics as well as in fabrication technology [18]. The Nb/Al·Al_2O_3/Nb junctions now are widely used, not only for digital applications, but also for analog applications.

Another important development in all-refractory junctions is the fabrication of "high-T_c" junctions with NbN electrodes. The ETL group has developed all-NbN junctions using MgO as a tunnel barrier, as shown in Fig. 1(f) [19]. A typical I-V curve for a NbN/MgO/NbN junction is shown in Fig. 3. The gap voltage for this junction is 5.1 mV, which is much higher than that of a typical NbN/oxide/NbN

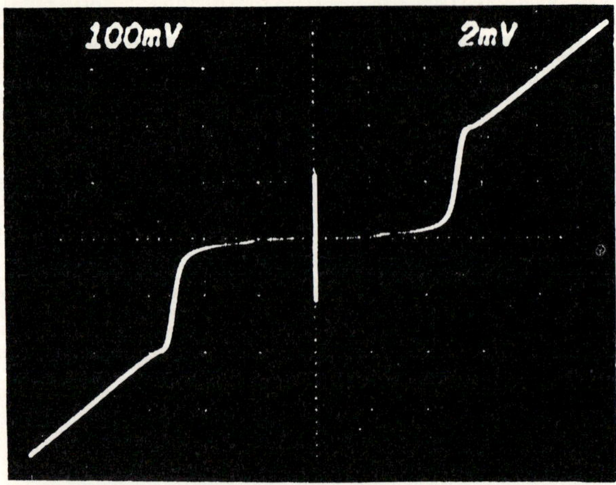

Fig. 3 A typical current-voltage curve for a NbN/MgO/NbN junction with dimensions of 2.5 μm x 2.5 μm. Vertical: 0.2 mA/cm. Horizontal: 2 mV/cm.

junction (Vg = 4 mV). This means that a reduction of T_C in the interface layer just above the barrier is prevented by the MgO barrier. In fact, Josephson current was observed in the NbN/MgO/NbN junction at temperatures up to 14.5 K, as shown in Fig. 4.

High-T_C Josephson tunnel junctions are important for digital applications. Although most Josephson digital circuits will probably be operated at 4.2 K, circuits with high-T_C junctions will exhibit improved operational stability against thermal fluctuations. Moreover, a high-T_C junction would generate a higher voltage when it switches, yielding a smaller propagation delay for an inductive load. Although Josephson digital circuits employing only NbN/MgO/NbN junctions have not yet been made, such junctions will grow in importance. Tunneling parameters for various fabricated all-refractory junctions are summarized in Table 2.

2-3 Junction Fabrication Process

The development of all-refractory junctions has changed the junction fabrication process. In Pb-alloy junctions, a lift-off process was utilized for

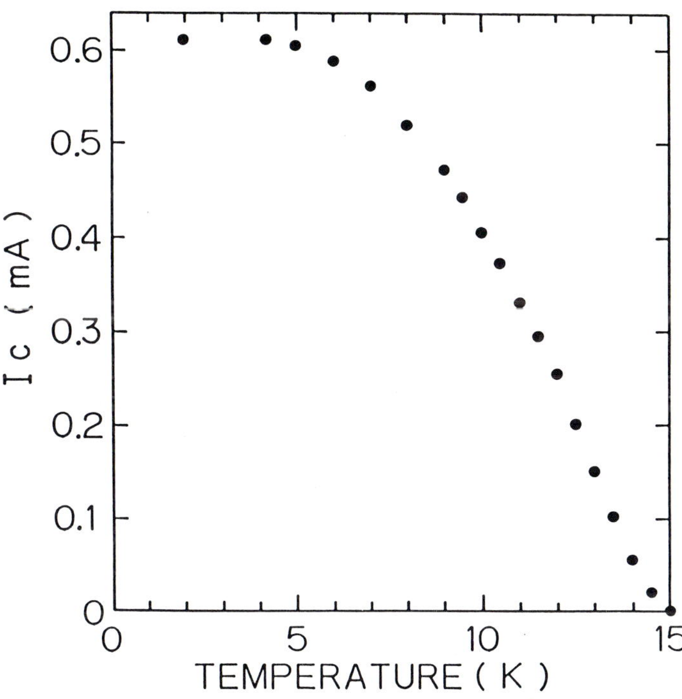

Fig.4 Temperature dependence of the Josephson critical current for a NbN/MgO/NbN junction. The Josephson current can be observed up to a temperature of 14.5 K [19].

Table 2 Characteristic physical quantities of all-hard junctions

Junction	V_g(mV)	I_cR_n (mV)	V_m (mV)	C_J(μF/cm^2)	Ref
Nb/Nb oxide/Nb	2.8	1.2	6~8	12~14	11
Nb/a-Si/Nb	2.7	1.3~1.6	11	2.5	14
Nb/a-Si a-Si(H) a-Si/Nb	2.8	1.3~1.6	28	2.5	15
Nb/Nb oxide Au/Nb	2.65	-	26	7.5	16
Nb/Al·Al_2O_3·(Al)/Nb	2.85	1.68	40~70	6	17,18
NbN/Nb oxide/NbN	4.0~4.4	1.9	15~25	10	13
NbN/MgO/NbN	5.1	3.2	45	8	19

patterning junctions. The most important change in the fabrication process for the all-refractory junctions is the introduction of the whole-wafer or full-wafer process developed by the Sperry group [14]. In the full-wafer process, a junction sandwich is formed over the entire area of a Si substrate without breaking vacuum, after which small junctions are isolated. This process can avoid contamination introduced from sequential processes, resulting in high uniformity of tunneling characteristics across the wafer. Moreover, the process is quite compatible with semiconductor processing, allowing a variety of fabrication processes developed in the semiconductor field. This process was first applied to fabricating Nb/a-Si/Nb junctions in which Nb was anodized to isolate each junction. This process was called SNAP for Selective Niobium Anodizaiton Process. Figure 5 shows the outline of the SNAP process.

In the SNAP method, the anodized layer usually grows laterally, making it difficult to define accurately junction area, especially when the junction size becomes small. In circuit applications where a large number of junctions are integrated, the uniformity of junction areas across a chip is very important to keep the spread of tunneling characteristics as small as possible.

In order to meet this requirement, a new process was developed [12] in which an etching process replaced the anoidization process to produce junction isolation. Figure 6 illustrates the etching process called SNIP for Self-aligned Niobium (Nitride) Isolation Process. As shown in Fig. 6, small junctions are isolated from a junction sandwich by reactive-ion etching, after which an insulating layer (SiO) is evaporated. After removing the resist masks on top of the junction, insulation of junction peripheries is completed in a self-aligned manner. This process was first used for NbN/oxide/NbN junctions [12], and then for Nb/Al$\cdot$$Al_2O_3$/Nb junctions [17].

When the etching process was adopted for fabricating Nb/Al$\cdot$$Al_2O_3$/Nb junctions, it was found that strains in sputtered Nb films have an important effect on the junction quality. This phenomenon was first noticed by ETL researchers [21]. They found that V_m's of Nb/Al$\cdot$$Al_2O_3$/Nb junctions fabricated by the etching process deteriorated significantly when the junction size became smaller, typically less than 5 μm. They speculated that this phenomenon is due to strain in the Nb electrodes, and that stress relaxation induced by etching causes damage to the tunnel barriers.

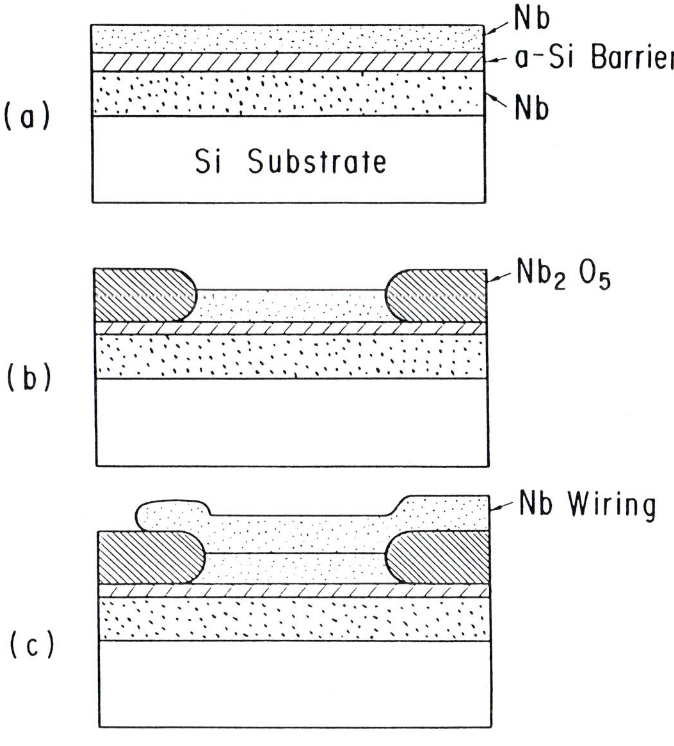

Fig. 5 An all-refractory junction process called SNAP in which an anodization of a counter Nb electrode is used for junction isolation [14].

The NTT and Fujitsu groups measured strain in sputtered Nb films and found that strain in the films depends on Ar gas pressure during Nb sputtering [21,22]. In the lower pressure region, the stress is tensile, changing to compressive with increasing Ar pressure, as shown in Fig. 7. By selecting the Ar pressure at which stress in the film becomes zero, it is possible to eliminate deterioration in tunneling characteristics due to stress relaxation.

In order to operate an LSI circuit within reasonable margins, it is important to make the spread of critical currents of the Josephson junctions on the chip as small as possible. The accurate definition of junction area due to etching, and the uniform barrier formation by employment of the full-wafer process, result in excellent

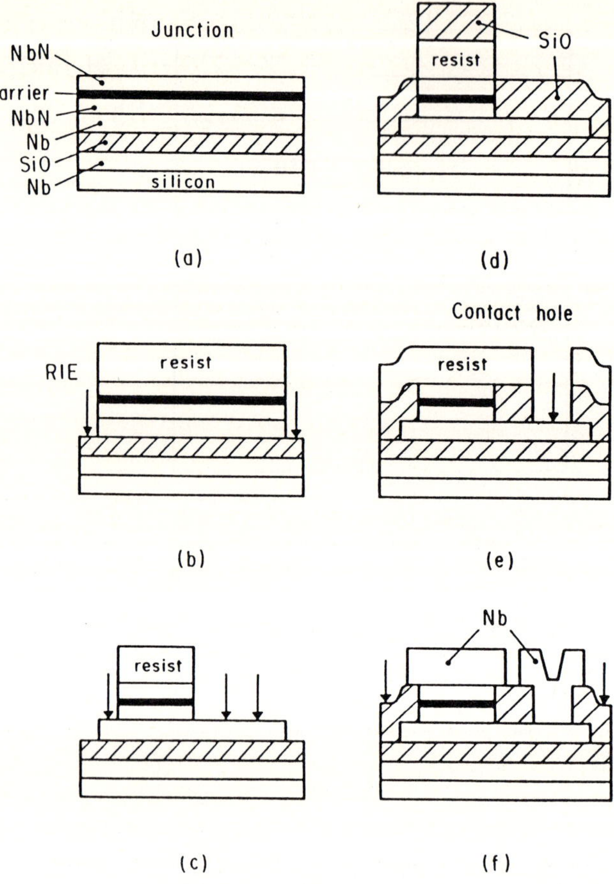

Fig.6 An all-refractory junction process called SNIP. The individual junction is isolated by a reactive-etching process applied to a junction sandwich [12].

junction uniformity. Figure 8 shows an example of an I-V curve for a series array of 1024 NbN/oxide/NbN junctions (2.5 μm x 2.5 μm) [23]. The trace of the critical currents is very flat, indicating good junction uniformity with a standard deviation of only 1.6%. Usually junction uniformity worsens when junction size becomes small. This is due to the fact that the precision of lithography is decreased as pattern size is decreased. The measured standard deviations of critical current versus junction size are plotted in Fig. 9 [23]. The use of a more accurate resist process, i.e. , a

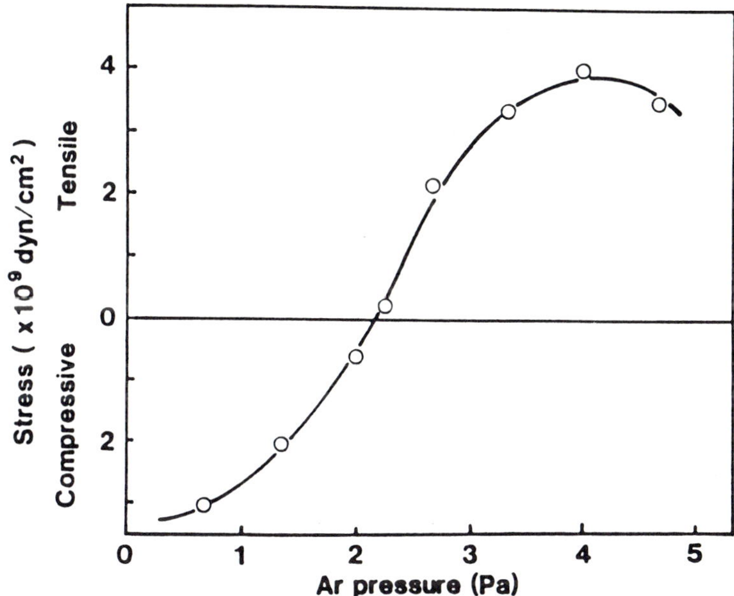

Fig. 7 Stress in a Nb film as a function of Ar pressure during sputter deposition. The stress in the film changes from compressive to tensile as the Ar pressure is increased [21].

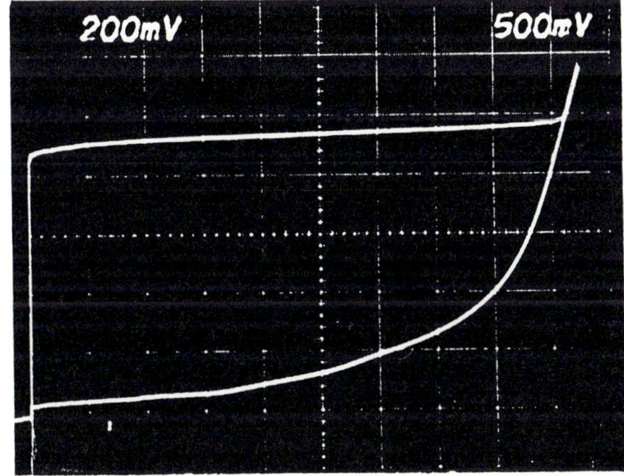

Fig. 8 A current-voltage curve of a series array of 1024 NbN/oxide/NbN junctions. The junction uniformity is evaluated from the trace of Joesphson critical current.

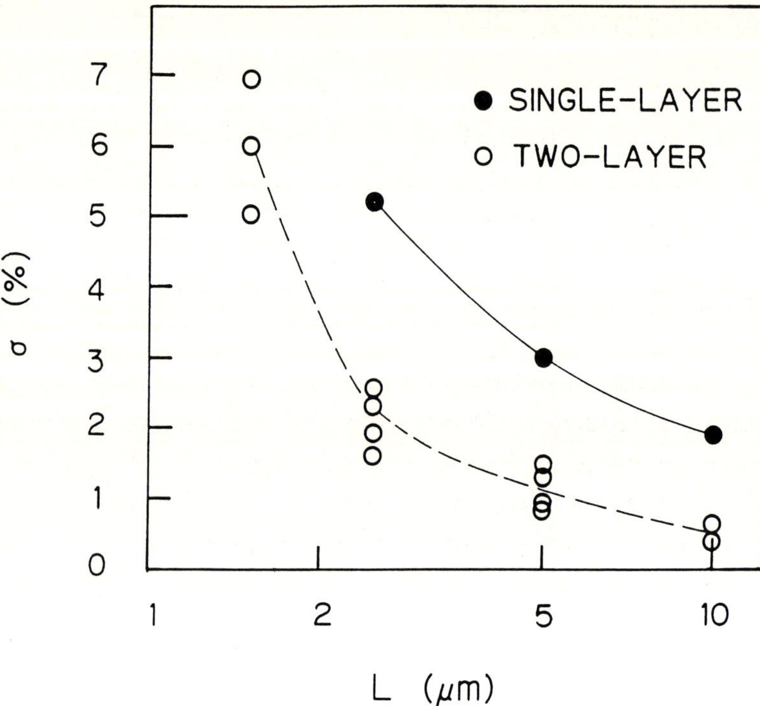

Fig. 9 Standard deviations in the critical currents of junctions on a chip as a function of junction sizes [23]. By the more accurate resist process, i.e., a double-layered resist mask, junction uniformity is much improved when junction size becomes smaller.

double-layered resist process, improves the uniformity of critical currents, especially for smaller junctions.

All-refractory junctions and improved junction fabrication processes have combined to make junction technology more reliable then it once was. Innovation in junction technology has played an important role in the realization of Josephson LSI circuits described in the following sections.

2-4 LSI Circuit Fabrication Process

Josephson LSI circuits have become a reality since the development of all-refractory junctions. As the integration scale increases, it becomes more important to have reliable circuit fabrication processes. Many new technologies have been introduced in fabrication processes to increase chip yields. As a result, a number of Josephson LSI circuits have been developed using the all-refractory circuit process and have been operated successfully.

Processing improvements which have been introduced to increase chip yields of all-refractory LSI circuits relate to planarization technologies and improvements in insulating layers. Figure 10 shows a cross-sectional view of a typical all-refractory Josephson circuit. As seen in the figure, Josephson circuits are made in a layered structure deposited on a substrate, usually of Si. Etching processes, mainly reactive-ion etching (RIE) processes, are used for patterning these layers: ground planes, ground-plane insulation layers, resistors, junctions,

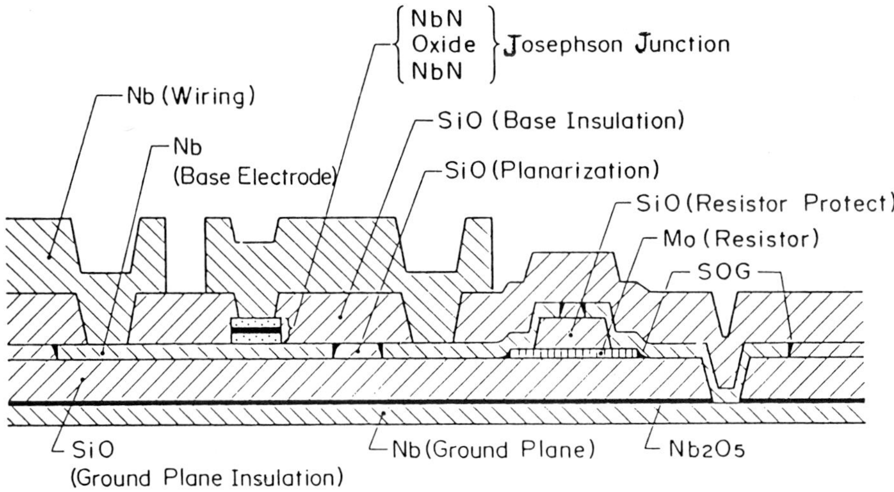

Fig. 10 A cross-sectional view of an all-refractory circuit.

interconnects, etc. Many steps appear after each layer is patterned. The edge profile of these steps (etched by the RIE) is so sharp that failures due to open and short circuits in or between overlaid metal lines can occur when these lines cross steps.

In order to avoid these failures, it is important to planarize the steps in each layer. The first introduction of planarization technology into circuit fabrication was made by ETL [13]. They used spun-on-glass (SOG) to control edge profiles. The SOG is a liquid glass which can be solidified easily by annealing. The planarization process called SAILOP {Self-Aligned Insulation Lift-off Process) is shown in Fig. 11. A Nb wiring layer is patterned by RIE. Before resist masks are removed, a SiO insulating film is deposited. After SiO layers on the surface of the Nb film are lifted off by removing the resist masks, the surface of the SiO layer is at the same level as the Nb layer; thus, the planarized insulating layers are formed in a self-aligned manner. However, crevasses are usually formed along edges between the Nb and the insulation layers. The spun-on-glass is used to fill these crevasses. By spin coating the SOG and annealing it at about 200 °C, crevasses are filled with the solid SOG. Then, the SOG is etched back to the surface level of the Nb film, completing the planarization.

ETL used this planarization process for fabricating a 4x4-bit multiplier with 652 gates based on NbN/oxide/NbN junction technology, which was the first all-refractory Josephson circuit with LSI complexity [13]. Figure 12 shows the outline of the integration process. As seen in the figure, the planarization process is used in every layer, resulting in greatly improved reliability.

Fijitsu has applied another planarization technique to circuit fabrication [24]. It utilized a bias-sputter deposition of SiO_2 films for growing planarized insulation layers. In the bias-sputtering, a low RF power is applied to the substrate during sputter deposition, so that a sputter etching of the deposited film simultaneously takes place with film deposition. By choosing appropriate bias sputtering conditions, the deposition rate at edges can be much slower than that at the other planar sections, resulting in a film with a flat surface. Fujitsu used this method for the deposition of planarized SiO_2 insulation layers in fabricating circuits based on $Nb/Al{\cdot}Al_2O_3/Nb$ junctions.

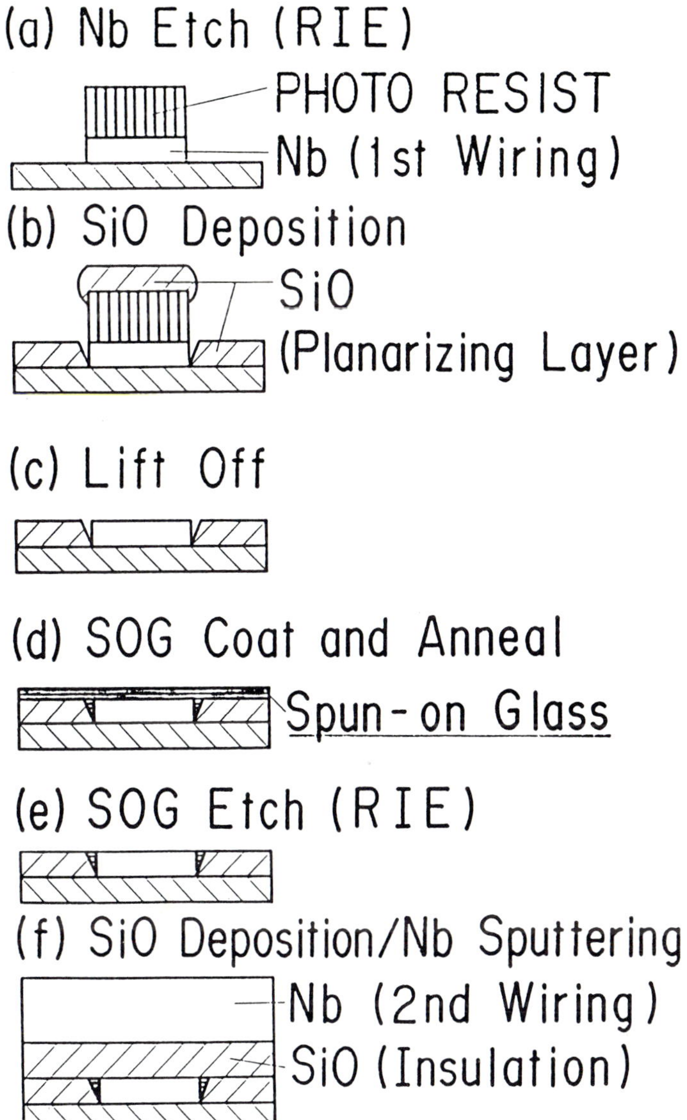

Fig. 11. A planarization process called SAILOP [13].

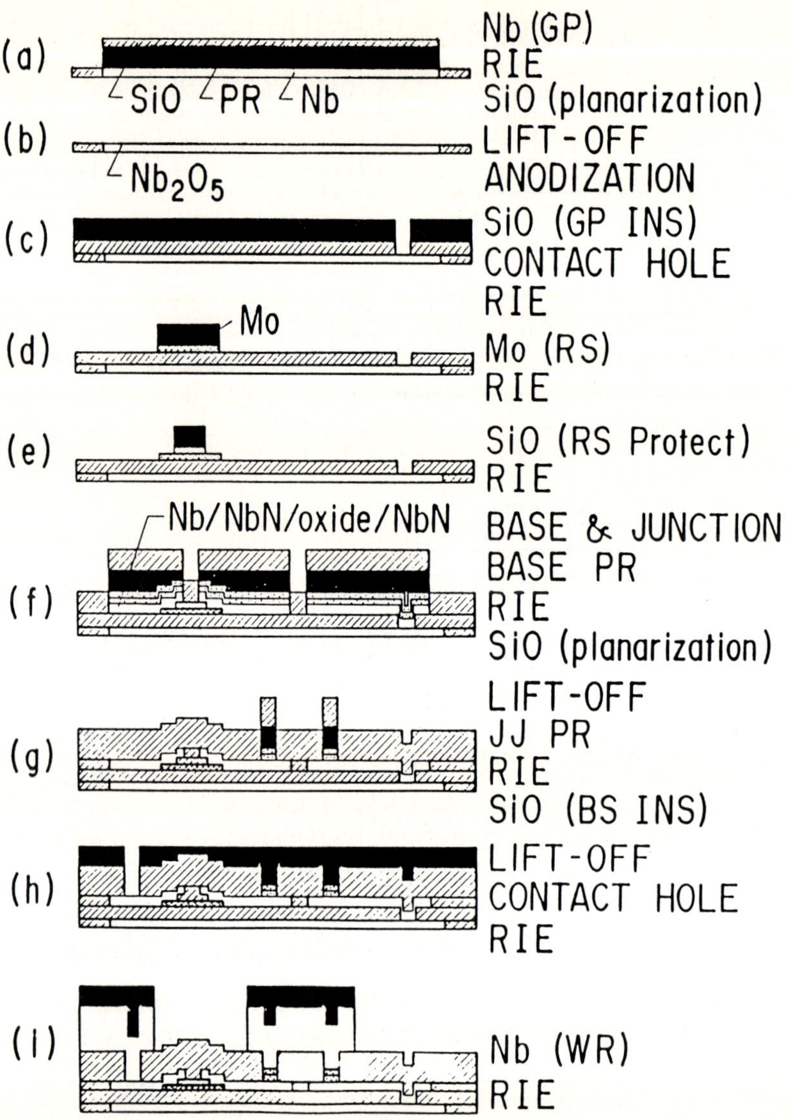

Fig. 12 A typical process flow for all-refractory LSI circuits [13]. In this case, NbN/oxide/NbN junctions are used.

Recently, the planarization technique using bias-sputtered SiO_2 insulation has been applied successfully to realize a three-dimensional Josephson circuit in which two layers of a circuit are integrated vertically [25]. Each circuit contains 5000 gates and the Josephson critical currents of the gates in both the lower and upper circuits have been found to be well controlled.

Another important factor to increase the production yield of LSI circuits is the selection of insulation layers. Earlier SiO films were used as insulation layers because these films can be deposited easily by vacuum evaporation. However, the SiO films usually were found to have a large density of pinholes, causing shorts between wiring layers.

In contrast to SiO films, SiO_2 films are expected to have a lower density of pinholes and higher breakdown voltages. SiO_2 films grown by plasma-enhanced chemical vapor deposition (PECVD) and bias-sputter deposition (described above) have been used as insulating layers in all-refractory circuits and were found to have pinhole densities much lower than those of SiO films. In Fig. 13, the measured pinhole densities in SiO_2 films deposited by PECVD, and SiO films deposited by vacuum evaporation, are compared [26].

The all-refractory circuit fabrication process recently has been made reliable enough to produce both LSI logic and LSI memory circuits. Moreover, performance of these circuits also has been much improved because of the increased uniformity of junctions. Based on progress in all-refractory integration technology, recent research efforts involve not only more complex circuits, but also small-scale Josephson computers in which logic and memory are included.

3 Josephson LSI Circuits

3-1 Logic Gates

A number of logic gates have been proposed. These logic gates are classified into two types: one is the magnetically-coupled gate, and the other is the direct-coupled gate.

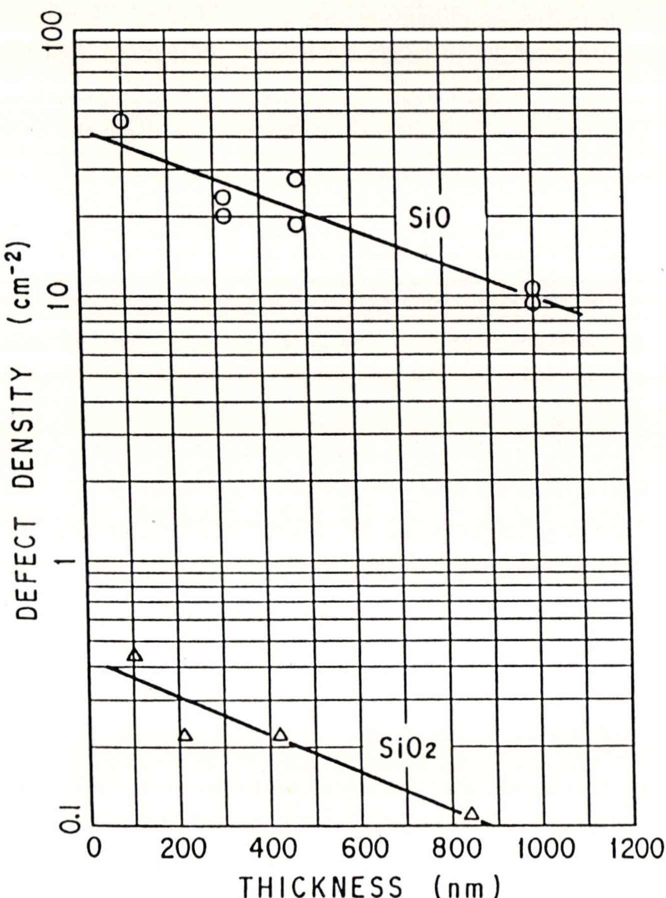

Fig.13 Pinhole densities in SiO_2 and SiO films are plotted as a function of film thickness [26]. Pinhole densities in SiO_2 films are much smaller than those in SiO films.

Magnetically Coupled Gate

The magnetically-coupled gate is based on SQUIDs which are controlled by the magnetic field produced by overlaying control lines. A typical gate of this type is the so-called three-junction interferometer logic (JIL) gate [27]. Figure 14 shows the equivalent circuit of a JIL gate and its threshold characteristics. When a sufficiently large input current is applied to the control line, the gate switches to the voltage state. This corresponds to an operation in which the operating point moves along the direction of the arrow in Fig. 14(b).

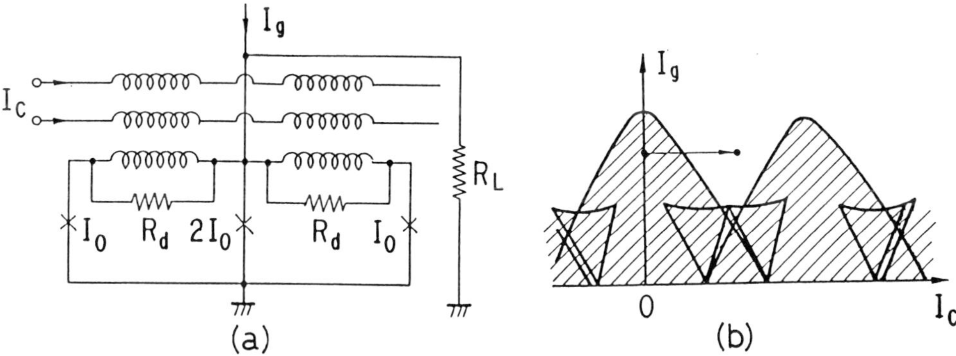

Fig. 14 The equivalent circuit configuration of a three-junction interferometer logic gate (a) and its threshold characteristics (b). The R_d's are damping resistors. The matched region of the threshold curve is the superconducting state.

Two-junction SQUIDs also can be used for logic gates. However, three-junction SQUIDs are more widely used than two-junction SQUIDs because of their wider operating margins. By choosing the critical current of the three junctions as I_0, $2I_0$ and I_0, side lobes between the main lobes in the threshold characteristics can be suppressed easily so that a wider operating window is obtained when the gate goes into the voltage state. The loop inductance L is typically chosen to be $LI_0 = 0.2 - 0.3\ \Phi_0$.

A SQUID gate with a serial tunnel junction whose equivalent circuit consists of a series combination of a loop inductance L and a junction capacitance C_j easily induces an LC resonance oscillation with a frequency of $\omega_r = 1/\sqrt{LC_j}$ when the gate goes into the voltage state. This resonance oscillation prevents the gate from achieving the full output voltage when it switches. In order to suppress the resonance oscillation, damping resistors are inserted to shunt the loop inductances as shown in Fig. 14(a) [28.].

Another important gate based on the SQUID is the so-called Current Injection Device (CID) [29]. The equivalent circuit and the threshold curve of the CID are shown in Fig. 15. The CID is operated by direct-current injection to the SQUID loop. The threshold characteristic becomes symmetric with respect to the input currents if the circuit parameters in Fig. 15(a) are chosen so that I_{01} and I_{02} are critical currents of junctions J_1 and J_2, respectively. In the CID, the gate switches into the voltage state with a large operating window if both inputs are simultaneously applied to the gate as shown in Fig. 15(b). This means that the CID is useful for operation as a two-input AND gate. However, the CID has no current isolation function, so that the CID must be used in combination with the current isolation gate. For example, the JIL's are used in front of the CID as the current isolation gates.

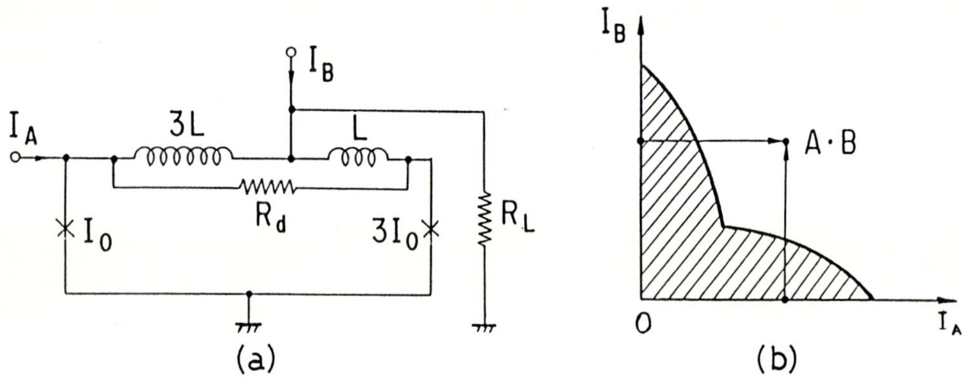

Fig. 15 The equivalent circuit configuration for a Current Injection Device (CID) (a) and its threshold characteristic (b). The threshold characteristic is symmetric for two inputs I_A and I_B, suggesting that the CID is useful as an AND gate.

Direct-Coupled Gates

Direct-coupled gates are controlled by the direct injection of input currents. The CID described above also is controlled by the injected current. However, the CID is based on a SQUID, and thus a magnetic field produced by the current flowing in the loop plays an important role in switching the gate. For this reason, the CID is classified as a magnetically-coupled gate in this chapter.

An important feature of the direct-coupled gates is that inductances can be eliminated from the gates, resulting in smaller gate size. In the direct-coupled gates, it is important to call attention to the current-isolation function. The first device with the current-isolation function was made of two junctions and a resistor, and was called JAWS (Josephson Atto-Weber-Switch) [30]. Figure 16 shows the equivalent circuit and the threshold characteristic of the JAWS gate. The operating principle of the JAWS gate is as follows. A gate current biases the gate somewhat below the critical current of the junction J_1 (so the gate remains in the superconducting state). When an input current I_c is applied to the gate through J_2, the current through J_1 is increased to $I_g + I_c$, which is large enough to switch J_1 into the voltage state. After J_1 switches, the gate current I_g flows to ground through J_2 and a small resistor r, which results in switching J_2 into the voltage state. At this stage, since both J_1 and J_2 switch into highly resistive states, I_g is transferred to a load resistor, and I_c flows to ground through r, completing the current isolation.

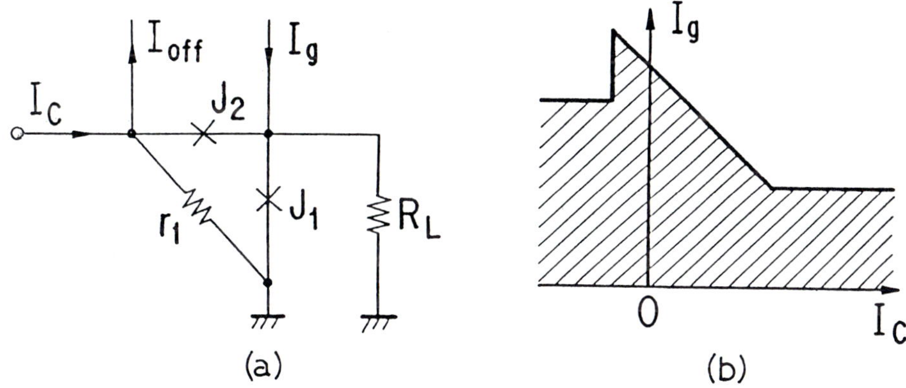

Fig. 16 Thin-film configurations of a three-junction interferometer gate known as a Josephson Atto-Weber-Switch (JAWS) (a) and its threshold curve (b). In order to make device size small, the effective loop inductance is increased by making holes in the ground plane.

The gate illustrated in Fig. 17, called a DCL (Direct-Coupled Logic) gate is another example of the current-isolation gate [31]. The operating principle of the DCL gate is quite similar to that of JAWS.

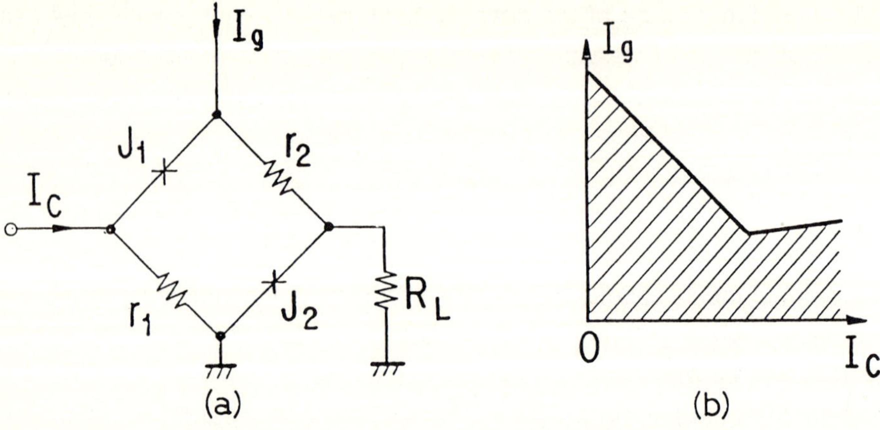

Fig. 17 The equivalent circuit of a Direct-Coupled Logic (DCL) gate (a) and its threshold curve (b).

In the JAWS and DCL current-isolation gates, the threshold curves separating the superconducting state from the voltage state have a slope of -1, i.e., $|\Delta I_g/\Delta I_c| = 1$. If the slope becomes steeper, i.e., $|\Delta I_g/\Delta I_c| > 1$, higher sensitivity can be obtained. Gates shown in Fig. 18 were developed to obtain higher sensitivity by adding another junction branch to the current isolation gates described above. These gates are called RCJL (Resistor Coupled Josephson Logic) [32] or RCL (Resistor Coupled Logic) [33].

Gates with modified threshold curves also were designed. Figures 19 (a) and (b) show the RCJ-AND and RCJL-2/3 gates [32]. As shown in Fig. 19, the threshold curves are modified to produce symmetric characteristics for input currents. This is useful for achieving an AND or a 2/3 majority function.

Another direct-coupled type logic family called 4JL for (4 Junction Logic) was demonstrated by ETL [34]. The 4JL gate is composed of four junctions which are closely coupled in a loop as shown in Fig. 20. The essential feature of the 4JL gate is that the loop inductance can be eliminated. Hence, the threshold characteristic is determined only by the phase relations of the junctions. The current isolation function is achieved by putting a small resistor at the input terminal and making junction (J_1) switches at the final stage of the switching sequence. In order

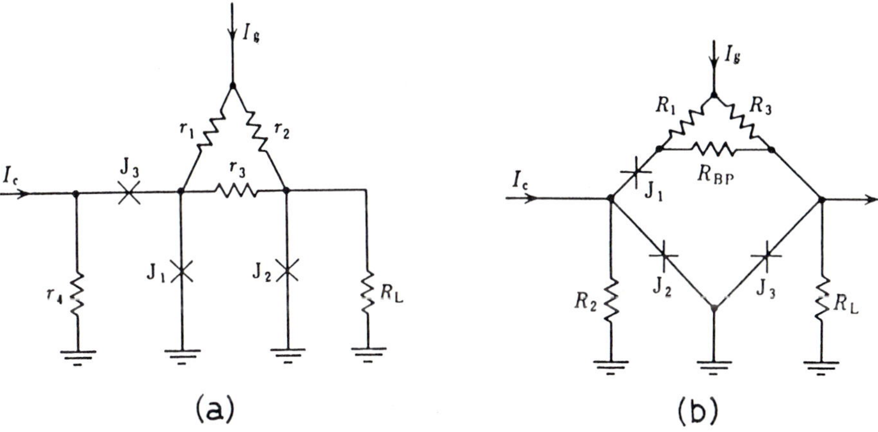

Fig. 18 The circuit configurations of the Resistor-Coupled Josephson Logic (RCJL) (a) and the Resistor-Coupled Logic (RCL) gates (b). By adding branches of Josephson junctions (J_2 for RCJL, J_3 for RCL), the sensitivities or the gains can be higher than those of the JAWS and the DCL gates.

to make the gate sensitivity high, the critical currents of junctions in the right branch are taken to be larger than those in the left branch (typically $I_{03,4} = 3I_{01,2}$). An AND gate based on the 4JL concept also was designed to make the threshold characteristic symmetric for the two input currents [33]. Figure 21 shows the 4JL-AND gate and its threshold characteristics.

The essential advantage of the direct-coupled gate is that the gate size can be smaller than that of the magnetically-coupled SQUID gates. This is because inductance loops can be eliminated in the direct-coupled gates. In fact, the 4JL gate fabricated with a 2.5 μm minimum rule has a gate area as small as 1200 μm^2. On the other hand, the JIL gate, fabricated using the same design rules, has an area of 4400 μm^2 [29].

Hybrid Gates

There is a series of gates which is not classified into the two categories described above. Recently, a gate family was developed which operates in the manner of a "hybrid" with both magnetic and the direct-coupled gates . Figure 22

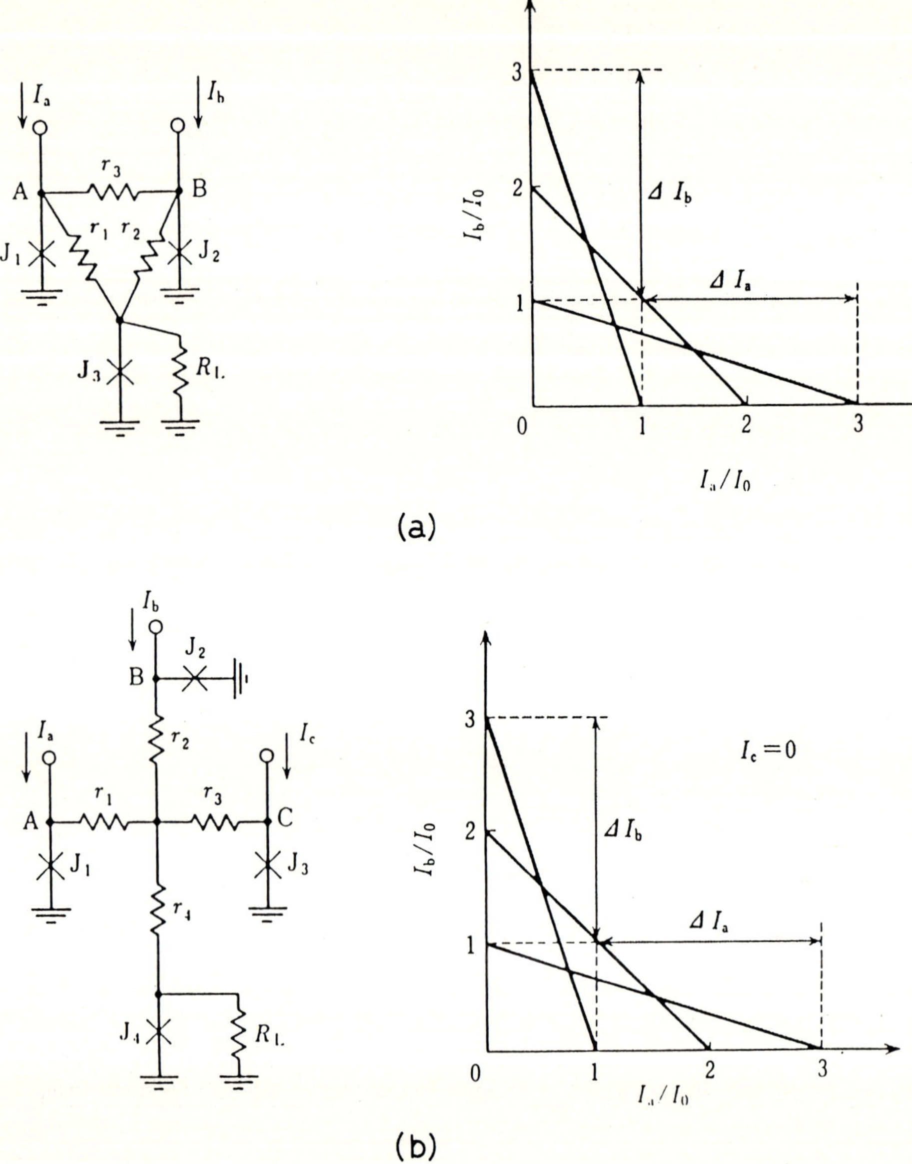

Fig. 19 The circuit configuration of the RCJL logic family. (a) An AND gate. (b) A 2/3 gate.

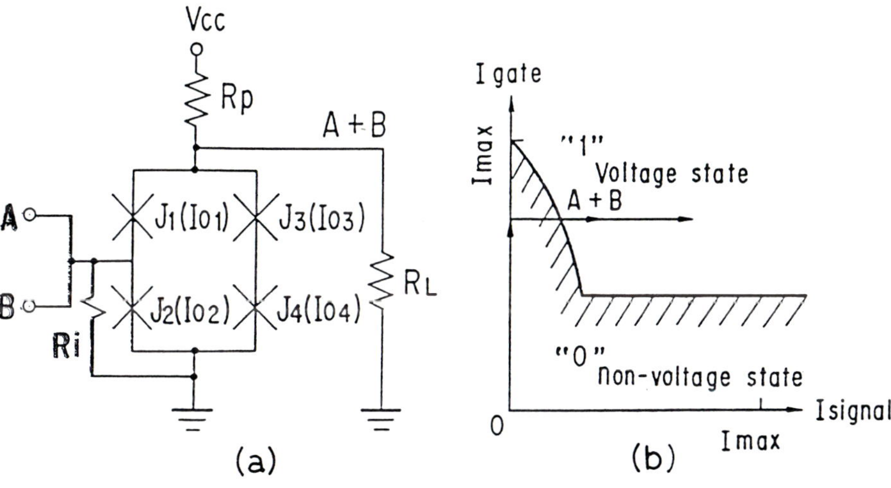

Fig. 20 The equivalent circuit configuration of the 4JL gate (a) and its threshold curve (b).

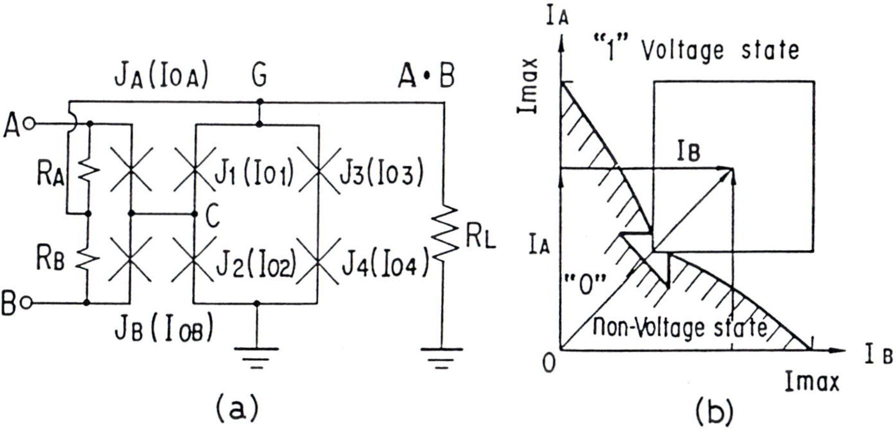

Fig. 21 The equivalent circuit configuration of the 4JL-AND gate (a) and its threshold curve (b).

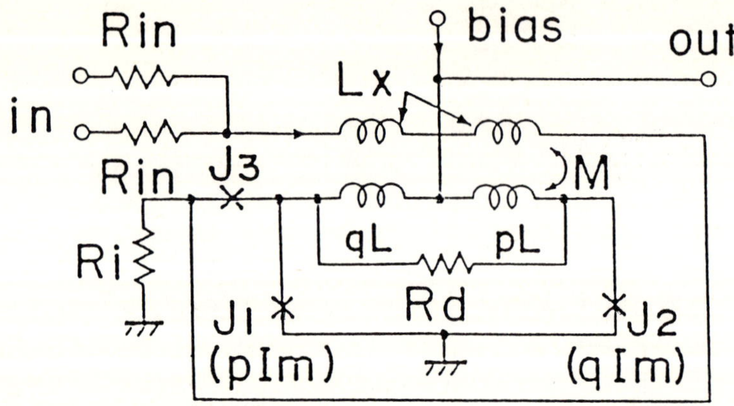

Fig. 22 The equivalent circuit configuration for the Modified Variable Threshold Logic (MVTL) gate.

shows a gate of this type, which is called an MVTL for Modified Variable Threshold Logic gate [36]. An input current is fed first into the control line of an asymmetric junction SQUID and then injected into the SQUID. The input-output current isolation function is obtained by placing a junction (J_3) and a small resistor (R_i) at the current injection terminal of the SQUID. This gate is controlled by both the magnetic field and the injected current so that the sensitivity of the gate can be high. Gates referred to as HTCID for (High-Tolerance Current-Injection Device) [37] and CCL (for Counter-Coupled Logic) [38] gates are similar to the CID-type of gate.

3-2 Switching Delay of Logic Gates

In Table 3 the logic delay of all gates measured until now are listed. The smallest delay obtained so far is 1.5 ps/gate in the MVTL gate fabricated with 1.5 μm $Nb/AlO_x/Nb$ junctions [43]. However, these data do not necessarily mean that the MVTL is superior to the other gates in speed of operation. Almost all of the delay measurements were made in the early stages of development, during which only Pb-alloy technology was available for fabricating gates. Recently, the fabrication technology based on all-hard junctions has been much improved, so that gate delays could be greatly reduced in any of the other gates if they were fabricated by more modern technology.

Table 3 Switching speeds for various logic gates

Gate	Rule (μm)	Switching Time (ps)	Power Dissipation	Junction Material	Ref
CIL	2.5	13	2 μW	Pb Alloy	29
JAWS	5	15	I_g = 90 μa	Pb-Alloy	39
RCJL	5	10.3	11.7 μW	Pb-Alloy	32
RCL	2	4.2	I_g = 0.72 mA	Pb-Alloy	40
4JL	2.5	7	4 μW	Pb-Alloy	41
DCL	1.5	5.6	4 μW	NbN/Pb-In	42
MVTL	1.5	2.5	4 μW	Nb/Alox/Nb	43

Power dissipation of gates depends on gate current. In actual Josephson logic circuits, almost all of the power is dissipated in the dropping resistors inserted between power lines and gates in order to assure a constant-current operation of the gates. Usually the dropping resistor should be chosen to be five times larger than the load resistor, i.e., $R_d \geq 5R_L$. Assuming I_g = 200 μA, R_L = 10 Ω and R_d = 50 Ω, the power dissipation of a gate becomes about 2.4 μW. The scatter in power dissipation for the various gates shown in Table 3 may be due to differences in gate current levels.

3-3 Memory Cells

In order to complete the discussion of superconducting digital systems, it is important to include a section on memories. In semiconductor devices, memories are made using charge stored in capacitors. In superconducting devices, on the other hand, persistient currents or magnetic flux in superconducting loops (i.e., inductances) are used for the storage of information. Josephson gates are used as switches to move magnetic flux in or out of the loops.

Various types of memory cells have been proposed, and can be divided into two categories: a nondestructive read-out (NDRO) memory and destructive read-out (DRO) memory.

Nondestructive (NDRO) Memory

The nondestructive read-out memory cell has the capability of reading stored information without changing the cell state. The NDRO memory is suitable for a cache memory, which communicates directly with the CPU, and, therefore, speed is vitally important.

The earliest memory cell of this type was proposed by Anacker [44] and experimentally demonstrated by Zappe [45]. Much effort has been made in trying to fabricate memory cells with wider operating margins. An improved NDRO memory cell is shown in Fig. 23 [46]. This cell makes use of a three-junction SQUID as a write gate and a two-junction SQUID as a sense gate. In order to write a "1", a

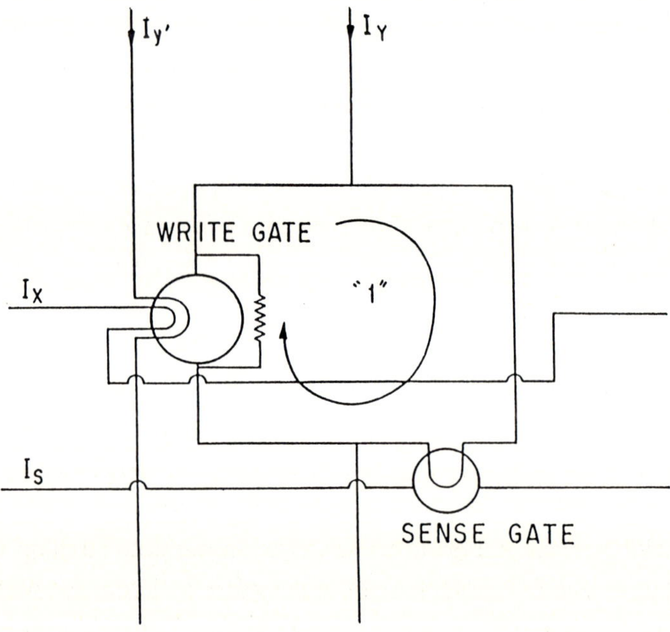

Fig. 23 The equivalent circuit configuration of the Nondestructive Read-Out (NDRO) memory cell. The three-junction interferometer gate is used for the write gate, and an asymmetric two-junction interferometer is used for the sense gate.

supply current I_y and a control current I_x are applied. In this sequence, the write gate switches to the voltage state, driving I_y to flow in the right branch of the loop. After I_y and I_x are turned off, the clockwise circulating current is maintained, leaving the cell in the "1" state.

In order to read the cell, the gate current of the sense gate, I_s, and I_y are applied. If the cell is in the "1" state, i.e., a circulating current is stored in the cell, the sense gate goes into the voltage state. If a "0" had been stored, the sense gate remains in the superconducting state. Thus, information in the cell can be read in a nondestructive manner. For the sense gate, an asymmetric-fed SQUID gate is used to widen the margin for sensing operations. This NDRO memory cell usually contains several flux quanta, typically about $3\Phi_0$.

Destructive Read-Out (DRO) Memory

A two-junction SQUID has a threshold characteristic with overlapping regions of vortex modes where either of two states is maintained. The DRO memory cell is designed to store a single flux quantum in the two-junction SQUID by using a mode overlapping region in the threshold characteristic [47]. This memory stores a single flux quantum so that the size of a cell can be smaller, which is desirable for construction of a main memory.

The structure of this type of memory cell is shown in Fig. 24. The memory operation performed using the mode-overlapping region of the threshold characteristic is shown in the lower part of Fig. 24. There are two types of transition modes on the threshold curve when the operating point moves across the curve. One is the so-called vortex transition, and the other is the voltage transition. For the vortex transition, the vortex state changes without generating a voltage except for a spike voltage when an operating point crosses a region of the threshold curve shown by dotted lines in Fig. 24. On the other hand, for the voltage transition, the cell generates a voltage when an operating point crosses a region of the threshold represented by the solid lines in Fig. 24. To write a "1", I_y and I_x are applied to move the operating point across the vortex transition region, as indicated by the arrow WR1 in Fig. 24. If the cell had been in the "0" state, the cell acquires one flux quantum as a result of this operation. To write a "0", I_y and I_x must be of opposite polarity to move the operating point across the vortex transition, as indicated by the

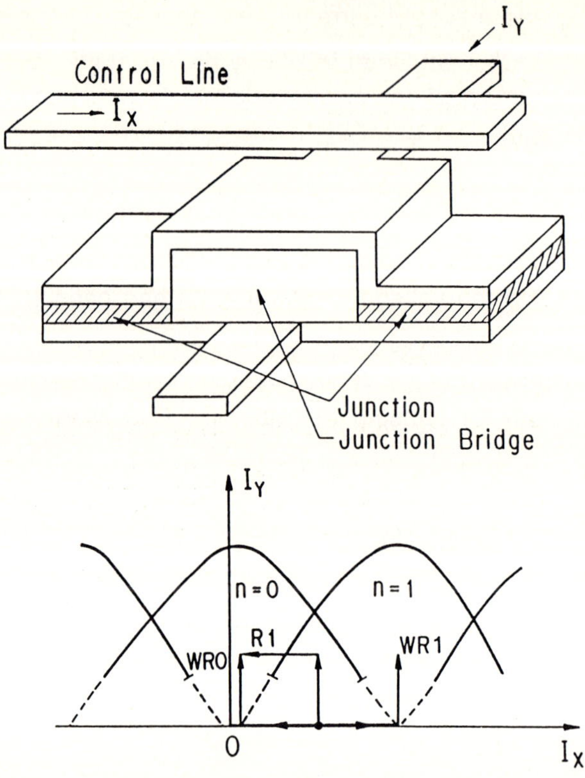

Fig. 24 The device configuration of a Destructive Read-Out (DRO) memory cell and its threshold characteristic. The DRO memory cell makes use of the mode overlap region in the threshold curve for writing and reading information.

arrow WR0 in Fig. 24. To read the cell, I_y and I_x also are applied so that the voltage-transition region is crossed, as indicated by the arrow R1 in Fig. 24. In this operation, the cell generates a voltage, assuming the cell had been in the "1" state. By reading the cell, the information stored in the cell usually is destroyed, thus requiring a "refresh" operation to rewrite information into the cell after each reading operation.

Variable Threshold Memory Cell [48]

This memory cell has a structure similar to the two-junction SQUID in which one flux quantum is stored. The equivalent circuit of the cell is illustrated in Fig. 25. As shown in Fig. 25, one junction in a two-junction SQUID is replaced by a three-junction SQUID gate to which a control current I_X is coupled. In this configuration, the vortex mode changes along the I_y axis, rather than along the I_X axis as shown in Fig. 25. To write a "1", I_X and then I_y are applied to produce a crossing of the mode boundary (0-A-B), as shown in Fig. 25. The dotted curves in the threshold curve also represent the vortex transition. To read the cell, I_y and then I_X are applied to produce a crossing of the boundary of the voltage transition (0-C-B). If the cell had been in the "0" state, it would generate a voltage. This reading sequence is destructive. However, the cell can be an NDRO cell by adding an external circuit

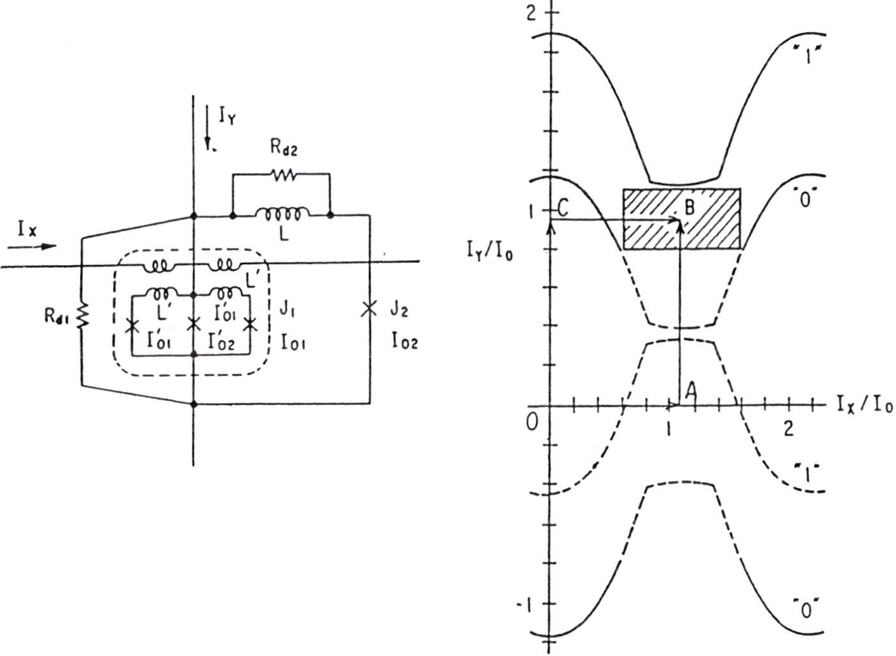

Fig. 25 The equivalent circuit of the Variable Threshold Memory cell and its threshold curve.

[49]. Figure 26 shows a bit-line arrangement of the cell for nondestructive readout. The bit-line consists of a large superconducting loop containing the cells and set-reset gates. In the reading sequence, a cell which had been in the "0" state generates a voltage, transferring I_y to the set gate. By this operation, I_y is automatically removed from the bit-line with the application of I_X, resulting in the rewriting of a "0" for the cell.

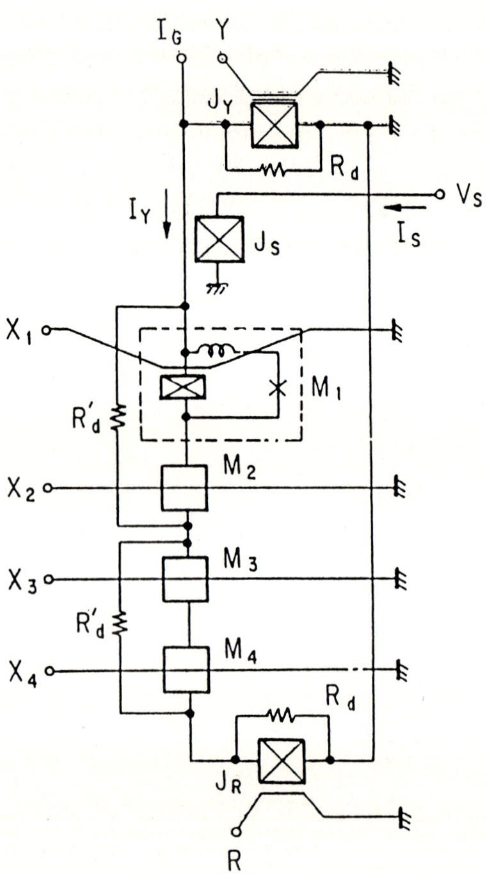

Fig. 26 The bit-line configuration of the Variable Threshold Memory cell. By using this configuration, the memory cell can have nondestructive readout.

3-4 Circuits

OR-AND Cell

Logic circuits are made using logic gates described in Section 3-2. Josephson logic gates are usually operated in a "latching" mode, making it difficult to use inverter gates in arbitrary places; thus, a dual-rail system is employed. In dual-rail logic, complementary signals always accompany true signals.

Using a dual-rail system, it is helpful to construct a unit cell composed of two OR gates and an AND gate, as shown in Fig. 27(a). In this configuration, the OR gates also act as the input-output isolation gates, since the AND gate usually has no isolation function. Using this unit cell, OR, AND and Exclusive-OR functions are easily obtained, so that all these logic functions can be performed as shown in Figs. 27 (b), (c), (d). Figure 28 shows a circuit configuration for the OR-AND unit cell based on the 4JL logic family [50].

Integrating these OR-AND unit cells, a number of logic circuits, such as adder and multiplier circuits, have been demonstrated. Logic circuits demonstrated thus far are summarized in Table 4.

These circuits are useful for evaluating the performance of logic gates. Recent results in a 16-bit x 16-bit multiplier fabricated by modern integration technology based on Nb junctions showed that a Josephson gate can be operated with a gate delay of under 10 ps (8.7 ps) in an LSI circuit [54].

AC-Power and Sequential Logic Circuits

Usually a Josephson logic gate which switches to the voltage state does not reset to the superconducting state when the input signal is turned off, i.e., this is a latching operation. In order to reset the gate, it is necessary to turn off the gate power. This means that an ac power supply is needed to operate a Josephson logic circuit, in contrast to the operation of semiconductor logic devices.

Two types of power supply have been suggested: one is the single-phase bipolar [55], and the other is the (2 or 3 phase) unipolar power system. These

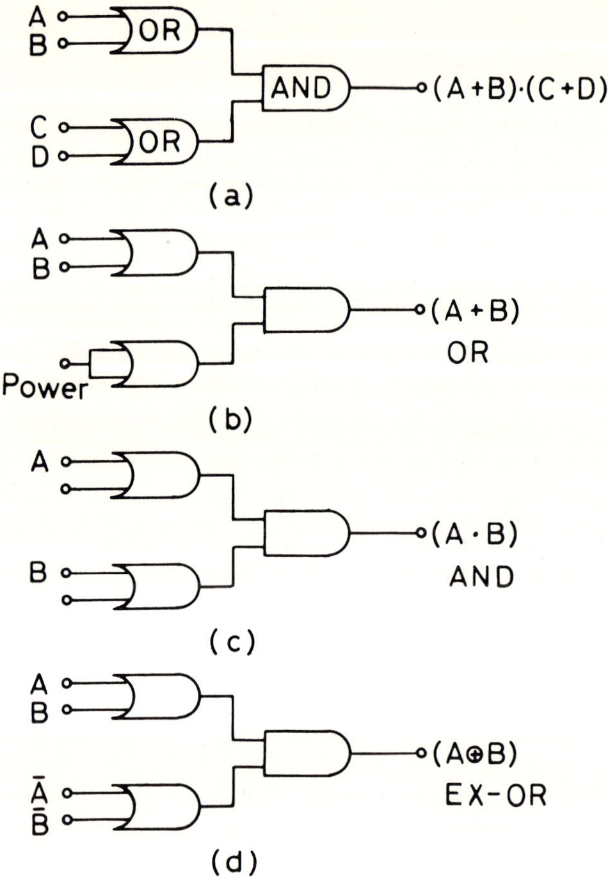

Fig. 27 The logic functions of the OR-AND unit cell. By using a dual-rail system, the OR-AND cell can be operated as an OR, AND, or Exclusive-OR gate.

power supply concepts are illustrated schematically in Fig. 29. In single-phase power systems (Fig. 29 (a)), logic operations are performed during both the positive and negative portions of the power cycle, and the gates reset during the transition time for changing polarities. As bipolar power is used, it is important to take into account the so-called punch-through phenomenon [56]. In latching gates there is a high probability that the gates will fail to set when the transition time becomes too fast. This effect is called "punch-through". In order to make the punch-through

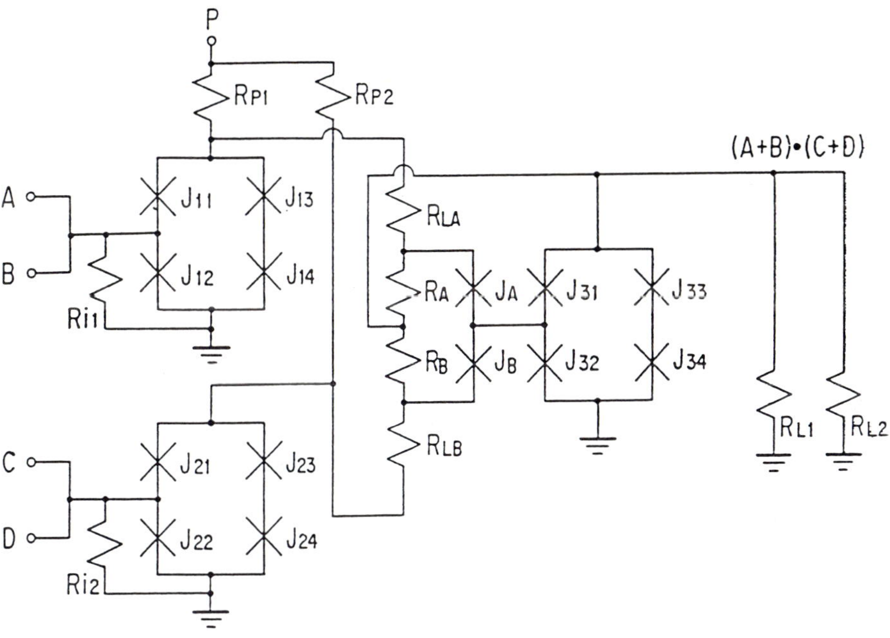

Fig. 28 The equivalent circuit for the OR-AND unit cell based on the 4JL logic family.

Table 4 Various logic circuits

Circuits	Gate Family	Junction	Gate Count	Performance	Ref
8-bit adder	4JL	PbAlloy	300	add time 300 ps	50
8-bit adder	4JL	NbNoxide/nbN	364	add time 700 ps	13
4-bit adder	RCJL	PbAlloy	56	add time 172 ps	51
4-bit multiplier	4JL	NbN/oxide/NbN	652	multiplication time 1 ns	13
4-bit multiplier	RCJL	PbAlloy	249	multiplication time 280 ps	52
4-bit multipler	JTHL	Nb/Aloxide/Nb	104	multiplication time 210 ps	53
16-bit multiplier	MVTJ	Nb/Aloxide/Nb	828	multiplication time 1.1 ns (8.7 ps/gate)	54

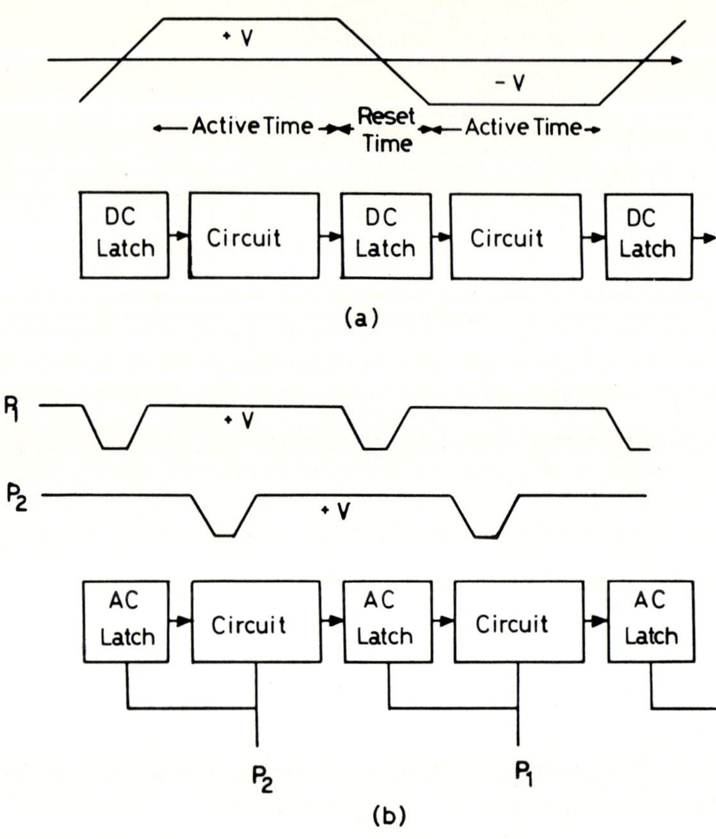

Fig. 29 Power system for Josephson logic circuits. (a) Single-phase bipolar power system. (b) Two-phase unipolar power system.

probability low, the transition time must be long enough to permit the gate to reset to zero before the other current polarity is applied. This limits the clock frequency.

During the reset time, all of the gates in the circuit reset to the superconducting state, so that a dc latch is needed to store the data in the preceding circuit. The dc latch is operated by another dc power supply. Figure 30 shows a typical dc latch circuit [57]. The latch circuit consists of a master latch and a slave latch. The calculated data in a circuit are stored in the superconducting loop of

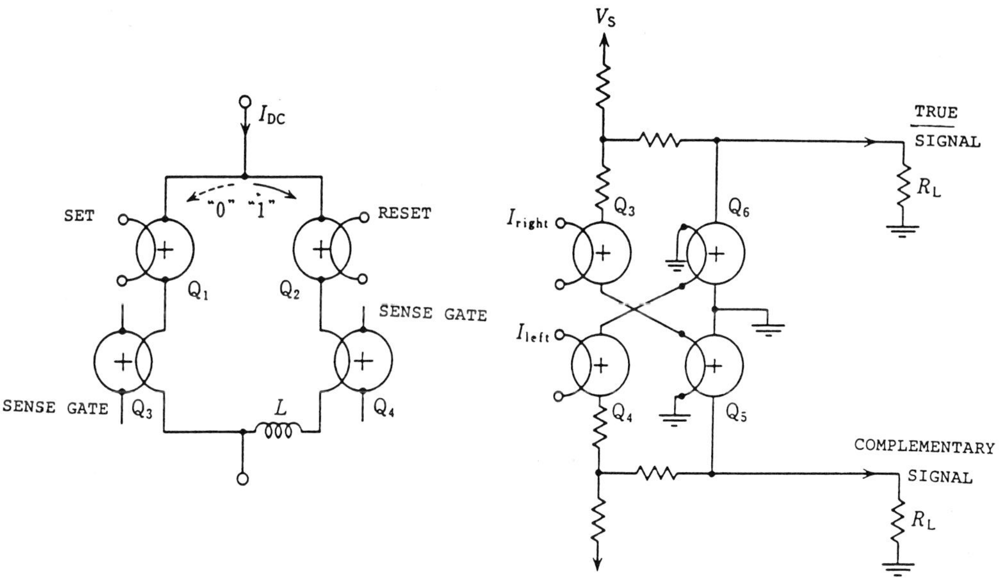

Fig. 30. DC-powered latches. The master latch is on the left, and the slave latch is on the right.

the master latch. The information stored in the master latch is sensed by the slave latch at the rise time of the next cycle of the power.

The multi-phase power system is another choice for powering Josephson logic. Figure 29(b) shows a two-phase unipolar-powered logic circuit [58]. As shown in the figure, two-phased overlapping power sources are fed to logic circuits alternately.

In this case, the dc-powered latch is not necessary to store data from the preceding circuit, since the data can be read out to the next circuit which is operated by the second power source before the first power source is turned off. A latching circuit for this type of power system can be composed of conventional latching gates. Figure 31 shows an example of the latch circuit based on the 4JL logic family for two-phase power [58]. One of the advantages of the two-phase power system is

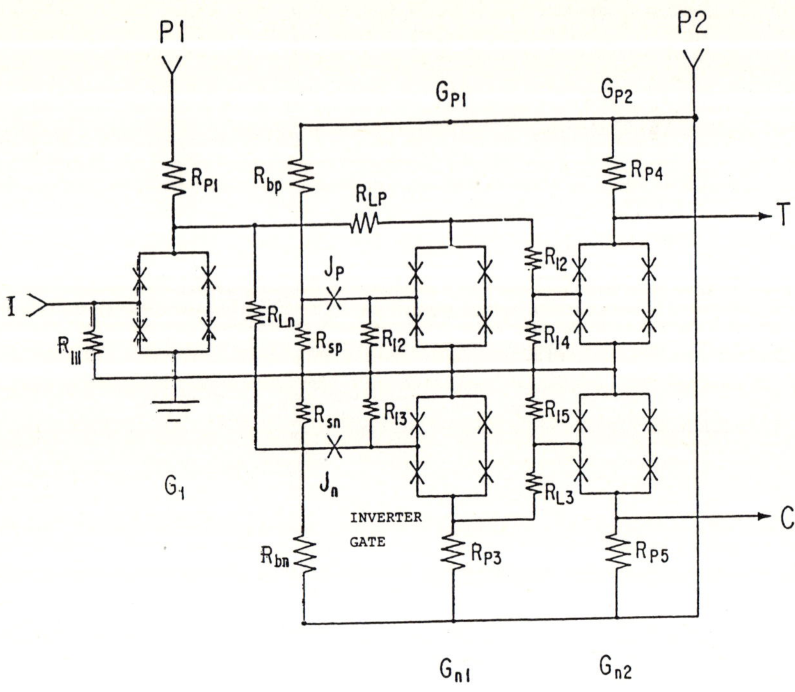

Fig. 31 A direct-coupled latch circuit based on the 4JL logic family. The latch is operated by a two-phase power system, so that the time inverter can easily be made.

that the timed inverter can be constructed easily as shown in Fig. 31. Two-phase power operation permits enough time to rese t the circuits, resulting in lowered punch-through probability without degradation of the effective cycle time. This is another advantage of the two-phase power system.

Taking advantage of a two-phase power system and using latching circuits, a number of sequential logic circuits such as shift registers [59], counters [60], and ALU's [61] have been fabricated and operated successfully. These circuits are important components for constructing computer CPU's.

Memory Circuits

In order to complete a fully decoded memory chip, one needs a well defined integration technology with the capability to fabricate circuits with LSI complexity. The memory circuit consists of a memory cell array and peripheral logic circuits, including address latches, X and Y decoders and a timing control circuit. A typical memory circuit arrangement is illustrated in Fig. 32 [62]. The address latches generate true and complementary signals. To write or read information, it is accessed by two crossing lines selected by the X and the Y decoders. The decoder is composed of several stages of AND gates from which one of the lines is selected.

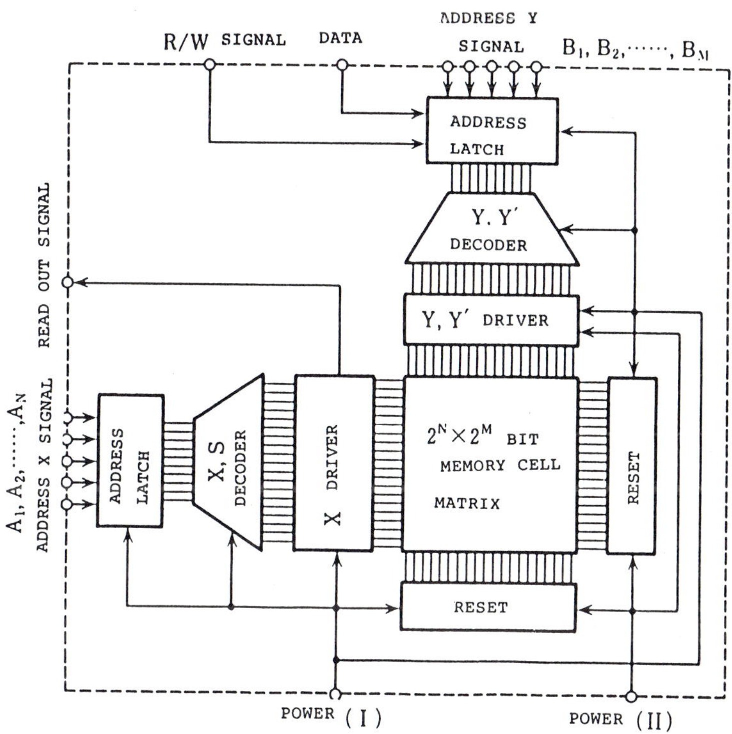

Fig. 32 A block diagram of a memory circuit.

Much effort has been directed to developing fully decoded memory circuits. Using Pb-alloy integration technologies, Yamamoto et al. fabricated a 1-kbit RAM (Random Access Memory) with an NDRO cell array built in 5 μm Pb-alloy technology [63]. The access time of the memory was found to be 3.3 ns. A 2.5 μm Pb-alloy technology has been adopted to design a 4-bit NDRO RAM chip [64]. A 1-kbit array cross-section in the 1 k x 4-bit design was operated. From measurement of individual circuit components, the access time of the 4-kbit RAM chip was estimated to be about 800 ps. A 1-kbit arrangement based on the variable threshold memory cell - see Section 3-3 - also has been integrated using 5 μm Pb-alloy technology [49]. This memory circuit is based on words rather than bits, resulting in a simplified peripheral circuit structure. The access time was found to be 1.1 ns.

Although much effort has been made to obtain full memory operation, only several bits were accessed in all of these memories based on Pb-alloy technology. This is due to non-uniformity of circuit parameters inherent in the Pb-alloy technology. Recently developed Nb-based integration technologies with increased uniformity and stability are expected to improve circuit performance for LSI memory chips.

Recently, a 1-kbit NDRO cell arrangement has been built by the NEC group using 3.3 μm Nb-based technology [65]. The access time was measured to be 570 ps at a power dissipation of 13 mW. Forty percent of the cells in the 1-kbit RAM were successfully operated, indicating that Nb technology is far superior to Pb-alloy technology. Figure 33 shows a photograph of the 1-kbit RAM chip based on Nb technology.

More recently, Kurosawa et al. of ETL have made a 1-kbit RAM based on a variable threshold memory cell using a 3 μm Nb junction process. They reported that 98% of the cells can be accessed [66].

<u>Small-Scale Computer Circuits-Microprocessors</u>

In order to evaluate the feasibilitiy of Josephson computer systems, it is important to operate processors in which circuit components for the computer system are arranged according to function .

The first attempt to operate this kind of circuit was made by Mukherjee of IBM [67]. He made a simple signal-processing circuit which included 12 latches and 16 AND gates based on the CIL logic family. This circuit processed data using 10-bit key codes and has been operated with a cycle time of 665 ps.

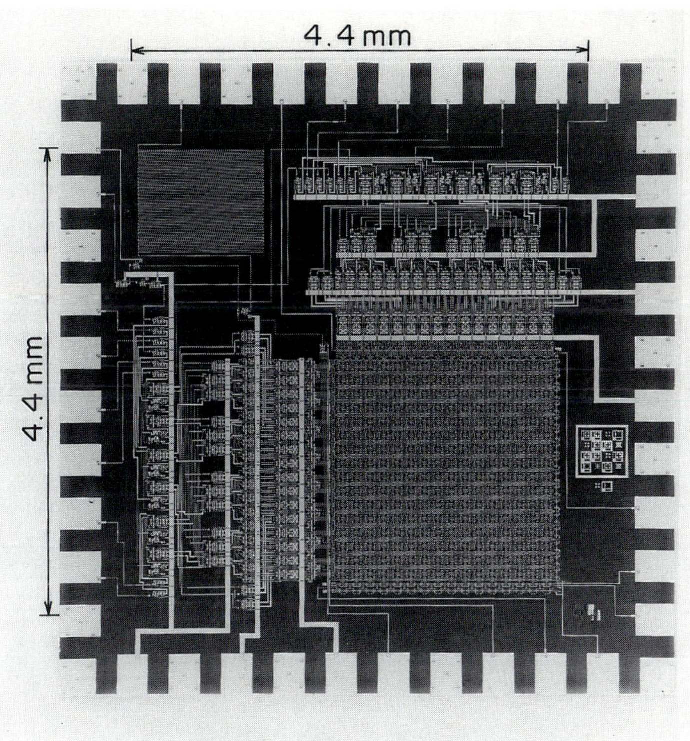

Fig. 33 Photograph of a 1-kbit NDRO memory circuit fabricated with 3.3 μm Nb/Al oxide/Nb junction technology.

As integration technologies based on Nb junctions have progressed, efforts have been made to build more complicated sequential circuits.

Kotani et al. of Fujitsu have made a 4-bit microprocessor operated with a three-phase power system [68]. The circuit consisted of a 64-bit NDRO RAM, an ALU and control circuits on a chip which was fabricated by integrating 1841 gates using a 2.5 μm Nb technology. The circuit block diagram is shown in Fig. 34. A critical path operation was performed with a maximum clock frequency of 770 MHz, which is 10 times faster than the GaAs version of the same processor. Figure 35 shows a photograph of the microprocessor chip. They also integrated another 4-bit

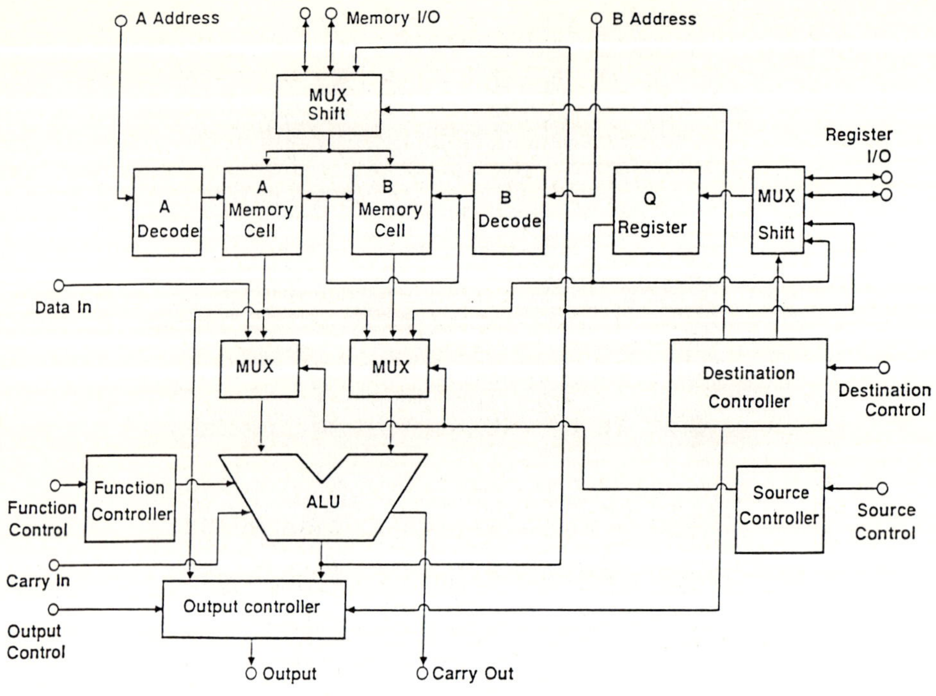

Fig. 34 Block diagram of a 4-bit Josephson microprocessor.

microprocessor comprised of 3056 gates in which an 8-kbit ROM plane and 4 x 4-bit multipliers were added to the processor described above. They reported that this newer processor operated at a maximum clock rate of 1.1 GHz [69].

Hatano et al. of Hitachi have operated a 4-bit Josephson data processor in which a decoded ROM (16 x 32 bits), ALU's (multiplier, adder, shifter, inverter, etc.) and a register file (16 x 4 bits) are integrated on the same chip using 2.5 μm Nb technology [70]. A critical path operation was performed, in which data taken from a register passes through the ROM and ALU units and finally is stored in a register. The maximum clock rate was 760 MHz. The ETL group has developed a 4-bit Josephson processor which can be extended to a small-scale Josephson computer system by adding a RAM and a ROM. The processor consists of 2 chips, one is a resistor ALU (RALU) [71] and the other is a sequence control unit (SQCU) [72]. The

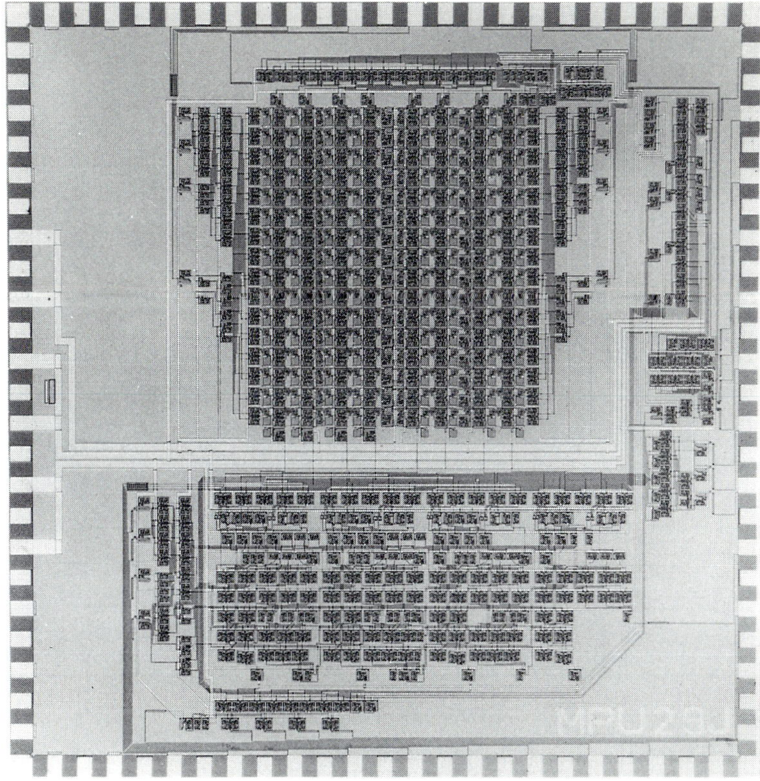

Fig. 35 Photograph of a 4-bit Josephson microprocessor chip.

RALU is composed of 1273 4JL gates fabricated using 3 μm Nb/Al oxide/Nb junction technology. The RALU chip contains a 10-bit instruction decoder, a 4-bit accumulator, an instruction decoder, a multiplexer and a memory and skip controller. The RALU can process 24 instructions using 10-bit instruction codes. The SQCU is a circuit to control data from an instruction ROM and a RAM. The circuit consists of 593 4JL gates with 2.5 μm NbN/oxide/NbN junctions. The processor operation was performed by combining the RALU and the SQCU chips. Full circuit operations were successfully confirmed in a quasi-static operation. They are planning to construct a Josephson computer system composed of 4 chips, namely, RALU, SQCU, ROM and RAM. At this time 3 of the 4 chips (RALU, SQCU, RAM) have been operated, so that a realistic Josephson computer is expected to appear in the near future. Recently developed Josephson processors are summarized in Table 5.

Table 5. Recently Developed Josephson Processors

Processor	Institute	Number of Gates	Configuration	Performances	Ref.
4-bit Microprocessor one chip	Fujitsu	1841	64 bit RAM ALU (8 functions)	Clock Rate 770 MHz	68
4-bit Microprocessor one chip	Fujitsu	3056	8-kbit ROM, 64 bit RAM ALU (8 functions) 4. x 4 Multiplier	Clock Rate 1.1 GHz Power 6.1 mW	69
4-bit Data Processor one chip	Hitachi	2066	Resistor File 16 x 4 bit ROM 16 x 32 bit ALU (8 functions)	Clock Rate 760 MHz Power 25 mW	70
4-bit Processor 2 chip Configuration	ETL	RALU 1273	RegistorAlu ALU (24 functions) 4 bit Accumulator 10 bit Inst. Registor Inst. Decoder Multiplexer	Full Operation at Low Clock Rate Power 2.6 mW	71
		SQCU 593	Sequence Control Unit Inst. Resistor Inst. Decoder Program Counter Multiplexer		72

4 Summary

The motivation to develop Josephson digital devices lies in the fact that Josephson junctions can switch with very high speed and very low power dissipation. As described in this chapter, Josephson gates fabricated with modern technology can be operated at delay times of less than 10 ps in an LSI circuit.

Moreover, the power dissipation is several μW, which is two or three orders of magnitude smaller than that for semiconductor gates.

The extremely low power dissipation of Josephson devices is important to obtain high computing speed. In modern computer systems a large part of the system delay is the wiring delay or the packaging delay rather than the switching delay of devices. In order to reduce system delay, it is essential to make the wiring length short by densely packing the devices. The low power dissipation of Josephson devices makes it possible to realize a three-dimensional high density packaging of devices, thus reducing the total interconnect length, and hence, the system delay. The packaging technology for Josephson devices has been investigated [73,74] and a cross-sectional model already has been tested [75].

The use of superconducting striplines for chip wiring and interconnects gives rise to another important advantage of Josephson devices. Signals can propagate along a matched superconducting stripline without loss or dispersion for frequencies less than the gap energy. This indicates that a superconducting stripline provides an ideal means to transmit fast pulses with a rise time of less than 10 ps.

Before IBM terminated its Josephson computer project, device technology was not reliable enough to operate logic and memory circuits with LSI complexity. Recent advances in all-refractory integration technology based on Nb and NbN junctions have made it possible to operate Josephson LSi circuits containing several thousand gates. Moreover, device performance in logic and memory has improved. Although interest in Josephson devices waned after the IBM decision, newly developed fabrication technologies have made Josephson digital technology appear attractive once again.

Recently, new high-temperature oxide superconductors whose T_C's are higher than liquid nitrogen temperature (77 K) have been discovered. Higher temperature operation is expected for devices fabricated from these high-T_C superconductors. The development of high-T_C tunnel junctions and new device structures will be very important. However, it should be noted that a sacrifice may be made when the operating temperature is raised. The power consumption of Josephson gates operated at 77 K will be increased compared to those operated at

4.2 K. The probability of operating failure increases exponentially with temperature due to the increase in thermal noise. In order to keep noise tolerance at acceptable levels at higher operating temperatures, the Josephson coupling energy must be increased, resulting in higher Josephson critical current and correspondingly higher power consumption. A simple calculation shows that the power consumption at 77 K will be two to three orders of magnitude larger than that at 4.2 K. This means that one loses one of the major advantages of Josephson devices by going to high temperature operation. It is important to recognize that much of the high performance in Josephson devices is the result of low temperature operation.

However, the discovery of high-T_C oxide superconductors is very important for Josephson devices in a different way. The high-T_C superconductors provide one with a means to connect Josephson devices operated at 4.2 K to semiconductor devices cooled to 77 K. This widens the applicability of Josephson devices. For example, a main memory system based on CMOS or HEMT devices operating at 77 K can be connected to a Josephson CPU operating at 4.2 K by means of high-speed striplines made of high-T_C superconductors, thus easing the heavy burden on the development of a large-scale Josephson computer.

As described in this chapter, newly developed Josephson technology has made Josephson devices much more reliable and desirable, and it promises to open a new era in the history of electronic technology.

REFERENCES

1. IBM achievements in Josephson digital technologies are summarized in the special issue of Josephson computer technology, IBM J. Res. Dev., 24 (1980).

2. J. H. Greiner, C. J. Kircher, S. P. Klepner, S. K. Lahiri, A. J. Warneche, S. Basavaiah, E. T. Yen, J. M. Baker, P. R. Brosious, H. C. W. Huang, M. Murakami and I. Ames: IBM J. Res. Dev. 24 (1980) 195.

3. J. H. Greiner: J. Appl. Phys. 45 (1974) 32.

4. M. Murakami: J. Appl. Phys. 52 (1981) 1309.

5. T. Imamura, S. Hasuo and T.Yamaoka: IEEE Trans. Magnetics MAG-21 (1985) 112.

6. H. Hoko, A. Yoshida, T. Imamura and S. Hasuo: Extended Abs. of the 18th Conf. on Solid State Devices and Materials, Tokyo, 1986, pp. 447-450.

7. S. Kosaka, F. Shinokia, S. Takada and H. Hayakawa: IEEE Trans. on Magnetics MAG-17 (1981) 314.

8. M. Hikaita, K. Takei, T. Iwata and M. Igarashi: Jpn. J. Appl. Phys. 21 (1982) 1724.

9. R. F.Broom, A. Oosenburg and W. Walter: Appl. Phys. Lett. 37 (1980) 237.

10. S. Yano, Y. Tarutani, H. Mori, H. Yamada, M. Hirano and U. Kawabe: IEEE Trans. on Magnetics, MAG-23 (1987) 1472.

11. R. F. Broom, R. B. Laibowitz, Th. O. Mohr and W.Walter: IBM J. Res. Dev. 24 (1980) 212.

12. A. Shoji, S. Kosaka, F. Shinoki, M. Aoyagi and H. H. Hayakawa: Appl. Phys. Lett. 41 (1982) 1097.

13. S. Kosaka, A. Shoji, M. Aoyagi, F. Shinoki, S. Tahara, H. Ohigashi, H. Nakagawa, S.Takada and H. Hayakawa: IEEE Trans. on Magnetics MAG-21 (1985) 102.

14. H. Kroger, L. N. Smith and D. W. Jillie: Appl. Phys. Lett. 39 (1981) 280.

15. H. Kroger, L.N. Smith, D. W. Jillie and J. B. Thaxter: IEEE Trans. on Magnetics MAG-19 (1983) 793.

16. E. E.Latta and M. Gasser: J. Appl. Phys. 54 (1983) 1115.

17. M. Gurvitch, M. A. Washington and H. A. Huggins: Appl. Phys. Lett. 42 (1983) 472.

18. S. Morohashi, F. Shinoki, A. Shoji, M. Aoyagi and H. Hayakawa: Appl. Phys. Lett. 46 (1985) 1179.

19. A. Shoji, M. Aoyagi, S. Kosaka, F. Shinoki and H. Hayakawa: Appl. Phys. Lett. 46 (1985) 1098.

20. H. Nakagawa, K. Nakaya, I. Kurosawa, S. Takada and H. Hayakawa: Jpn. J. Appl. Phys. 25 (1986) L70.

21. H. Kuroda, M. Yuda, J. Nakano and M. Ueki: IEICE Technical Report, SCE86-42 (1986) 59.

22. T. Imamura and S. Hasuo: IEICE Technical Report , SCE87-48 (1987) 49.

23. M. Aoyagi, A. Shoji, S. Kosaka, F. Shinoki and H. Hayakawa: Int. Cryogenic Materials Conf., Boston, 1985.

24. T. Imamura, H. Hoko and S. Hasuo: Extended Abs. 1987 Int. Superconductivity Conf., Tokyo, 1987, pp. 57-62.

25. H. Hoko, T. Imamura and S. Hasuo: Techn. Digest of Int. Electron Device Meeting, 1987, pp.385.

26. S. Kosaka: Int. Cryogenic Materials Conf., Boston, 1985.

27. M. Klein and D. J. Herrel: IEEE J. Solid-State Circuits, SC-13 (1978) 593.

28. H.H. Zappe and B. S. Landman: J. Appl. Phys. 49 (1978) 344.

29. T. R. Gheewala: IBM J. Res. Dev. 24 (1980) 130.

30. T A. Fulton, S. S. Pei and L. N. Dunkleberger: Appl. Phys. Lett. 34 (1979) 709.

31. T. R. Gheewala and A. Mukherjee: Tech. Digest International Electron Device Meeting (IEDM) (1979) 482.

32. J. Sone, T. Yoshida and H. Abe: Appl. Phys. Lett. 40 (1982) 886.

33. K. Hohkawa, M. Okada and A. Ishida: Appl. Phys. Lett. 39 (1981) 653.

34. S. Takada, S. Kosada and H. Hayakawa: Proc. 22th Conf. (1979) International Solid State Device, Tokyo, 1979; Jpn. J. Appl. Phys. Suppl. 19-1 (1981) 607.

35. H. Hayakawa: Jpn. J. Appl. Phys. Suppl. 22 (1983) 447.

36. N. Fujimaki, S. Kotani, S. Hasuo and T. Yamaoka: Jpn. J. Appl. Phys. 24 (1985) L1.

37. H. Beha and H. Jackel: IEEE Trans. Magnetics MAG-27 (1981) 3423.

38. S. Hasuo, H. Suzuki and T. Yamaoka: IEEE Trans. Magnetics MAG-17 (1981) 583.

39. S.S. Pei: Appl. Phys. Lett. 40 (1982) 739.

40. J. Nakano, Y. Mimura, K. Nagata, Y. Hasumi and T. Waho: Extended Abstracts 16th Conf. Solid State Devices and Materials, Kobe (1984) 636.

41. H. Nakagawa, T. Odake, E.Sogawwa, S. Takada and H.. Hayakawa: Jpn. J. Appl. Phys. 22 (1983) L297.

42. Y. Hatano, T. Nishino,Y. Tarutani and U.Kawabe: Appl. Phys. Lett. 44 (1984) 1095.

43. S. Kotani, T. Imamura and H.. Hasuo: IEEE IEDM Technical Digest (1987) 865.

44. W. Anacker: IEEE Trans. Magnetics MAG-5 (1969) 968.

45. H. H. Zappe: IEEE J. Solid-State Circuits, SC-10 (1975) 12.

46. S. M. Faris, W. H. Henkels, E. A. Valsmakis and H.H. Zappe: IBM J. Res. Dev. 24 (1980) 146.

47. H. H. Zappe: Appl. Phys. Lett. 25 (1974) 424.

48. I. Kurosawa, A.Yagi, H. Nakagawa and H. Hayakawa: Appl. Phys. Lett.43 (1983) 1067.

49. I. Kurosawa, H. Nakagawa, A. Yagi, S. Takada and H. Hayakawa: Extended Abstracts of 16th Conf. Solid State Device and Materials, Kobe (1984),p. 619.

50. H. Nakagawa, H. Ohigashi, I. Kurosawa, E. Sogawa, S. Takada and H. Hayakawa: Extended Abstracts of 15th Conf. on Solid State Devices and Material, Tokyo (1983), p. 137.

51. J. Sone, T. Yoshida, S.Tahara and H. Abe: Technical Digest Int. Electron Device Meeting (IEDM) (1982) 765.

52. J. Sone, J. S.Tsai and H. Abe: IEEE Solid-State Circuits SC-20 (1985) 1056.

53. H. Hatano, Y. Harada, K. Yamashita and U. Kawabe: ISSCC Digest of Tech. Paper (1986), p. 196.

54. S. Kotani, N. Fujimaki, S. Morohashi, S. Ohara and S. Hasuo: IEEE J. Solid-State Circuits SC-22 (1987) 98.

55. H. C. Jones and T. R. Gheewala: IEEE J. Solid-State Circuits SC-17 (1982) 1201.

56. R. E. Jewett and T. Van Duzer: IEEE Trans. Magnetics MAG-17 (1981) 599.

57. A. Davidson: IEEE J. Solid-State Circuits SC-13 (1978) 583.

58. H. Nakagawa, I. Kurosawa, S. Takada and H. Hayakawa: Extended Abstracts of 17th Conf. Solid State Devices and Materials, Tokyo (1985) 123.

59. N. Fujimaki, S. Kotani, T. Imamura and S. Hasuo: IEEE J. Solid-State Circuits SC-22 (1987) 886.

60. H. Nakagawa, I. Korosawa. S. Takada and H. Hayakawa: IEEE Trans. Magnetics MAG-23 (1987) 739.

61. H. Nakagawa, I. Kurosawa and S.Takada: Extended Abstracts. of the Int. Conf. on Solid State Devices and Materials , Tokyo, 1986, pp. 751-752.

62. Y. Wada, M. Hidaka, S. Nakagawa and I. Ishida: IEEE J. Solid-State Circuits SC-22 (1987) 892.

63. M. Yamamoto, Y. Yamauchi, K. Miyahara, K. Kuroda, F. Yanagawa and A.Ishida: IEEE Electron Device Lett. EDL-4 (1983) 150.

64. W. H.Henkels, K. H. Brown, T. V. Rajeerkumar, L. Geppert, J. W. Allan, Y. H. Lee and J. T. Yeh: Proc. Inf. Conf. Computer Design, New York (1983) 580.

65. Y. Wada, S. Nagasawa, I. Shida, M. Hidaka, H. Tsuge and S. Tahara: ISSCC Digest of Technical Papers (1988) 84.

66. I.Kurosawa, H. Nakagawa, S. Kosaka, M. Aoyagi and S. Takada: Extended Abs. 20th Int. Conf. Solid State Devices and Materials, Tokyo (1988) 605.

67. A. Mukherjee: IEEE Electron Device Lett. EDL-3 (1982) 29.

68. S. Kotani, N. Fujimaki, T. Imamura and S. Hasuo: IEEE ISSCC Digest of Technical Papers (1988) 150.

69. S. Kotani, T. Imamura and S. Hasuo: IEICE Technical Report SCE88-61 (1989) 47.

70. Y. Hatano, H. Mori, H. Yamada, H. Nagaishi,M. Hirano and U. Kawabe: 1989 Spring National Convention Record of IEICE, Higashi - Osaka (1989) 5-353.

71. H. Nakagawa, Y. Okada, Y. Hamazaki, M. Aoyagi,I. Kurosawa and S. Takada: IEICE Technical Report SCE88-36 (1989) 55.

72. S. Kosaka, H. Nakagawa, Y. Okada, Y. Hamazaki, M. Aoyagi, I. Kurosawa, A. Shoji and S. Takada: IEICE Technical Report SCE88-37 (1989) 61.

73. S. K. Lahiri, H. R. Bickford, P. Geldermans, K. R. Grebe, P. A. Moskowitz, M. Natan, W. M. Wang, C. T. Wu and T.Yogi: IEEE Trans. Components,Hybrids and Manufacturing Technology, CHMT-5 (1982) 271.

74. K. Aoki, Y. Tazoh and H. Yoshikiyo: IEEE Trans. Magnetics MAG-21 (1985) 741.

75. M. B. Ketchen, D. J. Herrell and J. Anderson: J. Appl. Phys. 57 (1985) 2550.

SUPERCONDUCTING FIELD-EFFECT DEVICES

T.M.Klapwijk, D.R.Heslinga and W.M.van Huffelen
Department of Applied Physics, University of Groningen
9747 AG Groningen, The Netherlands

1. INTRODUCTION

Superconducting electronics has evolved primarily as a result of the discovery of Giaever tunneling and the Josephson effects. Originally both phenomena were dealt with in the framework of two superconductors separated by a thin insulating layer. However, the Josephson effect is a much more general phenomenon which reveals the macroscopic quantum character of the superconducting state. Hence, the Josephson effect is found in a large variety of weak link structures such as superconductors connected by a narrow, short metallic contact (called a microbridge), by a normal metal, or by an insulator. In all cases the main features of the Josephson effect are present. The structures differ only with respect to the details of conduction of quasiparticles. The current-voltage characteristic is determined by the Josephson effect acting in parallel to a device-specific quasiparticle current. For various applications the response of either the quasiparticle current or the Josephson current is used. Which device is useful for which application depends on the required impedance, on an often needed nonlinearity and on the maximum supercurrent.

Superconducting tunnel devices have been investigated extensively for their potential as logic elements. The low power consumption should enable a high packing density which is needed for future main frame computers. In 1983 a major program on Josephson-junction logic was abandoned. One of the stumbling blocks has turned out to be the two-terminal character of a Josephson junction. Since that time, interest in possible three-terminal devices has increased substantially. For example (Fig.1), a superconducting field-effect transistor has been proposed by Clark, Prance and Grassie [1]. In the present paper, we review only various approaches for field-effect controlled superconductivity. Other proposals for three-terminal superconducting devices have appeared. The present status has been discussed in a recent summary by Kleinsasser [2], which also partly overlaps the present review. We will emphasize the physics of devices that try to explore the

NATO ASI Series, Vol. F 59
Superconducting Electronics
Edited by H. Weinstock and M. Nisenoff

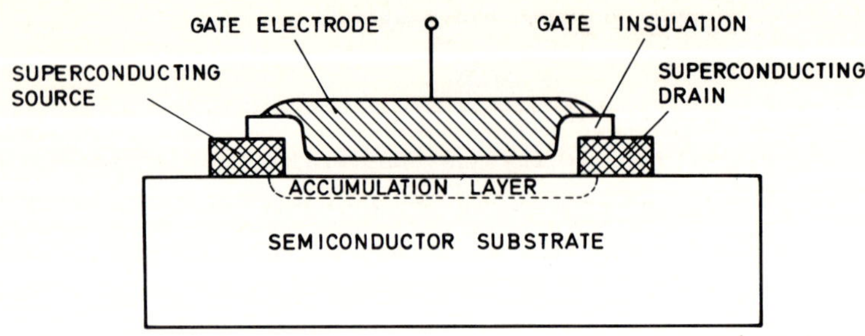

Fig.1 Cross section of a superconducting field-effect device [1]. A superconducting path from source to drain is controlled by a potential applied to the gate electrode.

effect of an electric potential on a superconductor. In conventional (semiconducting) field-effect transistors, the conductivity is modulated. Here, in principle, the **super**conductivity is changed when an electric field is applied.

Most of the discussion will center around metallic superconductors. We will show that combinations of semiconductors and superconductors may have a promising future, perhaps even more promising since the discovery of the new class of ceramic superconductors. It has been known for a number of years that many semiconductors achieve improved performance at the boiling point of liquid nitrogen (77K). At present, because of the aforementioned new ceramics, superconductors can, in principle, become a generic part of any semiconductor circuit operating at these temperatures. Our discussion will focus on fundamental problems that arise when metallic superconductors are combined with semiconductors. We believe that many of these problems will continue to be important when the ceramic superconductors are controlled well enough to be integrated with semiconductors. Section 2 discusses the effect of an electric field on a superconductor. Section 3 deals with the superconducting proximity effect, whereas Section 4 reviews results obtained on proximity-coupled devices.

2. FIELD EFFECT ON SUPERCONDUCTORS

Consider a thin metallic film serving as one plate of a parallel-plate capacitor. An electric field induces a surface charge density σ at the interface between the metal and the dielectric. The surface charge will act as a perturbation on the ambient electrons, changing the normal state conductivity σ_N of the metal, and, if the metal is a superconductor, its critical temperature T_c. The relative change in σ_N and in T_c may be expected to scale proportionally to the relative change in the number of electrons. For a film of thickness d and electron density n this number equals σ/ned. An upper limit on the induced surface charge is set by the breakdown field E_{br} of the dielectric. Taking SiO_2 as an example ($\varepsilon_r = 3.9$ and $E_{br} = 10^9 V/m$) with a common metallic superconductor such as Pb (thickness d = 10 nm, n = 13.2 10^{28} m^{-3}) we obtain a relative change of 1.6 10^{-4}. Hence, the effect of charging on the (super)conductivity will be very small indeed. Improvement only can be expected from atomically thin films and/or low-carrier-density materials.

2.1 Metallic superconductors

Field-effect induced changes in conductivity are conveniently realized with semiconductors. Unfortunately, very few semiconductors are known to be superconducting. Early experimental results on very thin metallic superconductors have been reported by Glover and co-workers [3]. They measured shifts of $\delta T_c = 10^{-4}$ K on 10 nm thick films of several superconductors on mica substrates. The observed shifts in critical temperature

Table I. Signs for normal and superconducting charging effects in response to an increase in electron number.

element	$\delta\sigma_N/\sigma_N$	$\delta T_c/T_c$
Sn	-	+
Al	-	+
Tl	+	-
Bi	+	-
Ga	-	-
Pb	-	-

and normal state conductivity were mostly of opposite sign and proportional to the induced surface charge. Furthermore, as expected, $\delta\sigma_N/\sigma_N$ scaled with 1/d. In an attempt to maximize the surface charge, Stadler [4] has used ferroelectric polarization of triglycine-sulfate substrates covered with Sn. This resulted in one order of magnitude higher T_c shifts.

Although most observations are in qualitative agreement with the naive model given above, significant deviations are observed. Based on a simple free-electron model, one expects a rise in both σ_N and T_c when the number of electrons is increased. Table I summarizes results obtained by Glover and co-workers [3] on films condensed on cold substrates. For some superconductors the conductivity increases (plus sign) by adding electrons, whereas the critical temperature decreases (minus sign). Others show the opposite behavior. It is difficult to find a clearcut correlation between the sign of the response of the conductivity and that of the critical temperature on charging. The signs for $\delta\sigma_N$ and δT_c are opposite for the weak-coupling superconductors, but are the same for the strong-coupling superconductors. Also, the response is smaller for the strong-coupling superconductors.

Bardasis [5] has made an attempt to explain these results. It is assumed, following Yu, Strongin and Paskin [6], that the critical temperature of a thin metallic film depends on the thickness of the film. The physical thickness of the film imposes restrictions on the electron wave function causing the T_c of the film to take on resonant values as a function of thickness. Bardasis focuses on the phase shift for electron wave functions at a metallic interface. A charged interface leads to a modification of the phase shift, which changes the resonant conditions and consequently the critical temperature. A charged surface leads to an effective thickness which is different from the actual physical thickness. Based on these assumptions, the following dependence is found:

$$\frac{\delta T_c}{T_c} = - \frac{1}{2Zg} \frac{\delta n}{N_A} \tag{1}$$

where g is the BCS coupling constant, N_A is the number of atoms, and Z is the valence. For a metal having electrons as charge carriers, one finds for positive δn (i.e., an addition of electrons), a decrease in the critical temperature. Of course, for positive δn, an increase in conductivity is expected.

This model is capable of explaining the existence of opposite signs for the normal and the superconducting field effect. Also, a dependence on the BCS coupling constant is found. However, the fact that the signs are the same for strong-coupling superconductors remains unexplained. In addition, in view of the proposed sensitivity to details of the metallic surfaces, one would expect comparable sensitivities to coverage by foreign atoms. Indeed, small T_c shifts have been observed due to oxidation, coverage with a dielectric, adsorption of noble gases, exposure to air, etc. One also would expect a sensitivity to the microstructure of the metallic films. Only layers with sufficiently large crystals support arguments which are based on the dependence of the electron wave functions on the thickness of the film. Quench-condensed films as used by Glover et al [3] are certainly not the best candidates.

The conductivity of polycrystalline thin films is primarily dependent on scattering from grain boundaries and the surface. The electrons added by charging the dielectric do not necessarily become free electrons and contribute to the conductivity. Due to screening, the extra charges will be confined to less than a nm from the metal-dielectric interface. Some of them will be trapped at the interface. Because surface scattering contributes substantially to the conductivity of thin metallic films, conditions at the interface are likely to affect the mobility as well. In particular, experiments on the field effect on epitaxial silver films have been interpreted as due to a change in surface scattering. In the Fuchs-Sondheimer[7] model for surface scattering, part of the scattering is assumed to be specular. A specularity parameter p is defined as the ratio of the number of electrons that are scattered specularly to the total number of scattered electrons. 1-p is the fraction that suffers diffuse scattering. Berman and Juretschke [8] explain their results on the electric field-effect in epitaxially grown silver films as indicative of a change of the specularity parameter due to charging.

Surprisingly little is known about the mechanism that causes both a change in conductivity and in critical temperature. It appears that an understanding could be derived from systematic experiments on well-characterized thin films with known crystallographic purity and surface conditions. Epitaxially grown niobium films have been proposed as a candidate system [9]. More recently, epitaxially grown silicides on silicon have been investigated [10]. Of course, for practical applications the effect will remain small.

2.2 Low carrier density superconductors

Substantial field effects may be expected from materials with a low carrier density. Systematic results, without the inconsistencies observed for pure metallic films, have been obtained by Hebard, Fiory and Eick [11] on In/InO_x films. These films are deposited by reactive ion-beam sputter deposition, and contain large amounts of oxygen. Most of the valence electrons are localized, although metallic conduction is still observed with electron densities as low as 10^{26} m^{-3}. Resistivities close to a critical value where superconductivity is rapidly suppressed with increasing disorder can be realized. The T_c becomes extremely sensitive to small changes in the normal-state sheet resistance. Any small field-induced variation in σ_N will have a large effect on T_c.

Application of an electric field normal to a conducting film (Fig.2) leads to an accumulation of carriers at the interface between the dielectric and the metal film. The depth will be set by the charge screening length L_s. Any change in conductivity due to the electric field will affect only the top layer. From this perspective a metallic film consists of a top layer of thickness L_s with a conductivity that changes upon application of a field and a bottom layer of thickness $d-L_s$ which is unaffected by the field. It is assumed that the conductivity can be described by the free-electron Boltzmann conductivity σ_N

$$\sigma_N = ne^2\tau/m = e^2k_F{}^2\ell/3\pi^2\hbar \tag{2}$$

with n the electron density, e the electronic charge, k_F the Fermi wave number, ℓ the mean free path and τ the elastic scattering time. It also is assumed that the scattering rate is the same throughout the whole film. As

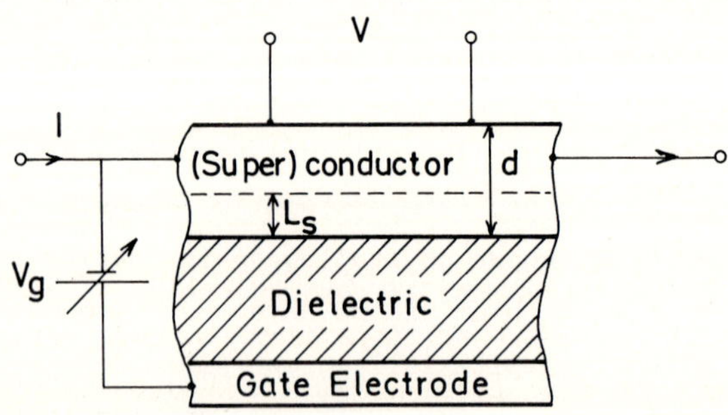

Fig.2 Field-effect arrangement for a superconducting film.

mentioned above, this is not a good approximation for polycrystalline thin films where surface scattering contributes substantially. The present materials are amorphous, and bulk scattering dominates. Furthermore it is assumed that the carriers induced at the interface are not trapped in a potential well, but have wave functions that extend throughout the film. The sheet conductance for the unperturbed layer of thickness d is $G = \sigma_N d$. Using the definitions $k_F = (3\pi^2 n)^{1/3}$ and $\ell = \mu k_F/e$, the mobility can be determined from:

$$\mu = \frac{3}{2}\,\frac{d}{L_s}\,\frac{\delta\sigma_N}{\delta ne} = \frac{3}{2}\,\frac{\delta G}{\delta\sigma ned}, \qquad (3)$$

where σ is the surface charge density. The measured field effect mobility can be used to find the mean free path, as well as other relevant normal-state parameters.

Hebard and Fiory [12] connect the normal-state properties to the superconducting properties using insights from localization theory. As shown by Fukuyama, Ebisawa and Maekawa [13], disorder attacks superconductivity by changing the BCS interaction. In the BCS gap equation:

$$T_c = 1.13\ \Theta_D \exp(-1/g'), \qquad (4)$$

the BCS coupling constant g' is of the form:

$$g' = g\ [1 - A(k_F\ell)^{-2} + \ldots] \qquad (5)$$

Here Θ_D is the Debije temperature and A is a constant. In Eq. 5 the expansion parameter is $k_F\ell$, which changes with electron density. Based on these equations one can show that the theoretical dependence of T_c on n is:

$$\frac{\delta T_c}{\delta n} = \frac{2gA}{3g'^2(k_F\ell)^2}\,\frac{T_c}{n}, \qquad (6)$$

in excellent quantitative agreement with experiment. Clearly, to maximize the field effect, it is not only necessary to have thin films with low electron densities, it also is advantageous to have sufficient disorder to make superconductivity a very sensitive probe to small changes in the normal-state sheet conductance. Experimental results reported by Hebard and Fiory are shown in Fig.3. Substantial shifts in T_c are observed with signs

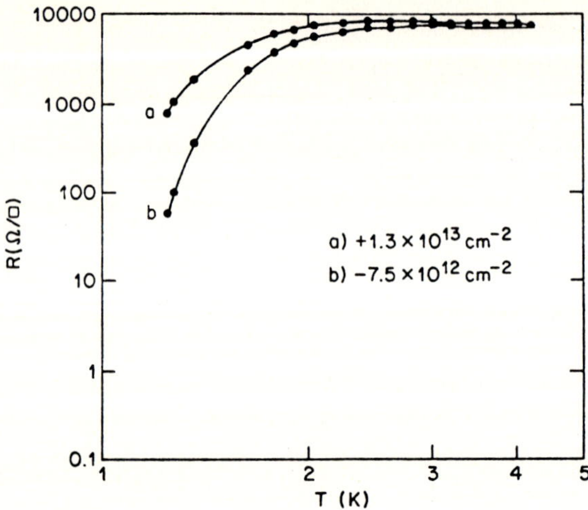

Fig.3 Resistive transition of In/InO_x film (6.3 nm thick) for two values of the surface-charge density (taken from Ref. 12). Note the logarithmic scales.

in agreement with the theory given above.

These results indicate that a consistent description of the field effect on disordered superconductors can be achieved. Within the framework of the model an attempt to optimize the materials parameters becomes feasible. The main factors that permit control are the thickness of the film and the electron density, at the expense of a further reduction of the critical temperature and the mobility. Despite assumptions in the model, it is very likely that the mobility in the conducting channel is dependent on the quality of the metal-dielectric interface.

Gurvitch and coworkers [14] have used Ta-doped surface layers of $SrTiO_3$ single crystals. This material is a superconductor with the lowest electron density known to date (3 10^{25} - 1 10^{26} m^{-3}). At the lower end of the doping range, where the material is not superconducting, 30-40 % changes in conductivity are observed. At the higher end, the conductivity changes only by a few percent, whereas substantial changes in T_c are reported. Even complete field-effect driven switching from the normal to the superconducting state is reported. However, many experimental difficulties, e.g., hysteresis effects, appear to be present.

A system that has received some interest is lanthanum sulfide. The composition can be La_3S_4, which is metallic with a T_c of 8.25 K and an electron density of 6.5 10^{27} m^{-3}, or La_2S_3, which is a semiconductor with a bandgap of 3 eV. By changing the sulphur content, the material can be tuned

Table II. Low carrier density superconductors

Superconductor	$T_c(K)$	$n(m^{-3})$
$SrTiO_3$	0.8	$5\ 10^{26}$
In/InO_x	2.5	$2\ 10^{26}$
LaS_x	8	$6\ 10^{27}$
$BaPbBiO_x$	12	$3\ 10^{27}$
$LaSrCuO_x$	40	$3\ 10^{27}$
$YBaCuO_x$	93	$6\ 10^{27}$

from metallic to insulating, without changing the crystal structure and the lattice parameter. Reports about successful thin film deposition by either coevaporation [15] or reactive sputtering [16] have appeared. In addition, Kent [17] has been able to deposit on La_3S_4 a thin, 3 nm thick layer of La_2S_3 as a tunnel barrier. Lead was used as a counter electrode. Reasonable quality tunnel data were obtained down to carrier densities of $3\ 10^{27}\ m^{-3}$.

At this point it also is of interest to recall the properties of the new ceramic superconductors. They have in common, as shown in Table II, with In/InO_x a low electron density and a relatively high specific resistance. Although much remains to be learned, they are certainly also very sensitive to changes in the normal-state properties. One may speculate about new superconducting devices based on the new ceramics employing the field effect.

3. PROXIMITY-INDUCED SUPERCONDUCTIVITY

An alternate approach to obtain superconductivity in low carrier density materials utilizes the proximity effect [18]. When a normal metal (N) is in good electrical contact with a superconductor (S), Cooper pairs easily will diffuse into the normal metal. As a result, the normal metal will become superconducting at the expense of a slight weakening of the strength of the superconductivity in S.

3.1 Basic theory

Consider an interface between a normal metal and a superconductor (Fig.4a). N_N and N_S are densities of states at the Fermi level of N and S, respectively. V_N and V_S are electron-electron interaction parameters. If

$V > 0$, the metal is a superconductor. The BCS coupling constant g, used in Section 2, equals NV. If g varies spatially, conventional BCS theory must be cast in the framework of the Gorkov theory. A Gorkov function F(r), called the condensation amplitude, is used. $|F(r)|^2$ is the probability of finding a Cooper pair at the position r. The pair potential, defined as $\Delta(r)=V(r)F(r)$ is analogous to the energy gap in the conventional BCS theory. In the following we assume that motion of electrons across the interface is diffusive. At the interface (with the x-coordinate running perpendicular to the interface) continuity of the condensation amplitude requires:

$$F_N(0)/N_N = F_S(0)/N_S \tag{7}$$

and

$$v_{FN}\ell_N \frac{dF_N(x)}{dx} = v_{FS}\ell_S \frac{dF_S(x)}{dx} \qquad \text{at } x = 0, \tag{8}$$

where ℓ and v_F are the mean free path and the Fermi velocity, respectively. Note that continuity in condensation amplitude implies discontinuity in energy gap. From the NS interface F(x) decays exponentially over the coherence length, ξ_N. Because it is a diffusion-limited process, the decay length ξ_N equals $(D\tau)^{1/2}$, with τ, the lifetime of a Cooper pair in the normal metal, given by $\hbar/2\pi k_B T$, and $D = v_F\ell/3$ is the diffusion constant. Above, it is assumed that the normal metal is in the dirty limit ($\ell < \xi_N$), the usual case of experimental interest. For the superconductor the same requirement applies, meaning $\ell < \xi_0$, with $\xi_0 = (\hbar D/\pi\Delta)^{1/2}$.

As shown in Fig.4a, a finite pair amplitude is induced in the normal metal, accompanied by a small decrease in the superconductor. Experimental consequences are that a thin film of normal metal in intimate contact with a superconductor develops an energy gap. This can be shown directly by measuring the tunnel conductance of a tunnel junction which has the NS layer as one electrode and a normal metal as the second. It also can be shown by measuring the transition temperature of NS bilayers as a function of thickness of the normal metal. The most dramatic demonstration is provided by a sandwich of superconductor-normal metal-superconductor (SNS) layers (Fig.4b). If the thickness of the normal metal is comparable to the normal-metal coherence length ξ_N, the system is capable of carrying a supercurrent. A measure of the strength of the proximity effect is the

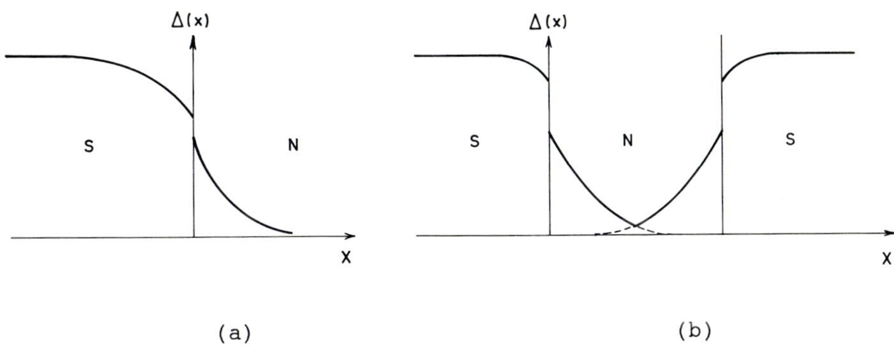

Fig.4 (a) Variation of the superconducting order parameter at an NS interface.
(b) A weak link established by the proximity effect.

maximum supercurrent that can be carried by an SNS sandwich. Likharev [19] finds for the critical supercurrent, assuming that the condensation amplitude in the superconducting electrodes is not suppressed,

$$I_c = \frac{4\Delta^2(T)}{R_N \pi k_B T} \frac{L}{\xi_N} \exp\left(-\frac{L}{\xi_N}\right), \tag{9}$$

where R_N is the normal-state resistance, and L is the thickness of the normal metal. The most striking aspect is the exponential decay as a function of thickness. This also will dominate the temperature dependence through the dependence on ξ_N.

3.2 Superconductor-semiconductor contacts

In recent years the proximity effect also has been used to induce superconductivity in doped semiconductors. The main target has been the development of a Josephson junction with a semiconducting barrier. The normal metal in an SNS junction must be replaced by a thin layer of semiconducting material. The thickness of the semiconducting layer will depend on the coherence length in the semiconductor. The decay length for a dirty three-dimensional semiconductor with mobility μ, carrier density n and effective mass m^*, is given by [20]:

$$\xi_N = (\hbar D/2\pi k_B T)^{1/2} = (\hbar^3 \mu/6\pi m^* e k_B T)^{1/2} (3\pi^2 n)^{1/3}. \tag{10}$$

With this definition one finds for silicon at 4.2 K, with a carrier density of 3 10^{25} m^{-3} and a mobility of 0.01 m^2/Vs, a value of 16 nm. For III-V semiconductors, higher values are attainable because of the lower effective mass. High carrier densities are needed to prevent carrier freeze-out at low temperatures. However, in bulk semiconductors an increase in carrier density leads always to a substantial reduction in mobility. Substantial improvement can be achieved in high mobility inversion layers where the carriers are spatially separated from the ionized impurities. For a two-dimensional electron gas (2DEG), using the relations $k_F=(2\pi N)^{1/2}$ and $D = v_F\ell/2$, one finds for the coherence length (in two dimensions):

$$\xi_N = (\hbar^3\mu N/2k_BTem^*)^{1/2}, \tag{11}$$

with N being the carrier concentration per unit area. This expression is numerically identical to the bulk coherence length, provided the identification $N = n^{2/3}$ is made. At 4.2 K in silicon MOSFETs, with mobilities on the order of 0.1 m^2/Vs at a carrier density of about 10^{15} m^{-2}, negligible improvement is obtained. However, with III-V semiconductors ,such as InAs, coherence lengths on the order of 0.1 μm are feasible.

3.3 Schottky barriers

Metal-semiconductor contacts are usually dominated by a Schottky barrier. At low temperatures and low voltages, carrier transport is dominated by tunneling. In this regime the resistance for an ohmic contact is given by:

$$R = \frac{\hbar^3\ 2\pi^2}{e^2m^*E_{00}}\ \exp[-E_{00}/E_B], \tag{12}$$

where E_B is the Schottky barrier height, and E_{00} equals $(e\hbar/2)(N_D/\varepsilon m^*)^{1/2}$.

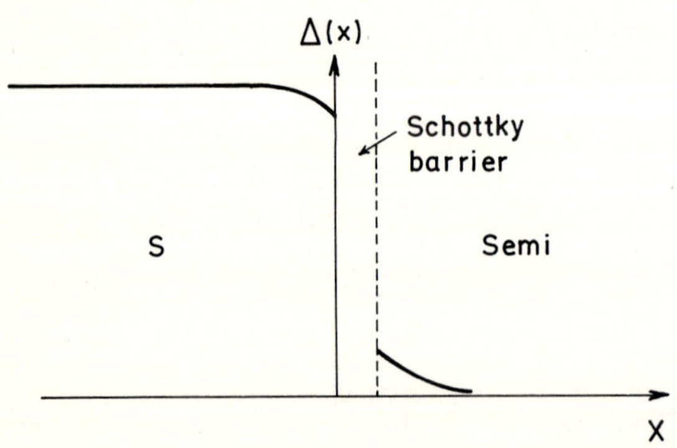

Fig.5 Assumed variation of the order parameter across the interface between a superconductor and a heavily doped semiconductor.

Here N_D is the doping concentration, and ε is the dielectric constant. In semiconductor technology, ohmic contacts are usually made by applying a high doping concentration. Experimentally observed values follow the behavior predicted by Eq.12 very well [21]. Satisfactory resistance values are on the order of 10^{-10} Ωm^2. However, in discussing superconducting devices these numbers can be put in a different perspective.

A typical tunnel junction, two superconducting electrodes separated by a thin oxide layer, has a comparable normal-state resistance. The current-voltage characteristics show, as is well known, all the salient features of tunneling behavior. Consequently, the probability for electrons to cross such a barrier is relatively small compared to the case of a clean interface between a normal metal and a superconductor (Fig. 5). A transition from clean metallic to tunneling contact has been well documented for NS contacts [22] with ample experimental support [23]. The striking aspect is that very little elastic scattering, i.e., a very thin oxide barrier, is sufficient to create tunneling behavior. In the same way, an insulating layer reduces the proximity effect considerably. Very little elastic scattering at the interface destroys the probability of finding a Cooper pair in the normal metal. Contact resistances on the order of 10^{-15} Ωm^2 are needed [24]. No systematic research has been done on the reduction of the proximity effect by a thin insulating layer, although it has been found that for a contact between aluminum and copper a resistance of only $4\ 10^{-13}$ Ωm^2 leads to a proximity effect of only a few percent [25]. Consequently, one expects, for contacts between a superconductor and a semiconductor, non-ohmic behavior on a mV scale comparable to observations on tunnel contacts between a normal metal and a superconductor. Contacts between superconductors and semiconductors have been investigated as mixers

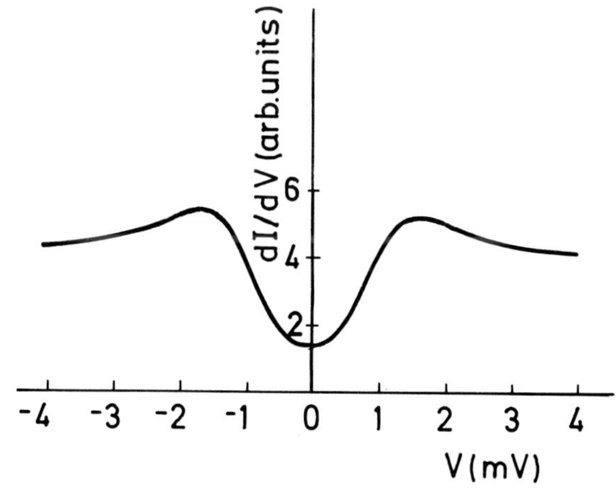

Fig.6 Conductance of a Nb/GaAs Super Schottky diode exhibiting Giaever tunneling (From Sugiyama et al [27]).

for microwave frequencies. Known as Super Schottky diodes, they have been replaced by the currently superior SIS mixers [26]. Super Schottky diodes were of interest because they showed, as expected from the above disussion, the nonlinearity characteristic of the superconducting energy gap. Contacts between Nb or Pb and GaAs have been used [27], as well as contacts between Nb and Si [28]. A steep rise of current at the energy gap of the superconductor is observed, indicative of tunneling transport (Fig.6).

On the other hand, as will be shown more extensively below, the proximity effect can be strong enough to allow supercurrents to flow through semiconducting layers. It also has been shown [29] that prior to deposition of the superconductor on a semiconducting substrate, a mild sputter cleaning is needed to obtain sufficiently low-resistance contact between the metal and the semiconductor. We recently have measured contact resistances and Schottky barrier heights between Nb and Si for various doping levels [30]. Results are summarized in Table III. Although some effect of doping and sputter cleaning is observed, the values for the contact resistances are too high to expect acceptable levels for the pair potential in the semiconductor. A comparison with the proximity effect would clearly be of interest, although it is evident that metal-semiconductor contacts at low contact resistances and on a millivolt scale are not very well understood. In this regime the discrete nature of the dopants, as well as the quality of the interfaces, will be important.

Table III. Schottky barrier height and contact resistances for Nb-Si contacts. N_D is the doping concentration, E_B the Schottky barrier height and R_C the contact resistance. The values marked with an asterisk are from samples without a sputter cleaning prior to deposition.

$N_D(m^{-3})$	$E_B^*(eV)$	$E_B(eV)$	$R_C^*(10^{-4}\Omega m^2)$	$R_C\ (10^{-4}\Omega m^2)$
$1\ 10^{21}$	0.60	0.48	--	--
$1\ 10^{22}$	0.58	0.48	--	--
$5\ 10^{22}$	0.57	0.46	--	--
$1\ 10^{23}$	0.56	0.46	7.6	0.17
$1\ 10^{24}$	0.50	0.41	0.93	0.035
$1\ 10^{25}$	--	--	0.12	0.025
$5\ 10^{25}$	--	--	0.037	0.002
$1\ 10^{26}$	--	--	0.003	0.002

3.4 Superconductivity observed in silicon

Some basic experimental facts about superconductivity induced in a semiconductor have been established. Hatano et al [31] measured the transition temperature of bilayers of Nb and a layer of heavily doped silicon on an insulating substrate (Fig.7a). Because of the contact between Nb and Si the critical temperature is suppressed. The suppression increases with increasing silicon thickness, with increasing doping level and decreasing niobium thickness. Using the De Gennes-Werthamer-Hauser[32] theory, values for ξ_N of 10 to 20 nm are inferred. They scale correctly with the dependence on carrier concentration predicted by Eq.10. Also, the absolute value agrees with theoretical expectations.

A direct demonstration of the superconducting gap in silicon through tunneling has been given by Nishino et al [33]. A thin membrane of boron-doped silicon backed by a superconducting lead film was used. A tunnel contact at the opposite side of the membrane enabled the measurement of the quasiparticle density of states of the semiconductor. The tunnel conductance (Fig.7b) shows the proximity-induced energy gap of about 0.3 meV in silicon, which in this case turns out to be about 20% of the energy gap of the lead electrode. The reverse experiment, the depression of the energy gap of lead backed by doped silicon, shows a systematic decrease of the gap for increasing doping concentration.

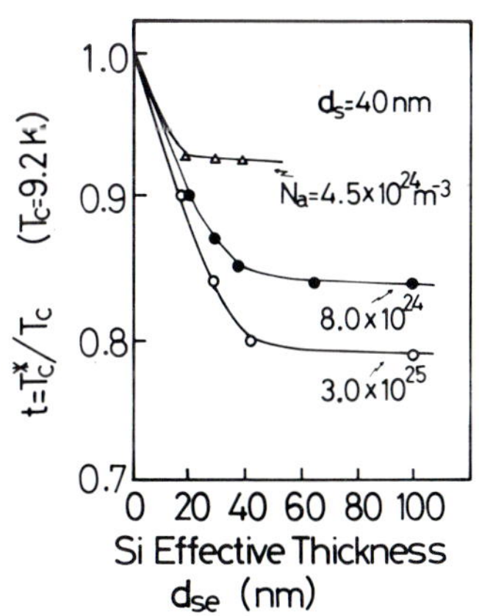

(a)

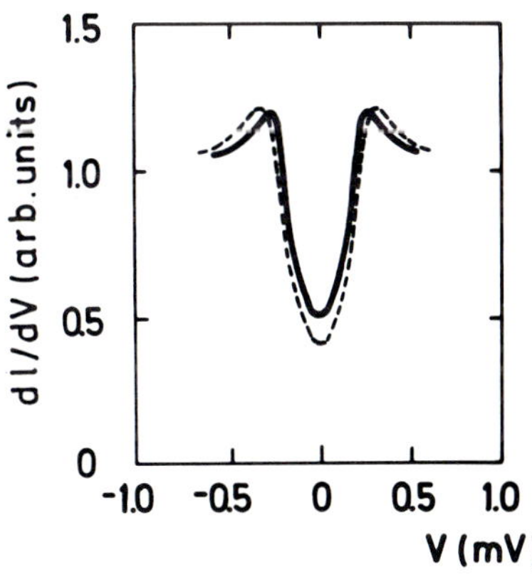

(b)

Fig.7(a) Depression of the critical temperature of a Nb film of thickness d_s as a function of the thickness (d_{se}) of a heavily doped silicon layer. N_a is the acceptor density.

(b) Tunnel measurement of the superconducting energy gap in silicon induced by a Pb film. Solid curve: N_a = 1 10^{25} m^{-3}; dashed curve: N_a = 8 $10^{25} m^{-3}$.

4. PROXIMITY-COUPLED SUPERCONDUCTING DEVICES

In Section 3.2 we have shown that the coherence length in a doped semiconductor is fairly short. Consequently a Josephson current through a single crystal semiconductor will be observable only if the semiconductor is thin enough. Two types of devices have been used succesfully. Huang and Van Duzer [34] used thin single-crystal membranes made by anisotropic etching. After the membrane is made, superconducting electrodes are deposited on the top and bottom of the membrane (Fig.8a). In a second approach, used for the first time succesfully by Ruby and Van Duzer [35], a doped surface layer of a semiconductor is covered by two superconducting films separated by a short distance (Fig.8b). The separation is usually defined by electron-beam lithography.

4.1 Josephson supercurrents

Experiments on supercurrents through semiconductors deal primarily with the critical value as a function of distance between the superconducting electrodes and of the temperature. The most extensive study has been made by Nishino and coworkers [36] on silicon. With lead on a (100)-oriented silicon wafer, using the geometry of Fig.8b, supercurrents have been observed for separations smaller than 0.25 μm. In Fig.9. results on the

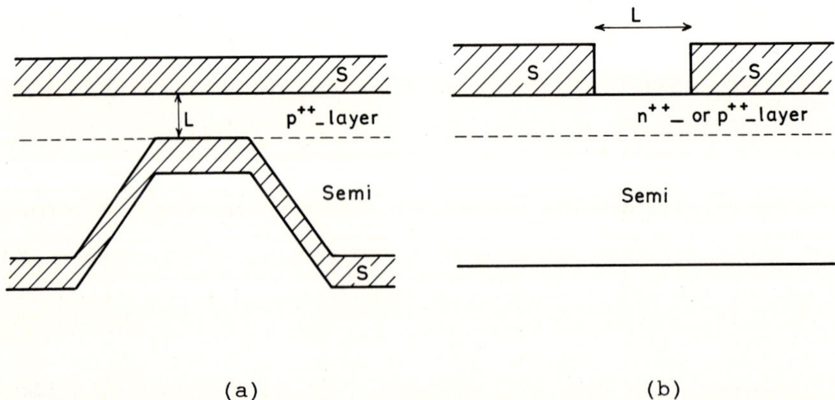

Fig.8 (a) Cross-sectional view of a thin silicon membrane sandwiched between two superconducting electrodes.
(b) Coplanar superconducting weak link with a semiconducting barrier.

critical current as a function of spacing, temperature and doping are reproduced. The solid line is the theoretical prediction of Eq.9. The results consistently can be modeled assuming 3-dimensional diffusion in the semiconductor (cf. Eq.10). The dependence on magnetic field also has been explored [37]. It is found that the supercurrent depends on magnetic field through two factors. The spatial dependence of the phase of the superconducting order parameter leads to the well known Fraunhofer diffraction pattern. In addition, the coherence length in Eq.9 is modified by the pair-breaking action of the magnetic field. Following Hsiang and Finnemore [38], ξ_n must be replaced by:

$$\xi_n = \xi_{n0}\ (1+B/b_0)^{-1/2}, \tag{13}$$

where ξ_{n0} is the coherence length for zero magnetic field, B is the strength of the magnetic field, and b_0 is a material-specific parameter related to the pair-breaking parameter.

Considerable work has been done on InAs, which has been reviewed recently by Kawakami and Takayanagi [39]. Compared to silicon, this material has a higher mobility because of its lower effective mass. It also has a negative Schottky barrier height [40], which in principle permits easy diffusion. Finally, InAs has a naturally inverted surface layer which leads to two-dimensional diffusion, and it is hoped, to enhanced response

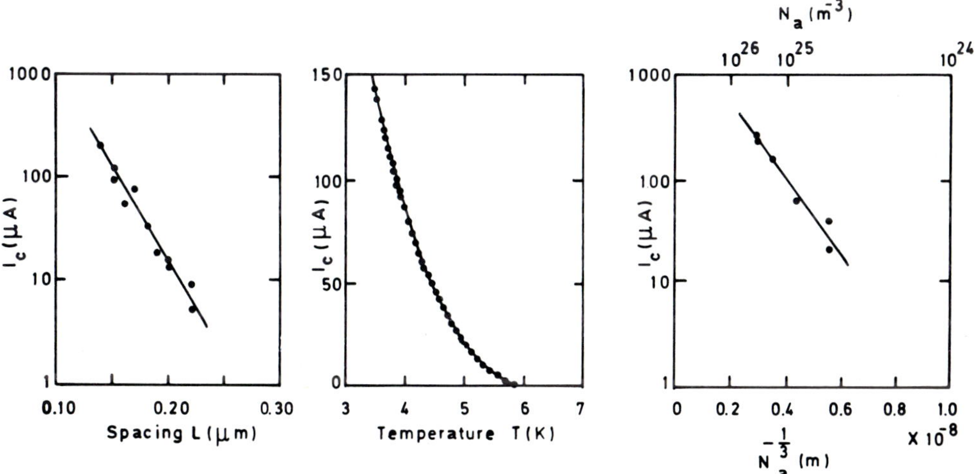

Fig.9 Critical current I_c of a silicon-coupled coplanar weak link as a function of spacing, temperature, and doping (N_A).

to an electric field [41]. As in the case of silicon, an exponential dependence on temperature and length is observed, but at substantially lower doping levels. A more advanced structure has been investigated by Kleinsasser et al [42]. A thin layer of InAs was grown on top of GaAs, and electrode spacings were defined by electron-beam lithography. Again, conventional proximity-effect theory seems to provide an excellent first-order description of the observed supercurrent as a function of temperature and length.

Despite its success so far, it is very unlikely that conventional proximity-effect theory can be used for low doping concentrations. The phenomenological theory of Section 3.1 assumes classical diffusion-limited transport, identical Fermi velocities, and **a priori** knowledge of the boundary conditions. An analysis of some idealized cases emphasizing the boundary conditions has been given by Tanaka and Tsukada [43] using the microscopic Gorkov theory. Because it is unlikely that the semiconductor can be treated as a normal metal throughout the whole range of doping concentrations, transport through the semiconductor has also been analyzed. In a series of papers, Aslamazov and Fistul' [44] have dealt with the transition from transport by tunneling (low doping), by resonant tunneling (intermediate doping level) and by diffusion (degenerately doped). It is claimed, based on a model proposed by Lifshitz and Kirpichenkov [45], that well before degeneracy sets in, resonant percolating trajectories along impurity states close to the Fermi level dominate the conduction. In this regime the supercurrent can no longer be described by Eq.9, although an exponential dependence on length remains. These theoretical predictions have not yet been tested experimentally, although Serfaty et al [29] have found excellent agreement in the degenerate case. Aslamazov and Fistul' [46] predict also a nontrivial dependence of the supercurrent on voltage. For example, the well-known logarithmic divergence of the supercurrent at $V = 2\Delta/e$, the Riedel singularity, is considerably enhanced.

In the degenerate case, as well as in the case of dirty normal metals, carrier diffusion is reduced by Anderson localization. This leads to a reduction of the coherence length as analyzed by Fukuyama and Maekawa [47]. A temperature-dependent diffusion constant enters Eqs.10 and 11. The correction is stronger for two-dimensional diffusion than for three-dimensional diffusion. The effect will be most prominent for semiconductors near the mobility edge, where the largest sensitivity to carrier concentration can be expected.

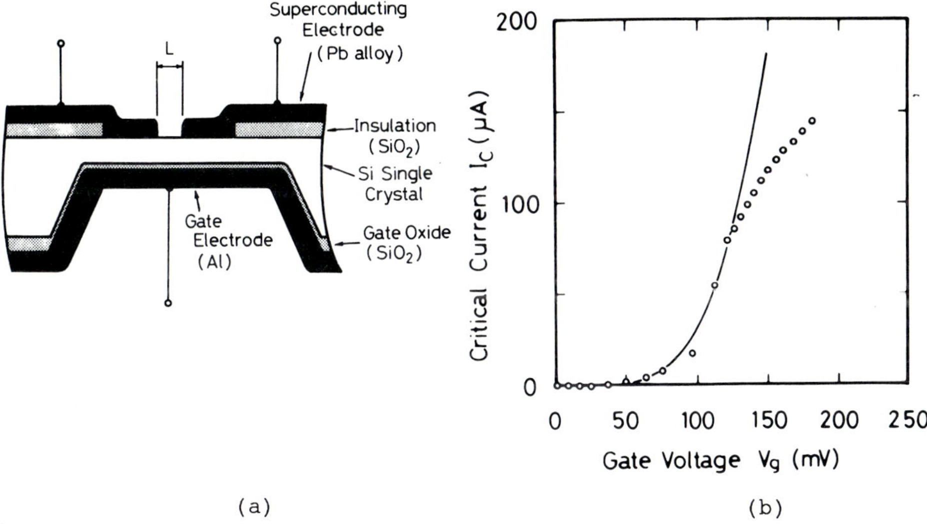

Fig.10 (a) Device configuration for field-effect controlled superconductivity in a silicon-coupled weak link.
(b) Critical Josephson current as a function of gate voltage.

4.2 Field-effect operation

Taking Eq.9 as a starting point, the effect of an electric field on a proximity-coupled Josephson junction can be understood [48]. The gate voltage will change the carrier concentration in the semiconductor. A change in carrier concentration leads to a change in coherence length (in the three-dimensional case):

$$\xi_N(V_G) = \left[\frac{\hbar^3 \mu}{6\pi m^* k_B T_c} \right]^{1/2} \{3\pi^2(n+\delta n)\}^{1/3} , \tag{14}$$

with δn being the change in carrier concentration, which is proportional to V_G, the gate voltage.
For small values of δn one finds:

$$\xi_N(V_G) = \xi_N(0)\ [1+\delta n/3n]. \tag{15}$$

From Eq.9, the dependence of the critical current on carrier concentration becomes:

$$I_c(V_G) = I_c(0) \exp\left[\frac{L\ \delta n}{\xi_N(0)\ 3n} \right] \propto \exp(V_G), \tag{16}$$

where $I_c(0)$ is the supercurrent for zero gate voltage. A similar expression can be found for two-dimensional diffusion. To maximize the field effect, a small value of n, as well as a large value of L/ξ_N, is needed. Unfortunately, this leads also to a reduced value of the prefactor $I_c(0)$.

Experimentally, gate controlled supercurrents (Fig.10 a and b) were observed for the first time in silicon-coupled, lead-lead junctions by Nishino et al [49]. Application of gate voltages up to 200 mV changes the critical current considerably. Samples that carry no supercurrent at all can be made to carry supercurrent above 75 mV. In Fig.11 the results are replotted to compare them with the predictions of Eq.16 [Cf. Ref. 48]. The exponential dependence on gate voltage is confirmed. However, a more quantitative evaluation of the factor in square brackets in Eq.16 shows unrealistically large values for $L/\xi_N(0)$. In other words, the supercurrent is much more sensitive to the gate voltage than can be expected on the basis of Eq.16 alone. At present, this is not understood and indicates that the gate voltage is doing more than influencing the supercurrent through the carrier concentration. An additional contribution might come from the value of the energy gap at the interface between the superconductor and the semiconductor. As we have seen in Section 3.3, the role of the Schottky barrier is unclear. It is less of a barrier than one would expect from a simple extrapolation from the low-doping situation. On the other hand, the precise conditions of the interface and its contribution to the reduction of the energy gap remain unclear.

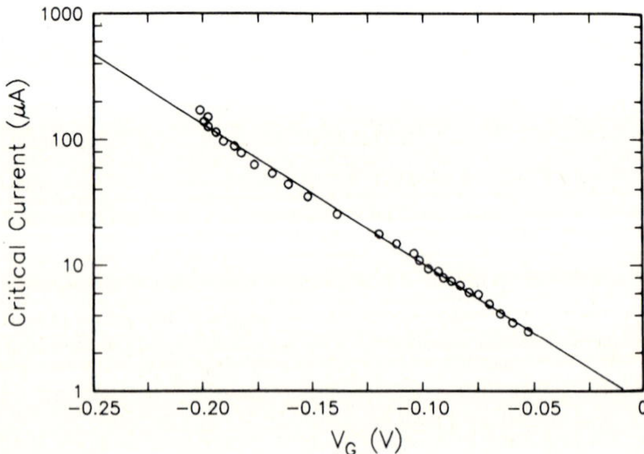

Fig.11 Exponential dependence of critical current on gate voltage (Taken from Ref.48).

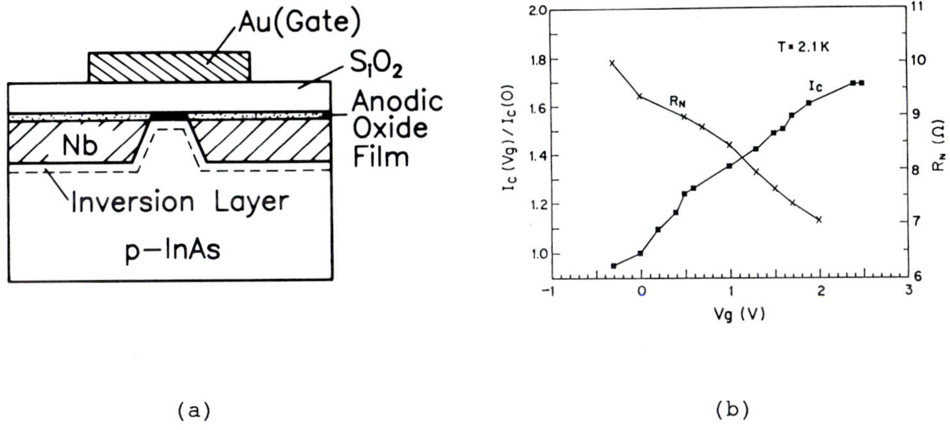

Fig.12 (a) Configuration for InAs coupled superconducting field-effect device.
(b) Critical current and resistance change on application of gate voltage.

InAs is a second system where a field-effect controlled supercurrent has been observed. Takayanagi and Kawakami [50] used the n-type inversion layer on a p-InAs substrate. In their experiment they were able to change the current by a factor of 3 by varying the gate voltage from +25 V to -60 V. These values were discouragingly high. Improved processing has brought the required gate voltages down to about a few volts [39]. Application of a gate voltage changes both the critical supercurrent and the normal-state resistance as in an ordinary field-effect transistor. Similar results have been obtained by Ivanov, Claeson and Andersson [51] on a GaAs accumulation layer. In both cases no attempt has been made to compare the results with theoretical predictions, such as in Eq.16. Further progress will require systematic experiments on well defined samples.

5. CONCLUSIONS

From the discussion presented, it is clear that a useful superconducting field-effect transistor is not yet available. It also is clear that theoretical understanding requires insight into systems with low carrier concentrations, where superconductivity is sensitive to small changes in carrier concentrations. In addition, better understanding of superconducting transport through semiconductors doped just below degeneracy is needed. Finally, more systematic knowledge about interfaces between superconductors and doped semiconductors is needed.

The present results deal exclusively with conventional superconductors. Taken together, they illustrate the fact that succesful applications of superconductivity are dependent on more than just the critical temperature of the material. There remains a need for systems in which superconductivity can be controlled by a third electrode. The sensational discovery of the new ceramic superconductors has not changed this situation. Nevertheless, integration of high-T_c superconductors with semiconductor barriers [52] should be explored as well.

Acknowledgment

This work is part of the research program of the Stichting voor Fundamenteel Onderzoek der Materie (FOM), with financial support from the Nederlandse Organisatie voor Wetenschappelijk Onderzoek (NWO).

REFERENCES

1. T.D. Clark, R.J. Prance and A.D.C. Grassie, J. Appl. Phys. 51, 2736 (1980).
2. A.W. Kleinsasser, **Superconductor-silicon heterostructures,** in NATO Advanced Study Workshop "Heterostructures on Si: One Step Further with Silicon", Cargese(Corsica),France, 16-20 May 1988.
3. R.E. Glover and M.D. Sherrill, Phys.Rev.Lett. 5, 248 (1960) W.Felsch and R.E.Glover, J.Vac.Sci.Technol. 9, 337 (1971).
4. H.L. Stadler, Phys. Rev. Lett. 14, 979 (1965).
5. A. Bardasis, Phys. Rev. B26, 1477 (1982).
6. M. Yu, M. Strongin and A. Paskin, Phys. Rev. B14, 996 (1976).
7. E.H. Sondheimer, Phys. Rev. 80, 401 (1950).
8. A. Berman and H.J. Juretschke, Appl. Phys. Lett. 18, 417 (1971).
9. T.H. Geballe, Materials Research Society Symposium on Thin Film Superconducting Materials, Dec.3-5 (1986).
10. P.A. Badoz, A. Briggs and E. Rosencher, **19th Int. Conf. on Physics of Semiconductors**, Warsaw, Poland, 1988, to be published.
11. A.F. Hebard, A.T. Fiory and R.H. Eick, IEEE Trans. on Magn. MAG 23, 1279 (1987).
12. A.F. Hebard and A.T. Fiory, Phys. Rev. Lett. 58, 1131 (1987).
13. H. Fukuyama, H. Ebisawa and S. Maekawa, J. Phys. Soc. Jpn. 53, 3560 (1984).
14. M. Gurvitch, H.L. Störmer, R.C. Dynes, J.M. Graybeal and D.C. Jacobson, Bull. Amer. Phys. Soc. 31, 438 (1986).
15. D.D. Berkley, J.H. Kang, J. Maps, J-C. Wan, A.M. Goldman, Thin Solid Films 156, 271 (1988).
16. A. Kapitulnik, A.D. Kent, T.H. Geballe and J.H. Kaufman, **Novel Superconductivity** (S.A. Wolf and V.Z. Kresin, Eds.) p.61.
17. A.D. Kent, Ph.D. Thesis, (Stanford University, 1988).
18. A. Gilabert, Ann. Phys. 2, 203 (1977).
19. K.K. Likharev, Sov.Phys.Tech.Letters 2, 12 (1976); see also K.K. Likharev, Rev.Mod.Phys. 51, 101 (1979).
20. J. Seto and T. van Duzer, in **Low Temp.Phys. LT13**, Vol.3, K.D. Timmerhaus, W.J. O'Sullivan and E.F. Hammel, Eds.(New York, Plenum 1974) p.328.
21. S.M. Sze, **Physics of semiconductor devices** (Wiley,New York, 1981) p.305.
22. G.E. Blonder, M. Tinkham and T.M. Klapwijk, Phys.Rev. B25, 4515 (1982).
23. G.E. Blonder and M. Tinkham, Phys.Rev. B27, 112 (1983); G.Voss, Thesis, (University of Cologne 1984).
24. T.Y. Hsiang and J. Clarke, Phys.Rev. B21, 945 (1980).
25. T.R. Lemberger and Y. Yen, Phys.Rev. B29, 6384 (1984).
26. J.R. Tucker, and M.J. Feldman, Rev.Mod.Phys. 57, 1055 (1985).
27. M. McColl, M.F. Millea and A.H. Silver, Appl.Phys.Lett. 23, 263 (1973); Y. Sugiyama, M. Tacano, S. Sakai and S. Kataoka, IEEE Electron Dev. Lett. EDL-1, 236 (1980).
28. L.B. Roth, J.A. Roth and P.M. Schwarz, in **Future Trends in Superconductive Electronics** B.S. Deaver, C.M. Falco, J.H. Harris and S.A. Wolf, Eds., American Institute of Physics, New York, 1978, p.384.
29. A. Serfaty, J. Aponte and M. Octavio, J. Low Temp. Phys.63, 23 (1986); Erratum J. Low Temp. Phys. 67, 319 (1987).
30. D.R. Heslinga and T.M. Klapwijk, to be published.
31. M. Hatano, T. Nishino and U. Kawabe, Appl. Phys. Lett. 50, 52 (1987).
32. P.G. de Gennes, Rev. Mod. Phys. 36, 225 (1964); J.J. Hauser, H.C.-Theuerer and N.R. Werthamer, Phys. Rev. 136, 637 (1964).
33. T. Nishino, M. Hatano and U. Kawabe, Jap.J.Appl.Phys. 26 Suppl-3, 1543 (1987).

34. C.L. Huang and T. Van Duzer, Appl. Phys. Lett. 25, 753 (1974).
35. R.C. Ruby and T. Van Duzer, IEEE Trans. Electron Devices ED-28, 1394 (1981).
36. T. Nishino, E. Yamada and U. Kawabe, Phys. Rev. B33, 2042 (1986).
37. T. Nishino, U. Kawabe, E. Yamada, Phys. Rev. B34, 4857 (1986).
38. T.Y. Hsiang and D.K. Finnemore, Phys. Rev. B22, 154 (1980).
39. T. Kawakami and H. Takayanagi, Jap. J. Appl. Phys. 26, Suppl-3, 2059 (1987).
40. K. Kajiyama, Y. Mizushima and S. Sakata, Appl. Phys. Lett. 23, 458 (1973).
41. V.Z. Kresin, Phys. Rev. B34, 7587 (1986).
42. A.W. Kleinsasser, T.N. Jackson, G.D. Pettit, H. Schmid, J.M. Woodall and D.P. Kern, Appl. Phys. Lett. 49, 1741 (1986).
43. Y. Tanaka and M. Tsukada, Solid State Comm. 61, 445 (1987); Phys. Rev. B37, 5087 (1988); Phys. Rev. B37, 5095 (1988).
44. L.G. Aslamazov and M.V. Fistul', Pis'ma Zh. Eksp. Teor. Fiz. 30, 233 (1979) [JETP Lett. 30, 213 (1979)]; idem, Zh. Eksp. Teor. Fiz. 81, 382 (1981) [Sov. Phys. JETP. 54, 206 (1981)]; idem, Zh. Eksp. Teor. Fiz. 83, 1170 (1982) [Sov. Phys. JETP, 56, 666 (1982)].
45. I.M. Lifshitz and V.Ya. Kirpichenkov, Zh. Eksp. Teor. Fiz. 77, 989 (1979) [Sov. Phys. JETP, 50, 499 (1979)].
46. L.G. Aslamazov and M.V. Fistul', Zh. Eksp. Teor. Fiz. 86, 1516 (1984) [Sov. Phys. JETP. 59, 887 (1984)].
47. H. Fukuyama and S. Maekawa, J. Phys. Soc. Japan 55, 1814 (1986); H. Fukuyama, Proximity effect and Anderson localization in **Novel Superconductivity** (S.A. Wolf and V.Z. Kresin, Eds.) p.51.
48. A.W. Kleinsasser, Phys. Rev. B35, 8753 (1987).
49. T. Nishino, M. Miyake, Y. Harada and U. Kawabe, IEEE Electron Device Lett. EDL-6, 297 (1985).
50. H. Takayanagi and T. Kawakami, Phys. Rev. Lett. 54, 2449 (1985).
51. Z. Ivanov, T. Claeson and T. Andersson, Jap. J.Appl. Phys. 26, Suppl-3, 1617 (1987).
52. V.Z. Kresin, Cryogenics 28, 409 (1988).

Cryogenics for Superconducting Electronics

C. Heiden

Institut fur Schicht- und Ionentechnik
Kernforschungsanlage Julich
P.O. Box 1913
D-5170 Julich, F.R.G.

1. Introduction

One major prerequisite for a wider acceptance of superconducting electronics is the availability of suitable cooling facilities. Certainly, the development of a Josephson technology utilizing the new high-T_C materials with transition temperatures well above the boiling point of liquid nitrogen will ease the situation to a considerable extent. Sensors with ultra-low instrument noise, nevertheless, may require working temperatures in the range of liquid helium or hydrogen. In order to become independent of the availability, the handling, and the storage of liquid cryogens, the development of small-scale cryocoolers was started in the past to meet the special requirements of superconducting devices. Such coolers should be quiet, i.e., their own electromagnetic and mechanical interference signals should be as small as possible. Furthermore, they should be reliable and easy to operate. Portability also may be required.

For many purposes one can tolerate a rather low cooling power, e.g., on the order of a milliwatt or less. This leads to comparatively long cool-down periods that may be on the order of one day. For practical purposes, it then may be advantageous to let the cooler run continuously over extended periods - months or even years. The cooler may not be serviced easily once placed in operation, for instance, as in space applications. Such coolers, therefore, should have a long mean time between failures (MTBF).

The development of coolers for superconductor electronics has been pursued in the recent past using, among others, refrigerators based on the Stirling, the Vuilleumier, and the Gifford-McMahon cycle. Pulse-tube refrigerators, Joule-Thomson coolers, and hybrid systems consisting of, for instance, the combination of a Gifford-McMahon with a Joule-Thomson cooler should be added to this list. Systems of the latter type may be an attractive choice to refill dewars with liquid

NATO ASI Series, Vol. F 59
Superconducting Electronics
Edited by H. Weinstock and M. Nisenoff

helium at appropriate intervals. In contrast to all these coolers, Peltier elements so far have not been able to reach temperatures as low as that of liquid nitrogen.

After a short discussion of cooling strategies in the following chapter, the focus will be on recent developments of systems that use gas for heat transport from the cold to the warm end of the cryocooler i.e., Stirling, pulse-tube, Joule-Thomson, and hybrid refrigerators.

2. Cooling Strategies

In order to achieve cooling or maintain a temperature below ambient, heat must be moved against the natural direction of heat flow. In all refrigerators using gas or vapor as a working fluid, this is done by a suitable compression/expansion cycle. For this purpose, the cooler consists mainly of three components (cf. Fig.1):

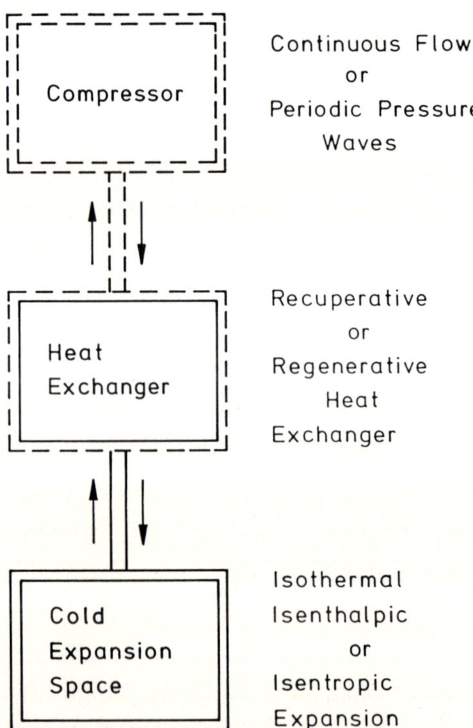

Fig.1 General cryocooler scheme. Indicated on the right side are ideal cases. A truly isothermal expansion can only be approximated.

compressor, heat exchanger, and cold expansion space. The working fluid, a gas such as N_2, Ar, H_2, He, or a suitable gas mixture, is compressed at the high temperature end of the cryocooler and then expanded at the other end, where it extracts heat, -Q, from its surroundings and cools it. In a closed cycle machine the cold gas then is returned to the compressor

There are many different ways to run the working gas through a complete thermodynamic cycle [1]. The compressor may produce a continuous high pressure gas stream or periodic pressure waves. The gas expansion in the ideal case may be performed at constant temperature, constant enthalpy, or constant entropy. An essential part of the refrigerator is the heat exchanger between compressor and cold space, taking heat from the downstream gas and rejecting it via the upstream flow to the hot end. Although there may be a considerable mass flow through the heat exchanger, it provides an effective thermal insulation between hot and cold end. Another way of looking at it is that the heat exchanger provides positive feedback to the system by cooling the downstream gas that, after expansion, cools the (new) downstream gas even further, and so on.

With regard to reliability and simplicity, the major development efforts have been on cryocooler schemes that incorporate as few moving parts as possible [2]. The most attractive systems, therefore, are Joule-Thomson units and pulse-tube refrigerators, where moving parts are needed only for the compressor; Stirling-cycle and some Vuilleumier-type cryocoolers that need no valves; and Gifford-McMahon refrigerators, where all valves are at room temperature and can be made rather reliable. Of course, combinations of Joule-Thomson units with other refrigerators used for precooling are also of much interest, since they allow the fabrication of small helium liquefiers. The development of periodically operating Simon liquefiers using isentropic expansion [3] also should be mentioned in this context, since it may prove to be suitable for cryoelectronic devices. For more details the reader is referred to the literature (e.g., [1].

3. Major Problem Areas of Cooler Development

The development of suitable cryocoolers for superconducting devices, in particular for sensors with ultra-low instrument noise (such as SQUIDs), has suffered from problems that are associated with each of the three major cooler components listed in Fig. 2.

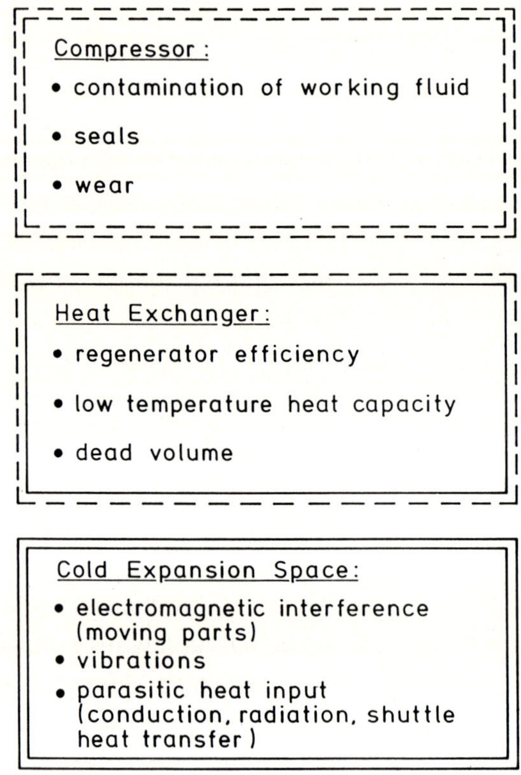

Fig. 2. Major problem areas in a cryocooler.

3.1 Compressor

One of the most stringent requirements for small cryocoolers is to keep the working gas free from impurities that may freeze out in the cold parts of the system thereby degrading the performance by clogging or increased friction. Lubrication and seals that may outgas and wear, can cause gas contamination problems. The contamination problem is most severe for microminiaturized Joule-Thomson coolers, whose fine capillaries are easily blocked by ice or other particles. This problem occurs also in the fine gas passages at the cold end of regenerators.

Desorption of water from walls of the system must be mentioned in this context, and the filling of the cooler with high purity gas must be done with extreme care.

3.2 Heat exchanger

One of the major problems encountered with regenerative heat exchangers is the reduction of their heat capacity at temperatures below about 20 K. This leads eventually to a saturation effect, i.e., the temperature at the cold end of the regenerator, due to insufficient heat capacity, starts to oscillate excessively with the oscillating gas flow. This limits the lowest attainable temperature in coolers using regenerative heat exchangers.

Another concern is the dead volume (inherent to this type of heat exchanger) which is needed to reduce the gas flow resistance to acceptable levels. For a given volume of the compressor the cooling power decreases with increasing dead volume [1]. It is important, especially for small cryocoolers, to make the dead volume as small as possible in order to have satisfactory cooling performance.

3.3 Cold expansion space

In coolers for SQUIDs, extensive precautions must be taken to keep electromagnetic interference as small as possible in the cold expansion space. Non-magnetic materials such as plastics or ceramics, therefore, are a good choice for displacers that move in the expansion space (of Stirling and Gifford-McMahon machines). Magnetic interference also can be generated by mechanical vibrations of the expansion volume in a non-zero external magnetic field. Such vibrations, therefore, should be reduced to a minimum.

Because of the small cooling power much attentions must be paid to parasitic heat that can enter the expansion space by conduction, radiation and (for coolers with moving displacers) due to shuttle heat transfer and friction.

All of these problems have been known for a long time and are discussed in the literature [4]. Progress in cooler development is closely related with solutions to these problem areas. Some recent developments will be reported below in dealing with particular cryocoolers.

4. Recent Cryocooler Developments

4.1 Stirling-cycle cryocoolers

Cryocoolers based on the Stirling cycle have been among the first candidate SQUID coolers. The thermodynamic Stirling-cycle ideally consists of two isothermal and two isochoric paths. The constant volume fractions of the cycle usually are generated by means of a displacer which moves in the cold expansion cylinder in fixed phase relation to the motion of the compressor piston, driven, for instance, by a common crankshaft (cf. Fig. 3). The regenerator can be incorporated in the displacer, heat being stored either in a metal or plastic screen inside the displacer cylinder or in the walls of the displacer and expansion cylinder (for a gap regeneration system).

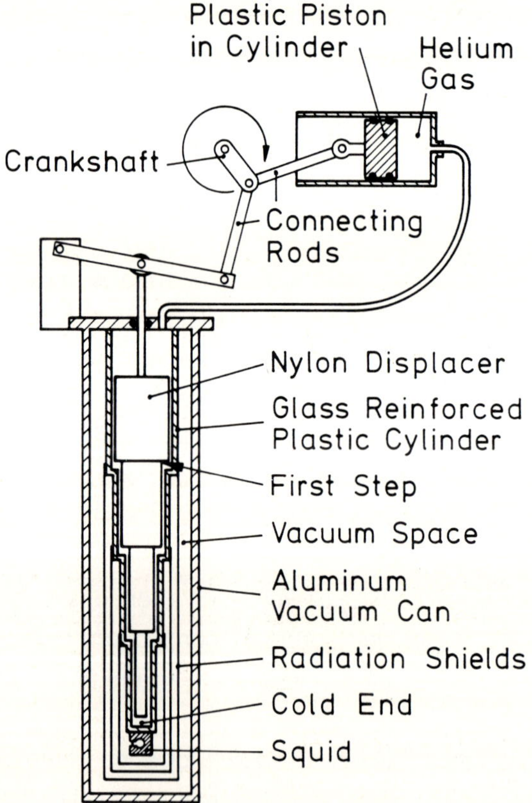

Fig. 3 Schematic of a small-scale Stirling cryocooler.

After the pioneering work of Zimmerman [5], in which he introduced a small Stirling cooler (cf. Fig.3) made of non- magnetic plastic material, there has been a continuous effort to improve this design and to eliminate weak spots [6]. Final temperatures below 7 K with a net cooling power on the order of a milliwatt have been reached using a displacer design with up to 5 stages. This leads to an effective distribution of cooling power along the cylinder from room temperature down to the coldest stage which can be used for cooling nested radiation shields.

A natural consequence of using an increasing number of stages is a conical design for the displacer unit that was first made by Myrtle et al. [7], achieving an ultimate temperature of 9 K. A special preparation technique for such conical displacer unit with optimized shape was developed by Lambert [8]. He also developed a technique to reduce helium diffusion through the warm section of the plastic unit by embedding a metal foil in the fiberglass-reinforced epoxy wall. An important point in this cryocooler design is the incorporation of a gas flow controller leading to better cooling performance. Placing lead particles in the low-temperature section of the plastic displacer in order to enhance the low temperature heat capacity of the regenerator resulted in temperatures approaching 7 K [9].

There also are several developments concerning the compressor unit. One is the use of pneumatically-driven diaphragm compressors by Zimmerman et al. [6]. To avoid diffusion of air and of water, a double membrane with a continuously-pumped spacing in between the membranes is used.

Another approach to a contamination free compressor has been followed by the author. Piston and cylinder of the compressor are made of alumina. In order to obtain a hermetic seal for the piston shaft, a stainless steel bellows is used. To avoid early fatigue fracture of such a seal, which otherwise can be a problem [10], the force transmission to the piston is effected by a lever leading only to a slight bending of the bellows (Fig. 4). Such a compressor has accumulated a total operating time of over 10,000 hours at a stroke frequency of 1 Hz. Similar to results obtained by Zimmerman [8] wear of the ceramic material was found to be negligible. The displacer unit can be driven from the crankshaft using a similar seal, thus eliminating O-rings and making it possible to hermetically seal the entire cryocooler unit.

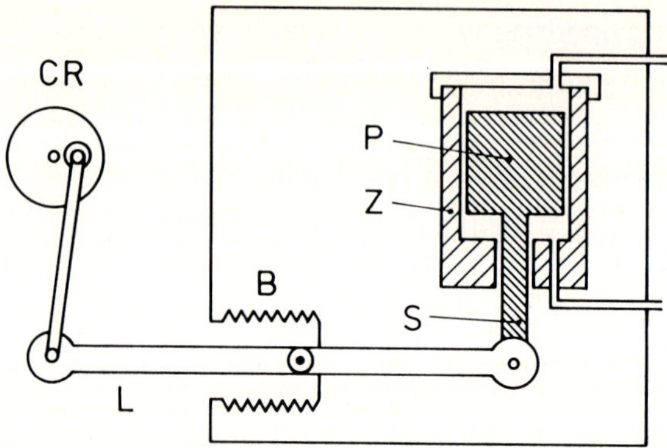

Fig. 4. Hermetically-sealed compressor unit. Piston, cylinder and piston shaft seal are made of alumina that is precision ground to a radial clearance of 4 micrometers. Two pressure waves are generated with a 180 degrees phase shift.

Recently, such a unit has been operated in the free-displacer mode, i.e., the displacer is not connected to the compressor crankshaft but is driven pneumatically by the pressure wave of the working fluid [11]. The weight of the displacer is balanced by a counterweight attached to a connecting lever. The latter is hermetically sealed using, again, a stainless-steel bellows in the bending mode. This design has the advantage that compressor and cold stage can be decoupled with the consequence that interference noise from the compressor at the cold stage can be reduced significantly. A length of connecting tubing of up to 6 m has been used without a significant change of end temperature, reaching 30 K for a three-stage displacer unit (cf. Fig.5). Further development of this unit is under way and includes the utilisation of a second compression volume for the ceramic cylinder. The corresponding pressure wave, being 180 degrees out of phase with that of the first compression volume, can be used to drive an additional displacer unit. This second displacer unit can be combined with the first one in the same cold stage, with the consequence that a counterweight and corresponding bellows seal are no longer needed.

4.2 Pulse-Tube Refrigerator

The pulse-tube refrigerator represents one of the simplest mechanical cooler designs, incorporating only one moving part, the compressor. It was described first

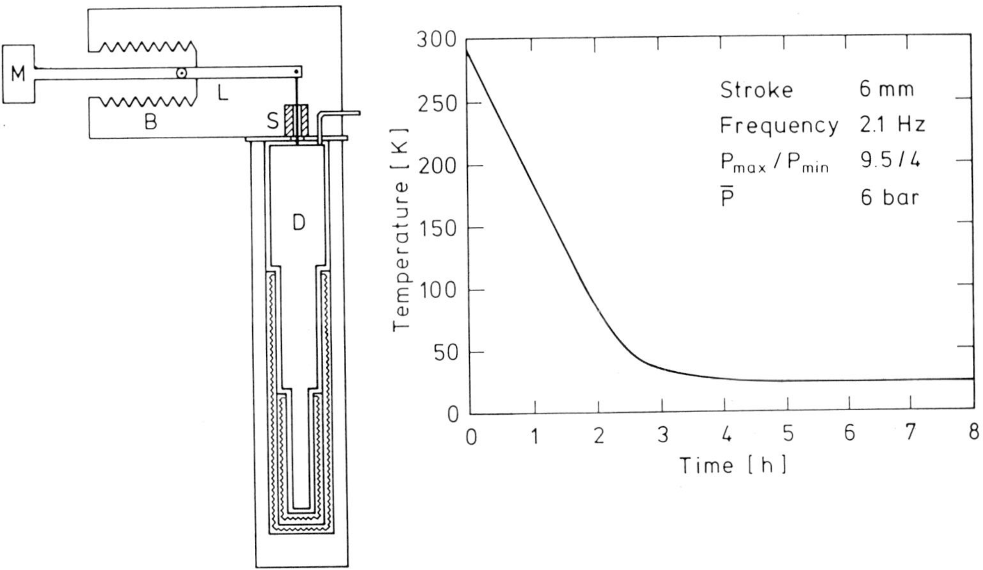

Fig. 5. Hermetically-sealed, three-stage plastic displacer unit driven pneumatically by a pressure wave. Also shown is its cool-down behavior.

by Gifford and Longsworth [12] in 1964, and has recently won renewed interest due to its potential for high reliability. Cooling is achieved at comparatively low gas pressures and pressure ratios. Figure 6 shows the essential parts: the reciprocating compressor which produces the pressure waves, the regenerator, and the pulse tube. The pulse tube consists of thin-walled stainless steel tubing with heat exchangers attached to it at both ends.

Although a definitive quantitative description of its operation still is not available [13], qualitatively the heat pumping in the pulse tube can be understood as follows: gas that initially is in the pulse tube (initial gas = i.g., cf. Fig. 7) is pushed during the compression phase of the cycle by the incoming additional gas (a.g.) toward the warm end of the tube. Both i.g. and a.g. warm inside the tube during this process and will give up heat by contact with the tube wall and the warm heat exchanger. In the ideal case this heat is carried away. During the following expansion phase the gas lowers its temperature and absorbs heat from the tube walls, and in particular, from the cold heat exchanger while passing through it.

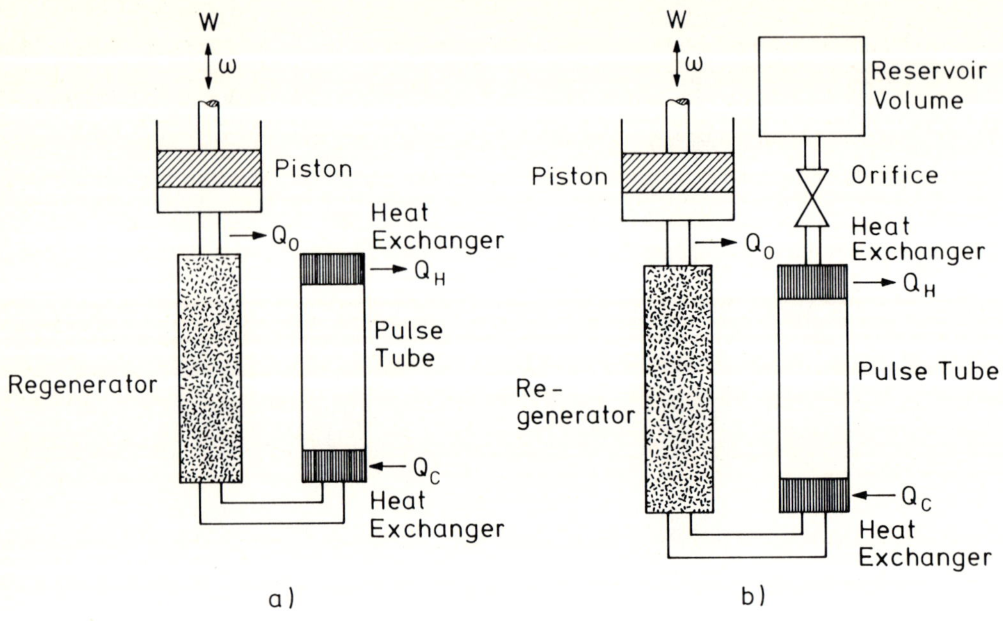

Fig. 6. a) Basic pulse-tube refrigerator; b) orifice-type pulse-tube refrigerator with additional expansion space, after [16].

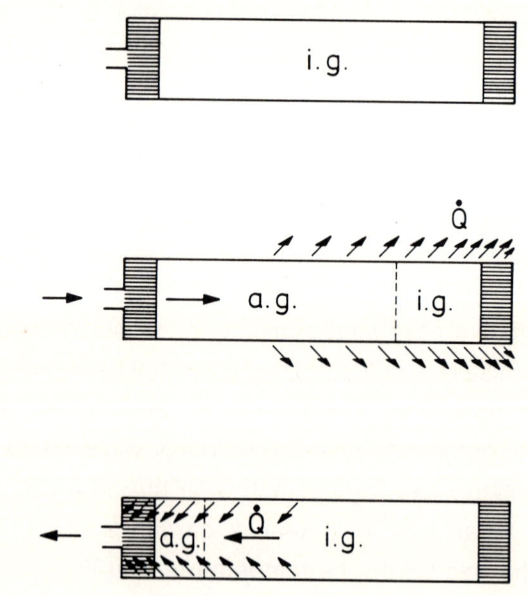

Fig. 7. Thermal cycle in a pulse tube. (i.g. = initial gas, a.g. = additional gas).

A temperature of 165 K was reached with this design by Gifford in 1967 [14]. With a modified design that uses a rotary valve to switch the pulse tube between the high and low pressure ports of a valved compressor a temperature of 124 K was achieved by Longsworth [15], also in 1967. Lower temperatures were obtained recently by introducing an orifice near the warm heat exchanger in order to let some of the gas go back and forth between the pulse tube and a larger reservoir (cf. Fig. 6b). According to an analysis by Radebaugh et al. [17] the effect of the orifice consists of introducing an additional phase shift between the gas velocity and dynamic pressure which in turn increases the enthalpy flow in the refrigerator, and thus the cooling efficiency. Mikulin et al. [16] achieved a temperature of 105 K, and 60 K was obtained by Radebaugh et al. [17], the lowest temperature so far for a one-stage pulse tube refrigerator.

It appears that the pulse-tube refrigerator requires further development, in particular, with regard to cascading several stages which would lower the achievable end temperature.

4.3. Joule-Thomson Coolers

With regard to vibrations and magnetic interference, a good choice for small reliable cryocoolers appears to be a system made completely of Joule-Thomson units. Except for the compressor it contains no moving mechanical parts (cf. Fig. 8). If short duration cooling is required, cylinders of compressed gas can be used. Cooling in a single-stage Joule-Thomson unit is achieved by expanding the real gas (when it is below its inversion temperature) through a throttle valve, a counterflow (recuperative) heat exchanger providing a positive feedback. Without external heat load, the working gas easily can be liquefied.

To obtain liquid helium, at least three stages are needed using, for instance, helium, hydrogen and nitrogen. The helium stage is precooled by the hydrogen stage, which is cooled by the nitrogen unit. Since the inversion temperature of nitrogen is above room temperature, it needs no precooling. For a more rapid cool down, and to reduce gas pressure, it is advantageous to add a forth stage operating, for instance with CF_4 or CH_4 as a working fluid.

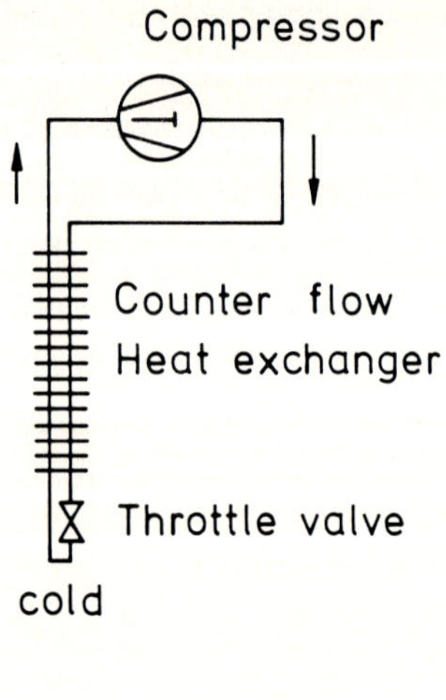

Fig. 8. Scheme of a one-stage Joule-Thomson cryocooler down, and to reduce gas pressure, it is advantageous to add a forth stage operating, for instance, with CF_4 or CH_4 as a working fluid.

Developments for such a system have been pursued at the Jet Propulsion Laboratory [18] (cf. fig. 9). Aside from the cold chambers and recuperative heat exchangers, the design is equipped with numerous filters in the downstream lines to remove any possible contaminant before the gas is expanded through the throttle valves. The entire unit is rather compact (some 30 cm in length) and has been connected to a compressor with individual compression spaces for the different gases. To reduce contamination, a compressor of the diaphragm type was used whose membranes are driven hydraulically on their back side. Such a system appears to be a good candidate for a SQUID cooler; however, much development is needed to bring it to the point that it can be manufactured commercially.

Instead of using mechanical compressors with all their contamination problems, it appears attractive to use sorption - desorption compressors for Joule-Thomson coolers. Considerable effort has been devoted recently to this possibility, and the feasibility of a long life cooler system has been demonstrated. Figure 10 shows a test assembly for a hydrogen liquefier using $LaNi_5$ sorption-desorption cells as a compressor unit. For details the reader is referred to the literature [19].

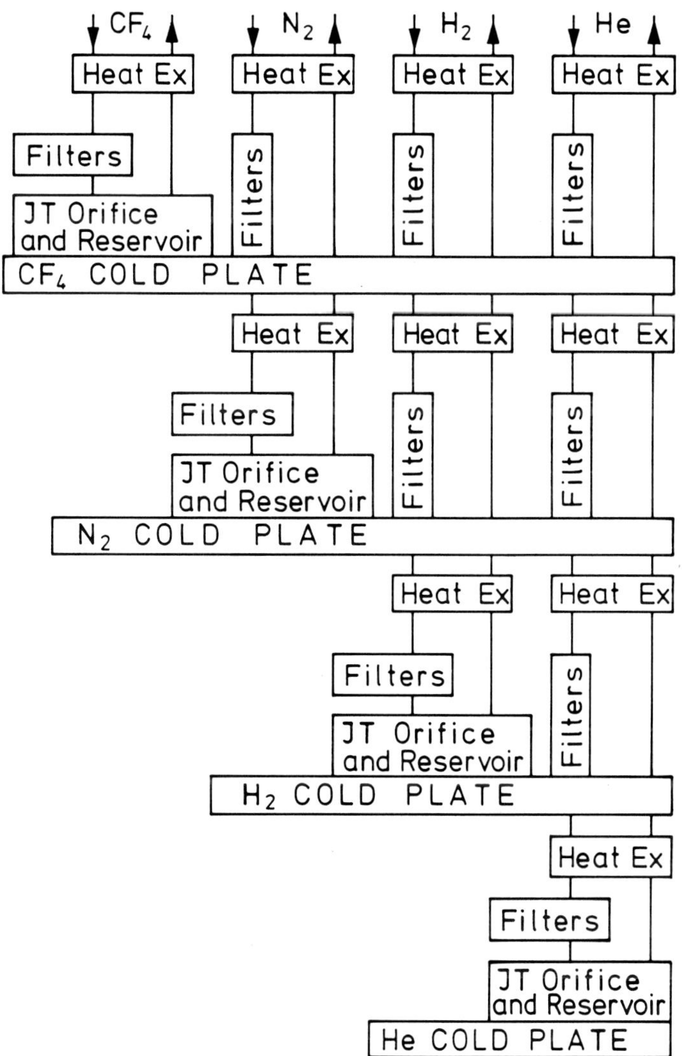

Fig. 9. Four-stage Joule-Thomson cryocooler for liquid helium temperatures, after [18].

Designs according to Fig. 8 allow rather compact construction, and small refrigerators were built many years ago [20 - 22]. Despite rather good performance, these refrigerators did not find widespread use because the throughput of high pressure gas was on the order of 10 barliters/min, limiting the operation time for a

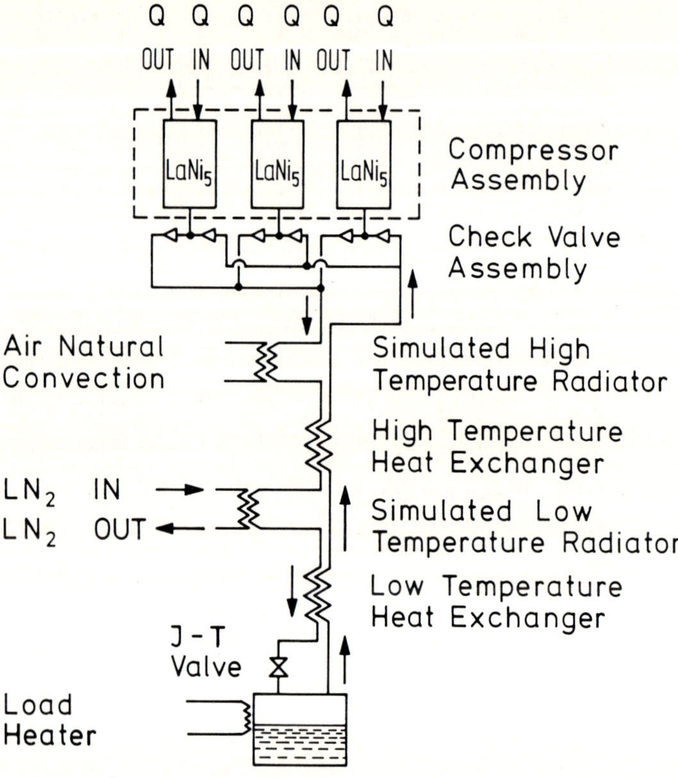

Fig.10 Joule-Thomson hydrogen liquefier with $LaNi_5$ sorption-desorption compressor, after [19]. The check valves are controlled so that there always is a (heated) cell in thedesorption mode while another cell is in the sorption mode.

standard cylinder of compressed gas to a few hours, and because high gas pressures on the order of 100 bar are required to obtain acceptable efficiencies. Closed-cycle operation was unattractive due to the lack of suitable high pressure compressors

This situation has been changed by a recent breakthrough in the development of microminiature Joule-Thomson coolers. In 1977 W.A. Little considered the possibility of reducing the size of Joule-Thomson coolers such that the cooling power was more appropriate for cryoelectronic devices, i.e., 100 mW or less, instead of a few watts. From simple scaling analysis [23] it became clear that the tubing diameter of the corresponding heat exchanger had to be reduced to the order of 0.1 mm or less. This led to the idea of using photolithography to produce the needed fine channels and nozzles.

In the following years, microminiature refrigerators were successfully developed along these lines [24]. In the fabrication process, channels for the heat exchanger, throttle, and the liquid reservoir are cut into a thin glass plate (cf. Fig. 11a). Instead of wet etching with the associated hazard of underetching, the channels are cut by a micro-sandblasting technique. Figure 11b shows an SEM photo of a portion of the heat exchanger. Glass is used because of its low thermal conductivity and high mechanical strength. The channels are sealed by a cover glass glued or fused to the base plate. The entire assembly is enclosed in a removable vacuum chamber which contains electrical feedthroughs for the devices to be cooled. A cool down curve for such a refrigerator operating with a 120-bar nitrogen pressure, a gas throughput of 1.2 barliter/min and a cooling power of 250 - 500 mW, is shown in Fig. 12. Lower flow rates of about 0.1 - 0.2 barliter/min also have been achieved, leading to a cooling power of 25 - 50 mW [24].

Although there is a possibility of clogging the fine capillaries in this design, using redundant channels and good filtering, sufficiently long continuous runs are obtainable in commercially available single stage N_2 or Ar units. Due to the small heat capacity of the entire unit, rather rapid cool-down and warm-up are achieved. Thus, if clogging does occur, it may be removed by a warm-up cool-down cycle.

Based on a Dutch patent [25], another significant improvement in performance recently was achieved by Little [26], who used a gas mixture in his miniature Joule-Thomson refrigerators instead of a single gas. With a mixture of N_2, Ar, and hydrocarbons such as ethylene and propylene, a much higher cooling efficiency was obtained than with N_2 or Ar alone. Figure 13 shows a comparison of the cool-down behavior of N_2, Ar, and a gas mixture consisting of equal parts of N_2, ethane, and methane, which was obtained recently in the author's laboratory. A small single-stage Joule-Thomson unit with a counterflow heat exchanger made of stainless steel tubing was used for this experiment. Comparing N_2 and Ar with the gas mixture, and starting from the same gas pressure (60 bar), there is a dramatic

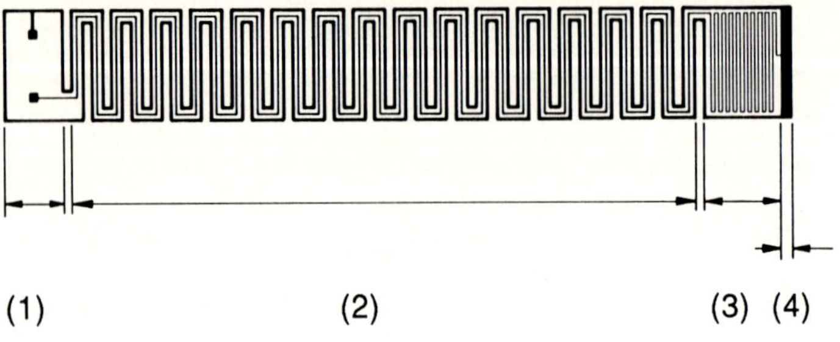

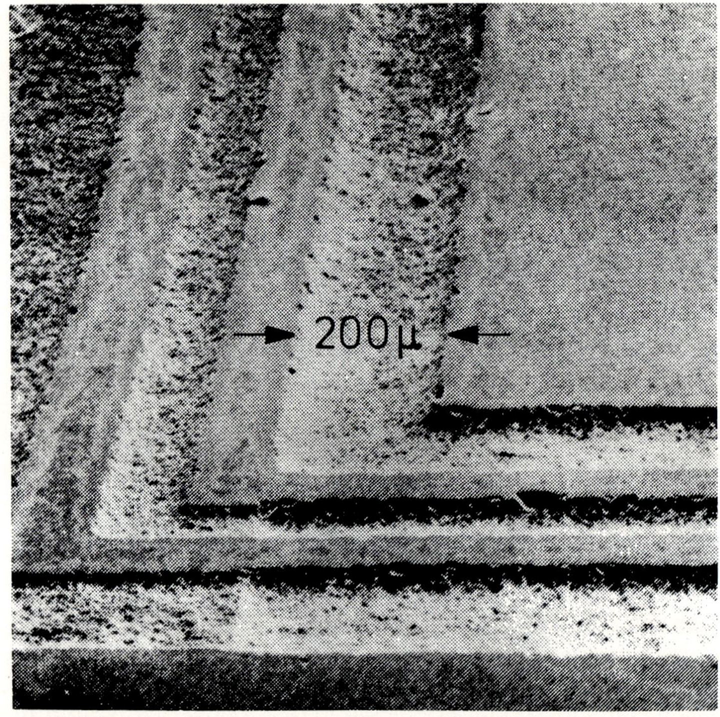

Fig. 11 a) Channel structure of a microminiature Joule-Thomson cryocooler [24]. (1) Gas inlet and outlet ports. (2) Counterflow heat exchanger. (3) Capillary expansion system. (4) Liquid reservoir.

b) SEM photo of a portion of the heat exchanger.

reduction of cool-down time for the mixture to reach the same end temperature. This enhanced cooling efficiency may be used to reduce the gas pressure prior to expansion through the throttle valve, thus relaxing significantly the requirements for

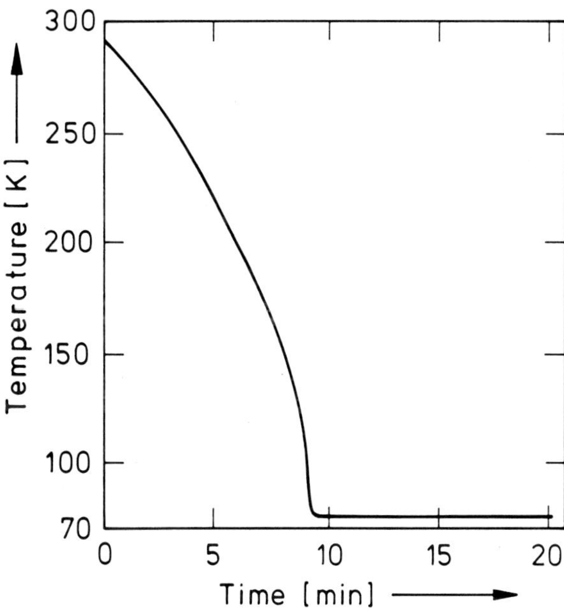

Fig. 12. Cool-down behavior of a microminiature Joule-Thomson cryocooler [24].

a compressor in a closed system. Starting from a 30-bar initial pressure for the above-mentioned ternary gas mixture, almost the same cool-down time is obtained as with 100 bars of N_2. Also, one can use this to cool down rapidly, starting with a higher pressure of the gas mixture and then reducing the pressure after reaching the final temperature to achieve stationary operating conditions. Thus, in the aforementioned experiment, it was possible to maintain the final temperature with only a 10-bar input pressure of the gas mixture. This development may have a considerable impact on miniature Joule-Thomson coolers for superconducting devices made of the new high-T_c oxide superconductors.

4.4 Hybrid Systems

Although cooling requirements may change due to the emergence of high-Cool down superconductor electronics, liquid helium temperatures, nevertheless,will be required in cases where very low noise is required. Hybrid systems that liquify helium for small volume cryostats, therefore, remain of interest. With an appropriate recovery system, closed-cycle operation also is feasible.

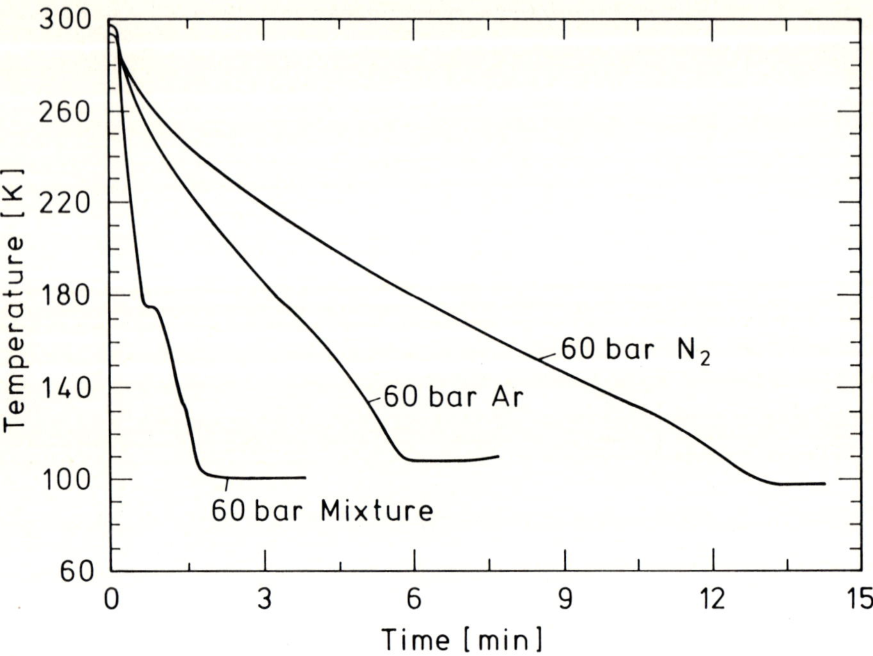

Fig. 13. Comparison of cool-down behavior of three working fluids: N_2, Ar, and a gas mixture of equal parts of N_2, CH_4 and C_2H_6.

Gifford-McMahon cryocoolers have been developed by many manufacturers of cryogenic equipment to a high degree of reliability. Combinations of such cryocoolers with a Joule-Thomson stage are currently available.

Another interesting variant [3] is the combination of a Gifford- McMahon cooler with an expansion space, in which precooled compressed helium gas can be expanded adiabatically, thereby liquefying some fraction of the helium. Provision is made so that this process is repeated at regular intervals. Adding a thermal diode to this periodic Simon liquefier [27] allows a small quantity of liquid helium to collect in a reservoir having temperature variations of only about 30 mK without any electronic regulation.

Finally , the integration of a Joule-Thomson stage into a plastic Stirling cooler [28] has been demonstrated to be a simple means to liquefy small amounts of helium, enough to cool, for instance, planar SQUIDs prepared on a chip.

An excellent survey of the status of cryocooler development can be found in a compilation that was made available a few years ago by Smith et al. [29].

References:

1. An excellent treatment of the entire field of cryocoolers can be found in G. Walker, Cryocoolers: Parts I and II, Plenum Press, New York (1983).

2. A more detailed discussion of this point can be found in J.E. Zimmerman, Cryogenics for SQUIDs. "SQUID 80: Superconducting Quantum Devices and their Applications," H. D. Hahlbohm and H. Lubbig, eds., Walter de Gruyter, Berlin, pp. 423-443 (1980)

3. C. Winter, S.Gygax, K. Myrtle and R. Barton, Proc. of the Third Cryocooler Conference, NBS Special Publication 698, U. S. Government Printing Office, Washington, DC p. 26 (1985).

4. J. E. Zimmerman an D. B. Sullivan, NBS technical note 1049, U. S. Government Printing Office, Washington, DC (1982).

5. J. E. Zimmerman and R. Radebaugh. NBS Special Publication SP-508 U. S. Government Printing Office, Washington, DC, p.59, (1978).

6. J. E. Zimmerman, D. E. Daney and D. B. Sullivan. NASA Conf. Publ. 2287, p. 95. (1983). See also [4].

7. K. Myrtle, C. Winter and S. Gygax. Cryogenics 22, 139 (1982).

8. N. Lambert, Third Cryocooler Conference, NBS Special Publications 698, U. S. Government Printing Office, Washington, DC., p. 119 (1985).

9. N.Lambert, Ph.D thesis, University of Delft, (1986).

10. C. Heiden, Festkorperprobleme (Advances in Solid State Physics), Vol. XXIV, Vieweg, Braunschweig p.331 (1985.)

11. C. Heiden and G.Reich, Advances in Cryogenics 33, 845 (1988).

12. W. E. Gifford and R. C. Longsworth, Pulse Tube Refrigeration, Trans. of the ASME, 63, 264 (1964).

13. R. N. Richardson, Cryogenics, 26, 331 (1986).

14. W. E. Gifford and G. H. Kyanka, Reversible pulse tube refrigeration, Adv. in Cryog. Eng. 12, 619 (1967).

15. R. C. Longsworth, An Experimental Investigation of Pulse Tube Refrigeration Heat Pumping Rates, Adv. in Cryog. Eng. 12, 608 (1967).

16. E. I. Mikulin, A. A.Tarasov, and M. P. Shkrebyonock, Low. Temp. Expansion Pulse Tubes, Adv. in Cryog. Eng. 29, 629 (1984).

17. R. Radebaugh, J. Zimmerman, D. R. Smith, and B. Louie, A comparison of three types of pulse tube refrigerators: New methods for reaching 60 K, Adv. in Cryog. Eng. 31, p. 779 1985).

18. E. Tward and R.Sarwinski, Proc. of the Third Cryocoolers Conference, NBS Special Publications, 698, U. S. Government Printing Office, Washington, DC, p. 220 (1985).

19. J. A. Jones and p. M. Golben, Cryogenics 25, 212 (1985).

20. G. K. Pitcher and F. K. du Pre', Adv.Cryog. Eng. 15, 447 (1970).

21. S. W. Stephens, Infared Phys. 8, 25 (1968).

22. J. M. Geist and P. K. Lashmet, Adv. Cryog. Eng. 6,73 (1961).

23. W. A. Little, Applications of Closed-Cycle Cryocoolers to Small Superconducting Device, J. E. Zimmerman and T M. Flynn, eds., NBS Special Publication 508, U. S. Government Printing Office, Washington, DC, p. 75 (1978).

24. W. A.Little, Rev. Sci. Instrum. 55, 661 (1984).

25. V. N. Alfeev, et. al., Dutch patent application No. 7106470 (Nov. 14, 1972).

26. W. A. Little, Proceedings of the Symposium Low Temperature Electronics and High Temperature Superconductors, The Electrochemical Society, Pennington, NJ, p. 251, 1988.

27. F. Simon, Phys. Z. 34, 232 (1932).

28. S.Barbanera, N. Lambert and J. E. Zimmerman, Cryogenics 26, 341 (1986).

29. J. L. Smith, Jr., G. Y. Robinson Jr., and Y. Iwasa Report prepared under Office of Naval Research, Contract N001483-K-0327 (1984). (Copies of this report can be obtained by writing to the U. S. Naval Research Laboratory, Code 6850.1, Washington, DC 20375-5000, USA.)

30. D. E. Daney, Refrigeration for Cryogenic Sensors and Electronic Systems, NBS Special Publications 607, U. S. Government Printing Office, Washington, DC, p. 48 (1981).

INTRODUCTION TO THE PHENOMENOLOGY OF TUNNELING IN HIGH-TEMPERATURE SUPERCONDUCTORS*

Antonio Barone
Dipartimento di Scienze Fisiche,
Universita di Napoli - P.le Tecchio,
Napoli, Italy
and
Istituto di Cibernetica del C.N.R., Via Toiano 6,
80072 Arco Felice, Napoli, Italy

1. INTRODUCTION

It is generally agreed that the new class of high-T_C (1,2) superconductors is revolutionary. There are unique problems associated with these materials which will require unique solutions and will require an undetermined period of time to achieve. However, the wide range of potential applications makes this a topic worthy of our consideration.

The aim of this chapter is to discuss various junction types, and to examine the similarities and differences between those comprised of oxide superconductors and those comprised of conventional metallic superconducting elements (3).

2. EFFECTS OF GRANULARITY

Two and three-dimensional arrays of superconducting weak links have been widely investigated both theoretically and experimentally (4).

A system of point-contact junction arrays was generated by Clark (5) and investigated in connection with the earlier work of Zimmer and of Saxena et al. See references quoted in Ref.(5). In Clark's work, a set of superconducting spheres, just touching each other at a point, produced the point-contact junction array. Attention was focussed on the response of these structures to microwave radiation, which resulted in a peculiar peaked distribution in dI/dV vs. V plots. Evidence of a cooperative-like effect was indicative of coherent behavior of the ordered array. Since that study, much progress has been made in the study of granular structures.

It is generally recognized that a YBCO sintered pellet can be modelled by a disordered three-dimensional array of superconducting junctions. Indeed, the

*Work partially supported by C.N.R. under the Project "Superconductive and Cryogenic Technologies

NATO ASI Series, Vol. F 59
Superconducting Electronics
Edited by H. Weinstock and M. Nisenoff

unpleasant feature of a relatively low critical transport current density for this material can be ascribed to the presence of Josephson junctions at the grain boundaries. It can be assumed, therefore, that such an array is characterized by a Josephson-like correlation established by a coupling energy $E_c = -E_{ik} \cos(\phi_i - \phi_k)$, where i (or k) labels quantities referred to the ith (or k^{th}) grain. The grains can be linked in different ways, ranging from tunneling to metallic paths. This variety of situations becomes even more complicated at the surface of the superconductor (where there may be effects due to oxygen deficiency) and, consequently, at the interface with another superconductor in a junction configuration.

When a junction device is considered, whether formed by two ceramics or by a ceramic and a conventional superconductor, the resulting behavior is far from ideal. The study of the current-voltage characteristics and the associated derivatives reflects this intriguing situation, and it is very difficult to interpret, in an unambiguous manner, the large variety of experimental results reported in the literature. The amount of structure present in the current-voltage characteristics is large, and an acceptable interpretation of the data can result only from a careful analysis for those features which are truly representative of the new oxide superconductors, if it is, indeed, possible to do so - see Ref.(6).

To investigate the consequences of the different types of links which can occur in actual junctions (either between a ceramic and a conventional superconductor or between two ceramics), we invoke the theory by Blonder, Tinkham and Klapwijk (BTK) (7). This theory provides an interpretation of the wide variety of effects which emerge from the inspection of the I vs. V and dV/dI vs. V characteristics of such devices. Indeed, BTK theory spans the entire range of "contacts", from the archtypical tunnel junction to the metal link. The essence of this is summarized in Fig. 1, where I-V curves (and their derivatives) are presented for different values of a characteristic dimensionless parameter $Z = H/hv_f$, which is a measure of the barrier strength.

Let us discuss now what can be learned from superconducting tunneling spectroscopy in junctions employing high-T_c ceramics. Figure 2 shows a dV/dI vs. V curve obtained from a point contact made by a Nb needle pressing on a YBCO pellet (8). It shows a dip (corresponding to a superconducting-normal path) which is a clear signature of a contact between the Nb and normal regions of the YBCO. Moreover, we observe a dip at about 20 mV, a value which can be interpreted as the sum of the gap of the YBCO, Δ , plus the gap of Nb, Δ_{Nb} (8,9), and deduce a

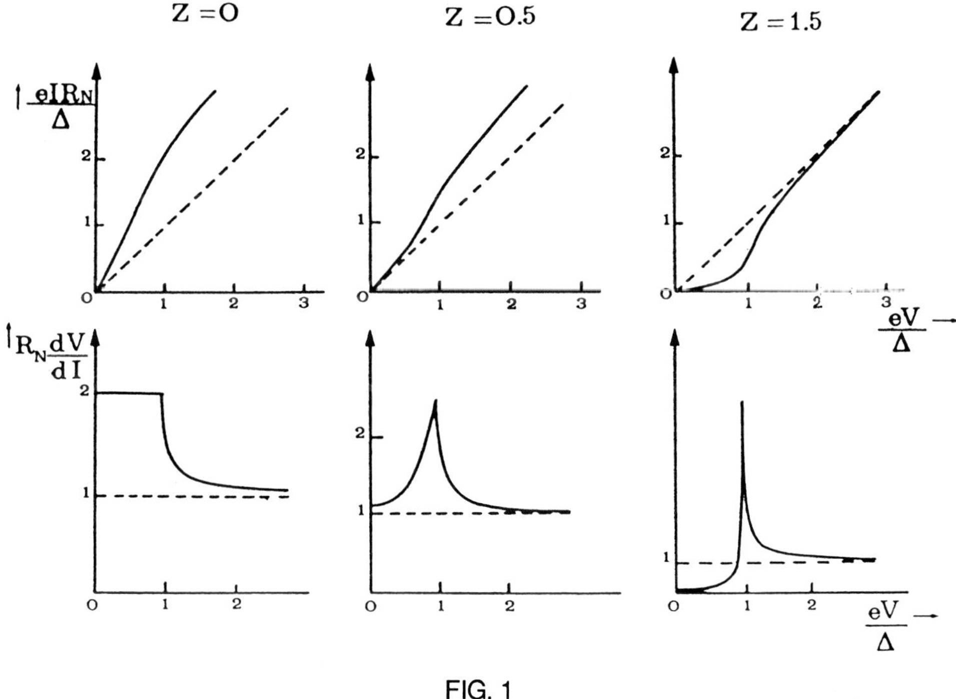

FIG. 1

Current-voltage characteristics and differential conductance vs.voltage for different barrier strengths Z at T = 0 K.

value for Δ of 19.5 ± 2.0 meV. In the literature, as mentioned above, there is a multitude of data, almost all of it characterized by such dips, together with a variety of resistance peaks and background current noise.

It seems likely that some of the characteristic features of these dV/dI vs. V curves can be ascribed to the granularity of the samples. To illustrate this, let us consider the experimental dV/dI vs. V curve shown in Fig. 3. The data refer to a Nb-Nb point contact junction obtained by using a sharp, pure Nb point on a Nb pellet counterelectrode (10). The pellet was formed by pressing Nb powder, roughly reproducing the characteristic size of the granularity of a YBCO pellet. In this case we observe , together with dips corresponding to the Nb gap, very pronounced resistance peaks and background noise current, reminiscent of the peculiar behavior observed for YBCO junctions. This serves as an example of how

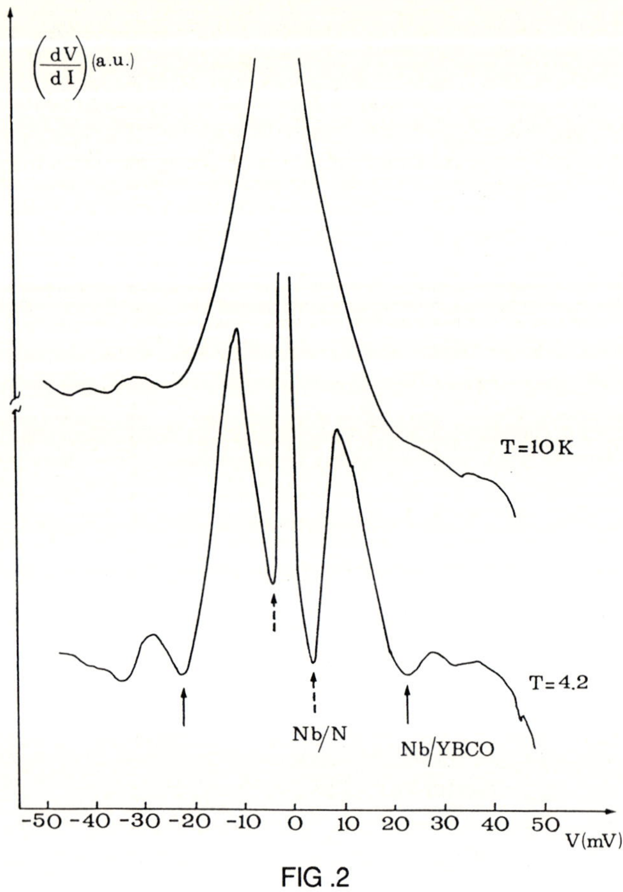

FIG .2

dV/dI vs. V curve of a YBCO-Nb point-contact junction (8).

granularity itself can be responsible for features which are not necessarily characteristic of intrinsic properties of the new class of ceramic superconductors. Let us discuss now possible differences.

At this point, let us consider possible differences between tunneling in ceramic and metallic superconductors. The model of Giaever and Zeller (11) for a granular system predicts that the dI/dV vs. V curve saturates at a voltage given by e/C, where e is the electron charge and C is the capacitance of a grain. This is observed in the data for the Nb-Nb structure of Fig. 3. However, in the case of high-T_C superconductors, a non-saturating behavior - see Fig. 4 - up to voltages on the order of hundreds of millivolts (e.g., in Ref. 10) has been observed. Such behavior

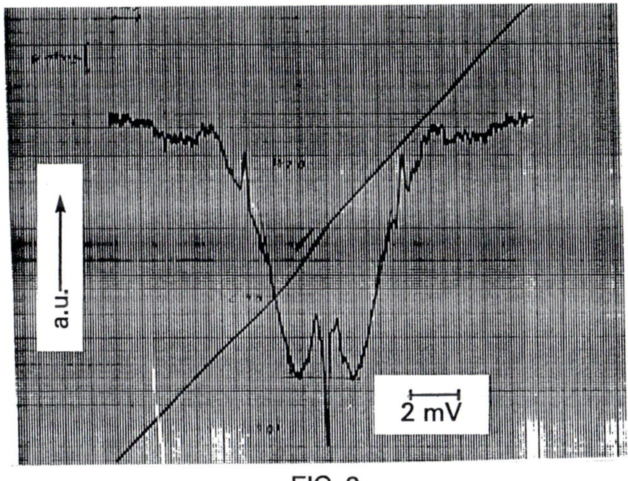

FIG .3

dV/dI vs. V curve of a granular Nb-Nb point-contact junction.

suggests that considerably lower values of capacitance could be involved, implying correspondingly smaller grain size. The possibility of a "subgranularity" has been considered by Rosenblatt (12). If an explanation in terms of the Zeller-Giaever model is ruled out, then the unusual tunneling behavior could be indicative of intrinsic properties of this new class of superconductors. In this respect we note the Anderson-Zou theory (13), which predicts the observed behavior. On the other hand, we can relate such phenomenology to the occurrence of a glassy state on a scale smaller than the grain size, as discussed by Deutscher and Muller (14). However, whether Josephson junctions are located at twin boundaries or at grain boundaries (or both) remains to be determined (15).

As mentioned previously, the granular nature of high-T_C superconducting oxides is responsible for the low critical-current density J_C. The values of J_C obtained for epitaxial films are orders of magnitude larger than for polycrystalline films or bulk ceramic samples. Recently, a striking demonstration of the limiting effect on the critical-current density by grain boundaries has been given (16). Large-grained YBCO oriented films were produced by using polycrystalline $SrTiO_3$ substrates with grain size exceeding 100 μm. Using excimer laser patterning, regions within a grain and across a grain boundary were obtained to probe the two different situations. The Josephson-like nature of the grain "network" has been

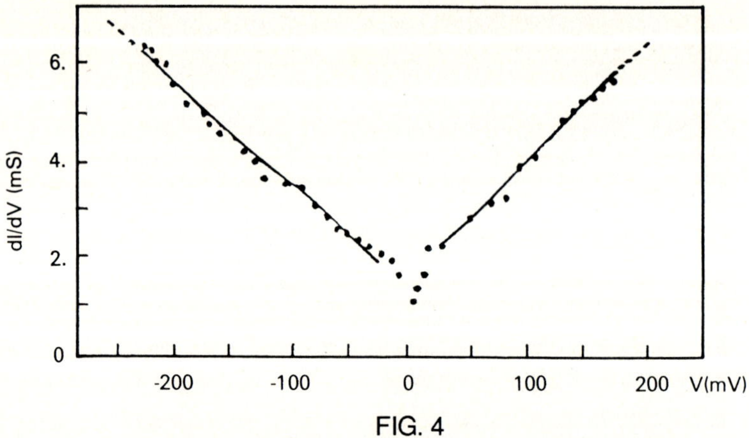

FIG. 4

dI/dV vs. V curve of a ceramic superconductor (BSCCO)-Nb junction.

extensively studied. The strength of intergrain coupling in YBCO also has been investigated measuring trapped magnetic flux by means of a Hall probe (17).

3. GAP-LIKE STRUCTURE

There is intense interest in discovering the fundamental mechanisms responsible for superconducting behavior in this class of ceramic materials. Accordingly, many experiments attempting to uncover an energy gap in these materials have been performed, mostly involving some form of tunneling spectroscopy. As previously noted, however, the results are far from conclusive.

It is worth mentioning that measurements on junctions made using bulk YBCO (or other high-T_C superconducting oxides), either in the form of a sintered pellet or a crystal, and covered by a film of a conventional superconductor (e.g. Nb), or using point contacts applied to crystals as a base layer, also produced ambiguous results. More recently thin-film, sandwich-type junctions have been made employing YBCO film as a base electrode, or as both base and counterelectrode. Actually, as reported in Ref.(6), the values obtained for the ratio $\gamma = 2\,\Delta(0)/k_BT$ show a considerable spread, although a value of $\gamma \sim 5$ appears to be found most frequently (e.g., in Ref.(18)). The temperature dependence of the energy gap has been found to exhibit BCS-like behavior in many cases (e.g., in Ref. (19)), although recently a temperature-independent "gap structure" also was observed (20).

The role of strong electron-phonon interactions in the new superconductors appears to be of importance (21). Application of Eliashberg theory provides information on the role of this interaction and has proven to be important in the study of a large class of conventional superconductors. An analysis based on Eliashberg theory can yield fundamental information on whether and to what extent the superconductive mechanism of the new high-T_c materials can be explained in terms of phonon exchange. A group at the Lebedev Physical Institute (22) is engaged in such an analysis.

We recall that Eliashberg theory considers the more general case of an energy-dependent order parameter $\Delta = \Delta(\omega)$. In this framework the phonon spectral distribution is written as $F(\omega)$ and the quantity $\alpha^2(\omega)F(\omega)$ is the matrix element of the electron-phonon interaction. In terms of these quantities

$$\lambda = 2\int_0^{\infty} \alpha^2(\omega)F(\omega)\frac{d\omega}{\omega} \quad \text{for } \lambda < E_F/\omega$$

where λ represents the phenomenological equivalent of the BCS electron-phonon interaction parameter N(0)V. In the work of the Lebedev group, tunneling data were obtained for various superconducting oxides ($La_{2-x}Sr_xCuO_{4-y}$, $EuBa_2Cu_3O_{7-}$ and $BiSrCaCu_2O_x$) with high values of the ratio $2\,\Delta(0)/k_BT$ (10, 11, 7.2, respectively). From the analysis in terms of Eliashberg theory, such experimental values were not expected. Indeed, unrealistically large values of λ would follow. The results suggest that a BCS-like electron-phonon interaction alone cannot explain completely the behavior of the new superconductors. This may be related to the fact that the theory was developed assuming isotropic behavior, which is certainly not the case for high-T_c oxide superconductors. Anisotropic and d-wave pairing models may prove to be more effective.

4. JOSEPHSON EFFECT

Evidence of the Josephson effect was demonstrated early in the new era of high-T_c superconductors. An example of I-V characteristics showing the Josephson effect is shown in Fig. 5, which is derived from the same sample used to generate Fig. 2 (8). A variety of junctions including point contacts, bulk junctions, break junctions, bridges and thin-film sandwiches have been investigated. Both the d.c. and a.c. Josephson effects have been observed. It is important to stress that only a small number of these many experimental results show clear-cut evidence of an energy-gap structure. Moreover, only a few experiments report a clear dependence of the Josephson current upon temperature (e.g., Ref.23).

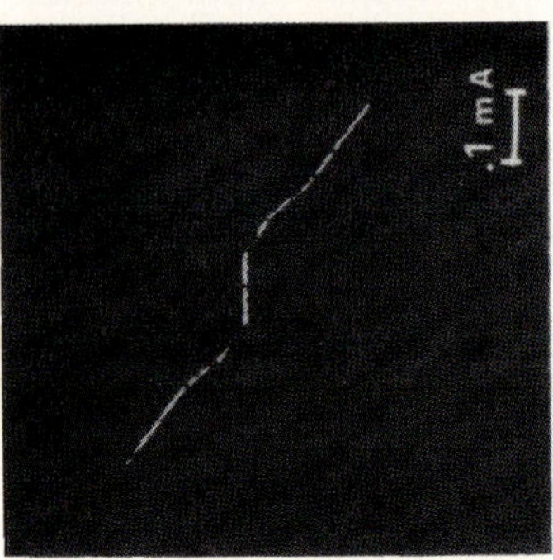

FIG. 5

I-V characteristics of a YBCO-Nb, point-contact Josephson junction.

A widely employed diagnostic tool for Josephson junctions, is the dependence of the Josephson critical current, I_c, upon the externally applied magnetic field H_c(3). High-T_c superconductors in the form of a sintered pellet are characterized by a high degree of disorder. Grains vary both in shape and in size. The resulting grain superstructure is not translationally invariant, and the links between the grains are randomly oriented. This circumstance could contribute to the observed deviations in the I_c vs H_c dependence from the expected pattern. Excellent results have been obtained in YBCO-Au-Nb thin-film junctions which closely reproduce the superconductor-normal-superconductor behavior (3) in both the small (24) and large (25) junction limit.

It is interesting to observe that, even in the presence of two identically-oriented uniaxial superconductors, the situation can be quite different from that occurring in conventional oxide-barrier junctions. Indeed, the sine-Gordon equation describing the dynamics of a Josephson junction (3) should be modified (in normalized units) as follows (26):

$$\mu_{ik} \frac{\partial^2 \phi}{\partial x_i \, \partial x_k} - \tau^2 \frac{\partial^2 \phi}{\partial t^2} = \sin \phi$$

with tensor components

$$\mu_{xx} = \lambda_j^2 (\sin^2\alpha + k\cos^2\alpha)$$

$$\mu_{yy} = \lambda_j^2 (\cos^2\alpha + k\sin^2\alpha)$$

$$\mu_{xy} = \mu_{yx} = \lambda_j^2 (1\text{-}k) \sin\alpha \cos\alpha$$

where α is the angle formed by the symmetry axis (which lies on the (x-y) plane of the barrier) and y, and k is the ratio of the London penetration depths. This analysis leads to a dependence on $I(\phi)$ (where ϕ= applied magnetic flux) which is no longer given by a Fraunhofer pattern, but rather by the expression:

$$I(\phi) = (I_1\, \phi_0^2 / \pi^2\, \phi_x\, \phi_y) \left|\sin\, (\pi\phi_x / \phi_0)\, \sin\, (\pi\phi_y / \phi_0)\right|$$

which is actually the product of two such patterns.

5. CONCLUDING REMARKS

In this chapter we have seen that the objective of making good quality junctions is still a challenge. Indeed, point-contact structures, as well as bulk-type junctions, cannot be considered ideal probes of superconductivity, nor as objects deserving the definition of "devices" with regard to application. A recent investigation has succeeded in yielding entirely YBCO thin-film tunnel junctions (27). However, the results of this effort are not conclusive. Also quite recently, $Tl_2Ca_2Ba_2Cu_3O_{10+x}$ has been employed as a base electrode for a Josephson junction (28).

Finally, we can say that there are still many problems to be solved in order to make good high-T_c junctions. The essential material constraints are represented by the extremely small coherence length ξ and the surface and interface degradation of films. It is interesting to observe that at this stage, paradoxically, some interesting results were obtained by using point-contact structures which, though they be rather primitive, allow one to probe the interior of the sample. Similarly, bridge-type junctions and SQUIDs, based on such weak links, do work despite the necessary condition $L < \xi$ (where L is a characteristic bridge dimension) due to the granularity of the material. On the other hand, recent results

(27) in sandwich structures and continuing improvements in film quality provide genuine cause for optimism.

References

1. J.G. Bednorz and K.A. Muller, A.. Phys. 64, 189 (1986).
2. M.K. Wu, J.R. Ashburn, C.J. Torng, P.H. Hor, R.L. Meng, L. Gao, Z.J. Huang, Y.Q. Zang, and C.A. Chu, Phys. Rev. Lett. 58, 908 (1987).
   R.J. Cava, R.B. Van Dover, B. Battlog and E.A. Rietman, Phys. Rev. Lett. 58, 408 (1987).
3. A. Barone and G. Paterno, "Physics and Applications of the Josephson Effect" John Wiley, New York (1982).
4. See for instance J. Rosenblatt et al in "Josephson Effect - Achievements and Trends" (A. Barone, Editor) World Scientific Publishing (1985).
5. T.D. Clark, Phys. Rev. B 8, 137 (1973).
6. A. Barone, "Tunneling Spectroscopy on High-T_C Superconductors: Certainties, Clues and Ambiguities" (invited paper) Proc. HTSC-M^2S Interlaken (1988), J. Muller and J.L. Olsen Eds., North-Holland, 1712
7. G.E. Blonder, M. Tinkham and T.M. Klapwijk, Phys. Rev. B 25, 4515 (1982).
8. A. Barone, A. di Chiara, G. Peluso, U. Scotti di Uccio, A.M. Cucolo, R. Vaglio, F.C. Matacotta and E. Olzi, in "Novel Superconductivity" (S.A. Wolf and V.Z. Kresin, Eds.) Plenum 1987.
9. A. Barone, A. di Chiara, G. Peluso, U. Scotti di Uccio, A.M. Cucolo, R. Vaglio, F.C. Matacotta and E. Olzi, Phys. Rev. B 36, 235 (1987).
10. A. Barone et al., to appear in Proceedings of the 1988 Applied Superconductivity Conference (as March 1989 issue of IEEE Transactions on Magnetics).
11. H.R. Zeller and I. Giaever, Phys. Rev. B 181, 789 (1969).
12. J. Rosenblatt (private communication).
13. P.W. Anderson and Z. Zou, Phys. Rev. Lett. 60, 132 (1988).
14. G. Deutscher and K.A. Muller, Phys. Rev. Lett. 59, 1745 (1987).
15. R.L. Peterson and J.W. Ekin, Phys. Rev. B 37, 9848 (1988).
16. P. Chaudhari et al., Phys. Rev. Lett. 60, 1653 (1988).
17. K. Yoshida, T. Kisu, N. Fuchigami and K. Empuku, preprint August 1988.
19. J.S. Tsai et al., 5th Int. Workshop Future Electron Devices (FED Hi$T_C$$S_C$-ED) June 1988, Miyagi-Zao, p. 219.

20. J. Geerk, X.X. Xi and G. Linker, preprint and references reported therein.
21. V.Z. Kresin, Sol. State Commun. 63, 725 (1987).
22. L.N. Bulaevskii et al., Superconductor Science and Technology (in press); J. of Modern Physics (in press); L.N. Bulevskii, O.V. Dolgov, M.O. Ptitsyn, V.A. Stepanov and S.I. Vedeneev (in press);
23. T. Yamashita et al., Jap. J. Appl. Phys. 27, L1107 (1988).
24. H. Akoh, F. Shinoki, M. Takahshi and S. Takada, Jap. J. Appl. Phys. 27, L519 (1988).
25. H. Akoh, F. Shinoki, M. Takahshi and S. Takada, preprint, August 1988.
26. R.G. Mints, preprint, 1988.
27. T. Shiota, K. Takechi, Y. Takai and H. Hayakawa, Proc. ISS 1988 (in press).
28. W. Eidelloth, F.S. Barnes, Z.Z. Sheng and A.M. Hermann (preprint).

NATO ASI Series F

Including Special Programme on Sensory Systems for Robotic Control (ROB)

Vol. 1: Issues in Acoustic Signal – Image Processing and Recognition. Edited by C. H. Chen. VIII, 333 pages. 1983.

Vol. 2: Image Sequence Processing and Dynamic Scene Analysis. Edited by T. S. Huang. IX, 749 pages. 1983.

Vol. 3: Electronic Systems Effectiveness and Life Cycle Costing. Edited by J. K. Skwirzynski. XVII, 732 pages. 1983.

Vol. 4: Pictorial Data Analysis. Edited by R. M. Haralick. VIII, 468 pages. 1983.

Vol. 5: International Calibration Study of Traffic Conflict Techniques. Edited by E. Asmussen. VII, 229 pages. 1984.

Vol. 6: Information Technology and the Computer Network. Edited by K. G. Beauchamp. VIII, 271 pages. 1984.

Vol. 7: High-Speed Computation. Edited by J. S. Kowalik. IX, 441 pages. 1984.

Vol. 8: Program Transformation and Programming Environments. Report on a Workshop directed by F. L. Bauer and H. Remus. Edited by P. Pepper. XIV, 378 pages. 1984.

Vol. 9: Computer Aided Analysis and Optimization of Mechanical System Dynamics. Edited by E. J. Haug. XXII, 700 pages. 1984.

Vol. 10: Simulation and Model-Based Methodologies: An Integrative View. Edited by T. I. Ören, B. P. Zeigler, M. S. Elzas. XIII, 651 pages. 1984.

Vol. 11: Robotics and Artificial Intelligence. Edited by M. Brady, L. A. Gerhardt, H. F. Davidson. XVII, 693 pages. 1984.

Vol. 12: Combinatorial Algorithms on Words. Edited by A. Apostolico, Z. Galil. VIII, 361 pages. 1985.

Vol. 13: Logics and Models of Concurrent Systems. Edited by K. R. Apt. VIII, 498 pages. 1985.

Vol. 14: Control Flow and Data Flow: Concepts of Distributed Programming. Edited by M. Broy. VIII, 525 pages. 1985.

Vol. 15: Computational Mathematical Programming. Edited by K. Schittkowski. VIII, 451 pages. 1985.

Vol. 16: New Systems and Architectures for Automatic Speech Recognition and Synthesis. Edited by R. De Mori, C.Y. Suen. XIII, 630 pages. 1985.

Vol. 17: Fundamental Algorithms for Computer Graphics. Edited by R. A. Earnshaw. XVI, 1042 pages. 1985.

Vol. 18: Computer Architectures for Spatially Distributed Data. Edited by H. Freeman and G. G. Pieroni. VIII, 391 pages. 1985.

Vol. 19: Pictorial Information Systems in Medicine. Edited by K. H. Höhne. XII, 525 pages. 1986.

Vol. 20: Disordered Systems and Biological Organization. Edited by E. Bienenstock, F. Fogelman Soulié, G. Weisbuch. XXI, 405 pages. 1986.

Vol. 21: Intelligent Decision Support in Process Environments. Edited by E. Hollnagel, G. Mancini, D. D. Woods. XV, 524 pages. 1986.

NATO ASI Series F

Vol. 22: Software System Design Methods. The Challenge of Advanced Computing Technology. Edited by J.K. Skwirzynski. XIII, 747 pages. 1986.

Vol. 23: Designing Computer-Based Learning Materials. Edited by H. Weinstock and A. Bork. IX, 285 pages. 1986.

Vol. 24: Database Machines. Modern Trends and Applications. Edited by A.K. Sood and A.H. Qureshi. VIII, 570 pages. 1986.

Vol. 25: Pyramidal Systems for Computer Vision. Edited by V. Cantoni and S. Levialdi. VIII, 392 pages. 1986. *(ROB)*

Vol. 26: Modelling and Analysis in Arms Control. Edited by R. Avenhaus, R.K. Huber and J.D. Kettelle. VIII, 488 pages. 1986.

Vol. 27: Computer Aided Optimal Design: Structural and Mechanical Systems. Edited by C.A. Mota Soares. XIII, 1029 pages. 1987.

Vol. 28: Distributed Operating Systems. Theory und Practice. Edited by Y. Paker, J.-P. Banatre and M. Bozyiğit. X, 379 pages. 1987.

Vol. 29: Languages for Sensor-Based Control in Robotics. Edited by U. Rembold and K. Hörmann. IX, 625 pages. 1987. *(ROB)*

Vol. 30: Pattern Recognition Theory and Applications. Edited by P.A. Devijver and J. Kittler. XI, 543 pages. 1987.

Vol. 31: Decision Support Systems: Theory and Application. Edited by C.W. Holsapple and A.B. Whinston. X, 500 pages. 1987.

Vol. 32: Information Systems: Failure Analysis. Edited by J.A. Wise and A. Debons. XV, 338 pages. 1987.

Vol. 33: Machine Intelligence and Knowledge Engineering for Robotic Applications. Edited by A.K.C. Wong and A. Pugh. XIV, 486 pages. 1987. *(ROB)*

Vol. 34: Modelling, Robustness and Sensitivity Reduction in Control Systems. Edited by R.F. Curtain. IX, 492 pages. 1987.

Vol. 35: Expert Judgment and Expert Systems. Edited by J.L. Mumpower, L.D. Phillips, O. Renn and V.R.R. Uppuluri. VIII, 361 pages. 1987.

Vol. 36: Logic of Programming and Calculi of Discrete Design. Edited by M. Broy. VII, 415 pages. 1987.

Vol. 37: Dynamics of Infinite Dimensional Systems. Edited by S.-N. Chow and J.K. Hale. IX, 514 pages. 1987.

Vol. 38: Flow Control of Congested Networks. Edited by A.R. Odoni, L. Bianco and G. Szegö. XII, 355 pages. 1987.

Vol. 39: Mathematics and Computer Science in Medical Imaging. Edited by M.A. Viergever and A. Todd-Pokropek. VIII, 546 pages. 1988.

Vol. 40: Theoretical Foundations of Computer Graphics and CAD. Edited by R.A. Earnshaw. XX, 1246 pages. 1988.

Vol. 41: Neural Computers. Edited by R. Eckmiller and Ch. v. d. Malsburg. XIII, 566 pages. 1988.

NATO ASI Series F

Vol. 42: Real-Time Object Measurement and Classification. Edited by A. K. Jain. VIII, 407 pages 1988. *(ROB)*

Vol. 43: Sensors and Sensory Systems for Advanced Robots. Edited by P. Dario. XI, 597 pages. 1988. *(ROB)*

Vol. 44: Signal Processing and Pattern Recognition in Nondestructive Evaluation of Materials. Edited by C.H. Chen. VIII, 344 pages. 1988. *(ROB)*

Vol. 45: Syntactic and Structural Pattern Recognition. Edited by G. Ferraté, T. Pavlidis, A. Sanfeliu, H. Bunke. XVI, 467 pages. 1988. *(ROB)*

Vol. 46: Recent Advances in Speech Understanding and Dialog Systems. Edited by H. Niemann, M. Lang, G. Sagerer. X, 521 pages. 1988.

Vol. 47: Advanced Computing Concepts and Techniques in Control Engineering. Edited by M.J. Denham, A.J. Laub. XI, 518 pages. 1988.

Vol. 48: Mathematical Models for Decision Support. Edited by Gautam Mitra. IX, 762 pages. 1988.

Vol. 49: Computer Integrated Manufacturing. Edited by I. Burhan Turksen. VIII, 568 pages. 1988.

Vol. 50: CAD Based Programming for Sensory Robots. Edited by B. Ravani. IX, 565 pages. 1988. *(ROB)*

Vol. 51: Algorithms and Model Formulations in Mathematical Programming. Edited by Stein W. Wallace. IX, 190 pages. 1989.

Vol. 52: Sensor Devices and Systems for Robotics. Edited by Alícia Casals. IX, 362 pages. 1989. *(ROB)*

Vol. 53: Advanced Information Technologies for Industrial Material Flow Systems. Edited by S.Y. Nof and C.L. Moodie. IX, 710 pages. 1989.

Vol. 54: A Reappraisal of the Efficiency of Financial Markets. Edited by R.M.C. Guimarães, B.G. Kingsman and S.J. Taylor. X, 804 pages. 1989.

Vol. 55: Constructive Methods in Computing Science. Edited by Manfred Broy. VII, 478 pages. 1989.

Vol. 56: Multiple Criteria Decision Making and Risk Analysis Using Microcomputers. Edited by B. Karpak and S. Zionts. VII, 399 pages. 1989.

Vol. 57: Kinematics and Dynamic Issues in Sensing-Based Control. Edited by G. Taylor, XI, 460 pages. 1989.

Vol. 58: Highly Redundant Sensing in Robotic Systems. Edited by J. T. Tou, J. G. Balchen. IX, 314 pages, 1989.

Vol. 59: Superconducting Electronics. Edited by M. Nisenoff, H. Weinstock. X, 441 pages. 1989.